AF780278

# Morphology, Palynology and Phytochemicals of Medicinal Plants

# Morphology, Palynology and Phytochemicals of Medicinal Plants

Editor

**Wajid Zaman**

MDPI

Basel • Beijing • Wuhan • Barcelona • Belgrade • Novi Sad • Cluj • Manchester

*Editor*
Wajid Zaman
Department of Life Sciences
Yeungnam University
Gyeongsan
Korea, South

*Editorial Office*
MDPI
St. Alban-Anlage 66
4052 Basel, Switzerland

This is a reprint of articles from the Special Issue published online in the open access journal *Horticulturae* (ISSN 2311-7524) (available at: www.mdpi.com/journal/horticulturae/special_issues/3PY0B8997Y).

For citation purposes, cite each article independently as indicated on the article page online and as indicated below:

| |
|---|
| Lastname, A.A.; Lastname, B.B. Article Title. *Journal Name* **Year**, *Volume Number*, Page Range. |

**ISBN 978-3-7258-0470-2 (Hbk)**
**ISBN 978-3-7258-0469-6 (PDF)**
**doi.org/10.3390/books978-3-7258-0469-6**

© 2024 by the authors. Articles in this book are Open Access and distributed under the Creative Commons Attribution (CC BY) license. The book as a whole is distributed by MDPI under the terms and conditions of the Creative Commons Attribution-NonCommercial-NoDerivs (CC BY-NC-ND) license.

# Contents

*horticulturae*

*Editorial*

# Morphology, Palynology and Phytochemicals of Medicinal Plants

**Wajid Zaman**

Department of Life Sciences, Yeungnam University, Gyeongsan 38541, Republic of Korea; wajidzaman@yu.ac.kr

The study of plant morphology and palynology not only enhances our understanding of plant biology, but also provides insights into the evolutionary adaptations and ecological dynamics that contribute to the medicinal properties of plants [1]. This Special Issue focuses on filling the gaps in existing research on medicinal plants by investigating new phytochemicals, changes in morphology and palynology across various plant families, and utilizing both traditional and sophisticated techniques for plant identification and analysis.

In both traditional and contemporary medicine, phytochemicals, which are the naturally occurring molecules that are produced by plants, play a key role [2,3]. The phytochemical study of a variety of medicinal plants is investigated in depth in this Special Issue, which aims to identify the molecules that are responsible for the beneficial effects that these plants have on human health. The great potential of plants as sources of new medications and therapies is highlighted by the wide variety of phytochemicals [4]. These phytochemicals range from anti-inflammatory compounds to antioxidant mediators [5].

Additionally, the relevance of integrative techniques in botanical research is highlighted in this Special Issue. Researchers can acquire a more comprehensive understanding of medicinal plants if they combine studies of plant morphology, palynology, and phytochemicals. It is essential to have this all-encompassing approach to investigate the entire potential of these plants, not only as possible sources of novel medications, but also as essential components in sustainable agriculture, environmental conservation, and world health [6,7]. Furthermore, the studies that are provided in this Special Issue highlight the significance of biodiversity for the research that is conducted in various medical fields. There are a variety of phytochemicals that are found in each plant species, and these phytochemicals have the potential to provide answers to some of the most important health problems that exist today. When viewed in this way, this Special Issue serves as a rallying cry for the preservation of plant diversity, bringing to our attention the intricate linkages that exist between the health of our planet and the well-being of the people who live on it.

When it comes to the multidisciplinary approach to the study of medicinal plants, this Special Issue represents a significant step forward. By doing so, it provides significant insights into the intricate relationship that exists between the chemical and physical properties of these plants, as well as the possible health advantages that they may bring. Not only does it demonstrate the significance of conducting an extensive study to unravel the mysteries that plants conceal, but it also shows the way for future discoveries that have the potential to completely transform the area of horticulture. This Special Issue is a goldmine of knowledge, inspiration, and hope for the future of health and healing, and it is a treasure trove for all those who are interested in the field, including practitioners, enthusiasts, and scientists.

check for updates

**Citation:** Zaman, W. Morphology, Palynology and Phytochemicals of Medicinal Plants. *Horticulturae* **2024**, *10*, 202. https://doi.org/10.3390/horticulturae10030202

Received: 14 February 2024
Accepted: 18 February 2024
Published: 22 February 2024

**Copyright:** © 2024 by the author. Licensee MDPI, Basel, Switzerland. This article is an open access article distributed under the terms and conditions of the Creative Commons Attribution (CC BY) license (https://creativecommons.org/licenses/by/4.0/).

**Conflicts of Interest:** The author declares no conflicts of interest.

**List of Contributions:** This Special Issue features 15 research articles and one review article, offering a comprehensive exploration of the structural, genetic, and chemical diversity of medicinal plants. These contributions delve into various aspects, such as the effects of light on plant metabolism, detailed morphological and anatomical studies, and the phytochemical composition affecting medicinal properties. The Guest Editors would like to thank every author for sharing their knowledge, and for providing engaging research findings in this Special Issue. Additionally, they express gratitude to *Horticulturae* for its invaluable assistance in making this Special Issue a reality.

1. Balilashaki, K.; Martinez-Montero, M.E.; Vahedi, M.; Cardoso, J.C.; Silva Agurto, C.L.; Leiva-Mora, M.; Feizi, F.; Musharof Hossain, M. Medicinal Use, Flower Trade, Preservation and Mass Propagation Techniques of Cymbidium Orchids—An Overview. *Horticulturae* **2023**, *9*, 690. https://doi.org/10.3390/horticulturae9060690.
2. Gangaram, S.; Naidoo, Y.; Dewir, Y.H.; Singh, M.; Magyar-Tábori, K. Micromorphology of *Barleria albostellata* (Grey Barleria) Flower and Pollen Grains. *Horticulturae* **2023**, *9*, 732. https://doi.org/10.3390/horticulturae9070732.
3. Idris, N.S.; Khandaker, M.M.; Rashid, Z.M.; Majrashi, A.; Alenazi, M.M.; Adnan, A.F.; Mahmud, K.; Mat, N. Discrimination of *Syzygium samarangense* cv. 'Giant Green' Leaves at Different Maturity Stages by FTIR and GCMS Fingerprinting. *Horticulturae* **2023**, *9*, 609. https://doi.org/10.3390/horticulturae9050609.
4. Idris, N.S.; Khandaker, M.M.; Rashid, Z.M.; Majrashi, A.; Alenazi, M.M.; Nor, Z.M.; Mohd Adnan, A.F.; Mat, N. Polyphenolic Compounds and Biological Activities of Leaves and Fruits of *Syzygium samarangense* cv. 'Giant Green' at Three Different Maturities. *Horticulturae* **2023**, *9*, 326. https://doi.org/10.3390/horticulturae9030326.
5. Kalendar, O.V.; Kostikova, V.A.; Kukushkina, T.A.; Erst, A.S.; Kuznetsov, A.A.; Kulikovskiy, M.S.; Vasilyeva, O.Y. Seasonal Development of *Paeonia obovata* and *Paeonia oreogeton* and Their Contents of Biologically Active and Reserve Substances in the Forest-Steppe Zone of Western Siberia. *Horticulturae* **2023**, *9*, 102. https://doi.org/10.3390/horticulturae9010102.
6. Khan, S.; Al-Qurainy, F.; Al-hashimi, A.; Nadeem, M.; Tarroum, M.; Salih, A.M.; Shaikhaldein, H.O. Comparative Study on Genome Size and Phytochemical Profile of Three Potential Species of *Acacia*: Threatened and Endemic to Saudi Arabia. *Horticulturae* **2022**, *8*, 994. https://doi.org/10.3390/horticulturae8110994.
7. Lee, G.O.; Jang, S.-N.; Kim, M.J.; Cho, D.Y.; Cho, K.M.; Lee, J.H.; Son, K.-H. Comparison of Growth Patterns and Metabolite Composition of Different Ginseng Cultivars (Yunpoong and K-1) Grown in a Vertical Farm. *Horticulturae* **2023**, *9*, 583. https://doi.org/10.3390/horticulturae9050583.
8. Lee, S.Y.; Kwon, H.; Kim, J.K.; Park, C.H.; Sathasivam, R.; Park, S.U. Comparative Analysis of Glucosinolate and Phenolic Compounds in Green and Red Kimchi Cabbage (*Brassica rapa* L. ssp. *pekinensis*) Hairy Roots after Exposure to Light and Dark Conditions. *Horticulturae* **2023**, *9*, 466. https://doi.org/10.3390/horticulturae9040466.
9. McVea, H.M.; Wood, L.J. Anatomical and Chemical Analysis of *Moringa oleifera* Stem Tissue Grown under Controlled Conditions. *Horticulturae* **2023**, *9*, 213. https://doi.org/10.3390/horticulturae9020213.
10. Parmar, G.; Zaman, W. Trichomes' Micromorphology and Their Evolution in Selected Species of *Causonis* (Vitaceae). *Horticulturae* **2022**, *8*, 877. https://doi.org/10.3390/horticulturae8100877.
11. Tuncay, H.O.; Akalın, E.; Doğru-Koca, A.; Eruçar, F.M.; Miski, M. Two New *Ferula* (Apiaceae) Species from Central Anatolia: *Ferula turcica* and *Ferula latialata*. *Horticulturae* **2023**, *9*, 144. https://doi.org/10.3390/horticulturae9020144.
12. Wu, Q.; Ye, R.; Duan, J.; Lin, D.; Jia, Y.; Dang, F.; Han, T. Physiological, Transcriptomic and Metabolomic Response of Basil (*O. basilicum Linn. var. pilosum* (Willd.) Benth.) to Red and Blue Light. *Horticulturae* **2023**, *9*, 1172. https://doi.org/10.3390/horticulturae9111172.
13. Wu, Z.; Li, Z.; Feng, W.; Meng, R.; Wang, X.; Wu, C. The Breeding of High-Quality Dandelions by NaCl Induced Callus Variation Combined with a Drosophila Tumor Cell Migration Test. *Horticulturae* **2022**, *8*, 1167. https://doi.org/10.3390/horticulturae8121167.
14. Yang, S.; Choi, G.; Song, J.-H. Morphology, Anatomy, Micromorphology, and Palynology of the Squirrel's Foot Fern, *Davallia mariesii* (Davalliaceae). *Horticulturae* **2023**, *9*, 939. https://doi.org/10.3390/horticulturae9080939.
15. Zemouri, T.; Chikhoune, A.; Benmouhoub, H.; Sahnoune, M. Taxonomic Comparison, Antioxidant and Antibacterial Activities of Three *Ebenus pinnata* Ait. ecotypes (Fabaceae) from Algeria. *Horticulturae* **2023**, *9*, 879. https://doi.org/10.3390/horticulturae9080879.
16. Zhang, D.; Xie, A.; Yang, X.; Shi, Y.; Yang, L.; Dong, L.; Lei, F.; Wu, J.; Sun, X. Study of 15 Varieties of Herbaceous Peony Pollen Submicroscopic Morphology and Phylogenetic Relationships. *Horticulturae* **2022**, *8*, 1161. https://doi.org/10.3390/horticulturae8121161.

## References


1. Zamira, D.; Khaydarov, K.; Zafar, M.; Ramadan, M.F.; Ahmad, M.; Aziza, N.; Ochilov, U.; Zebiniso, U.; Farzona, D. Comprehensive study of allergenic tree species: Palynological insights enhanced by HPLC and GC–MS profiling. *Biomed. Chromatogr.* **2023**, *38*, e5774. [CrossRef]
2. George, B.P.; Chandran, R.; Abrahamse, H. Role of phytochemicals in cancer chemoprevention: Insights. *Antioxidants* **2021**, *10*, 1455. [CrossRef] [PubMed]
3. Süntar, I. Importance of ethnopharmacological studies in drug discovery: Role of medicinal plants. *Phytochem. Rev.* **2020**, *19*, 1199–1209. [CrossRef]
4. Kaushik, J.; Kundu, N. Phytochemical screening, anti-oxidant and anti-microbial activity of polyphenolic flavonoids isolated from fruit of ananas comosus in various solvents. *Int. J. Sci. Res. Publ.* **2018**, *8*, 31–55.
5. Dhanani, T.; Shah, S.; Gajbhiye, N.A.; Kumar, S. Effect of extraction methods on yield, phytochemical constituents and antioxidant activity of *Withania somnifera*. *Arab. J. Chem.* **2017**, *10*, S1193–S1199. [CrossRef]
6. Harwood, R.R. A history of sustainable agriculture. In *Sustainable Agricultural Systems*; CRC Press: Boca Raton, FL, USA, 2020; pp. 3–19.
7. Smith, W. Understanding the changing role of global public health in biodiversity conservation. *Ambio* **2022**, *51*, 485–493. [CrossRef] [PubMed]


**Disclaimer/Publisher's Note:** The statements, opinions and data contained in all publications are solely those of the individual author(s) and contributor(s) and not of MDPI and/or the editor(s). MDPI and/or the editor(s) disclaim responsibility for any injury to people or property resulting from any ideas, methods, instructions or products referred to in the content.

*horticulturae*

Article

# Physiological, Transcriptomic and Metabolomic Response of Basil (*O. basilicum* Linn. var. *pilosum* (*Willd.*) Benth.) to Red and Blue Light

Qingfei Wu [1,†], Rigui Ye [2,†], Jingmian Duan [1], Duo Lin [1], Yuru Jia [1], Fengfeng Dang [1,*] and Tiantian Han [1,3,*]

1 China Shaanxi Key Laboratory of Chinese Jujube, Yan'an University, Yan'an 716000, China; feiqw1234@163.com (Q.W.); 0727mianmian@163.com (J.D.); jiayuruyydx@163.com (Y.J.)
2 Medical Innovation Center for Nationalities, Inner Mongolia Medical University, Huhhot 010110, China; 20220006@immu.edu.cn
3 State Key Laboratory of Plant Genomics, National Centre for Plant Gene Research (Beijing), Institute of Genetics and Developmental Biology, Chinese Academy of Sciences, Beijing 100101, China
* Correspondence: fdang@yau.edu.cn (F.D.); hantsthos@genetics.ac.cn (T.H.)
† These authors contributed equally to this work.

**Abstract:** Basil (*Ocimum basilicum* Linn. var. *pilosum* (*Willd.*) Benth.) is an aromatic plant with high nutritional and economic value, and the synthesis and regulation of its active ingredients have been studied in prior research. However, the mechanisms by which red and blue light—the most effective absorption spectra for photosynthesis—regulate the growth and metabolism of basil remain elusive. This study investigated the changes in phenotype, transcriptome, and metabolome in basil under red and blue light. The photosynthetic efficiency and biomass of basil under blue light (B) treatment were higher than those under white light (W), while red light (R) decreased photosynthesis and biomass. Metabolomic analysis showed that 491 significantly differentially accumulated metabolites were identified between the W and B groups, while 630 differentially accumulated metabolites were identified between the W and R groups. The DAMs were mainly enriched in pathways such as biosynthesis of secondary metabolites, monoterpenoid biosynthesis, limonene and pinene degradation, etc. In addition, transcriptomic analysis revealed that 34,760 and 29,802 differentially expressed genes were detected in the W vs. B pair and the W vs. R pair, respectively, while differentially expressed genes were divided into different unique subclasses, suggesting that they respond to light quality in specific ways. Overall, this work will not only enrich knowledge of the molecular mechanisms of light spectra's regulation of plant metabolism, but also provide a theoretical basis and guidance for the molecular improvement and quality cultivation of basil.

**Keywords:** basil; light spectra; metabolome; transcriptome; monochromatic

check for updates

**Citation:** Wu, Q.; Ye, R.; Duan, J.; Lin, D.; Jia, Y.; Dang, F.; Han, T. Physiological, Transcriptomic and Metabolomic Response of Basil (*O. basilicum* Linn. var. *pilosum* (*Willd.*) Benth.) to Red and Blue Light. *Horticulturae* **2023**, *9*, 1172. https://doi.org/10.3390/horticulturae9111172

Academic Editor: Wajid Zaman

Received: 5 September 2023
Revised: 25 October 2023
Accepted: 25 October 2023
Published: 26 October 2023

**Copyright:** © 2023 by the authors. Licensee MDPI, Basel, Switzerland. This article is an open access article distributed under the terms and conditions of the Creative Commons Attribution (CC BY) license (https://creativecommons.org/licenses/by/4.0/).

## 1. Introduction

Light is one of the indispensable environmental factors affecting plants' growth and development. It not only provides energy for photosynthesis, but also regulates multiple responses, including germination, seedling photomorphogenesis, circadian rhythms, phototropism, and flowering time [1,2]. In general, the light environment is divided into three main components: light photoperiod, light spectra, and light intensity. Specifically, light spectra play vital roles in many aspects, such as photosynthesis, chloroplast development, and plant morphogenesis, whose regulated processes are much more complicated than those caused by light photoperiod or intensity [3,4]. In higher plants, a series of photoreceptors can accurately perceive ambient light signals and regulate various developmental processes. Generally, photoreceptors in plants can be divided into five main categories: phytochromes (PHYA-E), which sense red/far-red lights (600–750 nm); phototropins (PHOT1 and PHOT2); F-box-containing flavin-binding proteins (e.g., ZEITLUPE, FKF1/LKP2); cryptochromes (CRY1-3), which sense blue/UV-A lights (320–500 nm); and UVR8, which senses UV-B light (280–320 nm) [5–9]. These photoreceptors play unique and/or redundant roles in regulating a wide variety of developmental processes and metabolic pathways.

As some of the most abundant photosynthetic pigments, chlorophyll a absorbs light in the wavelength between 400 nm and 662 nm, and chlorophyll b absorbs light in the wavelength between about 453 nm and 642 nm. Therefore, the red (600–700 nm) and blue (400–500 nm) light spectra are often considered to be the most effectively utilized wavelengths for driving photosynthesis [10]. It is interesting to note that plants grown solely under red light exhibit a lower photosynthetic rate than plants grown under white light, resulting from the fact that red light reduces stomatal density, stomatal conductance, and rubisco protein content [11]. In particular, red light induces substantial impairments in plant growth, such as causing reductions in chlorophyll content and leaf area, and provoking excessive stem elongation, which may eventually lead to the reduction in plant biomass [12–14]. Blue light is responsible for regulating chloroplast development, stomatal opening, and chlorophyll formation. Blue-light-grown plants contain higher amounts of Chl a/b, cytochrome (Cyt) F, and rubisco than plants grown under white light [15]. Moreover, blue light greatly affects the architecture of shoots, resulting in dense, compact plants, as well as increasing vegetative growth [16]. In addition to physiological and morphological effects, red and blue light have great effects on plants' metabolism, including primary metabolism and secondary metabolism. For example, it has been reported that red light inhibits the translocation of photosynthate from leaves, causing leaves to accumulate far greater starch and soluble sugar contents [17–19]. Greater anthocyanin content was accumulated in strawberries under red light [20]. In soybeans, red light can also promote the biosynthesis of lignin to improve cell wall firmness [21]. Blue light promoted glucosinolate accumulation in pak choi by increasing the precursor amino acid profiles [22], as well as promoting chlorogenic acid synthesis in strawberries [23]. Pepper plants produced greater amounts of epidermal flavonoids when exposed to enhanced blue light [24]. Thus, the effects of different light spectra on plants are distinctly different.

*Ocimum basilicum* (Basil) belongs to the *Lamiaceae* family, possessing important medicinal and culinary values, and has been widely grown in many warm and temperate countries [25]. Previous studies have suggested that basil has many potential pharmacological effects, such as antioxidant, antipyretic, antidiabetic, anticancer, and anti-stress effects, among others [26]. Biologically active ingredients such as essential oils, polysaccharides, flavonoids, and phenols are abundant in basil, contributing to its beneficial health properties [27]. Many basil species, especially sweet basil, are also good resources for the food industry. For example, basil's leaves can be used as a flavoring agent for food, and its extracts can be used in the manufacture of other foods [25]. Over the past few years, there has been intense research on different aspects of the herb, including its medicinal properties, chemical constituents, and organic cultivation. In contrast, the physiological response of basil to environmental changes has been poorly studied. Several studies have revealed that light spectra have significant effects on the growth and metabolic changes of basil. For example, postharvest irradiation with UV-B can promote the accumulation of flavonoids and phenolic acids in sweet basil's leaves, while also improving its pharmaceutical quality [28]. Though UV-C is filtered by the atmosphere and does not reach the Earth's surface/plants, UV-C has been shown to be effective in increasing the production of phenolic compounds and cellular antioxidant potential in callus cultures of purple basil [29]. Nadeem et al. (2019) also found that different light-emitting diodes (LEDs), especially those on the red and blue monochromatic spectra, stimulated the accumulation of biologically active ingredients in callus cultures of basil [30]. In addition, an RB ratio (red/blue LED lights) of three was optimal for indoor sweet basil cultivation, bringing about better performances in growth, resource use efficiency, and yield and quality formation [31]. A recent study showed that blue light did not induce the accumulation of antioxidants in green and purple basil cultivars at harvest [32]. However, the detailed molecular mechanisms by which red and blue light regulate the growth and quality of basil remain largely unknown.

Basil has many varieties with different leaf shapes and aromas, presenting distinct chemical profiles. China has abundant resources of basil, and *O. basilicum* Linn. var. *pilosum* (*Willd.*) Benth. is a variety that is widely distributed throughout China, used as a leafy vegetable and for prescriptions in folk medicine [33]. The effects of red and blue light on its growth and metabolism are still poorly understood. In this study, we explored the changes in phenotype, metabolome, and transcriptome in basil (*O. basilicum* Linn. var. *pilosum* (*Willd.*) Benth.) under continuous monochromatic red and blue light. These results should contribute to a better understanding of how light spectra affect the metabolic products of basil, as well as providing a theoretical basis for future cultivation management.

## 2. Materials and Methods

### *2.1. Plant Materials and Light Treatment*

Basil (*O. basilicum* Linn. var. *pilosum* (*Willd.*) Benth.) seeds were obtained from the Zhejiang Academy of Agricultural Science (Hangzhou, China). The seeds were sown in plastic flower pots with nutrient soil and cultivated under a 12 h/12 h light photoperiod at 30 °C (day)/24 °C (night) in a greenhouse. After germination, the seedlings were watered with half-strength Hoagland nutrient solution every five days. For light treatment, the plants were divided into three groups after being sown in the soil and placed under monochromatic red (R, 661 nm), blue (B, 449 nm), and white (W, peaks at 445 and 560 nm) LED light. The photon flux density of light in each group was set to 300 µmol photons $m^{-2}$ $s^{-1}$. After being cultivated for five weeks, leaves from each treatment sample were collected and stored for further analysis. In total, 40 seedlings in each group and three replicates were set.

### *2.2. Analysis of Photosynthetic Parameters and Chlorophyll Fluorescence*

Photosynthetic parameters and chlorophyll fluorescence were measured according to the manufacturer's instructions. Briefly, photosynthetic parameters, including net photosynthetic rate (Pn), intercellular $CO_2$ concentration (Ci), stomatal conductance (Gs), and transpiration (Tr), were measured using a portable photosynthesis system (LI-6400, LI-COR Lincoln, NE, USA). The white light intensity was set to 1000 µmol photons $m^{-2}$ $s^{-1}$ inside the leaf chamber for all measurements. Chlorophyll fluorescence was measured using a portable fluorometer (PAM 2500; Walz, Effeltrich, Germany). Before determining Fv/Fm, the plants were exposed to darkness for 30 min.

### *2.3. Determination of Photosynthetic Pigment and Antioxidant Activity of Leaf Extracts*

The photosynthetic pigment contents of leaves were determined using commercial kits (Shanghai Zhen Ke Biological Technology Co., Ltd., Shanghai, China). To measure antioxidant activity, 1 g of leaves was taken under different light quality conditions, and 20 mL of ethanol was added for ultrasonic extraction. Then, the antioxidant activities of the leaf extracts, including DPPH, ABTS, and hydroxyl-scavenging activities, were determined using the corresponding kits (Shanghai Enzyme-linked Biotechnology, Shanghai, China) according to the manufacturer's instructions. The measurement was repeated three times and averaged.

### *2.4. RNA Extraction and Illumina Sequencing*

Total RNA from frozen samples was extracted using the RNAprep Pure Plant Plus Kit (Tiangen, China), following the manufacturer's instructions. Then, contaminated genomic DNA in the extracted RNA was removed using DNase I (Takara, Dalian, China). RNA purity and quality were analyzed using a NanoPhotometer spectrophotometer (IMPLEN, Westlake Village, CA, USA) and an Agilent 2100 bioanalyzer (Agilent Technologies, Santa Clara, CA, USA). Samples with RIN (RNA integrity number) values $\geq$ 8.0 were used for sequencing library construction. The sequencing libraries were prepared with an NEB

Next Ultra RNA Library Prep Kit (NEB, MA, USA). In total, 9 libraries were prepared with different light spectra treatments, and 150 bp paired-end reads were sequenced using an Illumina HiSeq 4000 platform (Illumina, San Diego, CA, USA).

### 2.5. Transcriptomic Analysis

To retrieve clean reads from the raw data, the FastQC tool (http://www.bioinformatics.babraham.ac.uk/projects/fastqc/) (accessed on 13 November 2021) was used to remove reads containing adapters, poly-N, and low-quality reads. Transcriptome assembly was carried out using Trinity (v2.11.0) by employing the paired-end method, and Corset (https://code.google.com/p/corset-project/) (accessed on 22 November 2021) was used to realign the relevant transcripts into unigenes. Then, the assembled unigenes were functionally annotated with various databases, such as Nr (NCBI non- redundant protein sequences), Swiss-Prot (A manually annotated and reviewed protein sequence database), Trembl (a variety of new documentation files and the creation of TrEMBL), KEGG (Kyoto Encyclopedia of Genes and Genomes), GO (Gene Ontology), KOG/COG (COG: Clusters of Orthologous Groups of proteins; KOG: Karyotic Ortholog Groups), and Pfam (Protein Family). Using RSEM (RNA-Seq via Expectation Maximization) to estimate the gene expression levels, and calculating the FPKM (fragments per kilobase of transcript per million mapped reads) of each gene based on the gene length. DEG (differentially expressed gene) analysis was performed using the DESeq R package (DESeq2 v1.22.1), with an adjusted $p$-value of <0.05. The DEGs were then subjected to enrichment analysis of GO functions and KEGG pathways, as described in our previous study [34].

### 2.6. Extraction and Detection of Metabolites

The sample leaves were ground into powder in liquid nitrogen. We transferred 1 g of powder to a 20 mL headspace vial (Agilent, Santa Clara, CA, USA) with saturated NaCl solution, and we used crimp-top caps with TFE-silicone headspace septa (Agilent) to seal the vial. For SPME (solid-phase micro-extraction) analysis, we heated the vial at 60 °C for 5 min, and then we exposed a 120 μm DVB/CWR/PDMS fiber (Agilent) to the headspace of the sample at 100 °C for 15 min. Then, the GC apparatus (Model 8890; Agilent) was used to desorb VOCs from the fiber coating at 250 °C for 5 min in splitless mode. The identification and quantification of VOCs was performed by using an Agilent Model 8890 GC and a 7000 D mass spectrometer (Agilent) equipped with a 30 m × 0.25 mm × 0.25 μm DB-5MS (5% phenyl-polymethylsiloxane) capillary column. The operational parameters were set as described in a previous study [35].

### 2.7. Metabolomic Analysis

Metabolomic analysis was conducted according to a previously described method [36]. In brief, to study metabolite-variety-specific accumulation, the statistics function prcomp within R (www.r-project.org) (accessed on 25 November 2021) was used to perform unsupervised principal component analysis (PCA). The OPLS-DA model was produced using the R package MetaboAnalystR, which was used to analyze all of the comparison groups and to screen differential metabolites. Differentially accumulated metabolites (DAMs) between the compared samples were filtered with the criteria of absolute log2FC (fold change) ≥1 and VIP ≥ 1. Pathway annotation and enrichment analysis of DAMs were carried out based on the Kyoto Encyclopedia of Gene and Genomes (KEGG) database (http://www.genome.jp/kegg) (accessed on 29 November 2021).

### 2.8. Integrative Analysis of Metabolomic and Transcriptomic Datasets

Association analysis between the metabolomic and transcriptomic datasets was performed as previously described in [37]. The R package was used to normalize the data before establishing the relationships between the DEGs and DAMs. Correlation, two-way orthogonal partial least squares (O2PLS), PCA, KEGG pathway enrichment, and other analyses were conducted.

## 3. Results

### 3.1. Plant Morphology and Growth Characteristics

To determine the effects of different light quality on basil growth, phenotypes of basil planted under red light (R), blue light (B), and white light (W) for five weeks were analyzed. Compared with the W group, the B group showed increased biomass, with greater plant height (about 1.12 times that of the W group) and leaf area, while the R group showed the smallest plant height and leaves. In terms of branching between stems, compared to W, the levels of branching under R were increased, while they decreased under B (Figure 1A). Interestingly, there was no significant difference in root length between the W and B groups, while the root length of the R group decreased significantly (Figure 1B). Consistent with the phenotype, the leaf and stem fresh weight were the highest in the B group (Figure 1C). The chlorophyll contents and carotenoid contents were both significantly higher in the B group but lower in the R group (Figure 1D).

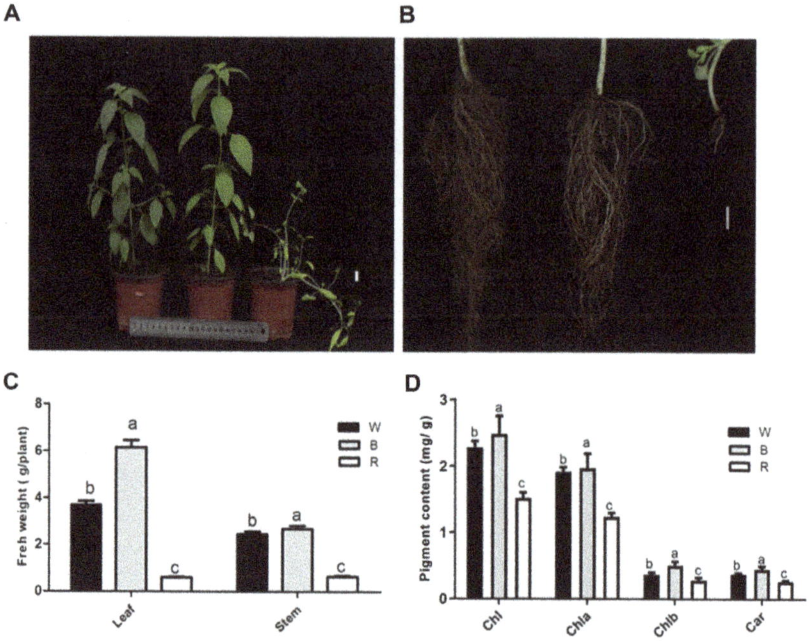


**Figure 1.** Phenotypes of basil under blue and red light: (**A**) Shoot phenotype of the W, B, and R groups. (**B**) Root phenotype of the W, B, and R groups. (**C**) Fresh weight of leaves and stems. (**D**) Pigment content of plant leaves. W, white light; B, blue light; R, red light. Scale bar: 1 cm. The data are shown as the mean ± SD ($n$ = 3), and different letters indicate significant differences between the data at $p < 0.05$.

As shown in Figure 2, compared to the W group, the photosynthetic parameters—including the net photosynthetic rate (Pn), stomatal conductance (Gs), intercellular $CO_2$ concentration (Ci), and transpiration rate (Tr)—were much higher in the B group. In contrast, the red group showed lower photosynthetic capacity, except for a higher Ci value than that of the W group. Chlorophyll fluorescence analysis revealed that the maximum photochemical efficiency of photosystem II (Fv/Fm) was reduced to 0.64 in the R group, while the values for the B and W groups remained around 0.8. Meanwhile, compared to the W group, the actual photochemical efficiencies (ΦPSII and ΦPSI) were significantly increased and reduced in the B and R groups, respectively (Figure S1).

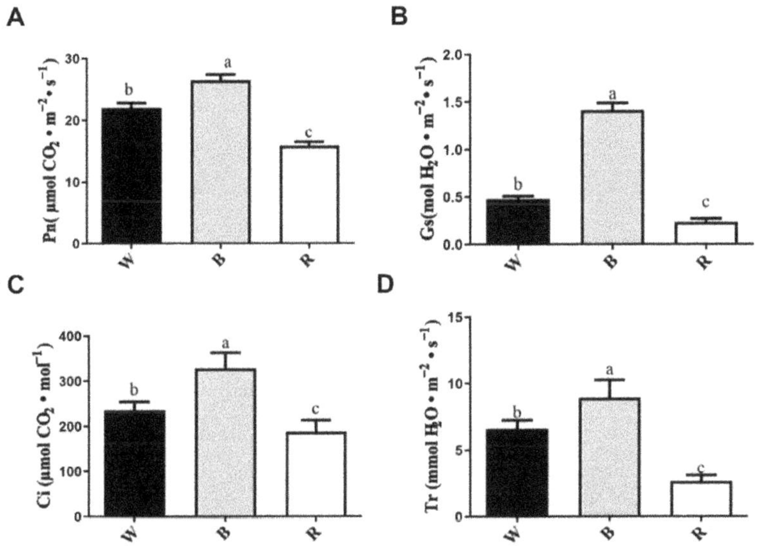


**Figure 2.** Measurement of photosynthetic parameters: (**A**) Pn, the net photosynthetic rate; (**B**) Gs, the stomatal conductance; (**C**) Ci, the intercellular $CO_2$ concentration; (**D**) Tr, the transpiration rate. The data are shown as the mean ± SD ($n = 3$), and different letters indicate significant differences between the data at $p < 0.05$.

*3.2. Antioxidant Activity Analysis of Leaf Extracts*

The antioxidant activity of the basil leaf extracts is shown in Figure 3. The DPPH radical scavenging activity under the blue light treatment was 1.25 times that under white light, while that under the red light treatment was only 0.3 times that under white light. Similarly, the ABTS radical scavenging activity under blue light treatment was 1.15 times that under white light, while that under the red light treatment was 0.75 times that under white light. There were no differences in hydroxyl radical scavenging activity under the three light treatments. Therefore, compared with white light, blue and red light were found to affect antioxidant activity, mainly by changing the activity of DPPH and ABTS.

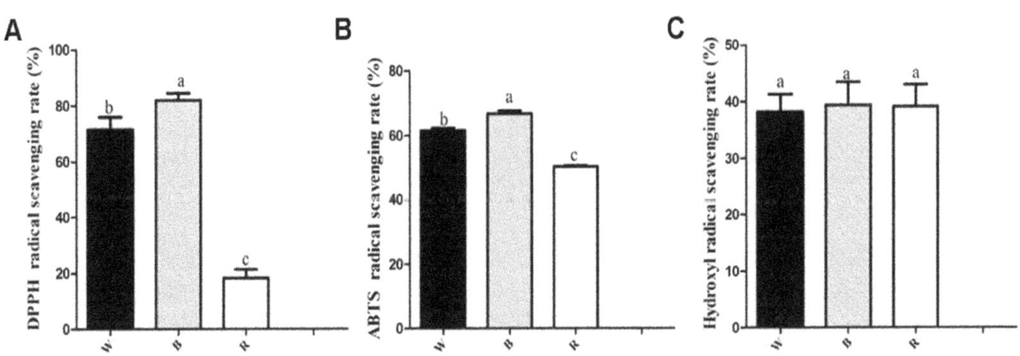


**Figure 3.** Antioxidant potential of leaf extracts: (**A**) DPPH radical scavenging activity. (**B**) ABTS radical scavenging activity. (**C**) Hydroxyl radical scavenging activity. The data are shown as the mean ± SD ($n = 3$), and different letters indicate significant differences between the data at $p < 0.05$.

### *3.3. Metabolite Profiling of the Leaves via GC-MS*

Basil is an aromatic plant containing many volatile organic compounds (VOCs) with important nutritional and medicinal values. To better understand the metabolic changes under different light qualities, VOCs of leaves from different samples were detected using GC-MS technology. A total of 1129 metabolites were identified in all samples, including 28 acids, 95 alcohols, 71 aldehydes, 27 amines, 75 aromatics, 174 esters, 1 ether, 4 halogenated hydrocarbons, 157 heterocyclic compounds, 90 hydrocarbons, 96 ketones, 10 nitrogen compounds, 29 phenols, 13 sulfur compounds, 251 terpenoids, and 8 others (Figure 4A, Table S1). The correlation analysis revealed that Pearson's correlation coefficients of all biological replicates were greater than or equal to 0.99 (Figure 4B), and biological replicates were clustered together in the heatmap (Figure 4C), suggesting that there was good homogeneity between the biological replicates, and that the data were reliable. The metabolites were divided into different clusters with specific expression trends in each group (Figure 4C), and principal component analysis (PCA) showed that the three groups were obviously distinguished from one another (Figure 4D). Hence, the above results indicate that different light qualities caused distinct metabolite profiles in basil.

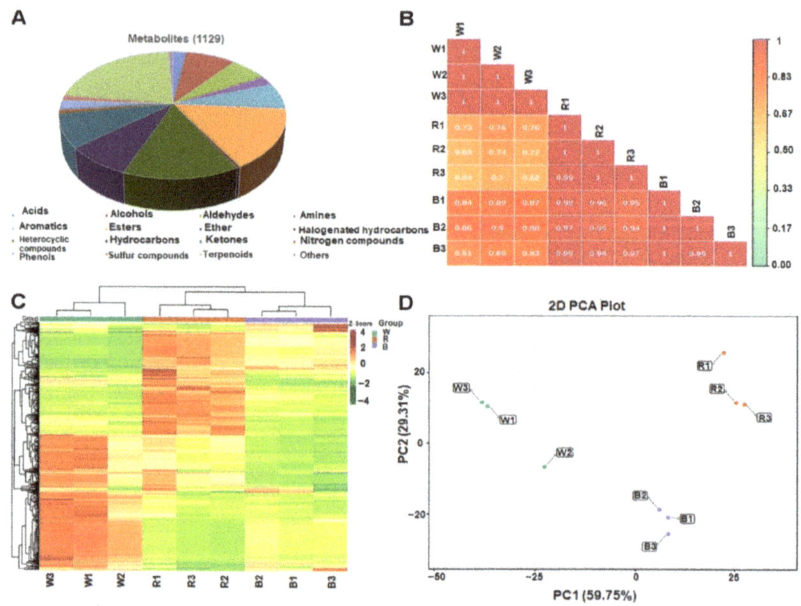


**Figure 4.** Statistical analysis of metabolome data: (**A**) Classification of identified metabolites. (**B**) Pearson's correlation analysis between samples. (**C**) Heatmap of the quantified identified metabolites. (**D**) Principal component analysis (PCA) score plot.

### *3.4. Pathway Enrichment Analysis of Differentially Accumulated Metabolites (DAMs)*

As a multivariate statistical analysis method, orthogonal partial least squares discriminant analysis (OPLS-DA) can maximize group differentiation, which is helpful for finding differentially expressed metabolites. As shown in Figure 5, pairwise comparisons were obtained based on the OPLS-DA model, and distinct metabolic differentiations were observed between three groups. Q2 is an important parameter for evaluating the model in OPLS-DA. The differences between W and R ($R^2X$ = 0.758, $R^2Y$ = 0.982, $Q^2$ = 0.964), W and B ($R^2X$ = 0.738, $R^2Y$ = 0.975, $Q^2$ = 0.957), and B and R ($R^2X$ = 0.702, $R^2Y$ = 0.993, $Q^2$ = 0.979) are shown in Figure S2. All $Q^2$ values in the three comparison groups were greater than 0.9, demonstrating that these models were meaningful. Then, the variable importance in projection (VIP) values of the OPLS-DA model and the fold change (FC) were combined to further screen the DAMs. An identification criterion of FC $\geq$ 2 or $\leq$0.5 together with VIP $\geq$ 1.0 was used to screen DAMs between the comparison groups. The screening results are shown in Supplementary Table S2. There were 491 DAMs between the W and B groups (322 downregulated, 169 upregulated), 630 DAMs between the W and R groups (288 downregulated, 342 upregulated), and 285 DAMs between the R and B groups (227 downregulated, 58 upregulated). The DAMs were divided into seven subgroups with different numbers using k-means cluster analysis (Figure S3). Thus, it can be concluded that the DAMs responded to specific light quality in unique ways.

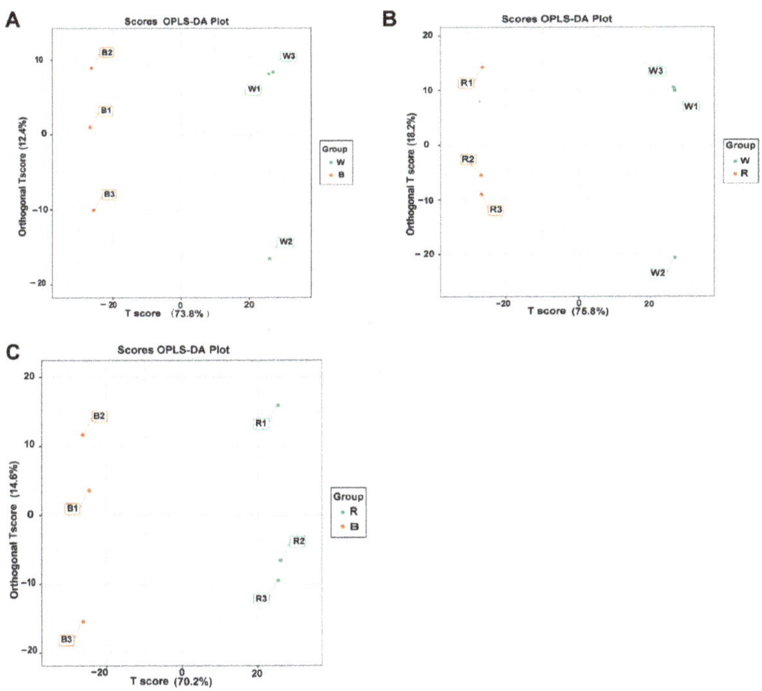


**Figure 5.** The score plots of orthogonal projections to latent structures discriminant analysis (OPLS-DA) pairwise comparisons of differentially accumulated metabolites (DAMs): (**A**) Comparison groups of W vs. B. (**B**) Comparison groups of W vs. R. (**C**) Comparison groups of B vs. R.

Furthermore, the DAMs from each comparison group were annotated using the Kyoto Encyclopedia of Genes and Genomes (KEGG) database. Most noteworthy, the DAMs were predominantly enriched in some overlapped KEGG pathways in the W vs. B, W vs. B, and B vs. R comparison groups, including the biosynthesis of secondary metabolites, monoterpenoid biosynthesis, phenylalanine metabolism, plant hormone signal transduction, limonene and pinene degradation, etc. (Figure 6).

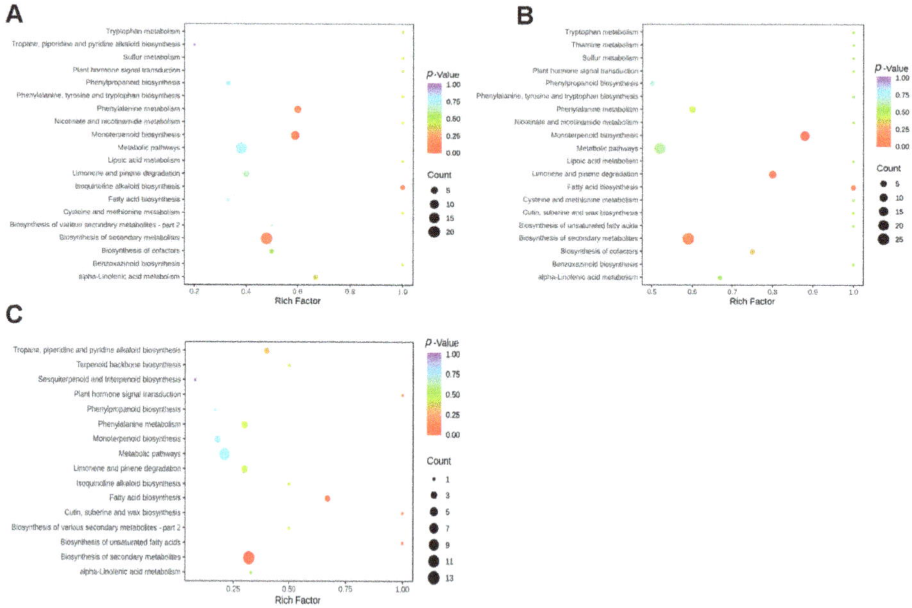


**Figure 6.** Kyoto Encyclopedia of Genes and Genomes (KEGG) pathway classification of DAMs in the comparison groups: (**A**) Comparison groups of W vs. B. (**B**) Comparison groups of W vs. R. (**C**) Comparison groups of B vs. R.

### *3.5. Annotation of Transcripts and Differentially Expressed Genes (DEGs)*

To investigate the global transcriptomic profile related to light quality responses in basil leaves, RNA-Seq analysis was carried out. A total of nine cDNA libraries were constructed from the W, B, and R groups. The cDNA sequencing generated 4.39–6.52 million raw reads and 4.24–6.32 million clean reads for each library. Approximately 97.97% of Q20, 94.18% of Q30, and 49.69% of GC contents were obtained for each library after filtering, which indicated that the RNA-Seq data were of high quality. Subsequently, 202,967 unigenes were produced by de novo transcriptome assembly, with a mean length of 1104 bp, an N50 length of 1623 bp, and an N90 length of 500 bp. Functional annotation analysis revealed that unigenes could be successfully annotated in at least one of the seven databases, including the COG, GO, KEGG, Swiss-Prot, NR, NT, and Pfam databases. Moreover, species distribution analysis showed that 42.19% of the unigenes were homologous with the sequence of *Sesamum indicum*, and 22.03%, 13.22%, and 2.91% of the unigenes were homologous with the sequences of *Handroanthus impetiginosus*, *Erythranthe guttata*, and *Olea europaea* var. *sylvestris*, respectively. FPKM values were used to estimate the gene expression levels. Pearson's correlation coefficients and PCA analysis showed high reproducibility among the biological replicates (Figure S4), suggesting that these sequencing data were reliable for differential gene expression.

### *3.6. Dynamic Transcriptome Analysis in Response to Light Spectra*

With the stand of $|\log2(\text{fold change})| > 1$ and FDR (false discovery rate) $< 0.05$, differentially expressed genes (DEGs) between the three comparison groups (W vs. R, W vs. B, and R vs. B) were screened. The results showed that 29,802 DEGs were detected in the W vs. R group, including 9690 downregulated DEGs and 20,112 upregulated DEGs. The W vs. B group showed the highest level of DEGs, at 34,760, of which 23,725 were upregulated and 11,035 were downregulated. For the R vs. B group, there were 16,417 DEGs, comprising 9189 upregulated and 7228 downregulated genes. Interestingly, 2898 common DEGs were detected in all three comparison groups, while the most specific DEGs (6952) were found in the W vs. B group (Figure 7).

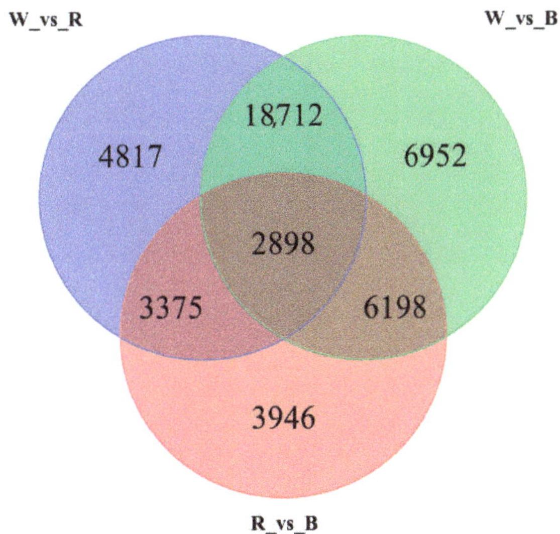


**Figure 7.** Venn diagram showing the overlapping and unique differentially expressed genes (DEGs) amongst the three comparison groups (W vs. R, W vs. B, and R vs. B).

In order to study the expression patterns of genes under different light spectra, the FPKMs of all different genes were normalized using the R language, and then k-means cluster analysis was performed. The expression patterns of genes in the W vs. B, W vs. R, and B vs. R groups could be divided into five subclasses (Figure S5, Table S3). In subclass 1 and subclass 4, the gene expression was highest under blue light and lowest under white light. In contrast, the gene expression was lowest under blue light and highest under white light in subclass 3. In subclass 2, a total of 3716 genes were expressed most highly under white light, followed by blue and red light. The 7530 genes in subclass 5 showed the highest expression under red light and the lowest under white light. In summary, the expression of DEGs was induced by different light spectra in specific ways.

### *3.7. GO and KEGG Pathway Enrichment Analysis of DEGs*

The DEGs were enriched in three GO categories: biological process, cellular component, and molecular function. In the W vs. B group, the top 50 most significantly enriched GO terms were as displayed in Figure S6A, among which the DEGs annotated in the biological process category mainly include the aminoglycan biosynthetic process, carbon fixation, cellular response to alkaline pH, cellular response to light intensity, and another 30 GO terms; the DEGs in the cellular component category contained photosystem, photosystem I, photosystem II, and plastoglobule; and the DEGs in the molecular function category involved adenosyl homocysteinase activity, chlorophyll binding, gibberellin binding, histone acetyltransferase activity, etc. The top 50 most significantly enriched GO terms in the W vs. R group are shown in Figure S6B; 28 biological process terms were most enriched, including mucilage extrusion from seed coats, negative regulation of anion channel activity, negative regulation of anion-channel activity by blue light, and negative regulation of anion transport; the DEGs in the cellular component category included amyloplast, DNA packaging complex, nucleosome, and another four terms; the DEGs in the molecular function category consisted of 15 terms, including 3−oxo−arachidoyl−CoA synthase activity, 3−oxo−cerotoyl−CoA synthase activity, adenosyl homocysteinase activity, trialkyl sulfonium hydrolase activity, etc. In the B vs. R group, 31 biological process terms, 11 cellular component terms, and 8 molecular function terms were enriched in the top 50 GO terms (Figure S6).

KEGG pathway enrichment analysis was performed to uncover the specific metabolic pathways that the DEGs participated in. In the W vs. B pair, the top 20 enriched pathways were mainly involved in carbon fixation in photosynthetic organisms, carbon metabolism, plant hormone signal transduction, MAPK signaling pathway–plant, terpenoid backbone biosynthesis, fatty acid elongation, etc. (Figure 8A). Pathways such as circadian rhythm–plant, aminoacyl-tRNA biosynthesis, selenocompound metabolism, isoflavonoid biosynthesis, and ubiquitin-mediated proteolysis were significantly enriched in the W vs. R group (Figure 8B). The top 20 enriched pathways in the B vs. R group are displayed in Figure 8C, including biosynthesis of secondary metabolites, metabolic pathways, plant–pathogen interaction, alpha-linolenic acid metabolism, and 16 other pathways.

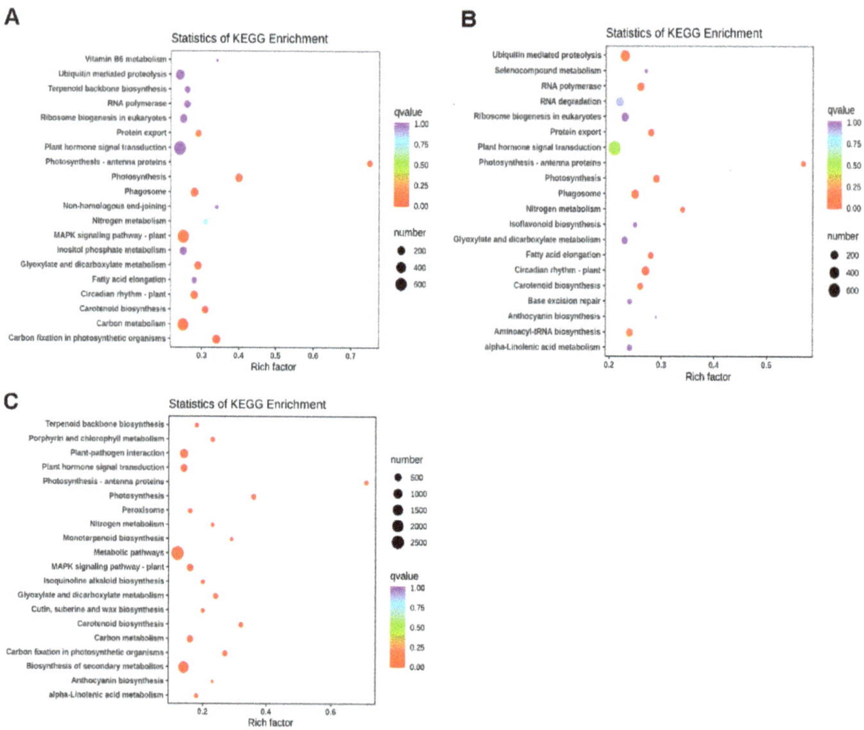


**Figure 8.** KEGG enrichment analysis of all DEGs: (**A**) Comparison groups of W vs. B. (**B**) Comparison groups of W vs. R. (**C**) Comparison groups of B vs. R.

### *3.8. Association Analysis of DAMs and DEGs*

Based on Pearson's correlation coefficients (PCCs) between the DEGs and DAMs, nine quadrant diagrams were constructed to reveal the variations in metabolites and their corresponding genes. The DEGs and DAMs in quadrants 3 and 7 had positive correlations, while the DEGs and DAMs in quadrants 1 and 9 had negative correlations (Figure 9). Additionally, the DEGs and DAMs were annotated to the KEGG pathway database to determine their common biological pathways. These significantly changed pathways in all of the comparison groups mainly included metabolic pathways, biosynthesis of secondary metabolites, phenylalanine metabolism, limonene and pinene degradation, and monoterpenoid biosynthesis (Figure S7). Interestingly, compared with the W vs. B group, more common pathways were enriched in the W vs. R group (Figure S7), suggesting that red light affects more pathways at the transcriptional and metabolic levels.

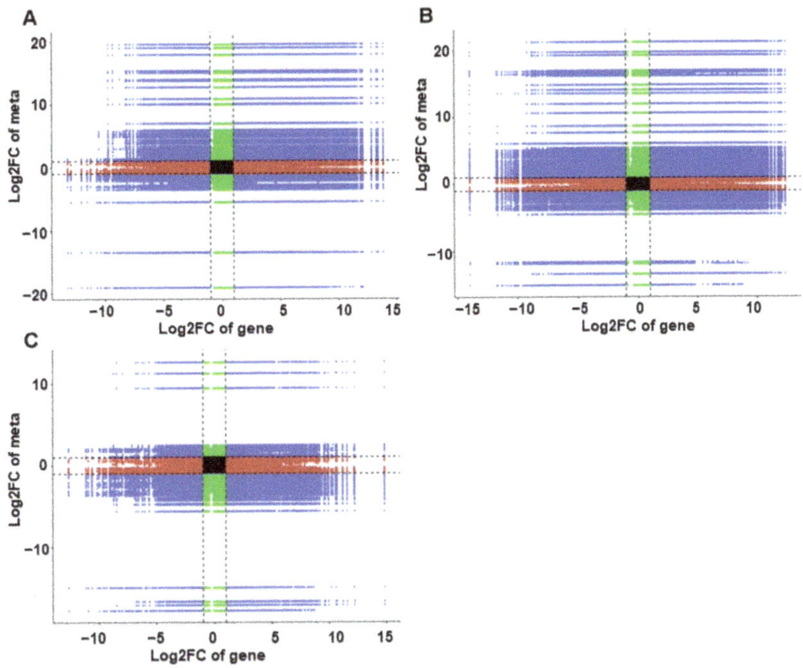


**Figure 9.** Quadrant diagram representing the association of transcriptomic and metabolomic variations in the W vs. B (**A**), W vs. R (**B**), and B vs. R (**C**) groups. The black dashed line indicates the difference threshold. Outside the threshold line, there are significant differences in genes/metabolites, while inside the threshold line the genes/metabolites are unchanged. Each dot represents a gene/metabolite. Black dots represent invariant genes/metabolites, green dots represent differentially accumulated metabolites with unchanged genes, red dots represent differentially expressed genes with unchanged metabolites, and blue dots represent both differentially expressed genes and differentially accumulated metabolites.

In addition, an O2PLS model was generated to determine the relationships between the DEGs and DAMs, which could be used to screen important variables from one set of data that influence the other. In the gene loading, 7 genes localized to chloroplasts were among the top 10 DEGs (Figure 10A), including carbonic anhydrase (Cluster-9688.94804), hydroxy-3-methylbut-2-enyl diphosphate reductase (Cluster-9688.72219), phosphoglycolate phosphatase 1A (Cluster-9688.101671), sugar partition-affecting protein (Cluster-9688.46987), WAT1-related protein (Cluster-9688.69111), PsbQ (Cluster-9688.96670) and plastid transcriptionally active 16 (Cluster-9688.82754), suggesting their importance in regulating metabolites. The top 10 DAMs displayed in Figure 10B, such as (2-bromoethyl)-oxirane (XMW0397), undecanoic acid (KMW0592), 1-dodecanol (KMW0629) and E-1-methoxy-4-hexene (XMW0279), were most closely associated with the DEGs.

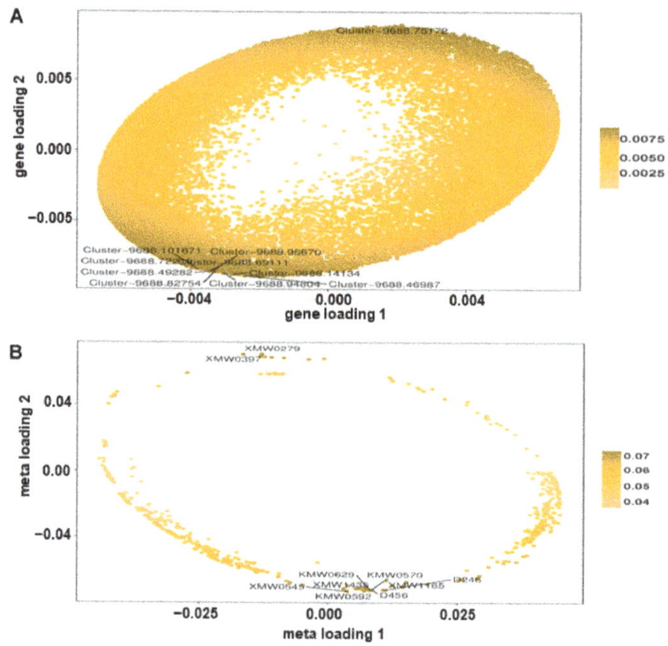


**Figure 10.** Integration analysis of metabolome and transcriptome by O2PLS: (**A**) The top 10 genes with transcriptomes strongly influencing the metabolome. (**B**) The top 10 metabolites substantially affecting the transcriptome.

## 4. Discussion

### *4.1. Changes in the Physiology and Antioxidant Capacity of Leaf Extracts*

As an important environmental factor, light quality plays irreplaceable roles in the growth and development of plants. Uncovering the mechanisms of plants in response to light quality is beneficial for developing crop improvement strategies. Red and blue light are generally considered to be the most efficient wavelengths to drive photosynthesis [10]. We found that the photosynthetic rate of basil increased significantly under blue light and decreased under red light compared to white light (Figure 2). At the same time, the stomatal conductance (Gs) under blue light was the largest, followed by white light and red light, indicating that the change in photosynthesis may be caused by stomatal opening, which is consistent with the findings of previous studies [38–40]. In addition, as one of the most important pigments in plants, chlorophyll is essential for capturing light for photosynthesis, which plays key roles in the interaction with light during the whole growth cycle of the plant. Many studies have revealed that light quality affects photosynthesis by changing the content and composition of chlorophyll. In some plants, including *Cattleya loddigesii* and *Triticum aestivum*, the chlorophyll content was highest under red light compared to other light [41].However, blue light was more effective than red light in promoting chlorophyll accumulation in lettuce, *Anoectochilus roxburghii*, and *Toona sinensis* [41,42]. Herein, we also found that blue light stimulated chlorophyll accumulation in basil more significantly than the other two types of light, which may further facilitate the light harvesting capacity to improve photosynthesis.

Light spectra affect plants' morphogenesis. For example, blue light significantly inhibited apples' plant growth and leaf extension, while red light treatment promoted apples' plant growth and root development [43]. In tomatoes, compared to white light, early photomorphogenesis was strongly promoted by red light and inhibited by blue light [44]. Conversely, it was found that blue light could promote plant growth and leaf extension better than red light in some species, such as *Petunia*, *Chrysanthemum*, and *Cucumber* [45–47], and similar phenotypes were found in our study, suggesting that red and blue light have unique effects on different plants. Plant hormones mediate various aspects of plants' growth and development, and some hormonal pathways are often influenced by light to regulate the developmental changes [48–50]. We found that 688, 783, and 465 genes involved in plant hormone signal transduction were differentially expressed in the W vs. R, W vs. B, and B vs. R groups, respectively (Figure 10). The altered expression patterns of plants' hormone signals may give rise to morphological changes in basil under different light quality conditions.

In addition to physiological and morphological changes, light spectra can also affect the synthesis of metabolites in herbs, thus modifying their biological activities. Blue LED irradiation increased the DPPH scavenging capacity of baby leaf lettuce and improved its antioxidant properties [51], and callus cultures of *Rhodiola imbricata* grown under blue light also displayed the highest antioxidant activity compared to other light conditions [52]. Consistently, in this study, antioxidant capacities—especially DPPH and ABTS scavenging activities—were significantly promoted by blue light and inhibited by red light. As the most abundant pigments in nature, chlorophylls and carotenoids are very potent natural antioxidants, which may be used for a range of health benefits in humans [53,54]. Sgherri et al. (2011) found that chlorophylls contributed about 40% to the bulk of the fast lipophilic antioxidants in basil extracts, whereas carotenoids could be ascribable to the slow antioxidant activity [55]. Since more chlorophylls and carotenoids were accumulated under blue light, and fewer under red light (Figure 1D), it can be inferred that the antioxidant activity of basil may be associated with these pigments' contents.

### *4.2. Involvement of Key Metabolites in Response to Light Spectra*

Due to its unique and pleasant aroma, volatile organic compounds (VOCs) of basils grown under different light spectra qualities were identified using GC-MS/MS analysis. A total of 1129 metabolites belonging to 16 classes were identified in all samples (Table S1), among which the terpenoids contained a maximum of 251 species and accounted for about 22%. To the best of our knowledge, previous studies have not identified so many VOCs in basil (*O. basilicum* Linn. var. *pilosum* (*Willd.*) Benth.) [33,56], and this study will therefore facilitate our understanding of the chemical components of basil. Furthermore, DAMs were screened to uncover the effects of light spectra on metabolites. All of the DAMs in the three comparison groups were mainly enriched in many of the same KEGG pathways, such as biosynthesis of secondary metabolites, monoterpenoid biosynthesis, phenylalanine metabolism, plant hormone signal transduction, limonene and pinene degradation, etc. (Figure 5), suggesting that both red and blue light have significant effects on these pathways. The k-means cluster analysis showed that all of the DAMs could be classified into seven subclasses with different numbers (Figure S3, Table S4). In subclasses 1, 2, and 7, the contents of 373 metabolites were highest under white light. In subclasses 4, 5, and 6, the contents of 378 metabolites were highest under red light. Interestingly, the highest accumulation of 35 metabolites in subclass 3 was observed under blue light. Taken together, it can be seen that metabolites respond to different light qualities in specific ways.

The top 10 most significant DAMs in each comparison group are listed in Figure 10. Among them, a total of seven significantly upregulated common metabolites were found in both the W vs. R and W vs. B groups, including debrisoquine, butanoic acid, 4-hexenyl ester, propanoic acid, 2-methyl-,phenylmethyl ester, (Z)-, (1aR,7R,7aR,7bS)-(+)-1a,2,3,5,6,7,7a,7b-Octahydro-1,1,7,7a-tetramethyl-1H-cyclopropa [a] naphthalen -3-one, ethyl tridecanoate, and 1H-pyrrolo [2,3-b]pyridine, 2-(1-methylethyl)-, 7-(2-hydroxypropan-2-yl)-1,4a-dimethyldecahydronaphthalen-1-ol, which indicated that both red and blue light significantly promoted the accumulation of these metabolites. Similarly, 2-octanone, 3-octanone, and 5-hexenal,4-methylene- were three common metabolites among the top ten most significantly downregulated in both the W vs. R and W vs. B groups, suggesting that the accumulation of these metabolites was seriously inhibited under both red and blue light. In terms of the R vs. B group, the top 10 most upregulated metabolites included cyclohexanecarboxylic acid, methyl ester, oleic acid, 2,6-piperidinedione, 3-ethyl-,2-hexenal, (E)-, and another 6 metabolites; it is worth noting that cyclohexanecarboxylic acid, methyl ester, oleic acid, and 3-octanol were among the top 10 most downregulated metabolites in the W vs. R group, indicating that these three key metabolites accumulated significantly under white and blue light but not under red light. The top 10 most downregulated metabolites in the R vs. B group consisted of iso-3-thujyl acetate, ketone, methyl 2,4,5-trimethylpyrrol-3-yl, fenchyl acetate, furfuryl pentanoate, pyrazine, 2,3-dimethyl-5-(1-methylpropyl)-, and another 5 metabolites; among them, 10-undecenal, (2E,4Z)-2,4-decadienal, and 3-cyclohexene-1-ethanol, beta.,4-dimethyl- were also included in the top 10 most upregulated metabolites in the W vs. R group, revealing that the accumulation of these three key metabolites was significantly strengthened by red light. Overall, these key metabolites could be used as the characteristic metabolites to distinguish the response of basil to blue, red, and white light.

### 4.3. *Key Genes Associated with the Light Response in Basil*

The phenotypic alterations induced by light are always accompanied by changes in gene expression; transcriptional regulation has emerged as an important regulation mode in organisms. In our transcriptomic data, DEGs with different quantities between the three comparison groups were identified, which were further enriched in many KEGG pathways (Figure 9). Transcription factors are central regulators of transcriptional reprogramming, and many transcription factors, such as HY5 and PIFs, have been proven to participate in light responses [57–59]. We found that a total of 2133 TFs were differentially expressed in the W vs. B group (Table S5), which included 586 downregulated genes and 1547 upregulated genes. In the W vs. R group, 1771 TFs were differentially expressed, including AP2/ERF (98), WRKY (76), bHLH (83), C3H (83), SNF2 (65), GRAS (65), C2H2 (62), and others. It was worth noting that the DEGs with the greatest influence on DAMs were screened using an O2PLS model. Interestingly, the top 10 DEGs closely associated with DAMs were screened (Figure 10), including 7 chloroplast localization genes, such as carbonic anhydrase (Cluster-9688.94804) and plastid transcriptionally active 16 (Cluster-9688.82754). Collectively, all of these DEGs could be associated with light signal pathways, and their specific regulatory mechanisms await further investigation through genetic modification combined with molecular biology experiments in future studies.

## 5. Conclusions

In this study, the response mechanisms of basil (*O. basilicum* Linn. var. *pilosum* (*Willd.*) *Benth.*) to blue and red light were revealed through physiological changes and multi-omics analysis. The physiological results demonstrated that blue and red light have different effects on photomorphogenesis by changing plants' height, leaf size, and branch number. Metabolomic and transcriptomic analyses showed that the DAMs and DEGs were annotated to common biological pathways such as metabolic pathways, biosynthesis of secondary metabolites, phenylalanine metabolism, limonene and pinene degradation, and monoterpenoid biosynthesis. These results contribute to understanding the mechanisms of plants' response to light spectra, which may also be useful in the cultivation of crops.

**Supplementary Materials:** The following supporting information can be downloaded at: https://www.mdpi.com/article/10.3390/horticulturae9111172/s1, Figure S1: Measurement of chlorophyll fluorescence parameters: (A) Maximum photochemical efficiency of PSII, Fv/Fm. (B) The actual photochemical efficiency of PSII, ΦPSII. (C) The actual photochemical efficiency of PSI, ΦPSI. The data are shown as the mean ± SD ($n$ = 3), and different letters indicate significant differences between the data at $p < 0.05$. Figure S2: Evaluation parameters of the OLPS-DA model: (A) Comparison groups of W vs. B. (B) Comparison groups of W vs. R. (C) Comparison groups of B vs. R. Figure S3: K-means cluster analysis of DEGs. Figure S4: Statistical analysis of transcriptome data: (A) Pearson's correlation analysis between samples. (B) Principal component analysis (PCA) score plot. Figure S5: K-means cluster analysis of DEGs. Figure S6: GO enrichment analysis of all DEGs: (A) Comparison groups of W vs. B. (B) Comparison groups of W vs. R. (C) Comparison groups of B vs. R. Figure S7:

Kyoto Encyclopedia of Genes and Genomes (KEGG) pathway classification of DAMs and DEGs in the comparison groups: (A) Comparison groups of W vs. B. (B) Comparison groups of W vs. R. (C) Comparison groups of B vs. R. Table S1: All metabolites identified in the samples. Table S2: DAMs in each comparison group. Table S3: DEGs in all k-means clusters. Table S4: DAMs in all k-means clusters. Table S5: Differentially expressed transcription factors.

**Author Contributions:** Q.W., F.D. and T.H. conceived the ideas and designed the methodology; R.Y. and J.D. collected the data; R.Y., D.L. and Y.J. analyzed the data; Q.W., F.D. and T.H. led the writing of the manuscript. All authors have read and agreed to the published version of the manuscript.

**Funding:** This research was funded by the Doctoral Research Project Foundation of Yan'an University (YAU202303794).

**Data Availability Statement:** Any data and codes used in this study are available upon request from the corresponding author.

**Acknowledgments:** Special thanks to Yan'an University for the financial support, and to the China Shaanxi Key Laboratory of Chinese Jujube for their help in the implementation of the experiment.

**Conflicts of Interest:** The authors declare that they have no conflict of interest.

## References

1. Deng, X.W.; Quail, P.H. Signalling in light-controlled development. *Semin. Cell Dev. Biol.* **1999**, *10*, 121–129. [CrossRef] [PubMed]
2. Yadav, A.; Singh, D.; Lingwan, M.; Yadukrishnan, P.; Masakapalli, S.K.; Datta, S. Light signaling and UV-B-mediated plant growth regulation. *J. Integr. Plant Biol.* **2020**, *62*, 1270–1292. [CrossRef] [PubMed]
3. Wang, J.; Lu, W.; Tong, Y.; Yang, Q. Leaf morphology, photosynthetic performance, chlorophyll fluorescence, stomatal development of lettuce (*Lactuca Sativa* L.) exposed to different ratios of red light to blue light. *Front. Plant Sci.* **2016**, *7*, 250. [CrossRef] [PubMed]
4. Wei, L.; You, W.; Gong, Y.; El Hajjami, M.; Liang, W.; Xu, J.; Poetsch, A. Transcriptomic and proteomic choreography in response to light quality variation reveals key adaption mechanisms in marine *Nannochloropsis oceanica*. *Sci. Total Environ.* **2020**, *720*, 137667. [CrossRef]
5. Suetsugu, N.; Wada, M. Evolution of three LOV blue light receptor families in green plants and photosynthetic stramenopiles: Phototropin, ZTL/FKF1/LKP2 and aureochrome. *Plant Cell Physiol.* **2013**, *54*, 8–23. [CrossRef]
6. Jenkins, G.I. The UV-B photoreceptor UVR8: From structure to physiology. *Plant Cell* **2014**, *26*, 21–37. [CrossRef]
7. Paik, I.; Huq, E. Plant photoreceptors: Multi-functional sensory proteins and their signaling networks. *Semin. Cell Dev. Biol.* **2019**, *92*, 114–121. [CrossRef]
8. Chaves, I.; Pokorny, R.; Byrdin, M.; Hoang, N.; Ritz, T.; Brettel, K.; Essen, L.-O.; van der Horst, G.T.J.; Batschauer, A.; Ahmad, M. The cryptochromes: Blue light photoreceptors in plants and animals. *Annu. Rev. Plant Biol.* **2011**, *62*, 335–364. [CrossRef]
9. Chen, M.; Chory, J. Phytochrome signaling mechanisms and the control of plant development. *Trends Cell Biol.* **2011**, *21*, 664–671. [CrossRef]
10. Liu, J.; van Iersel, M.W. Photosynthetic physiology of blue, green, and red light: Light intensity effects and underlying mechanisms. *Front. Plant Sci.* **2021**, *12*, 619987. [CrossRef]
11. Muneer, S.; Kim, E.J.; Park, J.S.; Lee, J.H. Influence of green, red and blue light emitting diodes on multiprotein complex proteins and photosynthetic activity under different light intensities in lettuce leaves (*Lactuca Sativa* L.). *Int. J. Mol. Sci.* **2014**, *15*, 4657–4670. [CrossRef] [PubMed]
12. Shen, Z.; Xia, K.; Wu, Q.; Su, N.; Cui, J. Effects of light quality on the chloroplastic ultrastructure and photosynthetic characteristics of cucumber seedlings. *Plant Growth Regul.* **2014**, *73*, 227–235.
13. Hoenecke, M.E.; Bula, R.J.; Tibbitts, T.W. Importance of "blue" photon levels for Lettuce seedlings grown under red-light-emitting diodes. *HortScience* **1992**, *27*, 427–430. [CrossRef]
14. Goins, G.D.; Yorio, N.C.; Sanwo, M.M.; Brown, C.S. Photomorphogenesis, photosynthesis, and seed yield of wheat plants grown under red light-emitting diodes (LEDs) with and without supplemental blue lighting. *J. Exp. Bot.* **1997**, *48*, 1407–1413. [CrossRef] [PubMed]
15. He, J.; Qin, L.; Chong, E.L.C.; Choong, T.-W.; Lee, S.K. Plant growth and photosynthetic characteristics of Mesembryanthemum crystallinum grown aeroponically under different blue- and red-LEDs. *Front. Plant Sci.* **2017**, *8*, 361. [CrossRef]
16. Sergejeva, D.; Alsina, I.; Duma, M.; Dubova, L.; Berzina, K. Evaluation of different lighting sources on the growth and chemical composition of lettuce. *Agron. Res.* **2018**, *16*. [CrossRef]
17. Shimizu, H.; Saito, Y.; Nakashima, H.; Miyasaka, J.; Ohdoi, K. Light environment optimization for lettuce growth in plant factory. *IFAC Proc. Vol.* **2011**, *44*, 605–609. [CrossRef]
18. Kobayashi, K.; Amore, T.; Lazaro, M. Light-emitting diodes (LEDs) for miniature hydroponic lettuce. *Opt. Photon. J.* **2013**, *3*, 74–77. [CrossRef]
19. Sæbø, A.; Krekling, T.; Appelgren, M. Appelgren light quality affects photosynthesis and leaf anatomy of birch plantlets in vitro. *Plant Cell Tissue Organ Cult.* **1995**, *41*, 177–185. [CrossRef]

20. Miao, L.; Zhang, Y.; Yang, X.; Xiao, J.; Zhang, H.; Zhang, Z.; Wang, Y.; Jiang, G. Colored light-quality selective plastic films affect anthocyanin content, enzyme activities, and the expression of flavonoid genes in strawberry (*Fragaria* × *Ananassa*) Fruit. *Food Chem.* **2016**, *207*, 93–100. [CrossRef]
21. Wang, C.; Qiu, H.; Chen, Y.; Xu, Y.; Shan, F.; Li, H.; Yan, C.; Ma, C. Red light regulates metabolic pathways of soybean hypocotyl elongation and thickening. *Environ. Exp. Bot.* **2022**, *199*, 104890. [CrossRef]
22. Mao, P.; Li, Q.; Li, Y.; Xu, Y.; Yang, Q.; Bian, Z.; Wang, S.; He, L.; Xu, Z.; Zheng, Y.; et al. The beneficial functions of blue light supplementary on the biosynthesis of glucosinolates in pakchoi (*Brassica Rapa* L. ssp. Chinensis) under greenhouse conditions. *Environ. Exp. Bot.* **2022**, *197*, 104834. [CrossRef]
23. Chen, X.; Cai, W.; Xia, J.; Yu, H.; Wang, Q.; Pang, F.; Zhao, M. Metabolomic and transcriptomic analyses reveal that blue light promotes chlorogenic scid synthesis in strawberry. *J. Agric. Food Chem.* **2020**, *68*, 12485–12492. [CrossRef] [PubMed]
24. Hoffmann, A.M.; Noga, G.; Hunsche, M. High blue light improves acclimation and photosynthetic recovery of pepper plants exposed to UV stress. *Environ. Exp. Bot.* **2015**, *109*, 254–263. [CrossRef]
25. Calderón Bravo, H.; Vera Céspedes, N.; Zura-Bravo, L.; Muñoz, L.A. Basil seeds as a novel food, source of nutrients and functional ingredients with beneficial properties: A review. *Foods* **2021**, *10*, 1467. [CrossRef]
26. Srivastava, S.; Lal, R.K.; Maurya, R.; Mishra, A.; Yadav, A.K.; Pandey, G.; Rout, P.K.; Chanotiya, C.S. Chemical diversity of essential oil among basil genotypes (Ocimum Viride Willd.) across the years. *Ind. Crops Prod.* **2021**, *173*, 114153. [CrossRef]
27. Prakash, P.; Gupta, N. Therapeutic uses of *Ocimum sanctum* Linn (Tulsi) with a note on eugenol and its pharmacological actions: A short review. *Indian J. Physiol. Pharmacol.* **2005**, *49*, 125–131.
28. Ghasemzadeh, A.; Ashkani, S.; Baghdadi, A.; Pazoki, A.; Jaafar, H.Z.E.; Rahmat, A. Improvement in flavonoids and phenolic acids production and pharmaceutical quality of sweet basil (*Ocimum Basilicum* L.) by ultraviolet-B irradiation. *Molecules* **2016**, *21*, 1203. [CrossRef]
29. Nazir, M.; Asad Ullah, M.; Mumtaz, S.; Siddiquah, A.; Shah, M.; Drouet, S.; Hano, C.; Abbasi, B.H. Interactive effect of melatonin and UV-C on phenylpropanoid metabolite production and antioxidant potential in callus cultures of purple basil (*Ocimum Basilicum* L. var.s Purpurascens). *Molecules* **2020**, *25*, 1072. [CrossRef]
30. Nadeem, M.; Abbasi, B.H.; Younas, M.; Ahmad, W.; Zahir, A.; Hano, C. LED-enhanced biosynthesis of biologically active ingredients in callus cultures of Ocimum basilicum. *J. Photochem. Photobiol. B.* **2019**, *190*, 172–178. [CrossRef]
31. Pennisi, G.; Blasioli, S.; Cellini, A.; Maia, L.; Crepaldi, A.; Braschi, I.; Spinelli, F.; Nicola, S.; Fernandez, J.A.; Stanghellini, C.; et al. Unraveling the role of red:blue LED lights on resource use efficiency and nutritional properties of indoor grown sweet basil. *Front. Plant Sci.* **2019**, *10*, 305. [CrossRef]
32. Larsen, D.H.; Li, H.; Shrestha, S.; Verdonk, J.C.; Nicole, C.C.S.; Marcelis, L.F.M.; Woltering, E.J. Lack of blue light regulation of antioxidants and chilling tolerance in basil. *Front. Plant Sci.* **2022**, *13*, 852654. [CrossRef] [PubMed]
33. Zhang, J.-W.; Li, S.-K.; Wu, W.-J. The main chemical composition and in vitro antifungal activity of the essential oils of *Ocimum basilicum* Linn. var. *pilosum (Willd.) Benth. Molecules* **2009**, *14*, 273–278. [CrossRef] [PubMed]
34. Wu, Q.; Han, T.; Yang, L.; Wang, Q.; Zhao, Y.; Jiang, D.; Ruan, X. The essential roles of OsFtsH2 in developing the chloroplast of rice. *BMC Plant Biol.* **2021**, *21*, 445. [CrossRef] [PubMed]
35. Wang, Y.; Zheng, P.-C.; Liu, P.-P.; Song, X.-W.; Guo, F.; Li, Y.-Y.; Ni, D.-J.; Jiang, C.-J. Novel insight into the role of withering process in characteristic flavor formation of teas using transcriptome analysis and metabolite profiling. *Food Chem.* **2019**, *272*, 313–322. [CrossRef] [PubMed]
36. Wang, H.; Hua, J.; Yu, Q.; Li, J.; Wang, J.; Deng, Y.; Yuan, H.; Jiang, Y. Widely targeted metabolomic analysis reveals dynamic changes in non-volatile and volatile metabolites during green tea processing. *Food Chem.* **2021**, *363*, 130131. [CrossRef]
37. Li, L.; Kong, Z.; Huan, X.; Liu, Y.; Liu, Y.; Wang, Q.; Liu, J.; Zhang, P.; Guo, Y.; Qin, P. Transcriptomics integrated with widely targeted metabolomics reveals the mechanism underlying grain color formation in wheat at the grain-filling stage. *Front. Plant Sci.* **2021**, *12*, 757750. [CrossRef] [PubMed]
38. Eduardo, Z.; Jianxin, Z. Role of zeaxanthin in blue light photoreception and the modulation of light-$CO_2$ interactions in guard cells. *J. Exp. Bot.* **1998**, *49*, 433–442.
39. Sharkey, T.D.; Raschke, K. Effect of light quality on stomatal opening in leaves of *Xanthium strumarium* L. *Plant Physiol.* **1981**, *68*, 1170–1174. [CrossRef]
40. Kinoshita, T.; Doi, M.; Suetsugu, N.; Kagawa, T.; Wada, M.; Shimazaki, K. Phot1 and phot2 mediate blue light regulation of stomatal opening. *Nature* **2001**, *414*, 656–660. [CrossRef]
41. Liu, Y.; Wang, T.; Fang, S.; Zhou, M.; Qin, J. Responses of morphology, gas exchange, photochemical activity of photosystem II, and antioxidant balance in *Cyclocarya paliurus* to light spectra. *Front. Plant Sci.* **2018**, *9*, 1704. [CrossRef] [PubMed]
42. Ye, S.; Shao, Q.; Xu, M.; Li, S.; Wu, M.; Tan, X.; Su, L. Effects of light quality on morphology, enzyme activities, and bioactive compound contents in *Anoectochilus roxburghii*. *Front. Plant Sci.* **2017**, *8*, 857. [CrossRef] [PubMed]
43. Li, Z.; Chen, Q.; Xin, Y.; Mei, Z.; Gao, A.; Liu, W.; Yu, L.; Chen, X.; Chen, Z.; Wang, N. Analyses of the photosynthetic characteristics, chloroplast ultrastructure, and transcriptome of apple (*Malus Domestica*) grown under red and blue lights. *BMC Plant Biol.* **2021**, *21*, 483. [CrossRef]
44. Izzo, L.G.; Mele, B.H.; Vitale, L.; Vitale, E.; Arena, C. The role of monochromatic red and blue light in tomato early photomorphogenesis and photosynthetic traits. *Environ. Exp. Bot.* **2020**, *179*, 104195. [CrossRef]

45. Miao, Y.; Chen, Q.; Qu, M.; Gao, L.; Hou, L. Blue light alleviates "red Light syndrome" by regulating chloroplast ultrastructure, photosynthetic traits and nutrient accumulation in cucumber plants. *Sci. Hortic.* **2019**, *257*, 108680. [CrossRef]
46. Zheng, L.; Van Labeke, M.-C. Chrysanthemum morphology, photosynthetic efficiency and antioxidant capacity are differentially modified by light quality. *J. Plant Physiol.* **2017**, *213*, 66–74. [CrossRef]
47. Fukuda, N.; Ajima, C.; Yukawa, T.; Olsen, J.E. Antagonistic action of blue and red light on shoot elongation in petunia depends on gibberellin, but the effects on flowering are not generally linked to gibberellin. *Environ. Exp. Bot.* **2016**, *121*, 102–111. [CrossRef]
48. Lau, O.S.; Deng, X.W. Plant hormone signaling lightens up: Integrators of light and hormones. *Curr. Opin. Plant Biol.* **2010**, *13*, 571–577. [CrossRef]
49. Kurepin, L.V.; Emery, R.J.N.; Pharis, R.P.; Reid, D.M. Uncoupling light quality from light irradiance effects in *helianthus annuus* shoots: Putative roles for plant hormones in leaf and internode growth. *J. Exp. Bot.* **2007**, *58*, 2145–2157. [CrossRef]
50. Kurepin, L.V.; Pharis, R.P. Light signaling and the phytohormonal regulation of shoot growth. *Plant Sci.* **2014**, *229*, 280–289. [CrossRef]
51. Samuolienė, G.; Sirtautas, R.; Brazaitytė, A.; Duchovskis, P. LED lighting and seasonality effects antioxidant properties of baby leaf lettuce. *Food Chem.* **2012**, *134*, 1494–1499. [CrossRef]
52. Kapoor, S.; Raghuvanshi, R.; Bhardwaj, P.; Sood, H.; Saxena, S.; Chaurasia, O.P. Influence of light quality on growth, secondary metabolites production and antioxidant activity in callus culture of *Rhodiola imbricata.* Edgew. *J. Photochem. Photobiol. B* **2018**, *183*, 258–265. [CrossRef]
53. Lanfer-Marquez, U.M.; Barros, R.M.C.; Sinnecker, P. Antioxidant activity of chlorophylls and their derivatives. *Food Res. Int.* **2005**, *38*, 885–891. [CrossRef]
54. Eggersdorfer, M.; Wyss, A. Carotenoids in human nutrition and health. *Arch. Biochem. Biophys.* **2018**, *652*, 18–26. [CrossRef] [PubMed]
55. Sgherri, C.; Pinzino, C.; Navari-Izzo, F.; Izzo, R. Contribution of major lipophilic antioxidants to the antioxidant activity of basil extracts: An EPR study. *J. Sci. Food Agric.* **2011**, *91*, 1128–1134. [CrossRef] [PubMed]
56. Qin, L.; Li, C.; Li, D.; Wang, J.; Yang, L.; Qu, A.; Wu, Q. Physiological, metabolic and transcriptional responses of basil (*Ocimum basilicum* Linn. var. *pilosum (Willd.) Benth.) to heat stress. Agronomy* **2022**, *12*, 1434.
57. Xiao, Y.; Chu, L.; Zhang, Y.; Bian, Y.; Xiao, J.; Xu, D. HY5: A pivotal regulator of light-dependent development in higher plants. *Front. Plant Sci.* **2021**, *12*, 800989. [CrossRef]
58. Fiorucci, A.-S.; Michaud, O.; Schmid-Siegert, E.; Trevisan, M.; Allenbach Petrolati, L.; Çaka Ince, Y.; Fankhauser, C. Shade suppresses wound-induced leaf repositioning through a mechanism involving *PHYTOCHROME KINASE SUBSTRATE (PKS)* genes. *PLoS Genet.* **2022**, *18*, e1010213. [CrossRef]
59. Yang, Z.; Liu, B.; Su, J.; Liao, J.; Lin, C.; Oka, Y. Cryptochromes orchestrate transcription regulation of diverse blue light responses in plants. *Photochem. Photobiol.* **2017**, *93*, 112–127. [CrossRef]


**Disclaimer/Publisher's Note:** The statements, opinions and data contained in all publications are solely those of the individual author(s) and contributor(s) and not of MDPI and/or the editor(s). MDPI and/or the editor(s) disclaim responsibility for any injury to people or property resulting from any ideas, methods, instructions or products referred to in the content.

*horticulturae*

Article

# Morphology, Anatomy, Micromorphology, and Palynology of the Squirrel's Foot Fern, *Davallia mariesii* (Davalliaceae)

**Sungyu Yang [1], Goya Choi [1] and Jun-Ho Song [2,*]**

[1] Herbal Medicine Resources Research Center, Korea Institute of Oriental Medicine, Naju 58245, Republic of Korea; sgyang81@kiom.re.kr (S.Y.); serparas@kiom.re.kr (G.C.)
[2] Department of Biology, Chungbuk National University, 1 Chungdae-ro, Seowon-gu, Cheongju 28644, Republic of Korea
* Correspondence: jhsong@cbnu.ac.kr; Tel.: +82-3-261-2298

**Abstract:** *Davallia mariesii* T. Moore ex Baker, a member of the section *Trogostolon* (Copel.) M. Kato and Tsutsumi (Davalliaceae M.R. Schomb.), is a lithophytic or epiphytic herb that grows on rocks and tree trunks in montane forests. This study analyzed the morphological, anatomical, micromorphological, and palynological characteristics of *D. mariesii* using a digital slide scanner and a field-emission scanning electron microscope and presented an expanded and updated description. A circumendodermal band was observed in the anatomical structure of the stipe, making *D. mariesii* the second species in the family Davalliaceae with such a band. The frond anatomical studies revealed that the epidermal cells of the indusium were thicker than those of the epidermis on both sides and that hypostomatic fronds with stomata chambers were present. Diacytic, anisocytic, and tetracytic stomatal complexes were observed on abaxial surfaces. The indusia covered numerous sporangia. Leptosporangium consisted of an apical cap, a basal cap, an annulus, and a stalk. The spore had an ellipsoidal outline, a monolete aperture, and verrucae with colliculate ornamentation. The obtained results provide systematic data for the phylogeny of Davalliaceae and establish a basis for future taxonomic delimitation of other taxa.

**Keywords:** circumendodermal band; Davalliaceae; *Davallia mariesii*; SEM; spore; stomata

check for updates

**Citation:** Yang, S.; Choi, G.; Song, J.-H. Morphology, Anatomy, Micromorphology, and Palynology of the Squirrel's Foot Fern, *Davallia mariesii* (Davalliaceae). *Horticulturae* **2023**, *9*, 939. https://doi.org/10.3390/horticulturae9080939

Academic Editor: Wajid Zaman

Received: 20 July 2023
Revised: 11 August 2023
Accepted: 16 August 2023
Published: 18 August 2023

**Copyright:** © 2023 by the authors. Licensee MDPI, Basel, Switzerland. This article is an open access article distributed under the terms and conditions of the Creative Commons Attribution (CC BY) license (https://creativecommons.org/licenses/by/4.0/).

## 1. Introduction

*Davallia* Sm. (Davalliaceae) is a genus of lithophytic or epiphytic ferns that are widely distributed from the Atlantic Ocean through Africa and southern and eastern Asia to Malaysia, Australia, and the Pacific Islands [1–5].

Morphology-based classifications traditionally divide the Davalliaceae family into from four to ten genera, containing approximately 49–130 species. These classifications are based on the organization of scales, hairs, frond texture, sori, and indusia characteristics [1,5,6]. However, a recent molecular phylogeny of the family using five combined DNA sequence datasets from plastid genes or intergenic spacers has indicated that the Davalliaceae family is divided into seven clades; therefore, phylogeny-based classifications have proposed the classification of Davalliaceae into a single genus, *Davallia*, with seven sections and approximately 65 species [6].

*Davallia* has a creeping rhizome covered with peltate scales, and its stems are densely and permanently covered with scales. These scales are often cordate or have peltate bases with serrated margins. Its fronds are monomorphic or dimorphic, with reduced or more dissected fertile blades and various venation segments, such as from bi-pinnate to four-pinnate pinnatifid. Their blades are deltoid- or pentagonal-shaped, leathery, or occasionally thickly herbaceous. The stipe has articulate to short phyllopodia that are terete or slightly winged. The fronds have various indusium, kidney-, or pouch-shaped structures that attach to the edge or side of the base of the sorus. The indusia are extrorse and elongated toward the margins [1,2,5,7].

On the Korean Peninsula, *Davallia* comprises only one species, *D. mariesii* T. Moore ex Baker [7]. Furthermore, this species is closely related to *D. griffithiana* Hook. [≡ *Humata griffithiana* (Hook.) C.Chr.] in the East Himalayas in Japan and Indochina [8]

The rhizomes of *D. mariesii* are commonly used in Asia (called "Gusuibu" by China and Taiwan with *Drynaria fortunei* (Kunze) J.Sm. (Polydiaceae)) as an indigenous herbal medicine, in combination with other herbs for various effects such as liver and kidney strengthening, musculoskeletal pain relief, rheumatoid arthritis progression control, anti-aging diets, and osteogenic activities [9,10]. Moreover, phenolic, flavonoid, and terpenoid components of *D. mariesii*, including gentistic acid, davallin, and dryocrassol, have been identified [11,12]. In addition, *D. mariesii* is grown as an ornamental plant and its associated microflora removes indoor air pollutants [13].

Anatomical characteristics of the stipe (petiole), such as the number and shape of vascular bundles, distribution of sclerenchyma, and presence of grooves, are helpful for fern systematics [14–16]. Frond (leaf) epidermal characteristics, including epidermal cell shape, stomatal size, stomatal type, and stomatal density, are widely used for the taxonomic analysis of many fern groups using scanning electron microscopy (SEM) [8,17–21]. In addition, spore morphology, such as shape and ornamentation, has been used in the taxonomy and classification of many fern species when exploring the relationships among closely related species [22–26]. Studies on frond anatomy [27,28], scale morphology [29], and spore morphology, including ornamentation [30], have focused on the Davalliaceae family. However, an integrative study comprising the stipe anatomy, frond anatomy, micromorphology, and palynological structure of *D. mariesii* has yet to be described in detail.

This study aimed to provide a detailed description of the anatomical, micromorphological, and palynological characteristics of the Korean *D. mariesii*. Moreover, we updated the morphology described in the Flora of Korea, and the taxonomic significance of the findings is discussed while considering previous studies on *Davallia*.

## 2. Materials and Methods

### 2.1. Study Species and Materials

Fertile frond samples of *Davallia mariesii* were collected during the mature period from five natural populations in South Korea (in Gui-myeon, Wanju-gun, Jeollabuk-do, 35°39′03.5″ N 127°07′05.7″ E, alt. 182 m, YSG_KIOM-2021-304; in Mt. Palbong-san, Hongcheon, Gangwon-do, 37°41′45.7″ N 127°41′58.4″ E, alt. 247 m, SJH_ KIOM-2021-435; in Seongsu-myeon, Imsil-gun, Jeollabuk-do, 35°38′06.7″ N 127°25′16.8″ E, alt. 580 m, SJH_KIOM-2021-476; in Mt. Geumsung-san, Naju, Jeollanam-do, 35°02′35.3″ N 126°42′24.4″ E, alt. 245 m, SJH_KIOM-2021-549; in Haman, Gyeongsangnam-do, 35°15′59.8″ N 128°25′35.1″ E, alt. 59 m, SJH_KIOM-2021-584).

Voucher specimens were stored at the Korean Herbarium of Standard Herbal Resources at the Herbal Medicine Resources Research Center (Index Herbariorum acronym: KIOM). Additional samples were collected from living plants and preserved as liquid specimens in a formalin-acetic acid-alcohol (FAA) solution [31]. Micromorphological, anatomical, and palynological investigations were performed using materials preserved in the FAA solution.

### 2.2. Morphological Analyses

Field photographs were taken using a digital camera (NIKON D850; NIKON, Tokyo, Japan). The measurements and optical observations of 20 individuals from each of the five collection sites (vouchers) were recorded for a total of 100 individuals. The sizes of the morphological structures were measured using a digital Vernier caliper (CD-15CP; Mitutoyo, Kawasaki, Japan). Fertile fronds (leaves) with sporangia were observed using a stereomicroscope (Olympus SZX16; Olympus, Tokyo, Japan).

### 2.3. Anatomical Analyses

Transverse sections of the stipe (petiole) and pinnule (subleaflet) were dehydrated in a tertiary butyl alcohol (TBA) series and embedded in paraffin using an automatic tissue processor (Leica EG1150H; Leica Microsystems, Wetzlar, Germany). The tissue blocks were sectioned using a manual rotary microtome (HistoCore MULTICUT; Leica Biosystems, Nussloch, Germany), and 5–8 μm sections were placed on glass slides. Sections were double-stained with Fast-Green FCF and safranin O solutions in an automatic slide stainer. Permanent slides were scanned using a digital slide scanner (3DHistech Pannoramic Desk II DW; 3DHistech Kft., Budapest, Hungary) and images were observed and captured using CaseViewer software (version 2.4.0; 3DHistech Kft., Budapest, Hungary).

### 2.4. Micromorphological and Palynological Analyses

For the observations using a field-emission scanning electron microscope (FE-SEM), fertile pinnae (leaflets) with sporangia and spores were dehydrated using an ethanol series (70, 90, 95, and 100%) at room temperature for one hour per ethanol concentration. The dehydrated material was immersed in liquid $CO_2$ for critical-point drying (CPD) (SPI-13200JE-AB; SPI Supplies, West Chester, PA, USA), and subsequently mounted on aluminum stubs using a double-sided adhesive conductive carbon disk (05073-BA; SPI Supplies, West Chester, USA). All samples were gold-coated using an ion-sputtering device (208HR; Cressington Scientific Instruments Ltd., Watford, UK) and observed using a low-voltage FE-SEM (JSM-7600F; JEOL, Tokyo, Japan) at an accelerating voltage of 5–10 kV and a working distance of 5–10 mm.

Epidermal cell terminology was based on Wilkinson [32], and spore terminology was based on Wang et al. [30]. Stipe (petiole) and frond (leaf) anatomical terminologies followed those of Sen et al. [33] and Hernández-Hernández et al. [34], respectively.

## 3. Results

### 3.1. Morphological Characteristics

***Herbs***: perennial, usually epipetric (lithophyte) or rarely epiphytic, 9.4–34.0 cm tall. ***Rhizomes***: long creeping, dark brown, 6.1–58 cm × 3.2–9.8 mm, densely scaly; scales persistent, black, dark to light brown or grayish orange, above a considerably broader base, evenly narrowed toward the apex, lanceolate, linear, or sickle-shaped ultimate lobes, peltate, 1.6–9.7 × 0.1–0.8 mm, margins denticulate. ***Fronds:*** monomorphic, alternate or opposite, symmetrical or asymmetrical, straight or bent, distantly spaced, petiolate; primary stipe grayish orange or yellowish green, 3.6–15.7 cm × 0.3–2.1 mm, adaxially grooved, sparse fugacious scales; blade rhomboid, triangular or triangular–ovate in outline, 4.5–22.8 × 5.2–26.3 cm, usually 4-pinnate, adaxial surface green or yellowish green, abaxial surface yellowish green, apex obtuse, acute or acuminate, glabrous; rachis winged; primary pinnae alternate or opposite, 5–17 pairs, triangular–ovate, petiolulate, basal pair largest, overlapping or not each pinna; primary upper and middle pinna ovate, oblong or elliptic, 1.0–5.2 × 0.5–2.4 cm; primary basal pinna asymmetry, lanceolate, oblong or ovate, 2.5–13.0 × 1.5–7.5 cm; ultimate segments oblanceolate, narrowly oblong or linear, 1.5–6.0 × 0.5–3.0 mm. ***Veins***: forked, branched, and did not reach the margins. ***Sori***: separate, terminal on veins, usually single on a segment at the forking point of veins, discoid, lateral opening; indusium linear, oblong, or elliptic pouch-shaped, 0.9–2.0 × 0.4–1.0 mm, membranous. ***Spores***: ellipsoidal with monolete apertures (Figure 1; Table 1).

**Figure 1.** Photographs of the *Davallia mariesii* T. Moore ex Baker in Korea. (**A**) Habit, lithophyte. (**B**) Adaxial side of a fertile frond. (**C**) Abaxial side of a fertile frond. (**D**) Rhizomes, under wet conditions. (**E**) Rhizomes, under dry conditions. (**F**) Abaxial side of a fertile pinna with sori.

**Table 1.** Morphological characters of *Davallia mariesii* T. Moore ex Baker based on our study and Flora of Korea [7].

| | Our Study | Flora of Korea |
|---|---|---|
| Plants | 9.4–34.0 cm tall | - |
| Rhizome | 6.1–58 cm × 3.2–9.8 mm | 3–5 mm in diam. |
| Scale | 1.6–9.7 × 0.1–0.8 mm | 5–8 mm long |
| Frond | 8.1–38.5 cm long | 15–35 cm long |
| Stipe | 3.6–15.7 cm × 0.3–2.1 mm | 5–15 cm long |
| Blade | 4.5–22.8 × 5.2–26.3 cm | 10–20 × 8–15 cm |
| Pinnae | 5–17 pairs | 6–12 pairs |
| Primary upper and middle pinna | 1.0–5.2 × 0.5–2.4 cm | - |
| Primary basal pinna | 2.5–13.0 × 1.5–7.5 cm | - |
| Ultimate segments | 1.5–6.0 × 0.5–3.0 mm | 1–2 mm wide |
| Indusium | 0.9–2.0 × 0.4–1.0 mm | - |

## *3.2. Anatomical Characteristics*

### 3.2.1. Stipe

The outline was circular, with a wing in the transverse section of the stipe. The size of the stipe (length of the ventral axis (VA) × length of the dorsiventral axis (DVA)) was 468.5–(559.2)–718.0 × 501.2–(585.6)–627.1 µm (Table 2). A thin cuticle layer covered the epidermis and consisted of uniseriate, oval, or rectangular cells (Figure 2A). The thickness of the abaxial side of the epidermal cell in the cross-section was 7.5–(10.5)–18.0 µm, and that of the adaxial side was 11.6–(15.2)–21.3 µm. The angular collenchyma tissue, located immediately under the epidermis, was composed of regular cells. The parenchyma cells of the cortex consisted of from four to seven oval, squashed oval, or almost rectangular layers and were isometrically arranged (Figure 2A). The stipe vascular bundle was open arc-shaped and its size (width of the main vascular bundle (WMV) × length of the main vascular bundle (LMV)) was 171.5–(213.5)–274.6 × 114.1–(133.5)–156.3 µm. The stipe vascular bundle was covered with a thick layer of the circumendodermal band (CB) (Figure 2B). The CB consisted of the continuity of the isodiametric cells formed by the band (ring) and was 6.9–(9.8)–11.7 µm thick (Table 2). The CB consisted of slightly square-shaped cells. Safranin histochemical tests for cell wall thickening in the CB were positive for red, which represented one-quarter to one-half of the cell forming U-shaped thickening (Figure 2B).

No individual had CB cell walls with thickened total occlusion of the cell lumina. The endodermis was 2.0–(3.7)–5.4 μm thick. The stipe surface was glabrous. No crystals were observed in any of the stipe cells.

**Table 2.** Anatomical characters of stipe and pinnule of *Davallia mariesii*.

| | ***Davallia mariesii*** |
|---|---|
| *Stipe* | |
| Outline of stipe | circular with wing |
| Length of ventral axis (VA) (μm) | 468.5–(559.2)–718.0 |
| Length of dorsiventral axis (DVA) (μm) | 501.2–(585.6)–627.1 |
| Epidermal thickness-abaxial surface (μm) | 7.5–(10.5)–18.0 |
| Epidermal thickness-adaxial surface (μm) | 11.6–(15.2)–21.3 |
| Arrangement of cortex | isometric |
| Outline of stipe vascular bundle | open arc |
| Width of main vascular bundle (WMV) (μm) | 171.5–(213.5)–274.6 |
| Length of main vascular bundle (LMV) (μm) | 114.1–(133.5)–156.3 |
| Circumendodermal band type | continuous ring |
| Circumendodermal band proportion | 1/4 to 1/2 |
| Circumendodermal band thickness (μm) | 6.9–(9.8)–11.7 |
| Endodermis thickness (μm) | 2.0–(3.7)–5.4 |
| *Blade (of ultimate segment)* | |
| Epidermal thickness-abaxial surface (μm) | 7.1–(13.1)–19.7 |
| Epidermal thickness-adaxial surface (μm) | 11.8–(15.1)–23.0 |
| Indusium epidermal thickness (μm) | 15.0–(21.3)–25.1 |
| Mesophyll thickness (non-indusium area) (μm) | 61.2–(99.4)–149.6 |
| Diameter of vascular bundle (μm) | 25.2–(44.3)–70.6 |

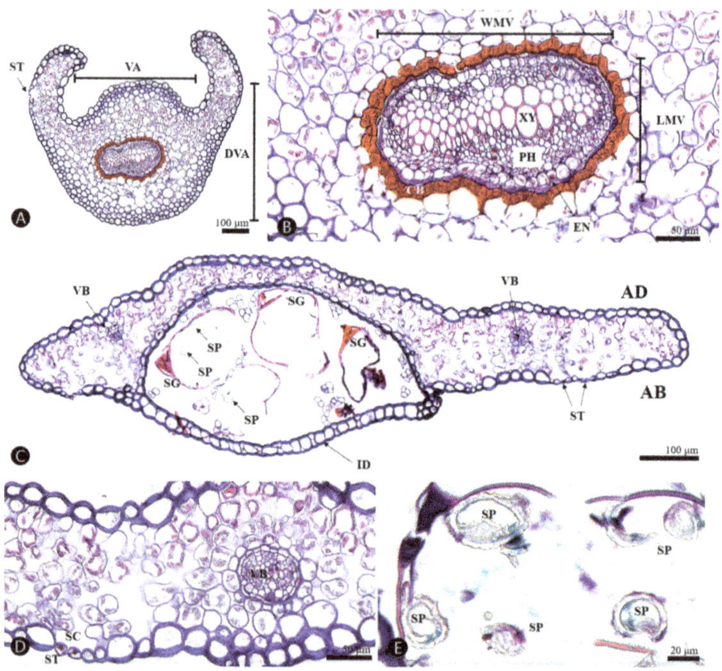


**Figure 2.** Micrographs of transverse section of the stipe and fertile frond in the *Davallia mariesii*. (**A**,**B**) Stipe. (**C**) Fertile frond. (**D**) Blade of fertile frond. (**E**) Sporangium with spore. AB, Abaxial side; AD, Adaxial side; CB, circumendodermal band; DVA, length of dorsiventral axis; EN, endodermis; ID, Indusium; LMV, length of main vascular bundle; PH, phloem; SC, stomata chamber; SG, Sporangium; SP, spore; ST, stomata; VA, length of ventral axis; VB, vascular bundle; WMV, width of main vascular bundle; XY, xylem.

3.2.2. Ultimate Segment (of Pinnule)

The adaxial and abaxial epidermises of the fronds were characterized by one layer of cells covered by a cuticle (Figure 2C). The epidermal thicknesses of abaxial and adaxial were similar [abaxial, 7.1–(13.1)–19.7 µm; adaxial, 11.8–(15.1)–23.0 µm]. The mesophyll contained 3–6 layers of mainly arm-cells that were 61.2–(99.4)–149.6 µm thick and slightly differentiated into palisade and spongy parenchyma (Figure 2C, Table 2). The upper mesophyll in the indusium was arranged more compactly. The epidermal cells of the indusium were thicker than those of the adaxial and abaxial epidermis [15.0–(21.3)–25.1 µm]. Stomata were found only in the abaxial epidermis (hypostomatic fronds), with a stomatal chamber (air space) (Figure 2D). The diameter of vascular bundles was 25.2–(44.3)–70.6 µm (Figure 2D, Table 2). All parts of the adaxial and abaxial epidermis of the fronds were glabrous. No crystals were observed in any of the epidermal cells.

*3.3. Micromorphological Characteristics*

3.3.1. Pinnule

Epidermal cell patterns were described separately for adaxial (AD) and abaxial (AB). Frond epidermal cells on both surfaces were arranged elongately and had striated, convex periclinal walls; however, their anticlinal walls differed (Figure 3, Table 3). AD had straight and slightly undulating anticlinal walls (Figure 3A,B), whereas AB had undulating sinuate walls (Figure 3C,D). Hypostomatic fronds were observed and three types of stomatal complexes were recognized: diacytic, anisocytic, and tetracytic (Figure 3C,D). The stomata were 35.6–(46.3)–51.4 µm long, 25.9–(31.4)–36.6 µm wide, and had an area of 912.5–(1178.5)–1410.9 µm$^2$, respectively (Table 3). The stomatal ledges were lip-shaped.

**Table 3.** Micromorphological characters of frond epidermal cells of *Davallia mariesii*.

| | **Adaxial Side** | **Abaxial Side** |
|---|---|---|
| Epidermal cell arrangement | Elongated | elongated |
| Anticlinal cell wall | straight, slightly undulate | undulate to sinuate |
| Periclinal cell wall | striate, convex | striate, convex |
| Stomata type | Absent | diacytic, anisocytic, tetracytic |
| Stomata length (µm) | Absent | 35.6–(46.3)–51.4 |
| Stomata width (µm) | Absent | 25.9–(31.4)–36.6 |
| Stomata area (µm$^2$) | Absent | 912.5–(1178.5)–1410.9 |
| Stomatal ledge | Absent | lip-shaped |

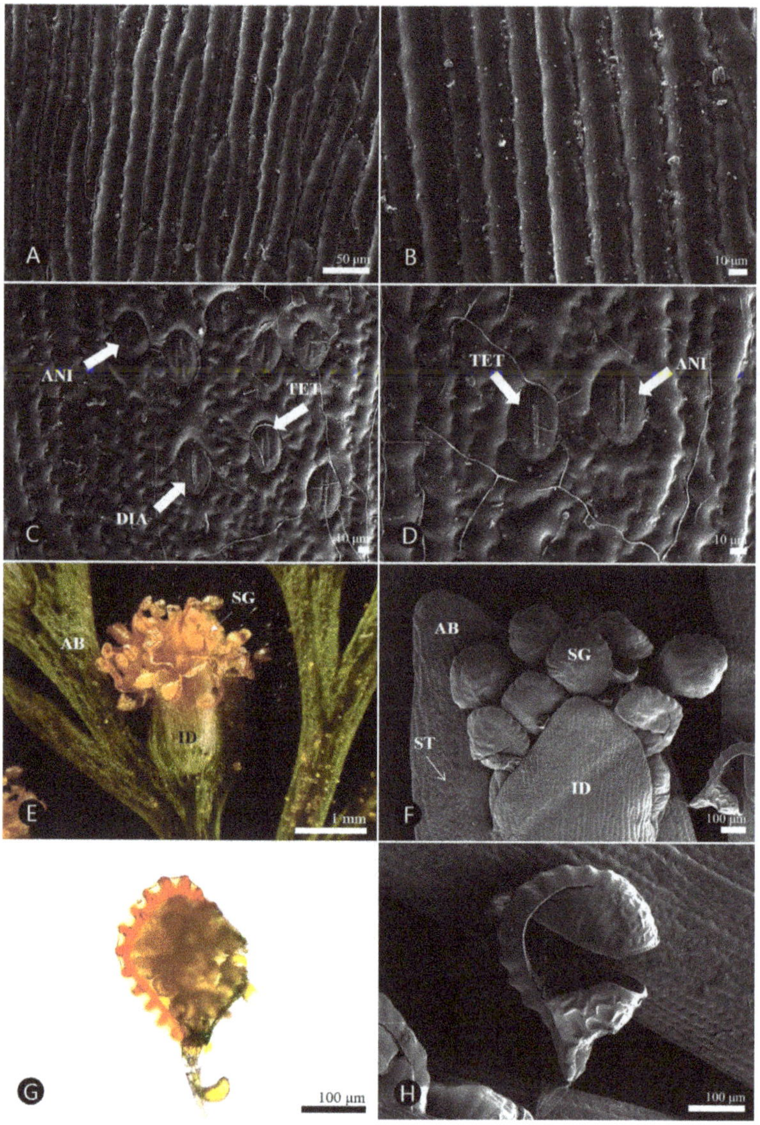


**Figure 3.** Stereo (SM), light (LM), and scanning electron (SEM) micrographs of the mature sporophyte in the *Davallia mariesii*. (**A**,**B**) Adaxial side of a fertile frond. (**C**,**D**) Abaxial side of a fertile frond with stomata. (**E**,**F**) Sorus with indusium. (**G**,**H**) Sporangium with spore. (**A–D**,**F**,**H**) SEM. (**E**) SM. (**G**) LM. AB, Abaxial side; ANI, Anisocytic, DIA, Diacytic; ID, Indusium; SG, Sporangium; ST, stomata; TET, Tetracytic.

3.3.2. Sorus

Each pinnule had many separate sori in fertile fronds, which were generally elongated (Figure 3E). The indusium covered numerous sporangia. The epidermal cells of the outer indusium were isodiametrically arranged, straight and elongated in the anticlinal wall, and striated and convex in the periclinal wall (Figure 3F). The leptosporangium consisted of an apical cap, a basal cap, an annulus, and a short stalk (Figure 3G,H). The annulus was arranged longitudinally and consisted of 12–16 thickened cells.

### 3.4. Palynological Characteristics

The spores were free and their sizes were 34.3–(38.8)–42.2 μm in polar axes, and 39.3–(50.8)–55.0 μm in equatorial axes. The exine was measured to be 2.3–(3.4)–5.1 μm thick (Table 4). The outline of the spore was transversely elliptical–reniform to ellipsoidal, bilaterally symmetrical, aniso-polar, and rounded–elliptical in polar view. The aperture was a monolete, which was a single linear mark designating the dividing axis of the proximal face (Figure 4). The ornamentation was a verrucae–colliculate type; the verrucae were numerous, slightly convex, densely packed, rounded, 2–(3.2)–5 μm diameter with radiating from laesura, forming a polygonal reticulate pattern (Figure 4D,E).

**Table 4.** Palynological characters of spore of *Davallia mariesii*.

| | *Davallia mariesii* |
|---|---|
| Polar axes (μm) | 34.3–(38.8)–42.2 |
| Equatorial axes (μm) | 39.3–(50.8)–55.0 |
| Shape | ellipsoidal |
| Exine thickness (μm) | 2.3–(3.4)–5.1 |
| Aperture | monolete |
| Exine ornamentation | verrucate colliculate |

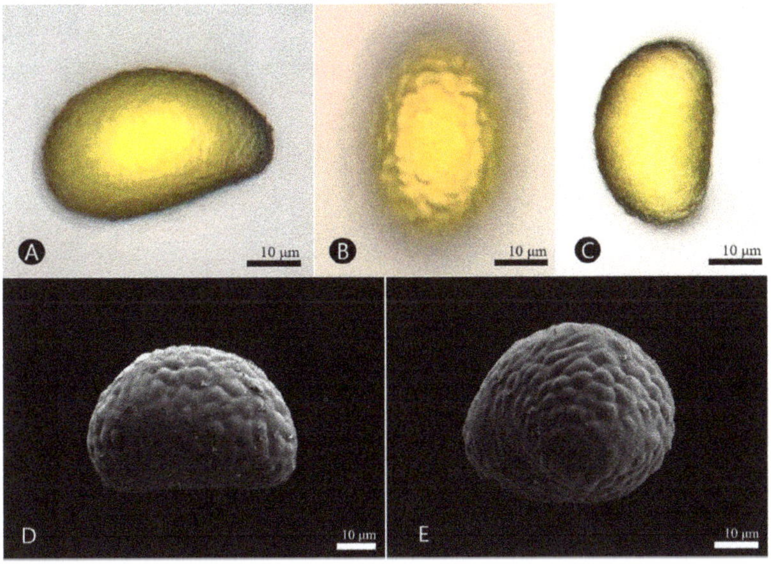


**Figure 4.** Light (LM) and scanning electron (SEM) micrographs of the verrucae-colliculate spore in the *Davallia mariesii*. (**A**) Equatorial side with transversely ellipsoidal shape (LM). (**B**) Proximal side with rounded–elliptical shape (LM). (**C**) Distal side with transversely elliptical–reniform shape (LM). (**D**) Equatorial side with rounded verrucae (LM). (**E**) Equatorial–distal side forming a polygonal reticulate pattern (SEM).

## 4. Discussion

This study is the first to comprehensively report the characteristics of the medicinal and economically important squirrel's foot fern, *D. mariesii*, and provide valuable information regarding its morphological, anatomical, micromorphological, and palynological characteristics.

Morphological characteristics such as the size of fronds and the shape and color of rhizome scales are taxonomically significant for identifying and delimiting species in the genus *Davallia* [5,35]. *D. mariesii* was closely related to *D. griffithiana* (=*D. tyermannii* (T. Moore) Baker), systematically [8,30,36]. *D. mariesii* was distinguished by its tri-pinnate or quadri-pinnate fronds (vs. bi-pinnate or tri-pinnate in *D. griffithiana*), ultimate segments 5–27 × 2–6 mm (vs. 2–5 × 2–3 mm), and pouch-shaped, oblong, longer than wide indusia (vs. semicircular, approximately as wide as long) [35]. Although the present results largely correspond with those reported for the Flora of Korea [7], several differences were found, and additional quantitative data were determined (Table 1). Characteristics such as plant height, rhizome length, scale width, stipes, primary pinna size, ultimate segments, and indusia were updated.

Stipe anatomical characteristics such as vascular bundle shape, number of vascular bundles, the presence or absence of the CB, and adaxial grooves are reflected in the promising results of taxonomic and systematic studies [15,37,38]. In particular, the stipe vascular bundles, which are conserved and stable, are valuable for taxonomy [37,38]. Our results indicated that the openly arc-shaped main vascular bundles of the studied species were similar to those of *D. hymenophylloides* (Blume) Kuhn (≡*Davallodes hymenophylloides* (Blume) M. Kato and Tsutsumi) [38].

In 89 species from 28 families of ferns, the second innermost layer of primary tissue next to the innermost endodermis, formed from thick-walled, tannin-rich cells, is present in the stipes, known as the CB [34]. To date, in Davalliaceae, a continuous CB with the total proportion of the cell lumen has only been observed in *D. canariensis* (L.) Sm. [15,16]. Our study reports an additional Davalliaceae species associated with CB. The degree of CB cell wall thickening in each species ranged from 25% to 100% of the cell lumina. Moreover, interspecific variations have been reported in 10 studied species of the genera *Elaphoglossum* and *Polypodium* [34]. In this study, total occlusion of the cell lumina of the CB in *D. mariesii* was not observed, in contrast to *D. canariensis*. Thus, the degree of cell wall thickening in the CB may be helpful for identification at the species level in the genus *Davallia*. However, further studies are required to evaluate the taxonomic significance of these findings.

CBs have been referred to as protective bands [39] and sclerenchymatic rings [15,40,41]. However, previous histochemical tests and developmental investigations revealed that CB cells were negative for lignin and positive for the cytoplasm and nuclei, indicating that CB cells were not lignified sclerenchyma [34,42]. Several studies have shown that CB cells contain numerous tannins [33]. Although this study showed positive results for safranin, more precise histochemical tests and chemical analyses are required to determine its specific molecular composition, including cellulose, tannins, lignin, and suberin.

In *D. mariesii*, CB may have two possible functions: a protective function similar to that in the endodermis [43] and a biomechanical function [34]. In particular, to resist bending and twisting stresses, the central tissues must be elastic, such as living parenchyma, and not rigid, such as lignified sclerenchyma [34,44]. The CB of *D. mariesii*, which is an epipetric (lithophyte) species with mid- to large-sized pinnate fronds [4.5–22.8 × 5.2–26.3 cm] and stipes [3.6–15.7 cm × 0.3–2.1 mm], probably experiences higher bending shear forces than other ferns with smaller ones.

Recent micromorphological studies have suggested that the characteristics of the frond epidermis are key features for the classification of Davalliaceae at the genus or species levels [8,17–21]. In particular, the frond (the ultimate segment of the pinnule) micromorpho–anatomical results revealed that the epidermal cell thickness and arrangement of the abaxial and adaxial walls were similar; however, those of the anticlinal walls were different. Most fern leaves have similar anticlinal walls on both epidermal surfaces [8,26,45]. Additional expanded frond cell data are required to evaluate their taxonomic importance in *Davallia* using the phylogenetic target-sampling strategies. The stomatal characteristics observed in this study disagreed with those from a previous survey. Sen [46] described that Japanese *D. mariesii* (voucher no. 0035-38 in Naturalis Biodiversity Center, The Netherlands) has polocytic, copolocytic, anomocytic, and staurocutic stomata of 31.5–36 × 22.5–31.5 μm;

however, our results revealed that those of Korean *D. mariesii* have diacytic, anisocytic, and tetracytic stomata with larger size (35.6–51.4 × 25.9–36.6 μm). Understanding the degree of variation in stomatal characteristics of this species requires frond micromorphological studies based on geographical, ecological, and developmental differences.

Palynological data have been extensively used to determine the systematic relationships between fern groups [22–26]. In this study, the spore morphology of *D. mariesii* was consistent with previous data from the genus *Davallia* [45,47–50]. The *D. mariesii* spore size includes the size variations in the *Davallia* clade, as described by Wang et al. [30]. All *Davallia* species have verrucate–colliculate ornamentation patterns [30], including macrofossils [51] such as *D. mariesii*. The verrucate–colliculate ornamentation type appears to be the most common in the genus and may represent a plesiomorphic state in the genus *Davallia*. Thus, spore characteristics, especially surface ornamentation in this genus, may be stable and helpful for taxonomy, systematics, paleobotany, and palynology at the generic level.

This study contributes to the knowledge of the anatomical, micromorphological, and palynological diversity of Davalliaceae plants. Nevertheless, further studies are necessary to evaluate species-level variations in the stipe, frond, and spore morphology in this family based on taxonomic and phylogenetic contexts using an expanded and targeted sampling strategy.

## 5. Conclusions

The present study provides valuable information regarding the morphology, anatomy, micromorphology, and palynology of the medicinally and economically important fern, *Davallia mariesii*, which is distributed in Korea. In particular, the morphological description of the species has expanded based on a large amount of living materials. This is the first comprehensive study of the stipe and frond anatomy, and the frond and spore micromorphology of *D. mariesii* in Korea. Our investigations aim to contribute to the comprehensive understanding of the taxonomy, systematics, paleobotany, and palynology of the genus *Davallia*, including the family Davalliaceae. However, additional molecular frameworks are required to address the challenges of the relationships and taxonomy of this genus. Further studies involving a broader range of *Davallia* taxa and extensive sampling would enhance our understanding of the taxonomic and systematic implications associated with the stipe, frond, and spore microanatomy.

**Author Contributions:** Conceptualization, J.-H.S.; methodology, S.Y., G.C. and J.-H.S.; software, S.Y. and J.-H.S.; validation, S.Y. and J.-H.S.; formal analysis, S.Y. and J.-H.S.; investigation, S.Y., G.C. and J.-H.S.; resources, S.Y. and J.-H.S.; data curation, S.Y., G.C. and J.-H.S.; writing—original draft preparation, S.Y. and J.-H.S.; writing—review and editing, J.-H.S.; visualization, S.Y., G.C. and J.-H.S.; supervision, J.-H.S.; project administration, J.-H.S.; funding acquisition, G.C. and J.-H.S. All authors have read and agreed to the published version of the manuscript.

**Funding:** This research was supported by the Korea Institute of Oriental Medicine, Naju, South Korea (grant numbers KSN1822320) and the National Research Foundation of Korea (NRF) grants funded by the Korea government (MSIT) (NRF-2020R1A2C1100147 and RS-2023-00208589) to J.-H. Song.

**Data Availability Statement:** All data are fully available without restriction. The plant images and micrographs can be made available upon requests addressed to correspondence (Jun-Ho Song).

**Acknowledgments:** The authors would like to thank the five anonymous reviewers whose comments and corrections have improved the work.

**Conflicts of Interest:** The authors declare that they have no competing interest or conflict of interest.

## References

1. Kramer, K.U. Davalliaceae. In *The Families and Genera of Vascular Plants: Pteridophytes and Gymnosperms*; Kubitzki, K., Ed.; Springer: Heidelberg/Berlin, Germany, 1990; Volume 1, pp. 74–80.
2. Nooteboom, H.P. Davalliaceae of China. *Acta Phytotax. Sin.* **1996**, *34*, 162–179.

3. Wu, S.H. Davalliaceae. In *Flora Reipublicae Popularis Sinicae*; Wu, S.H., Wang, C.H., Eds.; Science Press: Beijing, China, 1999; Volume 6, pp. 161–199.
4. Von Konrat, M.J.; Braggins, J.E.; de Lange, P.J. *Davallia* (Pteridophyta) in New Zealand, including description of a new subspecies of *D. tasmanii*. *N. Z. J. Bot.* **1999**, *37*, 579–593. [CrossRef]
5. Xing, F.W.; Wang, F.G.; Nooteboom, H.P. Davalliaceae. In *Flora of China (Pteridophytes)*; Wu, Z.Y., Raven, P.H., Hong, D.Y., Eds.; Science Press: Beijing, China; Missouri Botanical Garden Press: St. Louis, UK, 2013; pp. 749–757.
6. Tsutsumi, C.; Chen, C.W.; Larsson, A.; Hirayama, Y.; Kato, M. Phylogeny and classification of Davalliaceae on the basis of chloroplast and nuclear markers. *Taxon* **2016**, *65*, 1236–1248. [CrossRef]
7. Sun, B.Y.; Davalliaceae, M.R. Schomb. In *Flora of Korea: Pteridophytes and Gymnosperms*; Flora of Korea Editorial Committee, Ed.; The National Institute of Biological Resources: Incheon, Republic of Korea, 2015; Volume I, p. 148.
8. Ma, X.D.; Wang, A.H.; Wang, F.G.; He, C.M.; Liu, D.M.; Gerstberger, P.; Xing, F.W. A revised classification of Chinese Davalliaceae based on new evidence from molecular phylogenetics and morphological characteristics. *PLoS ONE* **2018**, *13*, e0206345. [CrossRef] [PubMed]
9. Ni, L.J.; Wang, N.N.; Zhang, L.G.; Guo, Y.Z.; Shi, W.Z. Evaluation of the effects of active fractions of Chinese medicine formulas on IL-1beta, IL-6, and TNF-alpha release from ANA-1 murine macrophages. *J. Ethnopharmacol.* **2016**, *179*, 420–431. [CrossRef] [PubMed]
10. Lin, Y.T.; Peng, S.W.; Imtiyaz, Z.; Ho, C.W.; Chiou, W.F.; Lee, M.H. In vivo and in vitro evaluation of the osteogenic potential of *Davallia mariesii* T. Moore ex Baker. *J. Ethnopharmacol.* **2021**, *264*, 113126. [CrossRef] [PubMed]
11. Cui, C.B.; Tezuka, Y.; Kikuchi, T.; Nakano, H.; Tamaoki, T.; Park, J.H. Constituents of a fern, *Davallia mariesii* Moore. I. Isolation and structures of davallialactone and a new flavanone glucuronide. *Chem. Pharm. Bull.* **1990**, *38*, 3218–3225. [CrossRef] [PubMed]
12. Cui, C.B.; Tezuka, Y.; Yamashita, H.; Kikuchi, T.; Nakano, H.; Tamaoki, T.; Park, J.H. Constituents of a fern, *Davallia mariesii* Moore. V. Isolation and structures of davallin, a new tetrameric proanthocyanidin, and two new phenolic glycosides. *Chem. Pharm. Bull.* **1993**, *41*, 1491–1497. [CrossRef]
13. Gunasinghe, Y.H.K.I.S.; Rathnayake, I.V.N.; Deeyamulla, M.P. Plant and plant associated microflora: Potential bioremediation option of indoor air pollutants. *Nepal J. Biotechnol.* **2021**, *9*, 63–74. [CrossRef]
14. White, R.A. Comparative anatomical studies of the ferns. *Ann. Mo. Bot. Gard.* **1974**, *61*, 379–387. [CrossRef]
15. Lin, B.L.; DeVol, C.D. The use of stipe characters in fern taxonomy I. *Taiwania* **1977**, *22*, 91–99.
16. Lin, B.L.; DeVol, C.D. The use of stipe characters in fern taxonomy II. *Taiwania* **1978**, *23*, 77–95.
17. Lee, C.-S. A taxonomical study of Korean Pteridaceae on the morphology of leaf epidermis. *Korean J. Plant Taxon.* **1988**, *18*, 275–290. [CrossRef]
18. Sen, U.; De, B. Structure and ontogeny of stomata in ferns. *Blumea* **1992**, *37*, 239–261.
19. Bondada, B.; Tu, C.; Ma, L. Surface structure and anatomical aspects of Chinese brake fern (*Pteris vittata*; Pteridaceae). *Brittonia* **2006**, *58*, 217–228. [CrossRef]
20. Chuang, Y.-Y.; Liu, H.-Y. Leaf epidermal morphology and its systematic implications in Taiwan Pteridaceae. *Taiwania* **2003**, *48*, 60–71.
21. Shah, S.N.; Ahmad, M.; Zafar, M.; Ullah, F.; Zaman, W.; Mazumdar, J.; Khuram, I.; Khan, S.M. Leaf micromorphological adaptations of resurrection ferns in northern Pakistan. *Flora* **2019**, *255*, 1–10. [CrossRef]
22. Giacosa, J.R.; Morbelli, M.; Giudice, G. Spore morphology and wall ultrastructure of *Anemia* Swartz species (Anemiaceae) from Argentina. *Rev. Palaeobot. Palynol.* **2012**, *174*, 27–38. [CrossRef]
23. Lee, S.J.; Park, C.W. Spore morphology of the genus *Dryopteris* Adans. (Dryopteridaceae) in Korea. *J. Plant Biol.* **2014**, *57*, 302–311. [CrossRef]
24. Giacosa, J.R. Spore morphology and wall ultrastructure of *Lomariocycas* (Blechnaceae) species from America. *Rev. Palaeobot. Palynol.* **2019**, *269*, 55–63. [CrossRef]
25. Yañez, A.; Marquez, G.J.; Morbelli, M.A. Spore morphology and ultrastructure of Dennstaedtiaceae from Paranaense Phytogeographic Province I.: Genus *Dennstaedtia*. *Rev. Palaeobot. Palynol.* **2016**, *224*, 181–194. [CrossRef]
26. Shah, S.N.; Ahmad, M.; Zafar, M.; Hadi, F.; Khan, M.N.; Noor, A.; Malik, K.; Rashid, N.; Iqbal, M. Spore morphology and leaf epidermal anatomy as a taxonomic source in the identification of *Asplenium* species from Malakand division Pakistan. *Microsc. Res. Tech.* **2020**, *83*, 1354–1368. [CrossRef] [PubMed]
27. Phillips, D.A.; White, R.A. Frond articulation in species of Polypodiaceae and Davalliaceae. *Am. Fern J.* **1967**, *57*, 78–88. [CrossRef]
28. Ma, X.; He, C.; Wang, F.; Wang, A.; Xing, F. Structural characteristics of leaf epidermis and their systematic significance in Davalliaceae. *Plant Sci. J.* **2015**, *33*, 438–447.
29. Tsutsumi, C.; Kato, M. Morphology and evolution of epiphytic Davalliaceae scales. *Botany* **2008**, *86*, 1393–1403. [CrossRef]
30. Wang, F.G.; Liu, H.M.; He, C.M.; Yang, D.M.; Xing, F.W. Taxonomic and evolutionary implications of spore ornamentation in Davalliaceae. *J. Syst. Evol.* **2015**, *53*, 72–81. [CrossRef]
31. Johansen, D.A. *Plant Microtechnique*; McGraw-Hill Book Company, Inc.: London, UK, 1940; 530p.
32. Wilkinson, H.P. The plant surface (mainly leaf). In *Anatomy of the Dicotyledons*, 2nd ed.; Metcalfe, C.R., Chalk, L., Eds.; Clarendon Press: Oxford, UK, 1979; Volume 1, pp. 97–165.
33. Sen, T.; Sen, U.; Holttum, R.E. Morphology and anatomy of the genera *Davallia*, *Araiostegia* and *Davallodes*, with a discussion on their affinities. *Kew Bull.* **1972**, *27*, 217–243. [CrossRef]

34. Hernández-Hernández, V.; Terrazas, T.; Mehltreter, K.; Angeles, G. Studies of petiolar anatomy in ferns: Structural diversity and systematic significance of the circumendodermal band. *Bot. J. Linn. Soc.* **2012**, *169*, 596–610. [CrossRef]
35. Nooteboom, H.P. Notes on Davalliaceae II. A revision of the genus Davallia. *Blumea* **1994**, *39*, 151–214.
36. Liu, H.; Schneider, H. Evidence supporting Davallia canariensis as a Late Miocene relict endemic to Macaronesia and Atlantic Europe. *Aust. Syst. Bot.* **2013**, *26*, 378–385. [CrossRef]
37. Noraini, T.; Ruzi, A.R.; Nadiah, N.; Nisa, R.N.; Maideen, H.; Solihani, S.N. Stipe anatomical characteristics in some *Davallia* (Davalliaceae) species in Malaysia. *Sains Malays.* **2012**, *41*, 53–62.
38. Tan, J.M.P.; Banaticla-Hilario, M.C.; Malabrigo, P.; Angeles, M.D.; Buot, I.E. Anatomical examination of the petiole of eupolypods I (Polypodiales). *Biodiversitas J. Biol. Divers.* **2020**, *21*, 1767–1777.
39. Russow, E. *Vergleichende Untersuchungen betreffend der Histologie (Histographie und Histogenie) der Vegetativen und Sporenbildenden Organe und die Entwicklung der Sporen der Leitbündel-Kryptogamen, mit Berücksichtigung der Histologie der Phanerogamen Ausgehend von der Betrachtung der Marsiliaceae;* Mémoires de l'Académie Impériale des Sciences de St. Petersbourg, sér. 7; Académie Impériale des Sciences: Saint Petersburg, Russia, 1872; Volume 19, 207p.
40. Ogura, Y. *Comparative Anatomy of the Vegetative organs of the Pteridophytes;* Borntraeger: Berlin, Germany, 1972.
41. Moran, R.C. Monograph of the neotropical fern genus Polybotrya (Dryopteridaceae). *Ill. Nat. Hist. Surv. Bull.* **1987**, *34*, 1–138. [CrossRef]
42. Hernández-Hernández, V.; Terrazas, T.; Stevenson, D.W.M. Ontogeny of *Ctenitis melanosticta* and *Diplazium expansum* fronds with emphasis on the circumendodermal band. *Feddes Repert.* **2009**, *120*, 426–442. [CrossRef]
43. Van Fleet, D.S. Histochemistry and function of the endodermis. *Bot. Rev.* **1961**, *27*, 165–220. [CrossRef]
44. Niklas, K.J. Research review: A mechanical perspectiveon foliage leaf form and function. *New Phytol.* **1999**, *143*, 19–31. [CrossRef]
45. Shah, S.N.; Celik, A.; Ahmad, M.; Ullah, F.; Zaman, W.; Zafar, M.; Malik, K.; Rashid, N.; Iqbal, M.; Sohail, A.; et al. Leaf epidermal micromorphology and its implications in systematics of certain taxa of the fern family Pteridaceae from northern Pakistan. *Microsc. Res. Tech.* **2019**, *82*, 317–332. [CrossRef]
46. Sen, T. The evidence of stomatal development on the relationships between *Davallia* and genera associated with it by recent authors. *Ann. Bot.* **1986**, *58*, 663–677. [CrossRef]
47. Huang, T.-C. *Spore Flora of Taiwan;* Meitai Color Print Co., Ltd.: Taipei, Taiwan, 1981.
48. Moy, C.J. Variations of fern spore ultrastructure as reflections of their evolution. *Grana* **1988**, *27*, 39–51. [CrossRef]
49. Tryon, A.F.; Lugardon, B. *Spores of the Pteridophyta;* Springer: New York, NY, USA, 1991.
50. Sofiyanti, N.; Isda, M.N.; Juliantari, E.; Suriatno, R.; Pranata, S. The inventory and spore morphology of ferns from Bengkalis Island, Riau Province, Indonesia. *Biodivers. J. Biol. Divers.* **2019**, *20*, 3223–3236. [CrossRef]
51. Conran, J.G.; Kaulfuss, U.; Bannister, J.M.; Mildenhall, D.C.; Lee, D.E. *Davallia* (Polypodiales: Davalliaceae) macrofossils from early Miocene Otago (New Zealand) with in situ spores. *Rev. Palaeobot. Palynol.* **2010**, *162*, 84–94. [CrossRef]


**Disclaimer/Publisher's Note:** The statements, opinions and data contained in all publications are solely those of the individual author(s) and contributor(s) and not of MDPI and/or the editor(s). MDPI and/or the editor(s) disclaim responsibility for any injury to people or property resulting from any ideas, methods, instructions or products referred to in the content.

*horticulturae*

MDPI

*Article*

# Taxonomic Comparison, Antioxidant and Antibacterial Activities of Three *Ebenus pinnata* Ait. ecotypes (Fabaceae) from Algeria

**Tassadit Zemouri [1], Amirouche Chikhoune [2], Hassina Benmouhoub [1] and Mohamed Sahnoune [1,*]**

[1] Laboratory of Ecology and Environment, Department of Environment and Biological Sciences, Faculty of Nature and Life Sciences, University of Bejaia, Targa Ouzemmour, 06000 Bejaia, Algeria; tassadit.zemouri@univ-bejaia.dz (T.Z.); hassina.benmouhoub@univ-bejaia.dz (H.B.)

[2] Laboratory of Microbial Ecology, Department of Food Science, Faculty of Nature and Life Sciences, University of Bejaia, Targa Ouzemmour, 06000 Bejaia, Algeria; amirouche.chikhoune@univ-bejaia.dz

* Correspondence: mohamed.sahnoune@univ-bejaia.dz; Tel.: 213-05-53-18-44-98

**Abstract:** *Ebenus pinnata* is not known as a traditional medicinal plant, but modern research has revealed its richness in components of medicinal value. Yet, the species remains understudied. Here, we assess the climate effect on its morphology, pollen grains size, chromosome numbers, pollen fertility, and antioxidant and antibacterial activities. Plant material was collected from the humid, sub-humid, and semi-arid areas of Northeastern Algeria. Data treatment by principal component analysis and/or cluster analysis and ANOVA post hoc tests revealed three significantly discriminated ecotypes correlated with the climate stage. Significant differences were detected for whole plant morphology, pollen size, and antioxidant activity. No differences were revealed for chromosome numbers, pollen fertility, and antibacterial activity. The studied material showed a chromosome number of $2n = 14$, high pollen fertility ($94.04 \pm 2.64$–$95.01 \pm 2.02$%), small pollen grains (polar axis: $17.95 \pm 1.10$–$19.47 \pm 1.27$ µm; equatorial axis: $12.80 \pm 1.18$–$13.03 \pm 0.99$ µm), high antioxidant activity (TPC: $50.79 \pm 0.51$–$56.89 \pm 0.46$ mg/g; DPPH: $71.18 \pm 2.24$–$95.67 \pm 2.02$ mg/g; RP: $11.09 \pm 1.24$–$25.88 \pm 0.26$ mg/g), and efficient antibacterial activity (Inhibition area diameter: $9.25 \pm 1.06$–$12.00 \pm 1.41$ mm). The climate seems to exert a significant impact on multiple aspects of the plant's biology. It would be interesting to assess the genetic basis of this phenomenon in *E. pinnata* and other species.

**Keywords:** *Ebenus pinnata*; morphology; pollen; Taxonomy; meiosis; chromosomes; antioxidants; antibacterial activity; climate

check for updates

**Citation:** Zemouri, T.; Chikhoune, A.; Benmouhoub, H.; Sahnoune, M. Taxonomic Comparison, Antioxidant and Antibacterial Activities of Three *Ebenus pinnata* Ait. ecotypes (Fabaceae) from Algeria. *Horticulturae* **2023**, *9*, 879. https://doi.org/10.3390/horticulturae9080879

Academic Editor: Wajid Zaman

Received: 14 July 2023
Revised: 27 July 2023
Accepted: 29 July 2023
Published: 3 August 2023

**Copyright:** © 2023 by the authors. Licensee MDPI, Basel, Switzerland. This article is an open access article distributed under the terms and conditions of the Creative Commons Attribution (CC BY) license (https://creativecommons.org/licenses/by/4.0/).

## 1. Introduction

The genus *Ebenus* comprises twenty species, fourteen of which are endemic to Turkey [1,2]. The six other species outside Turkey are *E. cretica* L. in Crete; *E. sibthorpii* DC. In Southeastern Greece and the Aegean Sea Islands; *E. stellata* Boiss. In Iran, Oman, Afghanistan, Pakistan, and India; *E. lagopus* Boiss. in Southern Iran; *E. armitagei* Schweinf. and Taubertin in Libya and Egypt; *E. pinnata* Ait. in Libya, Tunisia, Algeria, and Morocco [3–5].

Morphologically, the genus *Ebenus* distinguishes itself from other genera of the tribe Hedysareae mainly by its corolla, which is shorter than the calyx teeth, and its one-lomented pod enclosed within the calyx tube [3,6,7]. It constitutes a monophyletic group within the Hedysaroid clade [7], with its ancestral area of origin inferred in the Mediterranean Region [8] and its main center of diversity located in Turkey [9].

Pollen grains in *Ebenus* species are radially symmetrical, isopolar, tricolpate, prolate, and rarely perprolate; they have an intine of 0.5 µm and an exine of 1 µm, with reticulate ornamentations and more or less narrow lumina; they have a polar diameter of 24–40 µm and an equatorial diameter of 12.4–20 μm, with elliptical outlines on an equatorial view and circular to subcircular outlines on a polar view [10–12].

All *Ebenus* species so far investigated show a chromosome number of 2n = 2x = 14 [10,13–16]. The karyotypes are symmetric and include a pair of satellite chromosomes with chromosome lengths of 1.93–4.22 μm [10,14].

*Ebenus* plants are used in traditional folk medicine to treat various health disorders in Turkey [17,18]. *Ebenus haussknechtii* Bornm. ex Hub.-Mor. is used to prevent skin problems, hypertension, and stomach diseases [17]. The analysis of its chemical composition identified various natural compounds (including two flavonoid glycosides and a methylinositol) with very significant antimicrobial activity [17]. As to *E. hirsuta* Jaub. and Spach., it is used to treat kidney disorders [18,19]. Its aerial parts contain mainly hyperoside, rutin, hesperidin, tannic acid, and p-coumaric acid [20]. Its extracts display strong antigenotoxic effects and significant activity against bacteria and fungi [20]. Similarly, *E. laguroides* Boiss. and *E. macrophylla* Jaub. and Spach. have proven to be with significant antioxidant and antibacterial activities due to various chemical compounds, especially rutin, the dominant component in *Ebenus* species [21]. Analyses of roots and aerial parts extracts of *E. boissieri* Barbey have revealed immunomodulatory and antitumor activity inducing apoptosis in breast cancer cells [22], caspase-mediated apoptosis on the cervical cancer cell line Hela [23], and cytotoxic and apoptotic effects on the human lung cancer cell line A549 [24]. In addition to its osteoprotective role, *E. cretica* is very rich in flavonoids and isoflavones, such as formononetin, maesopsin glucoside (aurone), and other compounds [25]. *E. cretica* administration exerts a significant beneficial effect on bone density loss in ovariectomized rats [26]. Mitrocotsa et al. [27] report a long list of important components isolated from *E. cretica* and *E. sibthorpii* DC, including D-pinitol, quercetin, isorhamnetin glycosides, and, especially, rutin-7,4′-di-O-methyl ether, 8,4′-dimethoxy-7-hydroxy-isoflavone, and ionyl glycosides icaricide B1 and B2. Phytochemical tests of *E. stellata* extracts, known for their anticonvulsant and microbial activities, have revealed the presence of coumarins, alkaloids, cardiac glycosides, flavonoids, quinone, saponins, steroids, terpenoids, and tannins [28,29]. More information about the medicinal properties and antimicrobial activities of *Ebenus* plants can be found in Zemouri et al. [30] and references therein.

*Ebenus pinnata* is a spontaneous, uncultivated, and self-pollinated herbaceous species (pers. obs.), restricted to the northern parts of Morocco, Algeria, Tunisia, and Lybia [5]. The plant, as green fodder, is as palatable to livestock as other legumes (pers. obs.). No other traditional use is mentioned for it in literature. No specific data are found in the literature on its morphological and palynological variability. Two cytotypes have been mentioned for *E. pinnata*: one with 2n = 14 from Morocco and Algeria [13,15,16] and the second with 2n = 18 from Morocco [31]. A study of the chemical composition of *E. pinnata* plants from Tunisia has detected the presence of several secondary metabolites, including ombuoside, kaempferol 3-O-rutinoside, rutin, and catechin, which proved to be with significant antioxidant activity [32,33], conferring thus a potential medicinal value to the plant.

Despite its wide distribution all over the northern part of North Africa and its potential medicinal value, *E. pinnata* remains insufficiently investigated from both taxonomic and phytochemical points of view. In the present study, we compare three Algerian *E. pinnata* populations from different climate stages (humid, sub-humid, and semi-arid) using multivariate whole plant morphology, pollen grains size and shape, pollen fertility, karyology, antioxidant, and antibacterial activities. The aim is to highlight the climate impact on those aspects of plant biology. The results are statistically evaluated and discussed.

## 2. Materials and Methods

The taxonomic analysis, including morphometrics, multivariate analyses, cytogenetics, and pollen study, were achieved in the Laboratory of Ecology and Environment of University of Bejaia (Department and County of Bejaia, Northeastern Algeria). Antioxidant and antibacterial activities were assessed in the Laboratory of Microbial Ecology of the same university.

## 2.1. Plant Material

The whole plant morphology was studied on plants harvested from populations located in three different climate stages (humid, sub-humid, and semi-arid) [34]. The plants were kept fresh using plastic bags and moistened paper until study in the laboratory. More details are given in Table 1. Altogether, 126 plants were sampled: 44 specimens from the humid stage, 55 from the sub-humid, and 27 from the semi-arid.

**Table 1.** Characteristics, sampling dates, sample sizes and codes of the populations used in multivariate morphological study. H: Humid; SH: Sub-humid SA: Semi-arid.

| Climate stage | Humid | Sub-humid | Semi-arid |
|---|---|---|---|
| Locality name | Kherrata | Semaoun | Boudjelil |
| Population Code | H | SH | SA |
| GPS Localization | 36° 31′33.3″ N<br>5° 16′49.73″ | 36° 37′27.16″ N<br>4° 49′7.20″ E | 36° 22′11″ N<br>4° 26′48.30″ E |
| Altitude/<br>Exposure | 612 m/East | 195 m/West | 270 m/East |
| Soil | Red clay | Brown clay | White ground |
| Plant formation | Road bank | Garrigue, Grassland | Grassland, Sparse garrigue |
| Rainfall [1] (mm) | 800–1000 | 600–800 | 600–800 |
| Number of plants | 44 | 55 | 27 |
| Sampling date/<br>Plant codes | 08/06/2015/H01-H20<br>03/06/2015/H21-H35<br>15/07/2015/H236-H44 | 20/05/2015/SH01-SH15<br>09/06/2015/SH16-SH25<br>09/06/2015/SH26- SH34<br>20/05/2015/SH35-SH47<br>28/06/2016 SH48-SH55 | 13/05/2015/SA01-SA13<br>07/06/2015/SA14-SA27 |

[1] Mebarki [34].

In addition to fresh plants reserved for multivariate morphological analysis, inflorescences were collected from at least five plants at different stages of development: young floral buds for meiosis analysis, flowers just before anthesis for pollen fertility assessment, and pollen grains measurements. The material was in situ fixed in 10 mL tubes containing absolute ethanol–glacial acetic acid-chloroform (6:3:1) [35].

The plant material destined for chemical extractions was collected on March 2023 from the same populations as for morphology. A total of 400 to 500 g of fresh leaves and young stems (at the beginning of young inflorescences occurrence) were harvested from each climate stage. The plant material was spread on paper sheets for ten days under sun-free conditions. The sun-free dried plant material was then powdered and stored in sealed glass containers for further use [36].

## 2.2. Whole Plant Morphology Analysis

Forty-four quantitative morphological characters (Table 2, characters 1–44) were measured on the 126 fresh plants harvested from the three climate stages (see above and Table 1). Measurements were performed using a tape measure for stem height, a sliding caliper for stem diameter, and a ruler for the dimensions of internodes, leaves, inflorescences, etc. Graph paper was used to measure small features, such as hairs, flower parts, pods, and seeds, under a binocular magnifier. To minimize errors because of character misappreciation, the same researcher (T. Zemouri) performed all the morphometric scorings. Qualitative traits of color and hairiness are uniform, so they were not included in the analysis.

**Table 2.** Morphological characters used in the morphological multivariate study.

| No | Coding | Character Name | Type [1] | Unit |
|---|---|---|---|---|
| 1 | DBP | Diameter at the base of the plant | C | cm |
| 2 | LLS | Length of the longest stem | C | cm |
| 3 | NS | Number of stems | D | Stems |
| 4 | DLS | Diameter of the longest stem | C | mm |
| 5 | LIN | Length of the third internode | C | cm |
| 6 | NI | Number of inflorescences (racemes) | D | Raceme |
| 7 | LS1 | Length of stipules | C | mm |
| 8 | WS | Width of stipules | C | mm |
| 9 | LL | Length of the leaf (3rd node) | C | cm |
| 10 | WL | Width of the leaf (3rd node) | C | cm |
| 11 | LP | Length of the petiole | C | cm |
| 12 | NPL | Number of pairs of leaflets (3rd node) | D | Pairs |
| 13 | LNLP | Lowest number of leaflets pairs | D | Pairs |
| 14 | HNLP | Highest number of leaflets pairs | D | Pairs |
| 15 | LLB | Length of the leaflet blade | C | cm |
| 16 | WLB | Width of the leaflet blade | C | mm |
| 17 | LIP | Length of the inflorescence peduncle (3rd node) | C | cm |
| 18 | HI | Height of the inflorescence | C | cm |
| 19 | DI | Diameter of the inflorescence | C | cm |
| 20 | LNF | Lowest number of flowers | D | Flower |
| 21 | HNF | Highest number of flowers | D | Flower |
| 22 | LFB | Length of the flower bract | C | mm |
| 23 | WFB | Width of the flower bract | C | mm |
| 24 | LC1 | Length of the flower calyx | C | mm |
| 25 | LC2 | Length of the corolla | C | mm |
| 26 | TLW | Total length of the wing | C | mm |
| 27 | WWB | Width of the wing blade | C | mm |
| 28 | WBW | Width at the base of the wing | C | mm |
| 29 | WMW | Width at the middle of the wing | C | mm |
| 30 | TWK | Total width of the keel | C | mm |
| 31 | LWPK | Length of the widest part of the keel | C | mm |
| 32 | TLK | Total length of the keel | C | mm |
| 33 | TLS | Total length of the standard | C | mm |
| 34 | LSB | Length of the standard blade | C | mm |
| 35 | WSB | Width of the standard blade | C | mm |
| 36 | LSP | Length of the standard 'petiolule' | C | mm |
| 37 | LCM | Length of the calyx at maturity | C | mm |
| 38 | LCT | Length of the calyx tube at maturity | C | mm |

**Table 2.** *Cont.*

| No | Coding | Character Name | Type [1] | Unit |
|---|---|---|---|---|
| 39 | LHCT | Length of hairs at the base of the calyx teeth | C | mm |
| 40 | LPM | Length of the pod at maturity | C | mm |
| 41 | WPM | Width of the pod at maturity | C | mm |
| 42 | LS2 | Length of the seed | C | mm |
| 43 | WS | Width of the seed | C | mm |
| 44 | LR | Length of the radicle | C | mm |
| | | Extra variables (Not included in multivariate analyses) | | |
| 45 | P | Polar diameter of pollen grains | C | µm |
| 46 | E | Equatorial diameter of pollen grains | C | µm |
| 47 | P/E | Rate of P and E diameters of pollen grains | C | |
| 48 | PxE | Product of P * E. | C | $\mu m^2$ |
| 49 | PF | Pollen fertility rate | C | % |

[1] C: Continuous, D: Discrete.

### *2.3. Reagents Required for Pollen and Meiosis Study*

#### 2.3.1. Fixative Solution

Absolute ethanol (SIGMA-ALDRICH, 24103-2.5L-R, St. Louis, MO 63103 USA); Glacial acetic acid (BIOCHEM Chemopharma, 101132500, Europe Office, ZA Cosne-Sur-Loire, France); Chloroform (Carlo Erba, 438601, Val-De-Reuil, France).

#### 2.3.2. Lactopropionic Orcein Preparation

Orcein (BIOCHEM Chemopharma, 520280905, Europe Office: ZA Cosne-Sur-Loire, France); Lactic acid (80%) (PANREAC, 131034, E-08110 Montcada i Reixac (Barcelona) Spain); Propionic acid (BIOCHEM chemopharma, 116150500, Montreal, QC, Canada).

#### 2.3.3. Cotton Blue Preparation

Anilin blue (SCHARLAU, AZ 01000025, SCHALAB S.L., Barcelona, Spain); Glycerin (BIOCHEM Chemopharma, 201061000, Montreal, Quebec H2Y DA4); Lactic acid (85%): PANREAC, 131034, E-08110 Montcada i Reixac (Barcelona), Spain); Phenol cristals (SIGMA-ALDRICH, 242322, St. Louis, MO, USA).

### *2.4. Pollen Grains Size and Shape*

In situ fixed flowers just before anthesis (see above) were used to recover anthers on a microscope slide containing a drop of lactopropionic orcein prepared according to Dyer [37]. The anthers were dissected under a stereomicroscope to recover pollen grains. After eliminating anther debris, a cover glass was carefully applied over the stain drop. After 10 min, pollen grains are well stained. The observations were performed under an Optika B-353A (Optika, SN 373686, Ponteranica, BG, Italy) light microscope. Pollen grains were photographed at 40× magnification. Five flowers from different racemes were used for each population. Altogether, 235 pollen grains (humid: 116; sub-humid: 78; semi-arid: 41) were measured for their polar (P) and equatorial (E) axes; the rate P/E and the product PxE were calculated. Pollen grains shapes and number of colpi were scored.

### *2.5. Meiosis Analysis*

In situ fixed young floral buds were used (see above). In a drop of lactopropionic orcein [37] on a slide, after 5–6 min in a water bath (Memmert, L210.0187, Schwabach, FRG, Germany) at 60 °C, floral buds were dissected to recover the young anthers. Under a cover glass, the anthers were squashed with the thumb to eject the pollen mother cells. The observations were made under an Optika B-353A light microscope to search for under-division mother cells. The best metaphases, I and II, were photographed at 100x magnification.

### 2.6. Pollen Fertility Assessment

In situ fixed flowers (see above), just before anthesis, were used. In a small drop of distilled water on a slide, the anthers of a flower were recovered by dissecting and pressing them with needles to obtain the maximum of pollen grains. After drying the drop of water with slight heating on a hotplate, a small drop of cotton blue [35] was added to the dried spot. Pollen grains stain after 10 min at room temperature. The observations were made under a coverslip at 40× magnification on an Optika B-353A light microscope. A total of 10 to 30 flowers from different plants were used. A total of 700 to 900 pollen grains per flower were screened. Well-stained grains with uniform outlines were considered fertile, whereas the light-stained ones, with irregular outlines and relatively small sizes, were counted sterile. Pollen fertility rate (TF) is expressed as

$$\text{TF } (\%) = \frac{\text{Number of fertile grains}}{\text{Total number of fertile and sterile grains}} \times 100 \quad (1)$$

### 2.7. Determination of Antioxidant Activity

#### 2.7.1. Chemical Reagents

All chemicals were of analytical reagent grade. Folin–Ciocalteu reagent (catalog number 106060250), trichloroacetic acid ≥ 99.0% purity (catalog number 120130500), sodium carbonate anhydrous (catalog number 319060500) were from Biochem, Chemopharma (Montreal, QC, Canada. Potassium ferricyanide ≥ 99.5% purity (catalog number 316050250) was from Biochem (Chemopharma, GA, USA). Gallic acid (catalog number 91215-100MG); 2-2-diphenyl-1-picrylhydrazyl (catalog number 669237-1G) were from Sigma–Aldrich GmbH (St. Louis, MO, USA). Ferric chloride 97 % (catalog number 236489-100G), ethanol absolute (catalog number 1070172511-2,5L), and dimethyl sulfoxide (catalog number 472301-500ML) were from Sigma–Aldrich GmbH (Sternheim, Germany).

#### 2.7.2. Plant Extract Preparation

The dried and finely powdered plant material from each of the three climate stages (5 g for each assay) was extracted with 100 mL of absolute ethanol for 6 h or continued until the extract gave no coloration, using a Soxhlet apparatus (Behr Labor-Technik GmbH, Düsseldorf, Nordrhein-Westfalen, Germany). At the end of the extraction, the liquid extract was filtered and evaporated in a vacuum at 40 °C to complete dryness, using a Büchi rotavapor R-200 (BÜCHI Labortechnik Flawil, Switzerland) [36]. The extraction yield was calculated using the following equation:

$$\text{Extract yield } (\%) = \frac{\text{Mass of extract (g)}}{\text{Mass of dry leaves sample (g)}} \times 100$$

#### 2.7.3. Total Phenolic Content

The total phenolic content of plant extracts was measured by the Folin–Ciocalteu reagent GAEent assay, using the method described by Singleton et al. [38] with a few modifications. A diluted solution of each extract (200 µL) was mixed with 750 µL of Folin–Ciocalteu reagent (previously diluted with water 1:10 $v/v$). This mixture was maintained at ambient temperature for 5 min, after which 400 µL of sodium carbonate solution (75 g/L in water) was added. The mixture was left to stand for 1 hr at room temperature. The absorption was measured, in triplicate, at 765 nm against water blank, using ultraviolet-visible (UV-Vis) spectrophotometer (Shimadzu, China). The total phenolic contents of the extracts were calculated using the calibration curve of gallic acid standard. Results were given in mg of gallic acid equivalent (GAE)/g of dry extract.

2.7.4. DPPH Free Radical Scavenging Assay

Free radical scavenging activity was determined using the 2, 2-diphenyl-1-picrylhydrazyl free radical (DPPH) method [39] with some modifications. A total of 25 μL from each extract were added to 975 μL of 100 μM methanolic solution of DPPH. The mixture was shaken and left in the dark at room temperature. After 30 min, the absorbance was recorded, in triplicate, at 512 nm, using an ultraviolet-visible (UV-Vis) spectrophotometer), and compared to the absorbance of blank sample containing 25 μL of methanol and the same amount of DPPH solution. A standard calibration curve was obtained using different gallic acid concentrations. Antioxidant activity was expressed as mg gallic acid equivalent (GAE)/g of dry extract.

2.7.5. Reducing Power Assay

The method was based on [40] procedures with modifications. A total of 0.125 mL of each extracted sample was mixed with 2.5 mL phosphate buffer (0.2 M, pH 6.6) and 2.5 mL of potassium ferricyanide [$K_3Fe(CN)_6$] (1%). The mix was incubated in water bath at 50 °C for 10 min, followed by addition of 2.5 mL of trichloroacetic acid (10%) and then centrifuged at 1500 rpm for 10 min. Finally, 2.5 mL of the upper layer solution was mixed with distilled water (2.5 mL) and $FeCl_3$ (0.5 mL, 0.1%). The absorbance was recorded, in triplicate, at 700 nm. Reducing power assay was expressed as mg gallic acid equivalent (GAE)/g of dry extract.

*2.8. Screening of the Antibacterial Activity*

2.8.1. Bacterial Strains

The antibacterial activity test of *E. pinnata* extracts included six foodborne pathogen bacteria provided by Pasteur Institute (Algiers, Algeria) and identified with the ATCC number (American Type Culture Collection). The Gram-negative bacteria: *Escherichia coli* ATCC 25922, *Pseudomonas aeruginosa* ATCC 27853, *Vibrio cholerae* ATCC 14035, *Salmonella typhi* ATCC 14028, and the Gram-positive bacteria: *Staphylococcus aureus* ATCC 25923 and Methicillin-resistant *S. aureus* ATCC 43300 (MRSA) were studied. All strains were grown in nutrient agar (NA) and incubated at 37 °C for 18–24 h until the stationary growth phase was reached [41].

2.8.2. Antibacterial Test

The antibacterial activity was conducted using agar well-diffusion method in accordance with the National Committee for Clinical Laboratory Standards [41]. Inoculum containing $10^6$ colony-forming units per milliliter (CFU/mL) of each bacterial culture to be tested were evenly spread on the surface of Mueller Hinton agar plates using sterile swabs. Subsequently, wells of 6 mm diameter were punched into the agar medium and filled with 40 μL (5 mg/mL) of plant extract dissolved in dimethyl sulfoxide (DMSO) and allowed to diffuse at room temperature for 2 h. The plates were then incubated in the upright position at 37 °C for 24 h. A well containing the same volume of DMSO served as a negative control. After incubation, the diameters (mm) of the growth inhibition zones were measured. All tests were repeated in triplicate.

*2.9. Statistical Evaluation*

Whole plant morphology and pollen size data were treated using principal components analysis (PCA) and/or cluster analysis. Before performing cluster analysis, variable values were standardized by centering ($X_i = x_i -$ mean). Mean values were compared using Tukey's honest significant difference test for unequal-size samples (unequal N HSD test), Fisher's least significant difference (LSD), or the Student's *t*-test for paired (dependent) samples. Homogeneity of variances was checked using Levene's test. Statistical analyses and homogeneity tests were performed using Statistica 8.0 programs [42].

## 3. Results

### 3.1. Whole Plant Morphology

All the studied populations showed a hemicryptophyte habit (aerial parts dying after fructification and sprouting up from the base the next season) except in rare cases where some plant stems tend to be evergreen, especially in humid spots. All the plants appear the same for their qualitative traits: Greenish–brownish hairy stems with long upright white hairs; glaucous leaves, thickly hairy leaflets on both sides, appressed hairs; purple corolla parts.

The analysis of the quantitative characters (Table 2, characters 1–44; see Supplementary Materials S1 for the raw data matrix) by PCA yielded the scatterplot in Figure 1A (Planes 1–2), where the three studied groups are perfectly separated. On planes 1–3, 1–4, and 1–5 of the PCA (Appendix B), the semi-arid population was confirmed to be a separate group from the other two merged groups (humid and sub-humid). The cumulated explained variance for the five axes is 68.67%. The characters explaining the separation of the groups in Figure 1A are given in Table 3 and Appendix A.

**Table 3.** Morphological comparison of *Ebenus pinnata* plants from different climate stages. Values are expressed as Mean ± SD (See Appendix A for more details). Explanatory R values of PC1 and PC2 of Figure 1 are shown in bold. Mean values with different capital letters are statistically different (Unequal N HSD Test, $\alpha = 0.05$), with A < B < C. Lowercase letters a and b compare the polar (P) against the equatorial (E) axis of pollen grains (*t*-Test for dependent samples, $\alpha < 0.001$); with a < b. See abbreviation full names at the foot of the table [1].

| No | Trait Code | R1 | R2 | Humid | Sub-Humid | Semi-Arid |
|---|---|---|---|---|---|---|
| 1 | DBP | 0.49 | −0.24 | 0.73 ± 0.19 B | 0.73 ± 0.33 B | 0.55 ± 0.25 A |
| 2 | LLS | **0.80** | 0.10 | 68.22 ± 18.38 B | 89.28 ± 28.93 C | 44.41 ± 19.79 A |
| 3 | NS | 0.23 | −0.60 | 9.55 ± 6.52 B | 4.65 ± 2.47 A | 3.11 ± 2.41 A |
| 4 | DLS | 0.47 | −0.41 | 4.19 ± 1.12 B | 3.75 ± 1.01 AB | 3.16 ± 1.11 A |
| 5 | LIN | 0.67 | 0.16 | 5.27 ± 1.04 B | 6.53 ± 1.57 C | 4.36 ± 1.58 A |
| 6 | NI | 0.52 | −0.39 | 57.57 ± 28.25 B | 47.89 ± 32.23 B | 17.89 ± 23.80 A |
| 7 | LS1 | 0.66 | 0.17 | 10.98 ± 1.67 A | 13.60 ± 2.74 B | 10.10 ± 2.71 A |
| 8 | WS | 0.56 | −0.13 | 4.22 ± 0.75 B | 4.41 ± 0.74 B | 3.64 ± 1.00 A |
| 9 | LL | **0.77** | −0.05 | 10.01 ± 2.06 B | 10.85 ± 2.60 B | 6.91 ± 1.93 A |
| 10 | WL | 0.67 | −0.06 | 5.19 ± 0.80 B | 5.55 ± 0.88 B | 4.64 ± 0.66 A |
| 11 | LP | **0.73** | 0.03 | 4.56 ± 1.21 B | 5.23 ± 1.61 B | 2.96 ± 0.98 A |
| 12 | NPL | 0.43 | −0.07 | 4.73 ± 0.50 B | 4.78 ± 0.50 B | 4.11 ± 0.97 A |
| 13 | LNLP | −0.18 | −0.29 | 2.80 ± 0.55 B | 2.45 ± 0.50 A | 2.81 ± 0.68 B |
| 14 | HNLP | 0.43 | −0.23 | 4.98 ± 0.15 B | 4.91 ± 0.35 B | 4.52 ± 0.89 A |
| 15 | LLB | **0.70** | 0.03 | 2.45 ± 0.33 B | 2.66 ± 0.35 C | 2.15 ± 0.36 A |
| 16 | WLB | 0.22 | −0.63 | 7.77 ± 1.51 B | 6.14 ± 1.21 A | 6.18 ± 1.46 A |
| 17 | LIP | 0.56 | −0.31 | 22.00 ± 3.17 B | 21.29 ± 3.56 B | 18.20 ± 3.94 A |
| 18 | HI | 0.43 | −0.49 | 5.21 ± 1.41 B | 4.40 ± 0.75 A | 3.88 ± 1.61 A |
| 19 | DI | 0.33 | −0.16 | 2.44 ± 0.31 A | 2.40 ± 0.25 A | 2.30 ± 0.31 A |
| 20 | LNF | 0.24 | 0.24 | 11.84 ± 8.05 A | 16.64 ± 7.17 B | 13.56 ± 9.89 AB |
| 21 | HNF | 0.66 | −0.41 | 56.43 ± 11.83 C | 49.44 ± 9.47 B | 33.93 ± 17.42 A |
| 22 | LFB | 0.08 | 0.29 | 8.15 ± 0.62 A | 8.60 ± 0.89 B | 8.46 ± 1.04 AB |

**Table 3.** *Cont.*

| No | Trait Code | R1 | R2 | Humid | Sub-Humid | Semi-Arid |
|---|---|---|---|---|---|---|
| 23 | WFB | 0.06 | −0.12 | 3.29 ± 0.32 A | 3.25 ± 0.18 A | 3.19 ± 0.43 A |
| 24 | LC1 | −0.05 | 0.25 | 12.72 ± 0.94 A | 13.21 ± 0.90 B | 13.38 ± 1.06 B |
| 25 | LC2 | 0.13 | −0.25 | 8.21 ± 0.44 A | 8.05 ± 0.24 A | 7.98 ± 0.70 A |
| 26 | TLW | 0.40 | **0.87** | 1.79 ± 0.08 A | 3.54 ± 0.13 C | 2.40 ± 0.07 B |
| 27 | WWB | 0.45 | **0.85** | 1.37 ± 0.09 A | 2.45 ± 0.07 C | 1.67 ± 0.05 B |
| 28 | WBW | −0.67 | 0.23 | 1.10 ± 0.10 A | 1.14 ± 0.06 A | 1.38 ± 0.04 B |
| 29 | WMW | −0.02 | **0.89** | 0.43 ± 0.05 A | 0.89 ± 0.07 C | 0.83 ± 0.07 B |
| 30 | TWK | **−0.86** | −0.34 | 4.21 ± 0.23 B | 3.62 ± 0.10 A | 4.99 ± 0.09 C |
| 31 | LWPK | **−0.86** | −0.25 | 3.47 ± 0.26 B | 3.12 ± 0.08 A | 4.14 ± 0.08 C |
| 32 | TLK | **−0.81** | −0.29 | 6.36 ± 0.36 B | 5.94 ± 0.11 A | 6.99 ± 0.08 C |
| 33 | TLS | **−0.89** | −0.02 | 6.65 ± 0.45 B | 6.30 ± 0.23 A | 8.04 ± 0.08 C |
| 34 | LSB | **−0.86** | −0.29 | 5.13 ± 0.47 B | 4.49 ± 0.12 A | 6.06 ± 0.05 C |
| 35 | WSB | **−0.87** | −0.39 | 4.67 ± 0.36 B | 3.36 ± 0.16 A | 6.04 ± 0.10 C |
| 36 | LSP | −0.31 | 0.38 | 1.72 ± 0.47 A | 1.81 ± 0.16 A | 2.02 ± 0.10 B |
| 37 | LCM | −0.42 | 0.67 | 13.19 ± 0.48 A | 14.50 ± 0.89 B | 15.26 ± 0.18 C |
| 38 | LCT | 0.04 | **0.81** | 2.80 ± 0.14 A | 3.29 ± 0.12 C | 3.19 ± 0.12 B |
| 39 | LHCT | −0.15 | 0.22 | 3.27 ± 0.05 A | 3.28 ± 0.07 A | 3.29 ± 0.03 A |
| 40 | LPM | **−0.85** | 0.14 | 5.09 ± 0.09 A | 5.08 ± 0.11 A | 5.99 ± 0.10 B |
| 41 | WPM | **−0.71** | 0.40 | 3.04 ± 0.10 A | 3.15 ± 0.13 B | 3.46 ± 0.07 C |
| 42 | LS2 | **−0.87** | −0.06 | 2.35 ± 0.07 B | 2.19 ± 0.10 A | 3.00 ± 0.08 C |
| 43 | WS | **−0.85** | 0.06 | 2.05 ± 0.05 A | 2.02 ± 0.05 A | 2.34 ± 0.05 B |
| 44 | LR | **−0.77** | −0.34 | 2.14 ± 0.05 B | 2.02 ± 0.07 A | 2.25 ± 0.05 C |
| 45 | P | NA | NA | 17.95 ± 1.10 Ab | 19.47 ± 1.27 Bb | 18.43 ± 1.08 Ab |
| 46 | E | NA | NA | 12.80 ± 1.18 Aa | 13.03 ± 0.99 Aa | 12.84 ± 1.37 Aa |
| 47 | P/E | NA | NA | 1.41 ± 0.10 A | 1.50 ± 0.11 B | 1.45 ± 0.15 AB |
| 48 | PxE | NA | NA | 230.56 ± 31.60 A | 254.31 ± 30.81 B | 237.16 ± 32.74 A |
| 49 | PF | NA | NA | 94.04 ± 2.64 A | 95.01 ± 2.02 A | 94.97 ± 2.11 A |

[1] SD: Standard deviation; PC1 and PC2: Principal Components (axes) 1 and 2; R1 and R2: Pearson's coefficient of correlation with PC1 and PC2, respectively; NA: Not attributed. See Table 2 for trait code full names and measure units; Unequal HSD test: Honest significant difference for unequal size samples.

Table 3 details the morphological comparison of the three groups. There are significant differences for all morphological characters except DI (diameter of inflorescence), WFB (Width of the flower bract), LC2 (Length of the corolla), and LHCT (Length of hairs at the base of the calyx teeth). The morphological comparison of the three groups is better summarized by the cluster analysis in Figure 1B, where the "semi-arid" population shows up as a remote group in relation to the closer but different "humid" and "sub-humid" groups.

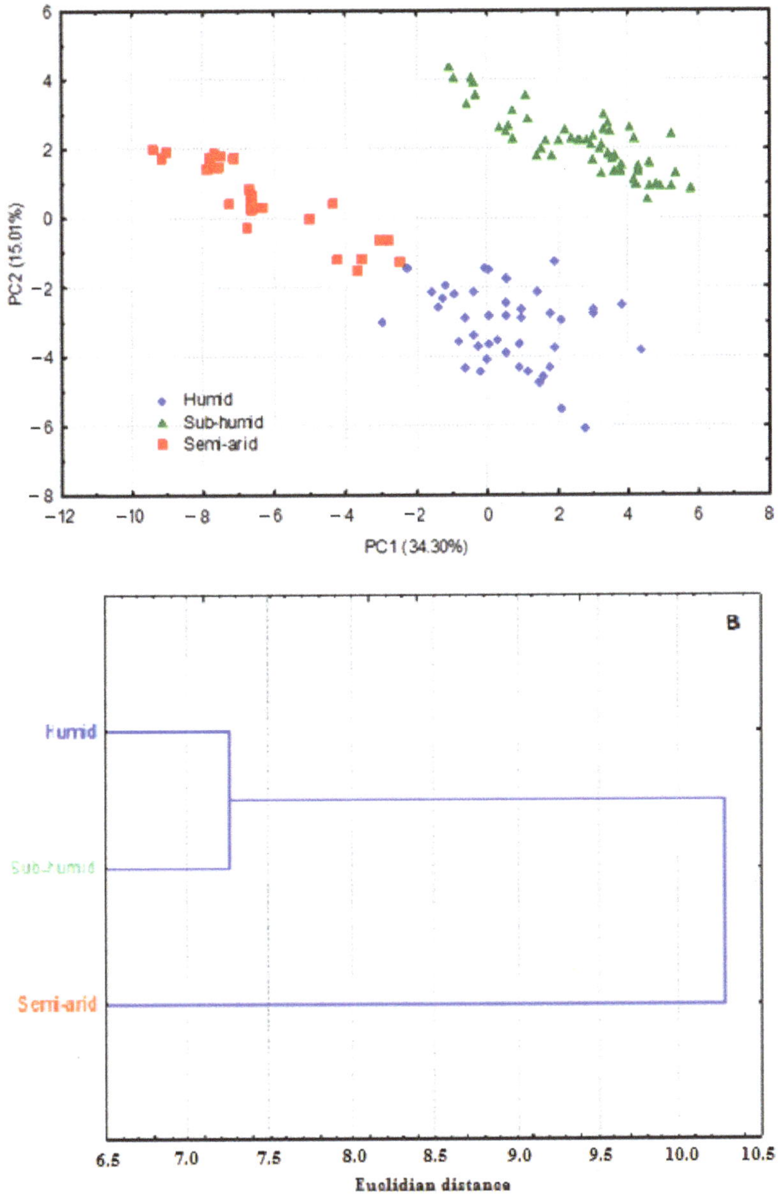


**Figure 1.** Multivariate analysis of morphological characters of three *Ebenus pinnata* ecotypes. (**A**): Principal Components Analysis based on 44 quantitative traits measured on 126 fresh plants (see Tables 1–3 for more details). (**B**). Cluster Analysis of centered mean values of the 44 quantitative traits using the unweighted pair group method with arithmetic mean (UPGMA).

### *3.2. Pollen Grains Size*

The results on pollen grains size are summarized in Table 3 (Characters 45–48). There are significant differences between groups for the polar axis (P), P/E, and PxE, but none were revealed for the equatorial axis (E). On the PCA scatterplot of Figure 2A (see Supplementary Materials S2 for raw data), the three groups appear to be merged in relation to Axis 1, explained by E and P/E. However, in relation to Axis 2, explained by P, it is clear that the "sub-humid" group has most of the highest values of P. Most of the lowest values of P are in the "humid" group, and those of the "semi-arid" group are mostly in the middle position between the two other groups. The dendrogram of Figure 2B shows that the "sub-humid" group is far distant from the other two groups. This topology is incongruent with that based on whole plant morphology (Figure 1B), where the "semi-arid" was the remote group. The cluster of the three groups of *E. pinnata* studied here appears as an outgroup of the other *Ebenus* species (Figure 2B).

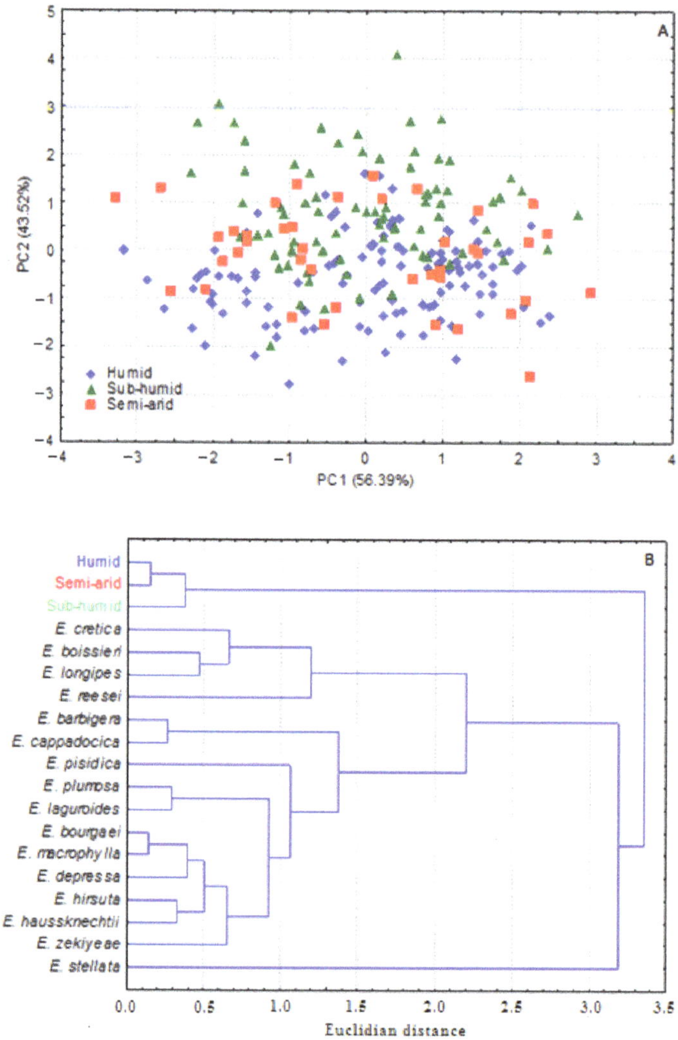


**Figure 2.** Multivariate analysis of pollen size. (**A**). Principal Components Analysis based on three pollen grains traits (polar axis P, equatorial axis E and their rate P/E) measured for the three *Ebenus pinnata* ecotypes. Axis 1 is explained by E and P/E with respectively R = 0.98 and R = −0.78; Axis 2 is explained by P with R = 0.94. (**B**). Cluster Analysis based on the centered mean values of three pollen traits (P, E and P/E).

In addition to this statistical evaluation, our observations showed that the pollen grains of the studied material are prolate (P = *c.* 1.5 × E), isopolaric, radially symmetrical, tricolpate with a circular outline on the polar view, and they have an elliptical outline on the equatorial view and reticulate ornamentations on their surface (Figure 3G–I) for all the studied material.

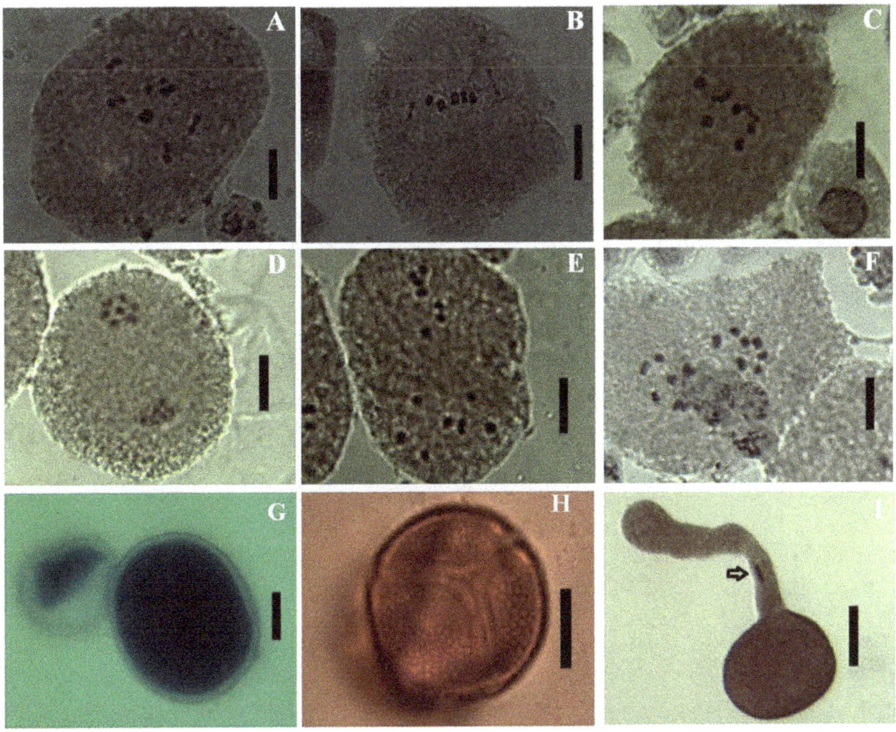


**Figure 3.** Haploid phase chromosomes and pollen grains in *Ebenus pinnata*. (**A–C**): Metaphase I of the "humid", "sub-humid", and "semi-arid" groups respectively (7 bivalents); (**D–F**): Metaphase II in the three groups respectively (7 chromosomes); (**G**): Cotton blue-stained fertile pollen grain (on the right) and sterile pollen grain (on the left); (**H**): Polar view of a lactopropionic orcein stained pollen grain showing three symmetrical colpi and reticulate ornamentations; (**I**): Lactopropionic orcein stained pollen grain with its pollen tube and nucleus (arrow). The scale bar corresponds to 10 µm.

### *3.3. Chromosome Numbers, Meiotic Abnormalities and Pollen Fertility*

For all three groups, seven bivalents in metaphase I and seven chromosomes in metaphase II were repeatedly counted (Figure 3A–F), which clearly confirms a chromosome number of 2n = 2x = 14, with a base number of x = 7. In metaphase I, there are often five bivalents with circular pairing and two with linear pairing. No multivalents were observed, and chromosome segregation at anaphase–telophase I was regular since metaphase II was always with n = 7.

Meiosis abnormalities were often absent, and cytomixis was observed only in very few cases. Pollen fertility rates were high for all the assessed flowers of the three groups (Table 3 and Supplementary Materials S3). Pollen fertility rates were 89.64–98.71% in the "humid" group, 89.01–98.07% in the "sub-humid" and 91.97–97.85% in the "semi-arid" (see Supplementary Materials S3 for pollen fertility raw data). No significant differences were detected among groups for pollen fertility (Table 3, line 49).

### 3.4. Antioxidant Activity

The antioxidant capacities of the studied extracts were evaluated by two in vitro methods: the free DPPH radical scavenging test and the reducing power assay. The results are shown in Table 4.

**Table 4.** Comparison of antioxidant and antibacterial activities of three samples of *Ebenus pinnata* ethanolic extracts from different climate stages. For each parameter, values are expressed as Min–Max (upper line) and Mean ± SD (lower line). All tests were repeated in triplicate. Different capital letters indicate significant differences between mean values according to the Fisher's least significant difference (LSD) test ($\alpha$ = 0.01 for antioxidant parameters, and $\alpha$ = 0.05 for antibacterial activity, with C > B > A).

| | Parameters/Strains | Sample 1 (Humid) | Sample 2 (Sub–Humid) | Sample 3 (Semi–Arid) |
|---|---|---|---|---|
| Antioxidant activity [1] | TPC (mg GAE/g EXT) | 50.32–51.33<br>50.79 ± 0.51<br>A | 50.93–53.02<br>52.04 ± 1.05<br>A | 56.49–57.40<br>56.89 ± 0.46<br>B |
| | DPPH (mg GAE/g EXT) | 68.62–72.74<br>71.18 ± 2.24<br>A | 82.93–88.56<br>86.39 ± 3.02<br>B | 93.53–97.54<br>95.67 ± 2.02<br>C |
| | RP (mg GAE/g EXT) | 10.27–12.52<br>11.09 ± 1.24<br>A | 16.45–17.94<br>17.21 ± 0.75<br>B | 25.64–26.16<br>25.88 ± 0.26<br>C |
| Antioxidant activity [2] | *Escherichia coli* | 10–13<br>11.50 ± 2.12<br>A | 11–13<br>12.00 ± 1.41<br>A | 10–13<br>11.50 ± 2.12<br>A |
| | *Staphylococcus aureus* | 10–10.4<br>10.20 ± 0.28<br>A | 8.5–10<br>9.50 ± 0.87<br>A | 9–12<br>10.50 ± 1.50<br>A |
| | *Pseudomonas aeruginosa* | 8.8–11<br>9.90 ± 1.56<br>A | 11.5–12<br>11.75 ± 0.35<br>A | 8.5–10<br>9.25 ± 1.06<br>A |
| | Methicillin-resistant *S.aureus* (MRSA) | 9–10.4<br>9.80 ± 0.72<br>A | 9–10<br>9.60 ± 0.53<br>A | 10–11<br>10.33 ± 0.58<br>A |
| | *Vibrio cholerae* | 9–10<br>9.50 ± 0.71<br>A | 8–11.2<br>9.60 ± 2.26<br>A | 9–14.5<br>11.75 ± 3.89<br>A |
| | *Salmonella typhi* | 9.6–11<br>10.53 ± 0.81<br>A | 8.9–13<br>11.30 ± 2.14<br>A | 10.3–13<br>11.77 ± 1.36<br>A |

[1] TPC: Total polyphenol content; DPPH: radical scavenging activity; RP: Reducing power assay. [2] Expressed as the diameter (mm) of the inhibition area.

In our results, *E. pinnata* extract from the semi-arid area (sample 3) had statistically ($p < 0.05$) the highest value of total phenolic content (56.89 ± 0.46 mg GAE/g dry extract), followed by samples one and two, with similar concentrations. Sample three also expressed the best scavenging activity (95.67 ± 2.02 mg GAE/g dry extract) and reducing power (25.88 ± 0.26 mg GAE/g dry extract), followed by the sub-humid sample and then the humid one. On cluster analysis (Figure 4), the "semi-arid' sample behaved as a remote group in relation to the closer "humid" and "sub-humid" ones.

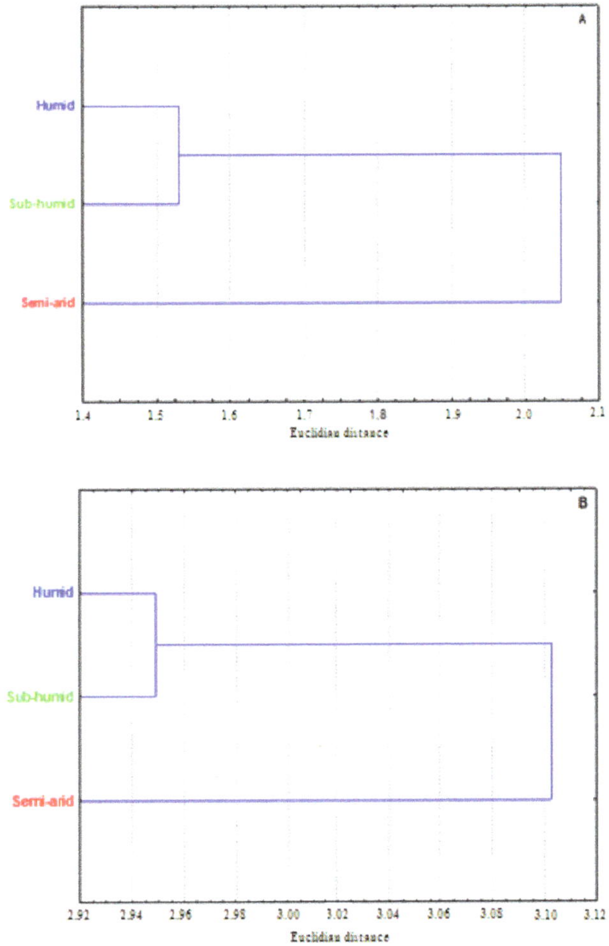


**Figure 4.** Cluster Analysis of antioxidant and antibacterial activities of three ecotypes of *Ebenus pinnata* from different climate stages. (**A**). Antioxidant activity; (**B**).Antibacterial activity. Data source in Table 4.

### *3.5. Antibacterial Activity*

The antibacterial activity results, displayed in Table 4, revealed that all studied extracts were potentially effective in inhibiting microbial growth. Although statistical analysis showed no significant difference between the three extracts, the semi-arid sample seems more active against the four microorganisms tested, namely *Staphylococcus aureus*, Methicillin-resistant *S. aureus*, *Vibrio cholera*, and *Salmonella typhi* with inhibition diameters (ID) of 10.50 ± 1.50, 10.33 ± 0.58, 11.75 ± 3.89, and 11.77 ± 1.36 mm, respectively. On the other hand, the sub-humid sample tends to be more efficient against *Escherichia coli* and *Pseudomonas aeruginosa*, with an ID of 12.00 ± 1.41 and 11.75 ± 0.35 mm, respectively. The cluster analysis dendrogram based on the data in Table 4 is shown in Figure 4, in which the semi-arid sample appears as a far distant group from the two other closer groups.

## 4. Discussion

Morphologically, the three ecotypes proved to be well-discriminated (Figure 1). They are significantly different for most characters, including those of the reproductive system (Table 3, Appendix A). The differences are slight but significant. The "semi-arid" ecotype behaved as a remote group in relation to the other close "humid" and "sub-humid" groups. Is this morphological divergence due to phenotypic plasticity alone? Or is it underpinned by a genetic divergence, too, due to differential adaption to ecological niches? The latter hypothesis seems supported by pollen data (see below) and preliminary karyological and molecular results. A thorough molecular genetic analysis is required to clarify the question.

Based on pollen data, the three groups discriminated well, too (Table 3, Appendix A). The "sub-humid" group showed larger pollen size than the two close "humid" and "semi-arid" groups. This situation is well rendered by the cluster analysis in Figure 2B. When compared to other *E. species*, using literature data, the three groups constituted a separate cluster within which distances between groups are of the same range as those separating other accepted *Ebenus* species (Figure 2B). This observation supports the hypothesis that the phenotypic differences between the three groups have a genetic basis. Moreover, the pollen grain size (as expressed by PxE) of the three groups is highly correlated (R = 0.89, $p$ = 0.30) with the total haploid chromosome length (THCL) of the three groups (unpublished data). PxE/THCL ($\mu m^2/\mu m$) values are, in increasing order, 229.76/23.82 (humid), 236.64/26.98 (semi-arid), and 253.69/28.33. This is not the case in other *Ebenus* species since their pollen grain sizes are not correlated (R = 0.22, $p$ = 0.72) with their haploid total chromosome lengths (Appendix C). Pollen size is not always positively correlated to genome size across taxa [43]. A pollen grain is not just plant cells with their nuclei, cytoplasm, membranes, and walls. It is, above all, the exine whose structure and thickness reflect evolutionary history and selective adaptation to various aspects of the environment. The variability of the exine structure and thickness seems to be at the origin of the lack of a positive correlation between pollen and genome size across taxa.

The chromosome number for all three ecotypes (Figure 4) is 2n = 2x = 14 (x = 7), the same as already reported from Morocco [13] and Algeria [15,16]. Preliminary results of an underway karyomorphological study show significant differences in chromosome lengths of material from the three climate stages involved here. On the karyogram reported by Gadoum and Hamma [15], there are two pairs of chromosomes with two pairs of large satellites. The four extra chromosomes of the 2n = 18 reported for a Moroccan population by Parra et al. [31] may correspond to those four large satellites mistaken for chromosomes.

According to Siddiqui and Alrumman [44] and the references therein, cytomixis is a phenomenon induced by both genetic and environmental factors; it occurs in mutants, hybrids, aneuploids, and stressed plants (heat, cold, drought, parasites, and pollution), leading to reduced pollen fertility. In our case, very scarce cytomixis and associated abnormalities were encountered while screening meiosis preparations, which is in accordance with the high pollen fertility rates assessed (89.01–98.71%) (Table 3 and Appendix A). From this perspective, it can be drawn that the local *E. pinnata* populations are in perfect equilibrium with their environment, without any aberrations occurring due to mutation, aneuploidy, or hybridization between diverging genotypes.

In the present study, the antioxidant and antimicrobial activities of *Ebenus pinnata* ethanol extracts revealed an interesting potential that could be used as an alternative medicinal source. Reports concerning *E. pinnata* phytochemicals and biological activities are very scarce. In the study by Abreu et al. [33], the measurement of the antioxidant activity of methanol extract from *E. pinnata* aerial parts collected from Tunisia has revealed an interesting potential. The authors linked this activity specifically to the presence of four phenolic compounds: ombuoside, kaempferol-3-O-rutinoside, rutin, and catechin, as evoked in the introduction. In another study from Algeria in relation to antioxidant parameters of methanol extracts of nine species, Nouioua and Gaamoune [45] have reported that *E. pinnata* extracts show a relatively high DPPH radical scavenging activity ($IC_{50}$ of 12.25 $\pm$ 2.80 $\mu$g/mL) compared to eight other taxa from Eastern Algeria. They have also reported that this plant contains 8.57 $\pm$ 0.16 mg GAE/g dry extract of TPC. The latter performances are very low compared with those obtained in the present study (Table 4).

The antioxidant potency of the semi-arid extract sample was found to be stronger when compared with the two other samples (humid and sub-humid) for both assays.

From these results, it is assumed that its high level of phenolics might have contributed to the observed antioxidant abilities. The present findings are in agreement with other studies reporting a high correlation between total phenolics and antioxidant activity [46]. However, the absence of statistically significant differences in the total phenolic content values of humid and sub-humid samples, combined with relatively high differences in DPPH and RP values between these samples, could be explained by the presence of other non-phenolic compounds, in particular carotenoids and terpenoids, which contribute to the antioxidant properties of the extracts. In addition, due to differences in the environmental growing conditions of these plants, the types of components and their proportions can vary considerably, resulting in a variation in the contribution of individual and synergistic activities and, consequently, a variation in biological properties [47].

In light of the antimicrobial results, it was observed that all extracts showed comparable power at the analyzed concentration. The main causes of the small differences observed in these results were the various bio-contents of the analyzed samples that were harvested from different bioclimatic stages (humid, sub-humid, and semi-arid). However, sometimes, the same species could have different bioactive values since many factors may be responsible for these changes, such as harvesting period, water availability, environmental factors (climate and altitude), and technological factors [48,49]. Indeed, Kabtni et al. [50] have investigated the influence of climate variation on the phenolic composition and antioxidant activity of *Medicago minima* populations selected from different provenances in Tunisia. They have concluded that the highest phenolic contents are observed in populations from the semi-arid area with a BSK climate and an altitude higher than 550 m, which agrees with the findings of this study. The accumulation of a higher level of phenolic compounds and the expression of the best antioxidant activity for *E. pinnata* grown under a semi-arid climate characterized by high temperature and low precipitation can be related to hydric and thermal stresses [51,52].

## 5. Conclusions

The three plant groups from the humid, sub-humid, and semi-arid climate stages are similar regarding stem and leaf color and hairiness as well as flower color. However, morphometrics revealed slight but significant differences for most quantitative morphological traits, with the semi-arid being far different from the two other closer humid and sub-humid groups. Pollen fertility rates were high for all the studied material (94.04 $\pm$ 2.64–95.01 $\pm$ 2.02%). All groups showed prolate, isopolaric, and tricolpate pollen grains with reticulate ornamentations. Their polar and equatorial axes were 17.95 $\pm$ 1.10–19.47 $\pm$ 1.27 µm and 12.80 $\pm$ 1.18–13.03 $\pm$ 0.99 µm, respectively, with the biggest pollen grains found in the sub-humid group. TPC, DFFH, and RP values (mg GAE/g EXT) were 50.79 $\pm$ 0.51–56.89 $\pm$ 0.46, 71.18 $\pm$ 2.24–95.67 $\pm$ 2.02, and 11.09 $\pm$ 1.24–25.88 $\pm$ 0.26, with the highest values found in the semi-arid group. The extracts showed activity against all the six bacterial strains tested; the diameters of activity zones ranged from 9.25 $\pm$ 1.06 mm (semi-arid extract vs. *Pseudomonas aeruginosa*) to 12.00 $\pm$ 1.41 mm (sub-humid extract vs. *Escherichia coli*). The same chromosome number of 2n = 2x = 14 was repeatedly counted. Pollen grain size seems positively correlated to karyotype length (genome size). In sum, the climate seems to exert a significant impact on the biology of the plant.

Experimental cultures and molecular analysis are required to check the genetic basis hypothesis of the phenotypic heterogeneity of the groups. It would be interesting to apply the present study to other species.

**Supplementary Materials:** The following supporting information can be downloaded at: https://www.mdpi.com/article/10.3390/horticulturae9080879/s1, Supplementary Materials S1: Whole plant morphology raw data; Supplementary Materials S2: Pollen grains size raw data; Supplementary Materials S3: Pollen fertility rates raw data.

**Author Contributions:** Conceptualization: M.S. and T.Z.; Plant material collect: M.S. and T.Z.; Morphometrics scoring: T.Z.; Meiosis and pollen grains analysis: T.Z. and H.B.; Antioxidant and antibacterial activities analysis: A.C. and T.Z.; Statistic evaluation: M.S.; Results interpretations, M.S., T.Z., A.C. and H.B., Original draft preparation: M.S. and A.C. (for antioxidant and antibacterial activities); Supervision: M.S.; Final version preparation M.S. and T.Z. All authors have read and agreed to the published version of the manuscript.

**Funding:** The study was carried out in the framework of research activities of the Laboratory of Ecology and Environment (University of Bejaia, Algeria) supported by Direction Générale de la Recherche Scientifique et du Développement Technologique (DGRSDT), Ministry of High Education and Scientific Research (MESRS), Republic of Algeria.

**Data Availability Statement:** The data used in this research are provided as Supplementary files attached to the article.

**Acknowledgments:** We would like to warmly thank all the technical staffs of the Laboratory of Ecology and Environment and the Laboratory of Microbial Ecology for their full availability during the preparation of this research.

**Conflicts of Interest:** The authors declare no conflict of interest.

## Appendix A

Detailed morphological comparison of Ebenus pinnata plants from different climate stages.

## Appendix B

Detailed whole plant morphology PCA results.

## Appendix C

Pollen grains size vs. karyotype length.

## References

1. Aytaç, Z. The genus *Ebenus* L. (Leguminosae/Fabaceae) in Turkey. *Karaca Arbor. Mag.* **2000**, *5*, 145–171.
2. Aytaç, Z.; Yildirim, H. *Ebenus zekiyeae* (Fabaceae), a new species from Turkey. *Ann. Bot. Fenn.* **2018**, *55*, 25–29. [CrossRef]
3. Aytaç, Z.; Suludere, Z.; Pınar, M. Examination of the leaflets hairs and stoma structures with the electron microscope of the genus *Ebenus* L. (Leguminosae) in Turkey. *Biodicon* **2015**, *8*, 32–50.
4. Boissier, E. *Flora Orientalis: Save Enumeratio Plantarum in Oriente a Græcia et Ægypto ad Indfiæ Fines Hucusque Observatarum*; NCSU Libraries: Genevae, Switzerland, 1872; Volume 2.
5. Dobignard, A.; Chatelain, C. *Index Synonymique de la Flore D'afrique du Nord. Dicotyledoneae: Fabaceae-Nymphaeaceae*; Conservatoire et Jardin botaniques de la ville de Genève et ECWP: Genève, Switzerland, 2012; Volume 4, 431p.
6. Huber-Morath, A. 1970: *Ebenus* L. In *Flora of Turkey and the East Aegean Islands*; Davis, P.H., Ed.; Edinburgh University Press: Edinburgh, UK; Volume 3, pp. 590–596.
7. Amirahmadi, A.; Kazampour-Osaloo, S.; Moein, F.; Kaveh, A.; Maassoumi, A.A. Molecular systematics of the tribe Hedysareae (Fabaceae) based on nrDNA ITS and plastid trnL-F and matK sequences. *Plant Syst. Evol.* **2014**, *300*, 729–747. [CrossRef]
8. Kaveh, A.; Kazampour-Osaloo, S. Estimation of *Ebenus* species divergence time based on nrDNA ITS and matK cpDNA sequences. In Proceedings of the 4th National Congress of Plants, Tehran, Iran, 12–13 May 2015.
9. Ghanavati, F.; Amirabadizadeh, H. Pollen grain morphology in Iranian Hedysareae (Fabaceae). *Crop Breed. J.* **2012**, *2*, 25–33. [CrossRef]
10. Aytaç, Z.; Ünal, F.; Pinar, M.N. Morphological, palynological, and cytotaxonomical study of *Ebenus longipes* Boiss. & Bal. and *E. argentea* Siehe ex Bornm. (Leguminosae) from Turkey. *Isr. J. Plant Sci.* **2000**, *48*, 321–326. [CrossRef]
11. Pinar, N.M.; Vural, C.; Zytac, Z. Pollen morphology of *Ebenus* L. (Leguminosae: Subfamily Papilionoideae) in Turkey. *Pak. J. Bot.* **2000**, *32*, 303–310.
12. Halbritter, H.; Auer, W.; Igersheim, A.; Ebenus cretica. In PalDat-A Palynological Database. 2020. Available online: https://www.paldat.org/pub/Ebenus_cretica/305777;jsessionid=B9B8A714C47DE9C0CB28F0595C4D0398 (accessed on 25 January 2021).
13. Molero, J.; Montserrat-Marti, J.M. Números cromosomáticos de plantas marroquíes. *Collect. Bot.* **1986**, *16*, 351–354.
14. Aksoy, H.; Ünal, F.; Aytaç, Z. Karyological study on four endemic *Ebenus* L. taxa (Leguminosae) in Turkey. *Caryologia* **2001**, *54*, 307–311. [CrossRef]

15. Gadoum, N.; Hamma, A. Etude Cytogénétique de L'espèce *Ebenus pinnata* Aiton (Fabaceae) du Golfe de Béjaïa et de la Vallée de la Soummam. Master's Thesis, Université Abderrahmane Mira de Bejaia, Algérie, North Africa, 2016; 58p. Available online: http://univ-bejaia.dz/dspace/123456789/10254 (accessed on 8 July 2023).
16. Hadawat, A.K.; Madani, S. Etude Cytogénétique de Quelques Populations d'*Ebenus pinnata* Ait. (Fabacées) de la Région de Béjaïa. Master's Thesis, Université Abderrahmane Mira de Bejaia, Algérie, North Africa, 2022; 49p. Available online: http://univ-bejaia.dz/dspace/123456789/21618 (accessed on 8 July 2023).
17. Uyar, Z.; Böke, N.; Türkay, E.; Koz, Ö.; Yas, A.I.; Kırmızıgül, S. Flavonoid glycosides and methylinositol from *Ebenus haussknechtii*. *Nat. Prod. Res.* **2006**, *20*, 999–1007. [CrossRef]
18. Özdemir, E.; Alpınar, K. An ethnobotanical survey of medicinal plants in western part of central Taurus Mountains: Aladaglar (Nigde–Turkey). *J. Ethnopharmacol.* **2015**, *166*, 53–65. [CrossRef] [PubMed]
19. Kültür, S.; Gürdal, B.; Sari, A.; Melikoğlu, G. Traditional herbal remedies used in kidney diseases in Turkey: An overview. *Turk. J. Bot.* **2021**, *45*, 269–287. [CrossRef]
20. Ceylan, R.; Katani, J.; Zengina, G.; Mati, S.; Aktumsek, A.; Boroja, T.; Stanic, S.; Vladimir Mihailovic, V.; Guler, G.O.; Boga, M.; et al. Chemical and biological fingerprints of two Fabaceae species (*Cytisopsis dorycniifolia* and *Ebenus hirsuta*): Are they novel sources of natural agents for pharmaceutical and food formulations? *Ind. Crops Prod.* **2016**, *84*, 254–262. [CrossRef]
21. Bektas, E.; Kaltalioglu, K.; Sahin, H.; Turkmen, Z.; Kandemir, A. Analysis of phenolic compounds, antioxidant and antimicrobial properties of some endemic medicinal plants. *Int. J. Second. Metab.* **2018**, *5*, 75–86. [CrossRef]
22. İmir, N.; Aydemir, E.; Şimşek, E.; Göktürk, R.; Yesilada, E.; Fişkin, K. Cytotoxic and immunomodulatory effects of *Ebenus boissieri* Barbey on breast cancer cells. *Genet. Mol. Res.* **2016**, *15*, 1–11. [CrossRef] [PubMed]
23. Simsek, E.; Imir, N.; Aydemir, E.R.; Gokturk, R.S.; Yesilada, E.; Fiskin, K. Caspase-mediated apoptotic effects of *Ebenus boissieri* Barbey extracts on human cervical cancer Cell Line HeLa. *Pharmacogn. Mag.* **2017**, *13*, 254–259. [CrossRef]
24. Aydemir, E.A.; Simsek, E.; Imir, N.; Göktürk, R.S.; Yesilada, E.; Fiskin, K. Cytotoxic and apoptotic effects of *Ebenus boissieri* Barbey on human lung cancer Cell Line A549. *Pharmacogn. Mag.* **2015**, *11* (Suppl. 1), S37. [CrossRef]
25. Kounadi, S.; Aligiannis, N.; Pongratz, I.; Lelovas, P.; Ismini, D.; Skaltsounis, A. Estrogenic activity of the methanolic extract of *Ebenus cretica* L. *Planta Medica* **2011**, *77*, PM168. [CrossRef]
26. Dontas, I.; Kounadi, S.; Aligiannis, N.; Galanos, A.; Skaltsounis, A.; Lelovas, P. Plant extract administration and mild daily exercise increase bone density of ovariectomized rats. In Abstracts of 14th FELASA congress 2019, PC41, p. 157. *Lab. Anim.* **2019**, *53*, 28–203.
27. Mitrocotsa, D.; Skaltsounis, A.-L.S.; Harvala, C.; Tillequin, F. Flavonoid and terpene glycosides from European *Ebenus* species. *Biochem. Syst. Ecol.* **1999**, *27*, 305–307. [CrossRef]
28. Khodaparast, A.; Sayyah, M.; Sardari, S. Anticonvulsant activity of hydroalcoholic extract and aqueous fraction of *Ebenus stellata* in mice. *Iran. J. Basic Med. Sci.* **2012**, *15*, 811–819.
29. Zameer, S.; Ali, S.; Gulmeena Tareen, R.B. Identification of volatile constituents and antimicrobial activity of *Ebenus stellata*. *Gu J. Phytosciences* **2022**, *2*, 214–222.
30. Zemouri, T.; Chikhoune, A.; Benmouhoub, H.; Sahnoune, M. Taxonomic Comparison, Antioxidant and Antibacterial Activities of Three Ecotypes of *Ebenus pinnata* Ait. (Fabaceae) from Algeria. *Preprints* **2023**, 2023071339. [CrossRef]
31. Parra, R.; Valdés, B.; Gordillo, I.; Venanzi, R. Meditenanean chromosome number reports. *Flora Mediterr.* **1999**, *9*, 323–387.
32. Braham, H.; Ben Jannet, H.; Castedo, L.; Mighri, Z. Isolation, for the first time, of a flavonoid glycoside and the (±)-catechin from the aerial parts of *Ebenus pinnata*. *J. De La Société Chim. De Tunis.* **2004**, *6*, 153–160.
33. Abreu, P.M.; Braham, H.; Ben Jannet, H.; Mighri, Z.; Matthew, S. Antioxidant compounds from *Ebenus pinnata*. *Fitoterapia* **2007**, *78*, 32–34. [CrossRef]
34. Mebarki, A. Hydrologie des Bassins de l'Est Algérien: Ressources en Eau, Aménagement et Environnement. Ph.D. Thesis, Université Mentouri de Constantine, Algérie, North Africa, 2005; 360p.
35. Mertens, T.R.; Hamnersmith, R.L. *Genetic Laboratory Investigations*; Eleventh, Ed.; Prentice Hall Inc.: Upper Saddle River, NJ, USA, 1998.
36. Herzi, N.; Bouajila, J.; Camy, S.; Romdhane, M.; Condoret, J.S. Comparison of different methods for extraction from *Tetraclinis articulata*: Yield, chemical composition and antioxidant activity. *Food Chem.* **2013**, *141*, 3537–3545. [CrossRef]
37. Dyer, A.F. The use of lactopropionic orcein in rapid squash. Methods for chromosome preparations. *Stain. Technol.* **1963**, *38*, 85–90. [CrossRef]
38. Singleton, V.L.; Orthofer, R.; Lamuela-Raventos, R.M. Analysis of total phenols and other oxidation substrates and antioxidants by means of Folin-Ciocalteu RGAEent. *Methods Enzymol.* **1999**, *299*, 152–178. [CrossRef]
39. Blois, M.S. Antioxidant determinations by the use of a stable free radical. *Nature* **1958**, *181*, 1199–1200. [CrossRef]
40. Oyaizu, M. Studies on products of browning reaction: Antioxidative activities of products of browning reaction prepared from glucosamine. *Jpn. J. Nutr. Diet.* **1986**, *44*, 307–315. [CrossRef]
41. NCCLS (National Committee for Clinical Laboratory Standards). *Performance Standard for Antimicrobial Disc Susceptibility Tests; Approved Standard*; Villanova, P.A., Ed.; Publication M2-A5; National Committee for Clinical Laboratory Standards: Wayne, PA, USA, 1993.
42. *Statistica*, version 8.0. Data analysis software system. Statsof, Inc.: Tulsa, OK, USA, 2007.
43. Knight, C.A.; Clancy, R.B.; Götzenberger, L.; Dann, L.; Beaulieu, J.M. On the relationship between pollen size and genome size. *J. Bot.* **2010**, *2010*, 612017. [CrossRef]

44. Siddiqui, S.; Alrumman, S.A. Methomyl, imbraclaobrid and clethodim induced cytomixis and syncytes behaviors in PMCs of *Pisum sativum* L: Causes and outcomes. *Saudi J. Biol. Sci.* **2022**, *29*, 103390. [CrossRef] [PubMed]
45. Nouioua, W.; Gaamoune, S. Antioxidant, antimicrobial and anti-inflammatory activities development of methanol extracts of some species growing in the massif of Boutaleb, Setif, Algeria. *Int. J. Pharm. Nat. Med.* **2018**, *6*, 15–20.
46. Zhang, Y.; Wang, Z. Phenolic composition and antioxidant activities of two *Phlomis* species: A correlation study. *Comptes Rendus Biol.* **2009**, *332*, 816–826. [CrossRef]
47. Martins, N.; Barros, L.; Ferreira, I.C. In vivo antioxidant activity of phenolic compounds: Facts and gaps. *Trends Food Sci.* **2016**, *48*, 1–12. [CrossRef]
48. Ahuja, I.; de Vos, R.C.; Bones, A.M.; Hall, R.D. Plant molecular stress responses face climate change. *Trends Plant Sci.* **2010**, *15*, 664–674. [CrossRef]
49. Rabeta, M.S.; Nur Faraniza, R. Total phenolic content and ferric reducing antioxidant power of the leaves and fruits of *Garcinia atrovirdis* and *Cynometra cauliflora*. *Int. Food Res. J.* **2013**, *20*, 1691–1696.
50. Kabtni, S.; Sdouga, D.; Bettaib Rebey, I.; Save, M.; Trifi-Farah, N.; Fauconnier, M.L.; Marghali, S. Influence of climate variation on phenolic composition and antioxidant capacity of *Medicago minima* populations. *Sci. Rep.* **2020**, *10*, 8293. [CrossRef]
51. De Abreu, I.N.; Mazzafera, P. Effect of water and temperature stress on the content of active constituents of *Hypericum brasiliense* Choisy. *Plant Physiol. Biochem.* **2005**, *43*, 241–248. [CrossRef] [PubMed]
52. Al-Huqail, A.; El-Dakak, R.M.; Sanad, M.N.; Badr, R.H.; Ibrahim, M.M.; Soliman, D.; Khan, F. Effects of climate temperature and water stress on plant growth and accumulation of antioxidant compounds in sweet basil (*Ocimum basilicum* L.) leafy vegetable. *Scientifica* **2020**, *2020*, 3808909. [CrossRef] [PubMed]

**Disclaimer/Publisher's Note:** The statements, opinions and data contained in all publications are solely those of the individual author(s) and contributor(s) and not of MDPI and/or the editor(s). MDPI and/or the editor(s) disclaim responsibility for any injury to people or property resulting from any ideas, methods, instructions or products referred to in the content.

*horticulturae*

MDPI

Article

# Micromorphology of *Barleria albostellata* (Grey Barleria) Flower and Pollen Grains

**Serisha Gangaram [1,*], Yougasphree Naidoo [1], Yaser Hassan Dewir [2], Moganavelli Singh [1] and Katalin Magyar-Tábori [3]**

[1] School of Life Sciences, Westville Campus, University of KwaZulu-Natal, P.O. Box X54001, Durban 4000, South Africa; naidooy1@ukzn.ac.za (Y.N.); singhm1@ukzn.ac.za (M.S.)
[2] Plant Production Department, College of Food and Agriculture Sciences, King Saud University, Riyadh 11451, Saudi Arabia; ydewir@ksu.edu.sa
[3] Research Institute of Nyíregyháza, Institutes for Agricultural Research and Educational Farm (IAREF), University of Debrecen, P.O. Box 12, 4400 Nyíregyháza, Hungary; mtaborik@gmail.com
* Correspondence: serishagangaram@yahoo.com

**Abstract:** *Barleria albostellata* C.B. Clarke (grey barleria, Acanthaceae) is an indigenous shrub to South Africa and has been relatively understudied. This shrub is a valuable medicinal plant with a wide spectrum of antibacterial and anti-inflammatory activities. Detailed studies on the floral and pollen morphology on *B. albostellata* are rare. This study was conducted to observe the morphology of the flower and pollen grains using stereomicroscopy and scanning electron microscopy (SEM). Morphological observations showed numerous non-glandular trichomes on the bracteoles and bracts of *B. albostellata*. Three types of trichomes were identified on these structures: I—unicellular, II—multangulate-dendritic branched non-glandular trichomes, and III—capitate glandular trichomes. A taxonomical description of the floral structures using stereo and SEM micrographs is provided. SEM micrographs revealed the pollen grains as globose tricolporate with a rough honeycomb exine, and small granules inside the lumina. The diameter of the pollen grains was 77.53 ± 5.63 μm, whereas the aperture of these grains was 14.31 ± 0.59 μm. This study provides insight into the floral biology of *B. albostellata*, and the results presented here will add to the body of knowledge and encourage further research on this species.

**Keywords:** capitate glandular trichomes microscopy; morphology; non-glandular trichomes; pollen grains; trichomes

check for updates

**Citation:** Gangaram, S.; Naidoo, Y.; Dewir, Y.H.; Singh, M.; Magyar-Tábori, K. Micromorphology of *Barleria albostellata* (Grey Barleria) Flower and Pollen Grains. *Horticulturae* **2023**, *9*, 732. https://doi.org/10.3390/horticulturae9070732

Academic Editors: Wajid Zaman and Alessandra Carrubba

Received: 17 May 2023
Revised: 14 June 2023
Accepted: 20 June 2023
Published: 21 June 2023

**Copyright:** © 2023 by the authors. Licensee MDPI, Basel, Switzerland. This article is an open access article distributed under the terms and conditions of the Creative Commons Attribution (CC BY) license (https://creativecommons.org/licenses/by/4.0/).

## 1. Introduction

Flower receptiveness plays an important role in pollination variability, reproductive success, and plant productivity [1,2]. These active traits include timing of the anther opening and pollen appearance, anther and stigma position, and flower receptiveness and morphology [3,4]. Pollination involves the transferal of pollen from the anther to the stigma of the flower [2,5], and the rate of success is highly dependent on pollen viability (the capability of pollen to induce seed set efficiently) [6,7]. Furthermore, the reproductive success of a plant may depend on its ability to attract flower visitors. These visitors that aid in pollination may exert selection on specific floral traits that are attractive to them [7–9]. The genus *Barleria* has approximately 300 species of shrubs and herbs that are distributed in the subtropical and tropical regions of the world [10–13]. Members of this genus originated from the Far East of Japan, through southern Asia, Arabia, India, Africa, Madagascar to as far west of Central America and Mexico [12,14]. Several species of *Barleria* are known for their floral diversity. Additionally, there are specialized structures on their surface for the synthesis, storage, and/or secretion of secondary metabolites with the ultimate goal of anti-herbivory tactics and protection against water loss [15,16]. These structures are known as trichomes and occur on the plant surface as hairs or external glands. Trichomes may be family or species-specific and may vary in their chemical composition [15]. Plant secretory structures can also be used as taxonomic characters, assisting in the identification of plant families [15].

*Barleria albostellata*, an evergreen shrub, thrives in semi-shade to full-sun woodland areas of South Africa, and under suitable conditions, grows up to 1.5 m in height (Figure 1). However, in colder regions, they can become deciduous to semi-deciduous [17]. In South Africa, *B. albostellata*, generally known as 'grey barleria' or in Afrikaans the 'Bosviooltjie', belongs to the family Acanthaceae [17]; and is widely distributed from Limpopo, Gauteng, and Mpumalanga to KwaZulu-Natal [17]. The genus name *Barleria* was derived from a French botanist and Dominican monk, Jacques Barrelier [17]. This shrub flourishes from September to May, with beautiful white flowers appearing sporadically (Figure 1). Flowers appear from a dense compound inflorescence and are surrounded by four leafy-bracts [18]. The blooming flowers are white in color and have a tinge of purple on the bracts. In contrast to the flowers, the leaves are grey-green and have an abundance of velvety hairs. This plant develops fairly quickly and reaches maturity in about three years [17]. *Barleria albostellata* contains medicinal properties which were verified by Amoo et al. [19]. It was found that several extracts from this plant exhibited excellent anti-inflammatory properties and a broad-spectrum of antibacterial activity. This plant has a relatively high flavonoid content, with an added effect from tannin and iridoid compounds [20]. Although *B. albostellata* has no recorded practice in traditional medicine, many species of *Barleria* have been used in traditional medicine and were confirmed to contain various compounds possessing biological effects such as analgesic, anti-inflammatory, antileukemic, antihyperglycemic, antitumor, anti-amoebic, antibiotic, and virucidal activities [21–26].

**Figure 1.** *Barleria albostellata* found along a pathway at the University of KwaZulu-Natal, Westville Campus. (**A**,**B**) White, tubular flowers emerge sporadically in spring and summer.

Acanthaceae is regarded as a eurypalynous family [27], with significant diversity in the pollen shape, size, exine structure, apertures, and ornamentation [28–32]. Plants in this family demonstrate a thriving diversity of ecological and morphological characteristics. These include a large range of pollinator relationships and floral morphologies [33–35]. Members of *Barleria* are easily recognized for their globose, tricolporate pollen with roughly reticulate (also referred to as the honeycomb-patterned) and inter-apertural exine [18,36,37]. Characterizing the morphology of pollen grains is useful in plant systematics and this can further add to the body of knowledge within the genus and family. Little is known on the floral and pollen morphology of *B. albostellata*, however a substantial amount of work has been done in other species within the family Acanthaceae [38–40] and in other species of *Barleria*. Previous studies have found that members of this genus are pollinated by moths [41,42], or attract various species of butterflies [17]. Additionally, it was noted that the flowers of *B. albostellata* were also pollinated by insects and butterflies [43]. On a regular basis, carpenter bees were also observed to visit the flowers of *B. albostellata*. Plants within this genus produce copious amounts of nectar which attract bumble bees [17]. Several morphological features of the *B. albostelata* flower and pollen grains have been largely unexplored. Secretory structures documented within *Barleria* include non-glandular and glandular peltate and capitate trichomes. Therefore, the present study aimed to describe the floral morphology and pollen of *B. albostellata* using stereo and scanning electron microscopy.

## 2. Materials and Methods

### 2.1. Plant Materials

Flowers of *B. albostellata* were collected from the University of KwaZulu-Natal, School of Life Sciences, Westville Campus (29°49′51.6″ S, 30°55′30″ E), Durban, South Africa. A voucher specimen (Accession no. 7973000) was deposited in the Ward Herbarium of the University of KwaZulu-Natal, Life Sciences, Westville Campus. Five replicates of flowers were analyzed using microscopy techniques.

### 2.2. Stereomicroscopy

Fresh flowers were examined using the Nikon AZ100 stereomicroscope (Nikon Corporation, Yokohama, Japan) equipped with a Nikon Fiber Illuminator and photographed using the NIS-Elements Software (NIS-elements D 3.00).

### 2.3. Scanning Electron Microscopy (SEM)

The micromorphology of chemically-fixed flowers of *B. albostellata* was examined in detail. The initial step of preparation involved dissecting the different parts of the flower: petal, stigma, style, anther, and filament and thereafter primary fixating the sections ($\pm 10$ mm$^2$) in 2.5% glutaraldehyde for 18–24 h. After primary fixation, samples were rinsed for 5 min each (thrice) with 0.1 M sodium phosphate buffer (pH 7.2) and then post-fixed in 0.5% osmium tetroxide for 3 h at room temperature. The samples were washed thrice (for 5 min each) with sodium phosphate buffer and dehydrated gradually with increasing concentrations of ethanol (30%, 50%, 75%, 100%) twice, for 5 min each, followed by exposure to 100% ethanol for two sessions, each of 10 min. Dehydrated samples were critically point-dried using the Quorum K850 Critical Point Dryer (Quorum Technologies Ltd., Laughton, East Sussex, UK) with a vertical chamber. Samples were mounted onto aluminum stubs using double-sided adhesive carbon tape and sputter coated with a layer of gold using the Quorum 150 RES (Quorum Technologies Ltd.), a combined system for carbon and sputter coating. The samples were then viewed and photographed using the LEO 1450 SEM at a working distance (WD) of 12–15 mm. Images were captured using the SmartSEM image software (Zeiss, Jena, Germany). (protocol was adapted from the microscopy and microanalysis unit, University of KwaZulu-Natal, Westville). With respect to the stigma of the flower, pollen grains were dusted from the stigma onto aluminum stubs using double-sided adhesive carbon tape and sputter coated with a layer of gold using the Quorum 150 RES (Quorum Technologies Ltd.), a combined system for carbon and sputter coating. Images of pollen grains were captured using the SmartSEM image software (Zeiss, Jena, Germany). Diameters of pollen grains were determined using ImageJ software Java 1.53e (Fiji, http://fiji.sc/Fiji, accessed on 10 June 2021) [44]

## 3. Results and Discussion

### 3.1. Analysis of Floral Structures via Stereomicroscopy

Bracteoles of *B. albostellata* vary from narrowly ovate to ovate, with glabrous or hairy surfaces and margins that are spiny, with scanty teeth (Figure 2A–D). Long white hairs (unicellular non-glandular trichomes) are prominent on the surface and margins of the floral bracts, (Figure 2A,B) upper, and lower bracteoles (Figure 2C,D), and a posterior lobe with sharp, curved apiculus (Figure 3A). Certain species of *Barleria* have characteristic non-glandular trichomes which are unicellular [45]. Non-glandular trichomes are recognized exclusively for their physical protection in plants against biotic and abiotic stresses [46,47], and to deter herbivores from feeding and ovipositing insects [48–50].

**Figure 2.** Stereomicrographs of the floral bracts of *B. albostellata*. (**A**,**B**) Floral bracts abaxial surface; (**C**) upper bracteole abaxial surface; (**D**) lower bracteole abaxial surface. Abbreviations: UT = unicellular non-glandular trichome.

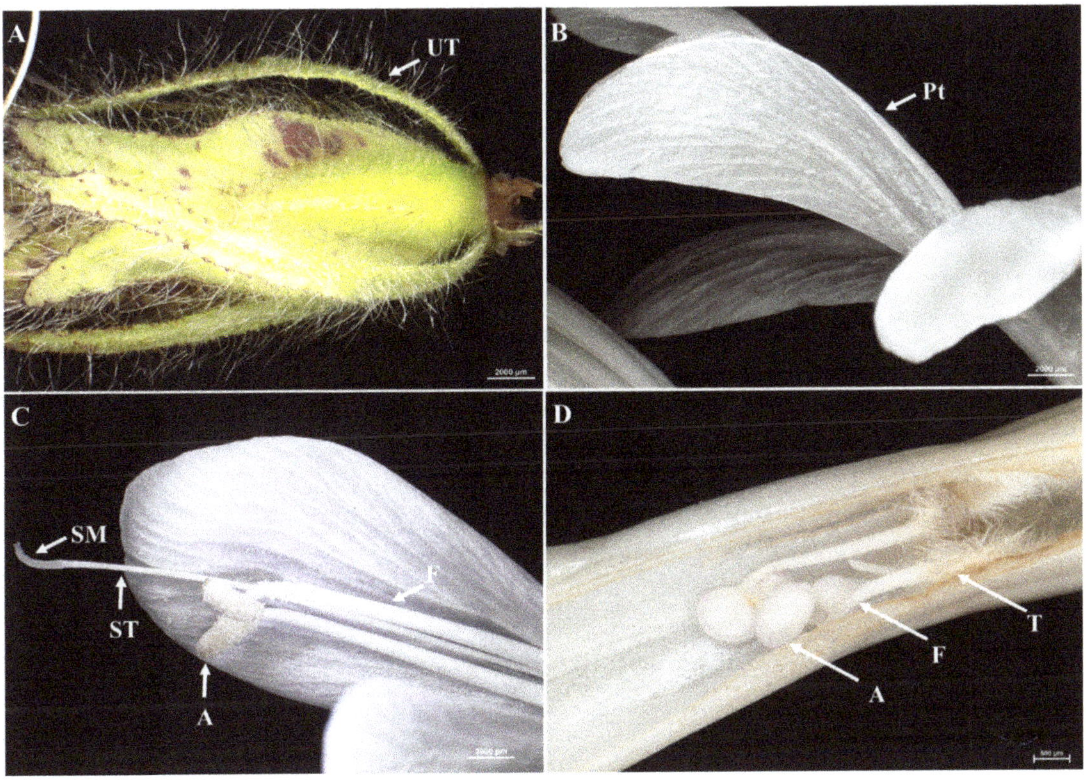


**Figure 3.** Stereomicrographs of the bracts and petals of *B. albostellata*. (**A**) Posticous calyx lobes of bracts, outer surface; (**B**) petals of the flower; (**C**) stamen, stigma, and style; (**D**) slit along the lower corolla lobe. Abbreviations: UT = unicellular non-glandular trichome; T = trichome; SM = stigma; ST = style; A = anther; F = filament; Pt = petal.

Bracts are highly modified, chartaceous, foliaceous, with reticulate venation being prominent; margins are entire, serrate, or irregularly dentate. The inflorescence is a compound, terminal synflorescence, capitate or strobilate with units of solitary flowers. Flowers of *B. albostellata* (2–4 flowers) are bisexual with a nectariferous disc, zygomorphic, and in cymes (a wide, flat-topped, distinct flower cluster in which the central flowers are opened first) [51,52]. Flowering is an important phenological event, which impacts the reproductive success of a species [53]. The flowers are enclosed by four leafy, hairy, purple-tinged bracts. Purple-tinged bracts are assumed to contain some sort of nectar (Figure 2A–D). The delimitation of the genus '*Barleria*' and specifically the taxonomic description of the leaves and flowers of *B. albostellata* have been only described by Balkwill and Balkwill [18].

The corolla (petals, 1 + 4) is irregular, thin, tubular, and gamopetalous (Figure 3B). The scent of the flowers of *B. albostellata* is produced nocturnally, with the strongest smell produced by mature, unopened buds, than with the open flowers. The floral visitors observed to interact with the flowers of *B. albostellata* were butterflies and bees. Similar observations were reported by Balkwill et al. [54] in flowers of *B. greenii*. Generally, flowers of *Barleria* are pollinated by butterflies, and they were described as large, white, and tubular, comprising deep nectaries/nectariferous discs, which function as nectar guides [55,56]. Bumblebees were observed to frequently visit *B. greenii* and remove nectar from outside of the flower, by creating a narrow slit at the base of the corolla tube [54]). Several trichome-derived compounds are used as attractants for species-specific pollination [57]. Trichomes are also involved in specialized mechanisms of insect capture for pollination [58]. The fertile stamens (anther + filament), inserted on the corolla, are usually present in pairs (Figure 3C) and are not didynamous. Filaments are long and may appear as twisted, can cross over each other, and are usually hairy at the base (Figure 3C). Anthers are basifixed and longitudinally dehisce, whilst the style is terete. The stigma is filiform and is found beyond the level of the dehisced anthers, while the style arches upward and is terete (Figure 3C). Similar morphological characteristics were observed in flowers of *B. greenii* [54] and in *B. saxatilis* [59]. They suggested that the position of the stigma above the anthers promotes autonomous self-pollination. The slit along the lower corolla lobe revealed the growing stamen with hairy trichomes attached to the lower region of the filament (Figure 3D), a characteristic of species within *Barleria* [60].

### 3.2. Floral Structures Observed via Scanning Electron Microscopy

Floral bracts were heavily pubescent with non-glandular and glandular trichomes (Figure 4A,B). Unicellular non-glandular trichomes were highly dense, long, pointed, and located on the serrated edges of the floral bracts or occurring along the mid-region. Similar observations were reported for *B. aristata* floral bracts [61]. Perhaps the edges of the floral bracts might have responded to insect damage, therefore increasing the trichome density. There were only a few glandular capitate trichomes scattered all over the floral bracts, while the multangulate-dendritic branched (MDB) non-glandular trichomes were predominant (Figure 4A,B). In certain cases, the MDB non-glandular trichomes were found to 'arch over' the glandular trichomes. Due to its proximity, the MDB non-glandular trichomes may provide some sort of physical protection to these glandular tichomes. The adaxial and abaxial surfaces of a petal contained several grooves and appeared as coarse and pitted (Figure 4C,D), with epidermal cells in an irregular shape. With high magnification, parallel striations can be seen on a section of the surface of the stigma (Figure 5B). The cap of the anther is curved and round (protection of pollen), with a slit in the middle (Figure 5C).

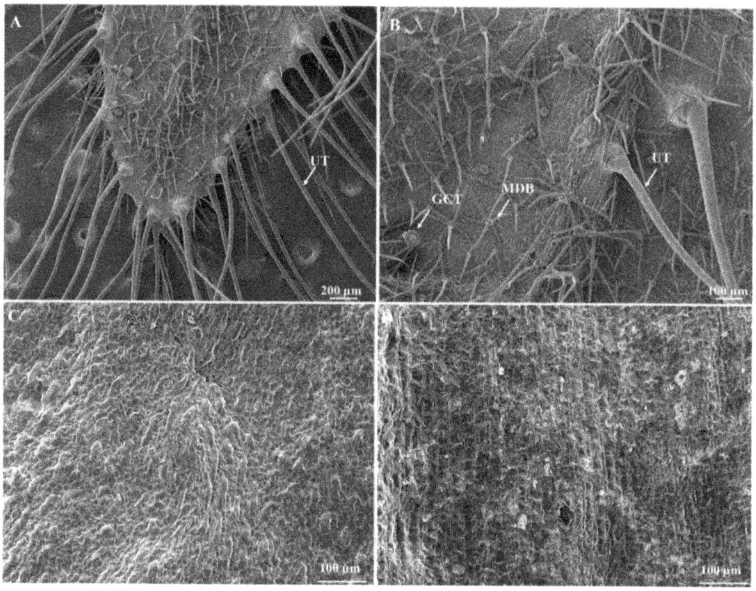


**Figure 4.** Scanning electron micrographs of the floral morphology of *B. albostellata*. (**A**) Floral bract; (**B**) glandular and non-glandular trichomes, on the floral bracts; (**C**) adaxial surface of a petal; (**D**) abaxial surface of a petal. Abbreviations: UT = unicellular non-glandular trichome; MDB = multangulate-dendritic branched non-glandular trichome; GCT = glandular capitate trichome.

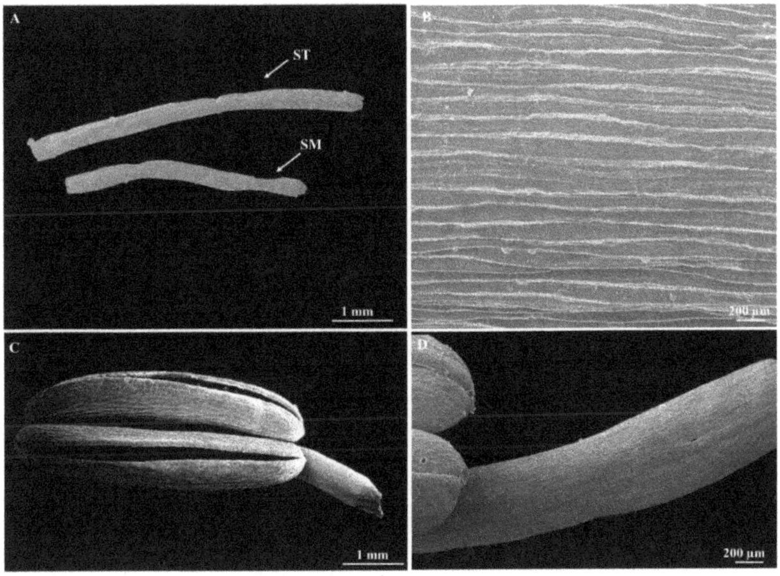


**Figure 5.** Scanning electron micrographs of the floral morphology of B. albostellata. (**A**) Low magnification image of a dissected section of the style and stigma; (**B**) high magnification image of a section of the stigma; (**C**) anther; (**D**) filament. Abbreviations: SM = stigma; ST = style.

### 3.3. Pollen Morphology

Pollen micromorphological features have contributed beneficial phylogenetic information in the accurate identification of species within Acanthaceae [13,39]. Scanning electron micrographs of pollen grains had an open reticulate tectum and appeared as globose tricolporate in equatorial view, honeycombed-shaped, with intense, coarse reticulation of the inter-apertural exine (Figure 6A); these characteristics are specific to pollen found in species of *Barleria* [11,18,36,37,60,62,63]. Pollen grains in *B. albostellata* had a diameter of 77.53 ± 5.63 µm (Figure 6).

This parameter varied from 60.5 ± 0.3 µm in *B. parviflora* to 81.5 ± 1 µm in *B. orbicularis*. Similar pollen grain sizes were found in *B. albostellata*, and almost the same diameters (74.9 ± 0.7 µm) were documented for *B. ventricosa* and *B. proxima* (79.1 ± 1 µm), respectively [13]. Before pollination at maturity, pollen grains are located inside the cap of the anther for protection (Figure 6B–D). Tiny granules are observed inside lumina of the pollen grain (Figure 6A). This was also noted in *B. parviflora*, *B. ventricosa* [13], *B. prionitis*, and *B. hochstetteri* [64]. The aperture of pollen grains in *B. albostellata* appeared circular in shape (Figure 6E,F), which was also noted in *B. bispinosa* [13]. The aperture width of pollen grains in *B. albostellata* was 14.31 ± 0.59 µm. The width varied from 9 µm in *B. acanthoides* and *B. aculeata*, 13 µm in *B. tetracantha*, 16 µm in *B. ventricosa* and *B. bispinosa*, to 23 µm in *B. prionitis* [13]. Pollen grains in various families are recognized by distinct morphological features represented in their exine [65]. Similar pollen grains characteristics to that of *B. albostellata* were found in *B. grootbergensis* [66] and *B. durairajii* [63]. Studies highlight that pollen viability is significantly reduced with increasing air humidity and temperature [67]. *Barleria albostellata* thrives in subtropical and tropical conditions, well-drained soils, and can grow under relatively cold conditions [43]. Pollen grains in *Barleria* are characteristic to the family Acanthaceae, however their reticulate ornamentation displays close resemblances with those found in their associated genera such as Lepidagathis, Crabbea, and Ruellia [68].

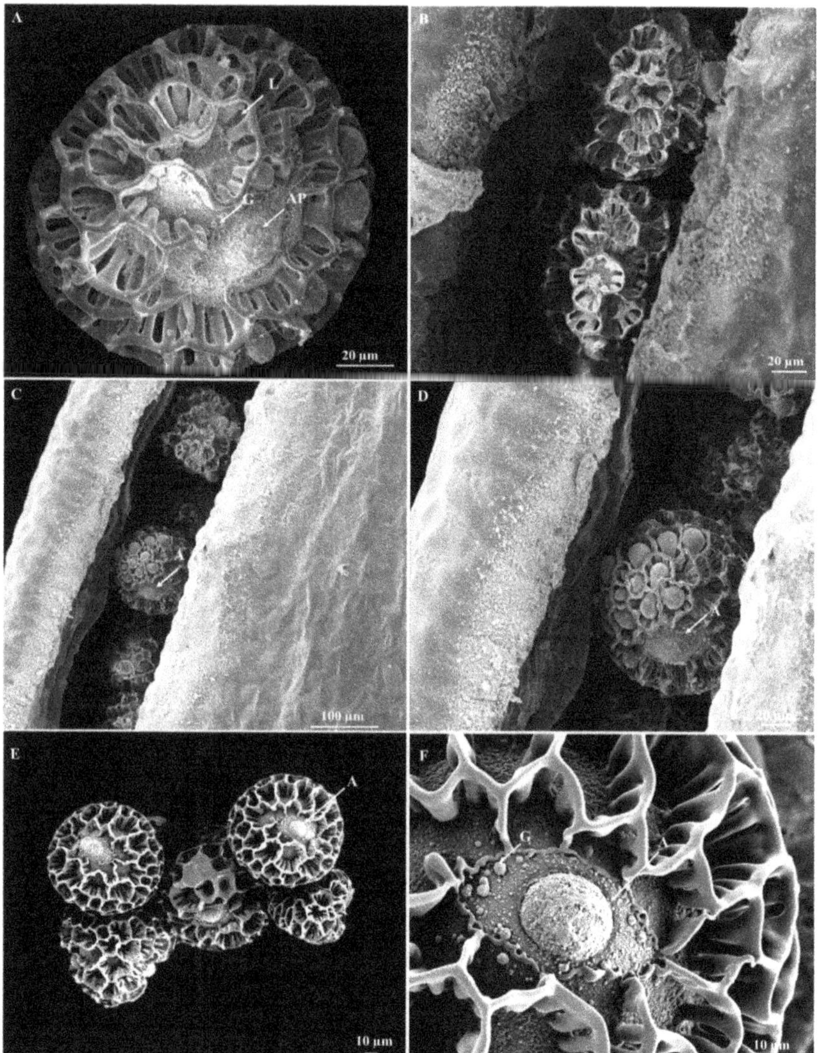


**Figure 6.** Scanning electron micrographs of the pollen micromorphology of *B. albostellata*. (**A**) Single pollen grain, equatorial view; (**B–E**) pollen grains found within the anther; (**F**) aperture of pollen grain. Abbreviations: AP = aperture; G = granules; L = lumina.

The morphology of the flower and pollen grains using various microscopic techniques showed numerous non-glandular trichomes on the bracteoles and bracts of *B. albostellata*. Three types of trichomes were identified on these structures and were found in other species of *Barleria* [54,59–61]. MDB non-glandular trichomes may provide some sort of physical protection to the glandular capitate tichomes. The pollen micromorphological features found are characteristic to species of *Barleria* [11,18,36,37,54,62,63].

## 4. Conclusions

The combination of stereo- and scanning electron microscopy facilitated the identification of the floral and pollen morphology of *B. albostellata*. Knowledge on the floral biology and pollen morphology of *B. albostellata* obtained by microscopy techniques has been very incomplete so far. Floral structures identified were compared to previously reported information in other species of *Barleria*. Pollen grains of *B. albostellata* are complex, intricate, and display reticulate sculpturing. Thus, the results presented in this study contribute significantly to our growing understanding of the floral and pollen biology of *B. albostellata*. In this regard, this study is novel, and results reported here will also assist taxonomists in identifying *B. albostellata* using SEM micrographs of their distinct pollen structures. Additional ultrastructural studies on the floral structures should be conducted to further examine the internal features of the cells and organelles. Further studies may also focus on evaluating the micromorphology of the seeds and roots of *B. albostellata*.

**Author Contributions:** Conceptualization, S.G. and Y.N.; methodology, S.G. and Y.N.; formal analysis, S.G., Y.N. and M.S.; investigation, S.G., Y.N. and M.S.; data curation, S.G., Y.N., Y.H.D. and M.S.; writing—original draft preparation, S.G. and Y.N.; writing—review and editing, Y.N., Y.H.D., M.S. and K.M.-T.; validation, M.S., Y.H.D. and K.M.-T.; visualization, M.S., Y.H.D. and K.M.-T.; supervision, Y.N., Y.H.D. and M.S.; project administration, Y.N.; funding acquisition, Y.N. and Y.H.D. All authors have read and agreed to the published version of the manuscript.

**Funding:** National Research Foundation (Grant No. 118897), South Africa and Researchers Supporting Project number (RSP2023R375), King Saud University, Riyadh, Saudi Arabia.

**Data Availability Statement:** All data are presented in the article.

**Acknowledgments:** Authors extend their appreciation to the National Research Foundation (Grant No. 118897), South Africa and the staff at the Microscopy and Microanalysis Unit at the University of KwaZulu-Natal for their assistance with the microscopy components of the research. The authors acknowledge Researchers Supporting Project number (RSP2023R375), King Saud University, Riyadh, Saudi Arabia.

**Conflicts of Interest:** The authors declare no conflict of interest.

## References

1. Zhang, J.; Lin, M.; Chen, H.; Zhu, Q.; Chen, X. Floral biology and pistil receptivity of the drumstick tree (*Moringa oleifera* Lam.). *Arch. Biol. Sci.* **2018**, *70*, 299–305. [CrossRef]
2. Liu, F.; Gao, C.; Chen, M.; Tang, G.; Sun, Y.; Li, K. The impacts of flowering phenology on the reproductive success of the narrow endemic Nouelia insignis Franch (*Asteraceae*). *Ecol. Evol.* **2021**, *11*, 9396–9409. [CrossRef] [PubMed]
3. Harder, L.D.; Williams, N.M.; Jordan, C.Y.; Nelson, W.A. The effects of floral design and display on pollinator economics and pollen dispersal. In *Cognitive Ecology of Pollination: Animal Behaviour and Floral Evolution*; Chittka, L., Thomson, J.D., Eds.; Cambridge University Press: Cambridge, UK, 2001; pp. 297–317.
4. Budečević, S.; Hočevar, K.; Manitašević Jovanović, S.; Vuleta, A. Phenotypic Selection on Flower Traits in Food-Deceptive Plant *Iris pumila* L.: The Role of Pollinators. *Symmetry* **2023**, *15*, 1149. [CrossRef]
5. McMullen, C.K.; Close, D.D. Wind pollination in the Galápagos Islands. *Not. Galápagos* **1993**, *52*, 12–17.
6. Dafni, A.; Firmage, D. Pollen viability and longevity: Practical, ecological and evolutionary implications. *Plant Syst. Evol.* **2000**, *222*, 113–132. [CrossRef]
7. Christopher, D.A.; Mitchell, R.J.; Karron, J.D. Pollination intensity and paternity in flowering plants. *Ann. Bot.* **2020**, *125*, 1–9. [CrossRef] [PubMed]
8. Kay, K.M.; Sargent, R.D. The role of animal pollination in plant speciation: Integrating ecology, geography, and genetics. *Annu. Rev. Ecol. Syst.* **2009**, *40*, 637–656. [CrossRef]
9. Sletvold, N.; Trunschke, J.; Smit, M.; Verbeek, J.; Ågren, J. Strong pollinator-mediated selection for increased flower brightness and contrast in a deceptive orchid. *Evolution* **2016**, *70*, 716–724. [CrossRef]
10. Mabberley, J.B. *The Plant—Book: A Portable Dictionary of the Vascular Plants*, 2nd ed.; Cambridge University Press: Bath, UK, 1997.
11. Darbyshire, I.; Vollesen, K.; Kelbessa, E. Acanthaceae, part 2. In *Flora Zambesiaca*; Timberlake, J.R., Martins, E.S., Eds.; Royal Botanic Gardens: London, UK, 2015; 8p.
12. Kumari, R.; Kumar, S.; Kumar, A.; Goel, K.K.; Dubey, R.C. Antibacterial, antioxidant and immuno-modulatory properties in extracts of *Barleria lupulina* Lindl. *BMC Complemen. Altern. Med.* **2017**, *17*, 484. [CrossRef]

13. Al-Hakimi, A.S.; Faridah, Q.; Abdulwahab, A.; Latiff, A. Pollen and seed morphology of *Barleria* L. (*Barlerieae*: Ruellioideae: Acanthaceae) of Yemen. *S. Afr. J. Bot.* **2018**, *116*, 185–191. [CrossRef]
14. Balkwill, M.J.; Balkwill, K. A preliminary analysis of distribution patterns in a large, pantropical genus, *Barleria* L. (Acanthaceae). *J. Biogeog.* **1998**, *25*, 95–110. [CrossRef]
15. Svoboda, K.P.; Svoboda, T.G.; Syred, A.D. *Secretory Structures of Aromatic and Medicinal Plants: A Review and Atlas of Micrographs*; Microscopix Publications: Knighton, UK, 2000; pp. 7–12.
16. Schmid, N.B.; Giehl, R.F.; Döll, S.; Mock, H.P.; Strehmel, N.; Scheel, D.; Kong, X.; Hider, R.C.; Von Wirén, N. Feruloyl-CoA 6′-Hydroxylase1-dependent coumarins mediate iron acquisition from alkaline substrates in *Arabidopsis*. *Plant Physiol.* **2014**, *164*, 160–172. [CrossRef] [PubMed]
17. Froneman, W.; Le Roux, L.N. Barleria albostellata. 2007. Available online: http://pza.sanbi.org/barleria-albostellata (accessed on 2 February 2019).
18. Balkwill, M.J.; Balkwill, K. Delimitation and infra-generic classification of *Barleria* (Acanthaceae). *Kew Bull.* **1997**, *52*, 535–573. [CrossRef]
19. Amoo, S.O.; Finnie, J.F.; Van Staden, J. In vitro pharmacological evaluation of three *Barleria* species. *J. Ethnopharmacol.* **2009**, *121*, 274–277. [CrossRef]
20. Amoo, S.O.; Ndhlala, A.R.; Finnie, J.F.; Van Staden, J. Antifungal, acetylcholinesterase inhibition, antioxidant and phytochemical properties of three *Barleria* species. *S. Afr. J. Bot.* **2011**, *77*, 435–445. [CrossRef]
21. Yosook, C.; Panpisutchai, Y.; Chaichana, S.; Santisuk, T.; Reutrakul, V. Evaluation of anti-HSV-2 activities of *Barleria lupulina* and *Clinacanthus nutans*. *J. Ethnopharmacol.* **1999**, *67*, 179–187. [CrossRef]
22. Wang, B.U.; Wu, M.; Perchellet, E.M.; Mcilvain, C.J.; Sperfslage, B.J.; Huang, X.; Tamura, M.; Stephany, H.A.; Hua, D.H.; Perchellet, J.P. Asynthetic triptycene bisquinone which blocks nucleoside transport and induces DNA fragmentation, retains its cytotoxic efficacy in daunorubicin-resistant HL-60 cell lines. *Int. J. Oncol.* **2001**, *19*, 1169–1178.
23. Jassim, S.A.A.; Naji, A.M. Novel antiviral agents: A medicinal plant perspective. *J. App. Microbiol.* **2003**, *95*, 412–427. [CrossRef]
24. Suba, V.; Murugesan, T.; Arunachalam, G.; Mandal, S.C.; Saha, B.P. Anti-diabetic potential of *Barleria lupulina* extract in rats. *Phytomedicine* **2004**, *11*, 202–205. [CrossRef]
25. Suba, V.; Murugesan, T.; Kumaravelrajan, R.; Mandal, S.C.; Saha, B.P. Antiinflammatory, analgesic and antiperoxidative efficacy of *Barleria lupulina* Lindl. extract. *Phytother. Res.* **2005**, *19*, 695–699. [CrossRef]
26. Chomnawang, M.T.; Surassmo, S.; Nukoolkarn, V.S.; Gritsanapan, W. Antimicrobial effects of Thai medicinal plants against acne-inducing bacteria. *J. Ethnopharma* **2005**, *101*, 330–333. [CrossRef] [PubMed]
27. Raj, B. Pollen morphological studies in the Acanthaceae. *Grana Palynol.* **1961**, *3*, 3–108.
28. Graham, V.A.W. Delimitation and infra-generic classification of Justicia (Acanthaceae). *Kew Bull.* **1988**, *43*, 551–624. [CrossRef]
29. Daniel, T.F. Pollen morphology of Mexican Acanthaceae: Diversity and systematic significance. *Proc. Calif. Acad. Sci.* **1998**, *508*, 217–256.
30. Bhatt, A.; Naidoo, Y.; Nicholas, A. The foliar trichomes of *Hypoestes aristata* (Vahl) Sol. ex Roem. & Schult var aristata (Acanthaceae) a widespread medicinal plant species in tropical sub-Saharan Africa: With comments on its possible phylogenetic significance. *Biol. Res.* **2010**, *43*, 403–409.
31. Choopan, T.; Grote, P.J. Cystoliths in the leaves of the genus *Pseuderanthemum* (Acanthaceae) in Thailand. *Int. J. Sci.* **2015**, *12*, 13–20.
32. House, A.; Balkwill, K. FIB-SEM enhances the potential taxonomic significance of internal pollen wall structure at the generic level. *Flora* **2017**, *236*, 44–57. [CrossRef]
33. McDade, L.A.; Masta, S.E.; Moody, M.L.; Waters, E. Phylogenetic relationships among Acanthaceae: Evidence from two genomes. *Syst. Bot.* **2000**, *25*, 106–121. [CrossRef]
34. Daniel, T.F.; McDade, L.A.; Manktelow, M.; Kiel, C.K. The "Tetramerium Lineage" (Acanthaceae: Acanthoideae: Justicieae): Delimitation and intra-lineage relationships based on cp and nrITS sequence data. *Syst. Bot.* **2008**, *33*, 416–436. [CrossRef]
35. McDade, L.A.; Daniel, T.F.; Kiel, C.A. The Tetramerium Lineage (Acanthaceae, Justicieae) Revisited: Phylogenetic relationships reveal polyphyly of many new world genera accompanied by rampant evolution of floral morphology. *Syst. Bot.* **2018**, *43*, 97–116. [CrossRef]
36. McDade, L.A.; Daniel, T.F.; Kiel, C.A. Toward a comprehensive understanding of phylogenetic relationships among lineages of Acanthaceae s.l. (Lamiales). *Am. J. Bot.* **2008**, *95*, 1136–1152. [CrossRef] [PubMed]
37. Darbyshire, I.; Vollesen, K.; Kelbessa, E. Acanthaceae, part 2. In *Flora of Tropical East Africa*; Beentje, H., Ed.; Royal Botanic Gardens: London, UK, 2010.
38. Scott, J. Dimorphism in *Eranthemum*. *J. Bot.* **1872**, *10*, 161–166.
39. Raza, J.; Ahmad, M.; Zafar, M.; Athar, M.; Sultana, S.; Majeed, S.; Yaseen, G.; Imran, M.; Nazish, M.; Hussain, A. Comparative foliar anatomical and pollen morphological studies of Acanthaceae using light microscope and scanning electron microscope for effective microteaching in community. *Microsc. Res. Techniq.* **2020**, *83*, 103–1117. [CrossRef] [PubMed]
40. Nilanthi, R.M.R.; Samarakoon, H.; Jayawardana, N.; Hathurusinghe, B.; Wijesundara, S.; Bandaranayake, P.C.G. *Strobilanthes glandulata* (Acanthaceae), a new species from Sri Lanka based on the morphological and molecular evidences. *Phytotaxa* **2022**, *573*, 1–14.

41. Pooley, E. *A Field Guide to Wild Flowers Kwazulu-Natal and the Eastern Region*; Natal Flora Publications Trust: Durban, South Africa, 1998.
42. Scott-Shaw, C.R.; Johnson, I.M.; Styles, D.; Makholela, T.; Von Staden, L. *Barleria greenii* (Balkwill, M., Balkwill, K.). National Assessment: Red List of South African Plants Version 2007, 1. Available online: http://redlist.sanbi.org/species.php?species=3909-26 (accessed on 10 October 2021).
43. Froneman, W.; Plants of South Africa. South Africa: Lowveld National Botanical Garden. 2010. Available online: http://www.plantzafrica.com/plantab/barleriapriondel.htm (accessed on 2 February 2019).
44. Schindelin, J.; Arganda-Carreras, I.; Frise, E.; Kaynig, V.; Longair, M.; Pietzsch, T.; Preibisch, S.; Rueden, C.; Saalfeld, S.; Schmid, B.; et al. Fiji: An open-source platform for biological-image analysis. *Nat. Methods* **2012**, *9*, 676–682. [CrossRef] [PubMed]
45. Ahmad, K.J. Epidermal hairs of Acanthaceae. *Blumea-Biodivers. Evol. Biogeogr. Plants* **1978**, *24*, 101–117.
46. Werker, E. Trichome diversity and development. *Adv. Bot. Res.* **2000**, *31*, 1–35.
47. Wagner, G.J.; Wang, E.; Shepherd, R.W. New approaches for studying and exploiting an old protuberance, the plant trichome. *Ann. Bot.* **2004**, *93*, 3–11. [CrossRef]
48. Levin, D.A. The role of trichomes in plant defense. *Q. Rev. Biol.* **1973**, *48*, 3–15. [CrossRef]
49. Baur, R.; Binder, S.; Benz, G. Non-glandular leaf trichomes as short-term inducible defense of the grey alder, *Alnus incana* (L.), against the chrysomelid beetle, *Agelastica alni* L. *Oecologia* **1991**, *87*, 219–226. [CrossRef]
50. Szyndler, M.W.; Haynes, K.F.; Potter, M.F.; Corn, R.M.; Loudon, C. Entrapment of bed bugs by leaf trichomes inspires microfabrication of biomimetic surfaces. *J. R. Soc. Interface* **2013**, *10*, 20130174. [CrossRef] [PubMed]
51. Palmer, E.; Pitman, N. *Trees of Southern Africa*; Balkema: Amsterdam, The Netherlands; Cape Town, South Africa, 1972.
52. Schmidt, S.; Lotter, M.; McCleland, W. *Trees and Shrubs of Mpumalanga and the Kruger National Park*; Jacana: Johannesburg, South Africa, 2002.
53. Fenner, M. The phenology of growth and reproduction in plants. *Perspec. Plant Ecol. Evol. Syst.* **1998**, *1*, 78–91. [CrossRef]
54. Balkwill, M.J.; Balkwill, K.; Vincent, P.L.D. Systematic studies in the Acanthaceae: A new species of *Barleria* from Natal. *S. Afr. J. Bot.* **1990**, *56*, 571–576. [CrossRef]
55. Burkhardt, D. Colour discrimination in insects. *Adv. Insect Physiol.* **1964**, *3*, 131–173.
56. Faegri, K.; Van der Pijl, L. A short history of the study of pollination ecology. In *Principles of Pollination Ecology*; Elsevier: Amsterdam, The Netherlands, 1979; pp. 1–77.
57. Caissard, J.C.; Joly, C.; Bergougnoux, V.; Hugueney, P.; Mauriat, M.; Baudino, S. Secretion mechanisms of volatile organic compounds in specialized cells of aromatic plants. *Recent Res. Dev. Cell Biol.* **2004**, *2*, 1–15.
58. Oelschlägel, B.; Gorb, S.; Wanke, S.; Neinhuis, C. Structure and biomechanics of trapping flower trichomes and their role in the pollination biology of *Aristolochia* plants (Aristolochiaceae). *New Phytol.* **2009**, *184*, 988–1002. [CrossRef]
59. Makholela, T.M.; Van der Bank, F.H.; Balkwill, K.; Manning, J.C. Allozyme variation in *Barleria saxatilis* (Acanthaceae) is lower than in two congeneric endemics. *S. Afr. J. Bot.* **2004**, *70*, 515–520. [CrossRef]
60. Obermeijer, A.A. A revision of the South African species of *Barleria*. *Ann. Transvaal Mus.* **1933**, *15*, 123–180.
61. Darbyshire, I. New species in *Barleria* sect. Stellatohirta (Acanthaceae) from Africa. *Kew Bull.* **2008**, *63*, 261–268. [CrossRef]
62. Gosavi, K.V.C.; Nalawade, A.D.; Yadav, S.R. Taxonomic identity, rediscovery and epitypification of *Barleria sepalosa* (Acanthaceae) from northern Western Ghats, India. *Rheedea* **2013**, *24*, 23–26.
63. Ravikumar, K.; Narasimhan, D.; Devanathan, K.; Gnanasekaran, G. *Barleria durairajii* (Acanthaceae): A new species from Tamil Nadu, India. *Rheedea* **2016**, *26*, 136–141.
64. Scotland, R.W.; Vollesen, K. Classification of Acanthaceae. *Kew Bull.* **2000**, *3*, 513–589. [CrossRef]
65. Tripathi, S.; Singh, S.; Roy, R.K. Pollen morphology of Bougainvillea (Nyctaginaceae): A popular ornamental plant of tropical and subtropical gardens of the world. *Rev. Palaeobot. Palynol.* **2017**, *239*, 31–46. [CrossRef]
66. Darbyshire, I.; Tripp, E.A.; Dexter, K.G. A new species and a revised record in Namibian *Barleria* (Acanthaceae). *Kew Bull.* **2012**, *67*, 759–766. [CrossRef]
67. Aronne, G.; Buonanno, M.; De Micco, V. Reproducing under a warming climate: Long winter flowering and extended flower longevity in the only Mediterranean and maritime Primula. *Plant Biol.* **2014**, *17*, 535–544. [CrossRef]
68. Shendage, S.M.; Yadav, S.R. Pollen Morphology of *Barleria* L. (Acanthaceae) from India. *Phytomorphology* **2009**, *59*, 121–126.

**Disclaimer/Publisher's Note:** The statements, opinions and data contained in all publications are solely those of the individual author(s) and contributor(s) and not of MDPI and/or the editor(s). MDPI and/or the editor(s) disclaim responsibility for any injury to people or property resulting from any ideas, methods, instructions or products referred to in the content.

horticulturae

Review

# Medicinal Use, Flower Trade, Preservation and Mass Propagation Techniques of Cymbidium Orchids—An Overview

**Khosro Balilashaki [1,*], Marcos Edel Martinez-Montero [2,*], Maryam Vahedi [3], Jean Carlos Cardoso [4], Catherine Lizzeth Silva Agurto [5], Michel Leiva-Mora [5], Fatemeh Feizi [1] and Mohammad Musharof Hossain [6]**

1 Department of Horticultural Science, Faculty of Agricultural Science, University of Guilan, Rasht 4199613776, Iran; f.feizi2013@gmail.com
2 Facultad de Ciencias Agrícolas, Universidad de Ciego de Ávila Máximo Gómez Báez, Ciego de Ávila 65200, Cuba
3 Department of Horticultural Science, Faculty of Agricultural Sciences and Engineering, College of Agriculture and Natural Resources, University of Tehran, Tehran 7787131587, Iran
4 Laboratory of Plant Physiology and Tissue Culture, Department of Biotechnology, Plant and Animal Production at Center of Agricultural Sciences at Universidade Federal de São Carlos, Rodovia Anhanguera, km 174, Araras 13600-970, Brazil
5 Laboratory de Biotecnología, Facultad de Ciencias Agropecuarias, Universidad Técnica de Ambato (UTA-DIDE), Cantón Cevallos Vía a Quero, Sector El Tambo-La Universidad, Cevallos 1801334, Ecuador; mleiva@uta.edu.ec (M.L.-M.)
6 Department of Botany, University of Chittagong, Chittagong 4331, Bangladesh; musharof20bd@yahoo.com
* Correspondence: khosrobali@alumni.ut.ac.ir (Kh.B.); cubaplantas@gmail.com (M.E.M.-M.)

**Abstract:** Cymbidium is an economically important genus in the orchid family (Orchidaceae) that has a pronounced medicinal and ornamental value. Medicinally, the plant is employed as a tonic to treat weakness in chronic diseases, dizziness, eye problems, burns, and wounds, etc. Cymbidiums are highly prized for their graceful flowers and sweet fragrance and are among the top ten most popular cut flowers. They are one of the most important commercial orchid groups and account for 3% of cut flowers in floriculture. Some orchid species in this genus are particularly threatened by excessive harvesting, so conservation measures are needed. Several enthusiastic organizations (e.g., The Cymbidiums Society of America, The Cymbidiums Club in Australia, The Golden Gate Cymbidiums Society, Alameda, CA, etc.) are dedicated to propagating, conserving, promoting, appreciating, and disseminating information about these beautiful and charming orchids. Through organogenesis (direct and indirect) and somatic embryogenesis, extensive propagation techniques for Cymbidiums have been developed to create protocols for synthetic seed production leading to large-scale propagation and long-term ex situ and in vitro conservation. This review highlights the medicinal uses, flower trade, conservation, and massive propagation techniques of Cymbidium orchids.

**Keywords:** Orchidaceae; traditional uses; health protection; conservation; micropropagation; horticulture

check for updates

**Citation:** Balilashaki, K.; Martinez-Montero, M.E.; Vahedi, M.; Cardoso, J.C.; Silva Agurto, C.L.; Leiva-Mora, M.; Feizi, F.; Musharof Hossain, M. Medicinal Use, Flower Trade, Preservation and Mass Propagation Techniques of Cymbidium Orchids—An Overview. *Horticulturae* **2023**, 9, 690. https://doi.org/10.3390/horticulturae9060690

Academic Editor: Wajid Zaman

Received: 29 April 2023
Revised: 30 May 2023
Accepted: 4 June 2023
Published: 11 June 2023

**Copyright:** © 2023 by the authors. Licensee MDPI, Basel, Switzerland. This article is an open access article distributed under the terms and conditions of the Creative Commons Attribution (CC BY) license (https://creativecommons.org/licenses/by/4.0/).

## 1. Introduction

The genus Cymbidium, also known as Boat Orchids, includes 75–80 species. As flowering plants in the Orchidaceae family, they are evergreens that bloom in winter and spring. They grow as epiphytes, terrestrial or lithophytes [1] in tropical and subtropical regions such as northeast India, eastern Asia and northern Australia.

Figure 1 shows multiple species of the genus Cymbidium that are predominantly epiphytic, but some species are also lithophytic and terrestrial or rarely leafless, saprophytic herbs, usually with pseudobulbs. Among orchids, Cymbidiums rank first, and in floricultural crops they account for 2.7% of the total cut flower production [2]. This genus has had medical applications for many years, particularly in the eastern part of Asia. Thus it serves as an important medicinal plant in the pharmaceutical industry [3].

**Figure 1.** Some natural and hybrid cymbidium orchids: (**A**) *C. giganteum;* (**B**) *C. iridioides* D. Don. (the iris-like cymbidium); (**C**) *C. insigne* 'Alba'; (**D**) Cymbidium 'Maluka Baby Pink'; (**E**) *C. lowianum,* (the Low's boat orchid, (**F**) *C. tracyanum* L. Castle; (**G**) *C. aloifolium* (L.) Sw. (the aloe-leafed cymbidium/boat orchid); (**H**) *C. bicolor* (L.) Sw. (the two-colored cymbidium); (**I**) Cymbidium 'Foxfire Mini Pharaoh Malcome'; (**J**) *C. tigrinum* C.S.P. Parish ex Hook. (the tiger-striped Cymbidium) [Photo plate prepared from Mohammad Musharof Hossain's unpublished photographs].

In the next sections, we describe these properties separately, evaluating the plant's medicinal use, its part in the flower trade, its preservation, and mass propagation techniques. We introduce this genus as a medicinal, economic and ornamental plant.

## 2. Medicinal Value of Cymbidiums

Orchids have a long history of traditional medicinal use. Some orchids have been utilized by Indians since the Vedic period (2000 BCE–600 BCE) for their healing and aphrodisiac properties [3]; the Chinese and Japanese also have an ancient cultural history with orchids. In legend, they advocated the medicinal fuction of some orchid species in the 28th century BCE [4].

Several Cymbidium species, namely *C. canaliculatum* R. Br, *C. madidum* Lindley, *C. eburneum* Lindl., *C. aloifolium* (L.) Sw., *C. devonianium* Lindl. ex Paxton, *C. iridioides* D. Don, *C. giganteum* Wall. Ex Lindl. and *C. sinense* Willd., are used as medicinal plants in the traditional medicine of many Asian countries [5]. So far, different compounds such as phenols, alkaloids, phenanthrenes, stilbenoid derivatives, and steroids have been extracted and identified from the Orchidaceae, and the molecular structures have been elucidated by various spectroscopic methods [3].

Of the various Cymbidium species, only a few have been critically studied for their ethnomedicinal, glycosidic, and other pharmaceutical properties. The *C. aloifolium* plant, for example, is said to have emetic and laxative properties. It yields salep, a nutritious drink enriched with starchy polysaccharides, which is consumed in traditional drinks and desserts. The root powder can relieve paralysis symptoms, and boiled-down aerial roots are used to bandage broken bones. The extract of the leaves is applied to treat fevers and boils. It is also used as an emetic, tonic, laxative; it can treat burns, wounds, and earaches. Crushed plant extracts with ginger are administered to treat eye weakness, dizziness, chronic diseases, and paralysis. It has two substituted bibenzyls, a phenanthraquinone (cymbinodin-A) and a dihy-drophenanthrene [3]. A decoction from the rhizome of *C. ensifolium* is used to treat gonorrhea, and a flower extract is used for eye inflammation [6]. The extract of leaves of Cymbidium (*C. giganteum*) has unique blood-clotting properties [7]. The pseudobulbs of *C. longifolium* are employed as a sedative, while an aqueous solution of dried and powdered pseudobulbs produces emesis when taken orally on an empty stomach [8]. The roots of *C. faberi* Rolfe. have been used in China for decades as an important herbal folk medicine to loosen phlegm and relieve cough, etc. [9]. The available literature demonstrates that phenanthrene compounds isolated from various orchids have shown various promising biological and antioxidant activities [10].

Recent reports state that extracts of Cymbidium roots have a high antimicrobial activity against *Staphylococcus aureus,* and the stem extracts contain phenolic compounds that exhibit a high antioxidant activity and cell cytotoxicity [11]. In another study, ephemeranthoquinone B, two phenanthrenes, and a phenanthrene glucoside were isolated from the roots of Cymbidiums along with six known phenanthrenes 5–10 [12]. The extracts from *C. kanran* Makino are enriched in flavone C-glycosides, including vicenin-2, vicenin-3, shaftoside, vitexin, and isovitexin [13]. The compound of 7-(4-hydroxybenzyl)-8-methoxy-9,10-dihydrophenane-threne-2,5-diol (HMD) was synthesized, together with five studied compounds [coelonin, 5,7-dimethoxy-phenanthrene-2,6-diol (DD), shancidin, 1-(4-hydroxybenzyl)-5,7-dimethoxy-phenanthrene-2,6-diol (HDP), and 2-methoxy-9,10-dihydrophenanthrene-4,5-diol (MDD)], from the roots of *C. faberi,* as reported by Lv et al. [14]. Except for HDP, other compounds dose-dependently suppressed the production of NO, tumour necrosis factor-alpha (TNF-$\alpha$), and interleukin-6 (IL -6) in lipopolysaccharide (LPS)-induced primary mouse peritoneal macrophages. Gigantol, a bibenzyl compound, has been isolated from *C. goeringii*, *C. aloifolium* and some other orchids and has shown anticancer activity [15]. It is a potent inhibitor of TNF-$\alpha$, IL-6 and IL-1 and affects the mRNA expression levels of these cytokines in a dose-dependent manner. The qualitative analysis of various organic extracts of *C. aloifolium* revealed eight different photochemical compounds, namely n-hexadecanoic acid, 9,12-octadecadienoic acid (Z,Z), 9,12,15-octadecatrienoic acid, (Z,Z,Z), octadecanoic acid, phytol, 2-butyn, 2-cyclopenten-1-one, and 1,4-benzenedicarboxylic acid [16]. Most of the identified compounds are biologically significant. In addition to the medicinal uses of Cymbidiums, endophytic fungi from orchid plants have been reported to secrete secondary metabolites containing bioactive antimicrobial siderophores [17].

## 3. Floristic Significance of Cymbidiums

Cymbidiums are a highly valued flower-growing plant. Because of their long-lasting inflorescences and large, attractive flowers, Cymbidiums are among the top ten commercial orchids. Among the 75–80 species, not counting the natural hybrids, are *C. floribundum* Lindl. (Golden-Edged Orchid), *C. devonianum* Lindl. ex Paxt, *C. elegans* Lindl. (Elegant Cymbidium), *C. eburneum* (Ivory-Edged Cymbidium), *C. mastersii* Griff. & Lindl, (Master Cymbidium), *C. erythraeum* Lindl., *C. iridioides* (Iris-Like Cymbidium), *C. lowianum* (Rchb.f.) Rchb.f. (Low's Boat Orchid) and *C. tracyanum* Rolfe. (Tracy's Cymbidium), *C. dayanum* (L.) Sw. (Day's Cymbidium or Phoenix Orchid), *C. suave* Sw. (Snake Orchid or Grassy Boat-Lipped Orchid), are the most beautiful, popular, and floristically important Cymbidium species. The greatest commercial use of this genus is associated with the splendor of its flowers and the splendor of its flowers and inflorescences. In floriculture worldwide, Cymbidiums hybrids are divided into three groups: miniature hybrids (e.g., Cymbidiums Autumn Emerald 'Royale'), large-flowered hybrids (e.g., Cymbidiums 'Kirby Lesh') and another commercial group called 'pending Cymbidiums'. Some Cymbidiums hybrids form clusters of up to 30 extravagant, multicolored flowers, including white, green, cream, mauve, and yellow [18]. The Tropical Cymbidiums Orchid is a well-known 'winter flower' with a flowering period of about two months, showing about 15 exquisitely beautiful and magnificent epiphytic flowers on the first inflorescence [19]. Undoubtedly, the commercial planting of this plant and the use of hybrid cultivars, despite having the advantages of a breeding cultivar, will lead to the elimination of native species and the reduction of genetic diversity and gene pool.

## 4. Reproductive Biology in Cymbidiums

Cymbidiums take 4–7 years to flower, but they are capable of triggering early flowering. This was demonstrated by Kostenyuk et al. [20] and suggests that the concerted action of phytohormone, as well as nutrient concentration and putative promoters/suppressors, determine the timing of the transition of the Cymbidium orchid from the vegetative to the reproductive stage [20]. The application of 6-benzylaminopurine, restricted nitrogen supplies with phosphorus enrichment, and root removal early induced the transition of a Cymbidium shoot from the vegetative to the reproductive stage for 90 days [20]. Furthermore, according to preview reports, the increase in leaf starch content during vegetative growth and soluble sugar in pseudobulbs and roots during the reproductive growth of Cymbidiums is critical for increasing the growth of the plant and thus promoting flowering [11]. Plants cultured at a high light intensity of photosynthetic photon flux density (PPFD) exhibited a lower time to flowering induction and development, alsoincreasing the number of inflorescences and flowers, in comparison with plants cultured at a low light intensity (PPFD) [11]. Moreover, shading treatments can significantly increase different inflorescence traits, such as quality, height, ratio, and spacing. The study by Zhou et al. [21] showed that inflorescence height, inflorescence ratio, and petal spacing of *Cymbidium* spp. increased significantly by 7.892 cm, 13.125 cm, and 0.484 cm, respectively, after shading [21].

The flowering of adult plants was influenced by several factors, such as fertilizer, light duration and quality, temperature, and plant growth regulators (PGRs), which affect flower induction, development, and flower characteristics. Flower diameter as well as the inflorescence length increased in response to increasing nitrogen and potassium fertilization during the adult vegetative stage in Cymbidiums grown at low light intensity and artificially induced inflorescence, while flower quality decreased in those grown at high light intensity [22]. Barman et al. [23] maintain that water-soluble fertilizers significantly affected the growth and reproductive stages of Cymbidiums. Their results showed that a maximum number of shoots (4.54), length of spikes (54.59 cm) and number of flowers per spike (10.19) were obtained when water-soluble fertilizer (N19P19K19 at 1 g/L concentration) was sprayed fortnightly. Day/night temperatures of 30/25 °C and 25/20 °C are best for vegetative shoot growth and flower bud formation in Cymbidiums, and intermediate flowers form most frequently in the just-growing young pseudobulbs and in the 1 to 2-year-old pseudobulbs [24]. Shoot growth rates and inflorescence numbers were lower at lower temperatures, such as 20/15 °C.

The molecular mechanisms underlying the regulation of the reproductive stage have been extensively studied in the long-day plant models such as Arabidopsis and the short-day plant such as Rice, and many processes of gene regulation during development, especially the reproductive biology of flowering, have been elucidated [25]. However, in orchids such as Cymbidiums, the understanding of the molecular mechanisms controlling flowering is just emerging [26]. Many genes are involved in the transition from the vegetative to the reproductive stage, so inducing these genes may be the best method for inducing the reproductive stage and flowering. In this sequence, researchers discovered some miRNAs, for example, the differential expression of two miRNAs, miR160 and miR396, targeting ARFs and the growth-regulatory factor (GRF), respectively [27]. Thus, genetic engineering is another approach to manipulate the switch from the vegetative to the reproductive stage of Cymbidiums.

The knowledge of reproductive biology is also limited in Cymbidiums. Four types of pollination are known: autonomous self-pollination, reward-based pollination, generalized food deception, and Batesian mimicry of the food source [28].

To date, autonomous self-pollination has been described in *C. macrorhizon*, *C. nipponicum*, and *C. nagifolium* Masam, all of which lack a rostellum that acts as a physical barrier between the anthers and the stigma [29]. In *Cym. mandidum*, pollination occurs by reward. The flowers are pollinated by the stingless bee *Trigona kockingsi* Cockerell., which collects viscous substance on the labellar surface [30]. The substance is probably used as nest-building material. A similar method is also thought to occur in *C. lowianum*, in which the labellar surface has proteinaceous papillae that may function as food hairs [31]. In some species, including *C. lancifolium* [32], *C. goeringii* Reichenbach Ill. [33], and *C. kanran* [33], a general feeding illusion has been observed. The nectarless flowers attract pollinating bees by visual and olfactory stimuli. Finally, a Batesian imitation of the food source occurs in *C. insigne*. The plant depends exclusively on the bumblebee (*Bombus eximius* Smith.), which also pollinates the nectar-producing flowers of *Rhododendron ciliicalyx* [34].

Pollination experiments were conducted on *C. macrorhizon*, *C. aberrans*, and *C. lancifolium* to study the breeding system. It was found that some rewarding myco-heterotrophic plants depend (at least in part) on an insect-mediated pollination system, and some myco-heterotrophic plants can attract pollinators without attractive materials [29].

## 5. Seed Biology of Cymbidiums

Orchid seeds are very small, extremely light, and produced in large numbers with the length range of from 0.05 to 6.0 mm, the range of longest and shortest known seeds in the family being 120 times. Known 1000-seed weights range from 310 μg to 24 mg (a 78-fold difference). The number of known seeds ranges from 20–50 to 4 million per fruit. The Testa are usually transparent, with outer cell walls that may be smooth or reticulate (Figure 2A). Embryos consist of 100–400 disorganized parenchymatous cells and seeds have no endosperm. The volume of the embryo is much smaller compared to the size of the seed. Orchid seeds are balloon-like with large internal air spaces and difficult to moisten so they can float in the air and on water for long periods of time, a characteristic that facilitates dispersal over long distances [35]. Orchid seeds can also be transported in and on terrestrial animals and birds (in fur, feathers or hair, mud on feet, and perhaps after ingestion) [2].

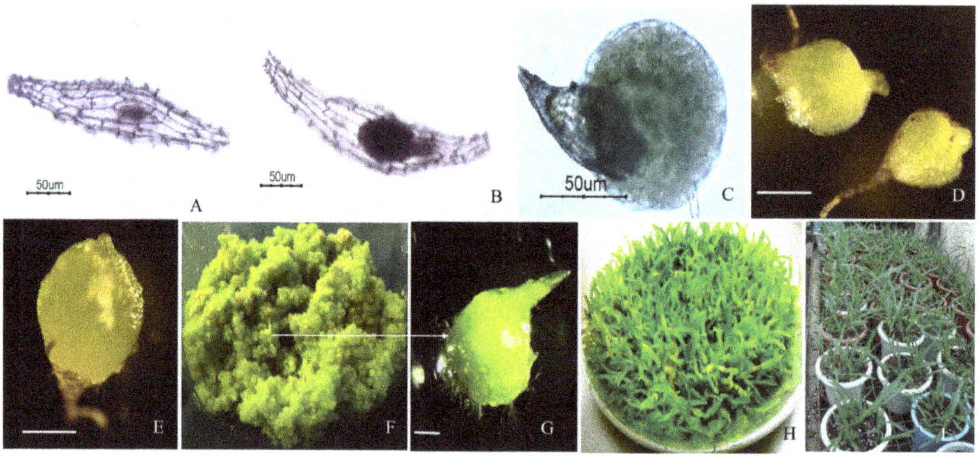


**Figure 2.** Embryo morphogenesis and seedling development in *C. aloifolium*: (**A**) seed with transparent and reticulate testa and small embryo, (**B**) embryos swell by uptake of nutrients and water, (**C**) cell number increases by repeated anticlinal and periclinal cell divisions, producing a parenchymatous cell mass called spherulite, (**D**,**E**) spherulites accumulated dense chloroplasts and formed a compact structure with bipolar character, (**F**) spherulites developed into protocorm, (**G**) germinated seeds, (**H**) young seedlings, (**I**) in vitro seedlings acclimatized in the outside environment. [Bar = 1 mm].

Cymbidiums produce large capsules that carry thousands to millions of long and fuzzy seeds with pointed ends [2,36].

The germination of orchid seeds under in vitro conditions depends mainly on their viability, which is determined by the pollination and fertilization process and the culture media used for germination. Direct and indirect methods are applicated for testing seed viability [6]. In the direct methods, the germinated orchid seeds on artificial media are counted [37], while in the indirect method, the metabolic activity of embryos is examined using chemicals such as fluorescein diacetate (FDA) ortetrazolium salts. To determine seed quality, germination tests are most commonly adopted to assess seed viability in plant production [38].

The seeds of orchids are orthodox, meaning they must be air dried to extend their longevity. The shelf life of the seeds can be extended by reducing the water content, temperature, and oxygen level in the storage room. Reducing the water content by up to 5 percent and then storing the seeds at a freezing temperature (−20 °C) is used in seed banks of orchids [39,40]. Orchid seeds stored at a low humidity and low temperature can survive for decades. On the other hand, they are resistant to sterilizing agents such as sulfuric acid or hypochlorite, which are used in the sterilization of seeds for in vitro germination [41].

Seed banks are recognized as the most efficient way to store large quantities of living plants in one location [42]. Orchid seed banks have the potential to make an invaluable contribution to orchid conservation [36]. According to studies, storing orchid seeds under cold conditions was the best conventional method to prolong seed viability.

## 6. Propagation of Cymbidiums

### *6.1. Conventional Methods*

Cymbidiums can be propagated by dividing the plant into parts, by taking cuttings from the mother plant, or by taking dorsal or pseudobulbs and placing them in clay pots or plastic bags with potting substrate. The most used substrates are small bricks/chunks of charcoal, moist sphagnum moss, coconut moss, or coconut peat. Interestingly, the conventional propagation methods are slow and laborious, and the propagation rate is very low [43], with only a few plants produced per year, making this method impractical for large-scale propagation to provide clonal plantlets or plants for commercial purposes.

### *6.2. Clonal Propagation of Cymbidiums*

In vitro methods of propagation are reliable, feasible and reproducible for the large-scale production of seedlings obtained by in vitro seed germination methods, or plantlets for clonal propagation of Cymbidiums using micropropagation. However, while seeds are most used for conservation or breeding, actual large-scale propagation of floricultural orchid varieties is realized preferentially by micropropagation due the clonal origin and industrialization of cultivation that aims at the scale-up and intensification of the cultivation [44].

### *6.3. Asymbiotic Seed Germination in Cymbidiums*

Although symbiotic germination is an important tool for the conservation and restoration of natural species, asymbiotic in vitro germination is the primary method for seedling production and the genetic improvement of commercial orchids, including many Cymbidium spp. and their hybrids. The success of seed germination by in vitro culture is influenced by several factors, including seed maturity and age [45], type of nutrient solution [46,47], type of carbohydrate additives [48], organic additives [49], and PGRs [43].

In nature, only 2–5% of Cymbidium seeds germinate because they require symbiosis with a mycorrhizal fungus and germinate only after infection with a mycorrhizal fungus [50]. Almost all Cymbidiums must form a mycorrhizal and symbiotic relationship with a partner corresponding to a particular fungus during their growth and development [48]. The nutrients stored by the cells for seed germination are very low, making germination very difficult [51]. Under certain conditions, different strains can significantly promote the germination of Cymbidiums. Further studies may be needed to verify the differences of this symbiosis under different ecological conditions [52].

Induced seed germination can be asymbiotic under in vitro conditions in a special culture medium or using symbiotic methods, which combines seed germination with the presence of a symbiont fungus [53]. Seed germination and propagation by symbiotic approaches are the preferred options for Cymbidiums' reintroduction into native habitats [52].

On the other hand, in vitro cultivation methods or asymbiotic orchid germination provides an interesting platform for the large-scale propagation of orchids without the requisite of fungal symbiosis. These methods facilitate the seeding and cultivation of seedlings on a different culture medium formulation, basically containing nutrients, carbohydrate sources, and agar, cultivated under aseptically controlled conditions [53].

The asymbiotic seed germination of orchids is a multipurpose technique that results in the production of high-quality seedlings and can be used for conservation, propagation, restoration, and breeding aiming at developing new hybrid cultivars [54,55]. In vitro germination and seedling development is the one of the most common methods also used for orchid propagation, mainly because it is technically less complicated than clonal propagation and allows for obtaining a large number of seedlings in a single culture medium, which contains only nutrients and a soluble carbohydrate source such as sucrose. Still, the clonal propagation by shoot propagation or protocorm-like bodies (PLBs) depends on a knowledge of the types, concentrations, and balance of phyto-regulators in the culture media used for each stage of plantlet induction, regeneration, proliferation and rooting/elongation [56]. As can be seen, there are opportunities for improvement in the successful propagation of Cymbidiums. These include selection of the optimal sterilant, improved seed recovery after sterilization, and selection of the optimal culture medium [57].

### 6.4. *Ex Vitro Seed Germination*

Seed propagation and seedling development is essential for the continuation of any plant population and Cymbidiums in particular. Due to the tiny seeds and undeveloped embryos, it has long been virtually impossible to observe the germination of Cymbidiums in vivo, which has been an obstacle to understanding seedling site requirements and fluctuations in Cymbidiums populations [58]. Because Cymbidium seeds lack an endosperm, they contain either low levels of food reserves or forms that are unlikely to be metabolized by Cymbidium embryos [58]. In nature, orchid seeds rely on Cymbidium mycorrhizal fungi to provide the necessary nutrients for germination, a process referred to as symbiotic seed germination [59]. Symbiosis, in its broadest sense, means the coexistence of two dissimilar organisms. A wide range of associations can be understood under this definition. The term germination mycobiont is used in this context to mean that a fungus participates in the mycorrhiza of orchid seedlings and supports seedling development. Seedlings and established plants may subsequently be attacked by other mycobionts and endophytes not involved in mycorrhiza. The function and effects of these endophytes in the adult stage on orchid physiology are often unclear [58].

The symbiotic view of germination can be summarized in six points: orchid roots are generally infected with a characteristic fungus that is not considered harmful; different orchid genera may have different strains or species of this fungus; seeds sown in pure culture on different growing media, especially soils containing starch or other insoluble organic matter, do not germinate unless the fungus is present; germination is apparently induced by some strains of fungus and not by others; a certain equilibrium between fungus and host must be maintained; although it is known that the germination of orchid seeds can be achieved by using sugar in the culture medium, germination under these conditions is abnormal and not common in nature [60,61].

A protocorm led to form shoots and roots at the beginning of germination in seeds without an endosperm [41]. Symbiotic germination, especially in vitro, is becoming increasingly popular because in many studies it promotes higher germination rates [62,63].

A study on the dependence of Cymbidiums on fungi during seed germination and seedling development provides a means of understanding the role of fungi in the developmental process [64]. Symbiotic seed germination is widely regarded as an effective tool for Cymbidium conservation and could be a cost-effective method to maintain the genetic diversity of reintroduced Cymbidium populations [65].

One of the major challenges is to obtain compatible fungi for symbiotic seed germination [26]. Nevertheless, the presence of compatible symbiotic fungi in Cymbidiums seedlings leads to greater adaptation to the environment, resulting in higher survival and faster growth than with asymbiotic seedlings [66].

To be recognized as an orchid mycobiont, it is generally necessary for the fungus to colonize orchid tissues with the formation of mycorrhizal features such as peloton formation (certain cells form cyclically dense hyphae), i.e., intracellular hyphae coils, and the absence of cortical tissue necrosis [58,67]. Many studies have been conducted to obtain compatible fungi for the germination of symbiotic seeds. On this subject, researchers reported that symbiotic seedlings with many strains show a higher stability than asymbiotic seedlings. For instance, Mahendran et al. [28] noted that mycorrhizal fungi significantly increases the number of roots and shoots [28]. Dong et al. [68] reported that the average increase in fresh weight of Cymbidium seedlings inoculated with strains CF1, CF3, and CF12 was 130.26%, 345.65%, and 153.34%, respectively, while the increase in the control was only 88.40%. The differences among the three treatments and the control are statistically significant. Liu et al. [69] investigated the effects of inoculated mycorrhizal fungi and non-mycorrhizal beneficial microorganisms on the plant characteristics, nutrient uptake, and root-associated fungal community composition of Cymbidium Hybrids Hort. They reported that all inoculants significantly increased the total dry weight of Cymbidiums. The mycorrhizal fungus positively affected the P, K, Ca, and Mg content in the shoots and the Zn content in the roots, suggesting that both mycorrhizal fungi and endophytic fungi have the potential to create a favorable symbiosis in orchid roots and stimulate growth [70]. Chand et al. [56] evaluated twelve isolated instances of the plant growth-promoting activity on *C. aloifolium* protocorms (84 days old). All of them showed growth-promoting activity [56].

As noted above, obtaining compatible fungi is a crucial step for symbiotic seed germination because mycorrhizal fungi have the ability to induce seed germination with varying efficiency and low specificity [71]. To this end, in situ/ex situ seed-baiting techniques have been proposed as an effective method to obtain efficient symbiotic seed germination-promoting fungi [72].

Ex-situ seed-baiting techniques developed in recent years are effective methods to study the compatible mycorrhizal fungi of Cymbidiums. Sheng et al. [73] noted that after 58 days of cultivation, seeds inoculated with a fungal strain showed a high germination rate, while seeds without the fungus did not germinate. Germination and protocorm production were higher in the dark (0/24 h light/dark) than under light conditions (12/12 h light/dark), while subsequent protocorm development was better under light [73].

After germination, the orchid seedling undergoes a long or short non-photosynthetic phase in which it is completely dependent on organic carbon from a mycobiont, so a suitable organic carbon source for the mycobiont is another important requirement.

### *6.5. Seedling Development*

Cymbidiums produce dust-like seeds of microscopic structure The cellular organization of the seeds is simple and consists of an undifferentiated mass of embryonic cells and a rudimentary endosperm covered by a transparent seed coat [74]. According to Arditti and Ghani [2], Cymbidium embryos are relatively small and simple, generally oval or spherical, and sometimes composed of only a few cells, usually without an endosperm (Figure 2A).

Regarding seed maturation, immature seeds from 3- to 5-month-old capsules could be successfully germinated on a culture medium. Immature seeds from 4- and 3-month-old capsules showed the highest and lowest germination rate [45]. In another study in relation to planting times, younger seeds germinated slowly in early planting but grew rapidly and formed more rhizomes in later planting. In contrast, older seeds germinated rapidly in early culture but formed fewer rhizomes the later planting [75].

In vitro germination in an artificial culture medium is one of the best solutions for increased germination rates and rapid seedling development. During seed germination, four to five different developmental stages are observed in Cymbidiums, and this is also more or less common in other Orchidaceae. Orchid seeds undergo a typical metamorphogenetic process during germination. Stage 1: In this stage, viable embryos swell only by the uptake of nutrients and water (Figure 2B); Stage 2: The number of cells increases by repeated anticlinal and periclinal cell divisions, resulting in the formation of irregularly shaped parenchymatous cell masses that emerge when the seed coat breaks open; this stage is called the globular stage (Figure 2C); Stage 3: The parenchymatous cell mass is enriched with dense chloroplasts and exhibits a bipolar character; the compact structures are referred to as spherulites; some rhizoids emerge from the posterior/basal part of the spherulites, while an appendage appears from the anterior/upper part (Figure 2D); Stage 4: In this stage, the spherulite enlarges and a protuberance appears at the anterior part, demarcating the meristematic zone for the development of a leaf primordium; this stage is called the protocorm stage, and is thought to be the effective germination of orchid seeds (Figure 2E); Stage 5: This is the final stage of germination, where roots emerge from the posterior part of the protocorm and gradually develop into young seedlings (Figure 2F). Protocorm development is considered a characteristic feature of post-seminal development in orchids, and the shape of protocorms is taxon-specific. They may be oval, round, disc-shaped, elongated, branched, spindle-shaped, or thorn-shaped. In Cymbidiums, the early protocorm is usually round, radially symmetrical, green, and gradually assumes an oval shape.

### 6.6. Culture Media

In terms of culture media, there are a variety of formulas that have different effects on seed germination [76]. Mohanty et al. [77] noted that the percentage of seed germination varied with the composition of the culture media and was highest in a full-strength MS basal medium; the number of secondary protocorms that developed from seed-derived protocorms increased with the addition of 5.0 μM 6-benzilaminopurine (6-BAP) and 2.5 μM α-naphthaleneacetic acid (NAA) [77]. Dep and Pongener [78] reported that immature embryos were successfully germinated 9 months after pollination on an MS medium (sucrose 2% (*w*/*v*) + NAA and benzyladenine (BA) (3 and 6 μM in combination, respectively) within 7 weeks of culture, recording 90% germination [78].

Mahendran et al. [28] cultured immature seeds of *C. bicolor* Lindl. on four basal media, namely Murashige and Skoog (MS), Knudson C (KC), Knudson C modified Morel (KCM), and Lindemann orchid (LO). The results were significantly higher and lower on the LO medium (96.6%) and KC (62.7%) media after 8 weeks, respectively [28].

Gogoi et al. [70] cultured *C. eburneum* Lindl. asymbiotically in different basal media, namely MS, Mitra, B5 and Nitsch. It was found that the medium MS, nourished with 15 μM each of BAP and NAA in combination, increased the number and length of shoots and the number and length of roots in the seedlings [70].

It is believed that nutrient requirements for orchid seed germination are species specific and that the nitrogen source plays an important role in orchid seed germination, which may explain the superior germination of seeds on a Murashige & Skoog medium [79]. It has been reported that nitrogen in the medium MS strongly influences cell growth and differentiation, and that ammonium nitrate in the medium MS is the most suitable source for seed germination and plantlet development. Other reduced nitrogen forms such as pyridoxine, thiamine and nicotinic acid as vitamins are also absent in KC and VW media but present in MS media [80].

The addition of different types and concentration of organic substances to in vitro culture should stimulate seed germination. These organic additives are natural sources of amino acids, vitamins, minerals, organic acids, sugars, proteins, and natural growth regulators that help in orchid propagation by stimulating development and morphogenesis in the asymbiotic seed culture [81].

For example, the additional presence of riboflavin, biotin and folic acid in a Mitra medium may have further promoted seed germination. Mitra medium has high concentration of phosphate ions that affects the seed germination of asymbiotic orchids [82]. Dutra et al. [82] also showed that nitrogen does not play an important role in the seed germination of asymbiotic orchids compared to other nutrients, especially those in a Mitra medium [82]. Hajong et al. [83] reported that the low response of orchid seeds to nitrogen in the B5 and Nitsch media could be due to the inhibitory effect of nitrogen in the form of ammonium sulfate on seedling growth in B5 media or vitamin mixtures present in both B5 and Nitsch media.

With respect to PGRs, the modification of free PGR culture media by the addition of specific PGRs and other components such as activated charcoal, as well as changes in culture conditions, have been reported to improve the germination percentage and subsequent protocorm development in many Cymbidium genotypes [84]. For the development of efficient germination and micropropagation protocols, the conventional tissue culture media must be modified by adding specific PGRs and various complex additives (peptone, yeast extract, banana pulp, etc.), and plant production must be automated by adjusting the bioreactor system and culture conditions. Recently, orchids have become the focus of new research areas, including genetic engineering, functional genomics, proteomics, and metabolomics, all of which require standardized micropropagation techniques. The successful application of new approaches will contribute to the further improvement of orchids and orchid products [85].

## 7. Clonal Propagation of Cymbidiums

The clonal propagation of Cymbidiums can be practically realized by division of rhizomes or pseudobulbs, but this conventional technique generates 2–4 plants per year and is applied for amateur cultivation. This technique not practicable in the large-scale production of cut and pot flowers. The actual and realistic technique used in large-scale floriculture is micropropagation by shoot proliferation or by the induction, proliferation and regeneration of protocorm-like bodies, also called IPR-PLBs [86].

### *7.1. Micropropagation of Cymbidiums*

Cymbidiums were the first orchid to be micropropagated by tissue culture. The first success of clonal propagation of *Cymbidium* sp. was reported by Morel [87] through the culture of shoot tips, which led to the in vitro production of millions of plantlets from tiny plant parts [87,88]. The first detailed protocol for in vitro propagation of Cymbidiums, which began with meristem culture, was published by Wimber [89]. The use of explants from plants grown outdoors is generally associated with the problem of a high contamination rate. Different types of explants can be used to start micropropagation, e.g., shoot tips [5,90], leaf segments [30,91], thin cross-sections of PLB [91,92], shoot segments [3], whole proto-buds [3], PLB segments [93], flower buds [94], leaf bases of axenic seedlings [95]; rhizome segments [94,95]; complete seedlings [96], transverse thin cell layers (*t*TCL) cut from stem internodes and nodes from the base of the shoot apex [92], and root tips [97]. Nevertheless, for each species or hybrid, a particular explant has been shown to be suitable for efficient micropropagation.

Plantlet propagation in orchids can be accomplished by direct shoot bud formation (direct organogenesis), formation of shoot buds via callus formation (indirect organogenesis), formation of secondary proto-buds or PLBs without an intervening callus phase (direct embryogenesis), and formation of PLBs via callus (indirect embryogenesis) [98–100]. Proto-buds (secondary embryogenesis) can be regenerated from the outer tissues of protocorms [99]. This is an independent pathway in which they develop without an intervening callus phase. The cells of young protocorms are highly meristematic in nature and can be used to rejuvenate and enhance plant regeneration in Cymbidiums [38].

The type of morphogenetic differentiation (organogenesis or embryogenesis, directly or indirectly) depends on a number of plant endogenous and exogenous factors, such as type and origin of explants, the culture media, culture media consistency, cultivation conditions, PGRs, complex additives, and even culture duration [3].

The use of liquid or gelled culture media influences the morphogenic formation. As an example, the new formation of PLBs (also called secondary protocorms) from seed-originated protocorms was observed in *C. aloifolium* and *C. giganteum* in either a liquid or gelled agar medium [96]. Although the liquid medium was more efficient for propagation of PLBs, this method failed to develop plantlets unless transferred to a gelled agar medium. The non-uniformity and hyperhydricity of the PLBs were also observed [3]. Hyperhydricity (development of physiologically abnormal tissues), high contamination rate, and production of etiolated plantlets are the common disadvantages of in vitro propagation using the liquid media system [93]. In general, hyperhydric tissues are unable to regenerate physiologically true plantlets. The failure of hyperhydric tissues to regenerate in a liquid medium could be related to tropisms or perturbations in polarity [92]. There is much evidence that the reduction of the agar concentration or the absence of agar in the culture medium leads to vitrification during tissue growth. However, a number of studies suggest that not all plant species exhibit vitrification when cultured in a liquid medium. Although regeneration of plantlets could be achieved in both a liquid and semisolid culture medium, a liquid medium generally proved to be better for propagation and growth of newly developed plantlets because the cultures are maintained under constant agitation.

PLBs were induced from protocorm sections, pseudostem segments, and even complete seedlings of *C. aloifolium* and *C. giganteum* on a semisolid PM agar medium supplemented with BAP and NAA. The frequency of PLB regeneration was strongly influenced by the concentrations and combinations of the two PGRs. Although histological and cytological analyses suggest that a PLB is indeed equivalent to a somatic embryo, this claim has never been made in the Orchidaceae [91,98].

### 7.2. Propagation by Artificial Seeds

Nowadays, the encapsulation technique to produce artificial seeds has become an important part of micropropagation [101]. Artificial seeds are artificially encapsulated vegetative parts that can be sown as seeds and transformed into a plant under in vitro or in vivo conditions. These parts include somatic embryos (usually), cell aggregates, shoot buds, auxiliary buds or other micropropagules [102]. Another study on in vitro germination and propagation of *C. aloifolium* (L.) Sw. was conducted by Pradhan et al. [101] using artificial seeds prepared in vitro by encapsulating PLBs with 4% sodium alginate and 0.2 mol·L$^{-1}$ calcium chloride solution. This study showed that artificial seeds are a good alternative for in vitro mass propagation and the short-term preservation of *C. aloifolium* [101]. Artificial seeds are also an excellent way to store orchid material at room temperature, under refrigeration, or even in cryopreservation for weeks, months, or even years while maintaining the clonal stability of the material [103]. Table 1 summarizes the research on some orchids with emphasis on Cymbidiums.

**Table 1.** Summary of research on some orchids with emphasis on Cymbidiums.

| Researchers | Year | Subject | Results |
|---|---|---|---|
| Deb and Pongener [104] | 2002 | Studies on the in vitro regenerative competence of aerial roots of two horticulturally important Cymbidium species | Of the three basal media tested, MS medium supported optimum regeneration and culture proliferation in both the species. In *C. aloifolium* ~12, shoot buds developed on medium nourished with sucrose 3% and benzyl adenine (BA) 3 µM but in *C. iridioides* optimum regeneration was achieved when medium supplemented with sucrose 3%, coconut water (CW) 15%, casein hydrolysate (CH) 100 mg/L and ~20 shoot buds formed per subculture |

**Table 1.** *Cont.*

| Researchers | Year | Subject | Results |
|---|---|---|---|
| Hossain et al. [105] | 2009 | Cost-effective protocol for in vitro mass propagation of *C. aloifolium* (L.) Sw.–a medicinally important orchid | Mitra medium supplemented with 2.0 g/L activated charcoal (AC) showed 100% seed germination and effective for induction of significantly large-size protocorms (1.64 mm in dia.) |
| Hossain et al. [106] | 2010 | Seed germination and tissue culture of *C. giganteum* Wall. ex Lindl | The effects of peptone, AC and two-plant growth regulators [6-benzylaminopurine (BAP) and 2,4-Dichlorophenoxyacetic acid (2,4-D)] were also studied. Both M and PM supplemented with 2.0 g/L peptone or 1.0 mg/L BAP resulted in ~100% seed germination. |
| Deb and Pongener [78] | 2011 | Asymbiotic seed germination and in vitro seedling development of *C. aloifolium* (L.) Sw.: a multipurpose orchid | Immature embryos of 9 months after pollination were successfully germinated on MS medium containing sucrose (2%) (*w*/*v*) and α-naphthalene acetic acid (NAA) and BA (3 and 6 μM, respectively, in combination) within 45 days of culture where 90% germination was recorded. The germinated seeds formed PLBs on the optimum germination medium within two passages. |
| Nahar et al. [107] | 2012 | Effect of different light and two polysaccharides on the proliferation of protocorm-like bodies of Cymbidiums cultured in vitro | The highest protocorm-like bodies (PLBs) formation, shoot formation rate (90%) and root formation rate (50%) were found among explants cultured on medium supplemented with 0.1 mg/L Chitosan H under green light. After 11 weeks of culture, fresh weight of PLBs was higher (241.3 mg) at HA9 (1 mg/L) treatment with green light. The average number of PLBs (5.7) was higher under green light at HA9 treatment. PLBs under white light showed the highest number of shoot (1.2) at Chitosan H treatment. |
| Parmar and Pant [108] | 2016 | In vitro seed germination and seedling development of the orchid *Coelogyne stricta* (D. Don) Schltr | MS medium supplemented with 1 mg/L BAP and 1 mg/L NAA was found to be the best condition for the development. The germination started after 7 weeks of culture and complete seedlings were obtained after 23 weeks of culture on the medium supplemented with 1 mg/L BAP and 1 mg/L NAA suggesting the usefulness of both hormones in root induction. In the hormone, free MS medium germination started after 5 weeks, but root initials were not developed even after 32 weeks of culture. |
| Pradhan et al. [109] | 2016 | Efficient plant regeneration of *C. aloifolium* (L.) Sw., a threatened orchid of Nepal through artificial seed technology | Full strength of MS medium without plant growth regulators was found to be the most favourable condition for efficient plantlet regeneration of *C. aloifolium* (9.83 shoot and 2.66 roots per culture). |
| Bhowmik and Rahman [84] | 2017 | Effect of different basal media and PGRs on in vitro seed germination and seedling development of medicinally important orchid *C. aloifolium* (L.) | Medium supplemented with hormones favored optimum condition for the germination (approx. 95%) of seeds followed by full strength and half strength on KC, MS, PM and VW media. MS medium supplemented with 0.5 mg/L BAP and 0.5 mg/L NAA showed comparatively better response within 6 weeks of culture than other conditions of MS medium as well as KC, PM and VW media. |
| Philip Robinson et al. [110] | 2017 | In vitro seed germination of *C. aloifolium* (L.) Sw., a potential medicinal orchid from Eastern Ghats of Tamil Nadu, India | The highest seed germination of 90% was observed KC basal media after30th days whereas germination percentages were 40% and 30% on 1/2 MS and VW media respectively. |

**Table 1.** *Cont.*

| Researchers | Year | Subject | Results |
|---|---|---|---|
| Paul et al. [111] | 2019 | In vitro mass propagation of *C. aloifolium* (L.) Sw | Developing an efficient protocol for rapid propagation of *C. aloifolium* starting with in vitro asymbiotic seed germination, leading to protocorm induction followed by plantlet development and successful ex vitro acclimation. |

## 8. Conclusions

Cymbidium orchids are the most important floricultural plants in the world. They are a potential source of special metabolites used in alternative and conventional medicine. The large number of species and their origin from different habitats allowed the development of varieties that flower in different parts of the world. However, in some tropical regions, the production of flowering plants continues to be concentrated in the autumn/winter because they require low temperatures for flowering. Breeding programs could use new genotypes to obtain hybrids that bridge this period and allow production of commercial plants at other times of the year. Propagation by seed is reliable and simple and can be done under in vitro and ex vitro conditions. However, the large-scale production of seedlings in industrial floriculture requires the development of efficient micropropagation protocols applied to a significant number of important cultivars on the market, producing millions of plants each year to match the number of seedlings commercially distributed in all of the world.

**Author Contributions:** Conceptualization, methodology and writing—original draft preparation, Kh.B. and M.V.; validation, investigation and formal analysis, Kh.B., M.V. and M.M.H.; resources, M.V. and M.M.H.; writing—review and editing, M.E.M.-M., J.C.C., C.L.S.A., M.L.-M., F.F. and M.M.H.; visualization, M.M.H.; supervision, M.E.M.-M. and M.V.; project administration, M.E.M.-M.; funding acquisition, M.E.M.-M., C.L.S.A. and M.L.-M. All authors have read and agreed to the published version of the manuscript.

**Funding:** This research received no external funding.

**Institutional Review Board Statement:** Not applicable.

**Data Availability Statement:** Data are contained within this article.

**Conflicts of Interest:** The authors declare no conflict of interest.

## References

1. Zotz, G. The systematic distribution of vascular epiphytes-a critical update. *Bot. J. Linn. Soc.* **2013**, *171*, 453–481. [CrossRef]
2. Arditti, J.; Ghani, A.K.A. Numerical and physical properties of orchid seeds and their biological implications. *New Phytol.* **2000**, *145*, 367–421. [CrossRef]
3. Hossain, M.M.; Sharma, M. Dual phase regeneration system for mass propagation of *Cymbidium aloifolium* (L.) Sw.: A High Value Medicinal Orchid. *Plant Tissue Cult. Biotechnol.* **2019**, *29*, 257–266. [CrossRef]
4. Kong, J.M.; Goh, N.K.; Chia, L.S.; Chia, T.F. Recent advances in traditional plant drugs and orchids. *Acta Pharm. Sin.* **2003**, *24*, 7–21.
5. Pant, B.; Swar, S. Micropropagation of *Cymbidium iridioides* Nepal. *J. Sci. Technol.* **2012**, *12*, 91–96. [CrossRef]
6. Singh, D.K. Morphological diversity of the orchids of Orissa/Sarat Misra. In *Orchids: Science and Commerce*; Pathak, P., Sehgal, R.N., Shekhar, N., Sharma, M., Sood, A., Eds.; Bishen Singh Mahendra Pal Singh: New Delhi, India, 2001; p. 35.
7. Teoh, E.S. India: Van Rheede, Caius and Others. In *Orchids as aphrodisiac, medicine or food*; Teoh, E.S., Ed.; Springer: Singapore, 2019; pp. 195–232. [CrossRef]
8. Jana, S.K.; Sinha, G.P.; Chauhan, A.S. Ethnobotanical aspects of Orchids in Sikkim. *J. Orchid Soc. India* **1997**, *11*, 79–84.
9. Wang, G.Q. *National Chinese Herbal Medicine Collection*; People's Medical Publishing House: Beijing, China, 2014.
10. Sujin, R.M.; Jeeva, S.; Subin, R.M. *Cymbidium aloifolium*: A review of its traditional uses, phytochemistry, and pharmacology. *Phytochem. Pharmacol. Asp. Ethnomedicinal Plants* **2021**, 363–371.
11. Kim, Y.J.; Lee, H.J.; Kim, K.S. Carbohydrate changes in *Cymbidium* 'Red Fire'in response to night interruption. *Sci. Hortic.* **2013**, *162*, 82–89. [CrossRef]

12. Yoshikawa, K.; Ito, T.; Iseki, K.; Baba, C.; Imagawa, H.; Yagi, Y.; Morita, H.; Asakawa, Y.; Kawano, S.; Hashimoto, T. Phenanthrene derivatives from *Cymbidium* Great Flower Marie Laurencin and their biological activities. *J. Nat. Prod.* **2012**, *75*, 605–609. [CrossRef] [PubMed]
13. Jeong, K.M.; Yang, M.; Jin, Y.; Kim, E.M.; Ko, J.; Lee, J. Identification of major flavone C-glycosides and their optimized extraction from *Cymbidium kanran* using deep eutectic solvents. *Molecules* **2017**, *22*, 2006. [CrossRef] [PubMed]
14. Lv, S.S.; Fu, Y.; Chen, J.; Jiao, Y.; Chen, S.Q. Six phenanthrenes from the roots of *Cymbidium faberi* Rolfe. and their biological activities. *Nat. Prod. Res.* **2020**, 1–12. [CrossRef]
15. Won, J.H.; Kim, J.Y.; Yun, K.J.; Lee, J.H.; Back, N.I.; Chung, H.G.; Chung, S.A.; Jeong, T.S.; Choi, M.S.; Lee, K.T. Gigantol isolated from the whole plants of *Cymbidium goeringii* inhibits the LPS-induced iNOS and COX-2 expression via NF-kappaB inactivation in RAW 264.7 macrophages cells. *Planta Med.* **2006**, *72*, 1181–1187. [CrossRef]
16. Rampilla, V.; Khasim, S.M. GC-MS analysis of organic extracts of *Cymbidium aloifolium* (L.) Sw. (Orchidaceae) leaves from Eastern Ghats of India. In *Orchid Biology*; Recent Trends & Challenges; Springer: Singapore, 2020; pp. 507–517.
17. Chowdappa, S.; Jagannath, S.; Konappa, N.; Udayashankar, A.C.; Jogaiah, S. Detection and characterization of antibacterial siderophores secreted by endophytic fungi from *Cymbidium aloifolium*. *Biomolecules* **2020**, *10*, 1412. [CrossRef]
18. Hinsley, A.; De Boer, H.J.; Fay, M.F.; Gale, S.W.; Gardiner, L.M.; Gunasekara, R.S.; Kumar, P.; Masters, S.; Metusala, D.; Roberts, L.R.; et al. A review of the trade in orchids and its implications for conservation. *Bot. J. Linn. Soc.* **2018**, *186*, 435–455. [CrossRef]
19. Park, P.H.; Ramya, M.; An, H.R.; Park, P.M.; Lee, S.Y. Breeding of *Cymbidium* 'Sale Bit'with bright yellow flowers and floral scent. *Korean Soc. Breed. Sci.* **2019**, *51*, 258–262. [CrossRef]
20. Kostenyuk, I.; Oh, B.J.; So, I.S. Induction of early flowering in *Cymbidium niveo-marginatum* Mak in vitro. *Plant Cell Rep.* **1999**, *19*, 1–5. [CrossRef] [PubMed]
21. Zhou, D.; Chen, G.; Ma, Y.P.; Wang, C.G.; Lin, B.; Yang, Y.Q.; Li, W.; Koike, K.; Hou, Y.; Li, N. Isolation, structural elucidation, optical resolution, and antineuroinflammatory activity of phenanthrene and 9,10-dihydrophenanthrene derivatives from *Bletilla striata*. *J. Nat. Prod.* **2019**, *82*, 2238–2245. [CrossRef] [PubMed]
22. An, H.R.; Kim, Y.J.; Kim, K.S. Flower initiation and development in *Cymbidium* by night interruption with potassium and nitrogen. *Hortic. Environ. Biotechnol.* **2012**, *53*, 204–211. [CrossRef]
23. Barman, D.; Bharathi, T.U.; Medhi, R.P. Effect of media and nutrition on growth and flowering of *Cymbidium* hybrid 'HC Aurora'. *Indian J. Hortic.* **2012**, *69*, 395–398.
24. Lee, N.; Lee, C.Z. Growth and flowering of *Cymbidium ensifolium var. misericors* as influenced by temperature. *Acta Horticulturae* **1991**, *337*, 123–130. [CrossRef]
25. Yu, H.; Goh, C.J. Molecular Genetics of Reproductive Biology in Orchids. *Plant Physiol.* **2001**, *127*, 1390–1393. [CrossRef]
26. Yang, W.K.; Li, T.Q.; Wu, S.M.; Finnegan, P.M.; Gao, J.Y. Ex situ seed baiting to isolate germination-enhancing fungi for assisted colonization in *Paphiopedilum spicerianum*, a critically endangered orchid in China. *Glob. Ecol. Conserv.* **2020**, *23*, e01147. [CrossRef]
27. Li, X.; Jin, F.; Jin, L.; Jackson, A.; Ma, X.; Shu, X.; Wu, D.; Jin, G. Characterization and comparative profiling of the small RNA transcriptomes in two phases of flowering in *Cymbidium ensifolium*. *BMC Genom.* **2015**, *16*, 1–17. [CrossRef]
28. Matsuda, Y.; Sugiura, N. Specialized pollination by honeybees in Cymbidium dayanum, a fall–winter flowering orchid. *Plant Species Biol.* **2019**, *34*, 19–26. [CrossRef]
29. Suetsugu, K. Autonomous self-pollination and insect visitors in partially and fully mycoheterotrophic species of *Cymbidium* (Orchidaceae). *J. Plant Res.* **2015**, *128*, 115–125. [CrossRef]
30. Du Puy, D.; Cribb, P. The genus *Cymbidium*. In *Surrey, Royal Botanic Gardens*, 2nd ed.; Kew Publishing: London, UK, 2007.
31. Davies, K.L.; Stpiczyńska, M.; Turner, M.P. A rudimentary labellar speculum in *Cymbidium lowianum* (Rchb. f.) Rchb. f. and *Cymbidium devonianum* Paxton (Orchidaceae). *Ann. Bot.* **2006**, *97*, 975–984. [CrossRef]
32. Thummavongsa, T. Taxonomy, Reproductive Biology and Seed Germination of Habenaria rhodocheila Hance complex (Orchidaceae). Ph.D. Dissertation, School of Biology Institute of Science, Suranaree University of Technology, Nakhon Ratchasima, Thailand, 2021.
33. Balilashaki, K.; Vahedi, M.; Ho, T.T.; Niu, S.C.; Cardoso, J.C.; Zotz, G.; Khodamzadeh, A.A. Biochemical, cellular and molecular aspects of Cymbidium orchids: An ecological and economic overview. *Acta Physiol. Plant.* **2022**, *44*, 24. [CrossRef]
34. Kjellsson, G.; Rasmussen, F.N.; Dupuy, D. Pollination of *Dendrobium infundibulum*, *Cymbidium insigne* (Orchidaceae) and Rhododendron lyi (Ericaceae) by Bombus eximius (Apidae) in Thailand: A possible case of floral mimicry. *J. Trop. Ecol.* **1985**, *1*, 289–302. [CrossRef]
35. Nanekar, V.; Shriram, V.; Kumar, V.; Kishor, P.K. Asymbiotic in vitro seed germination and seedling development of *Eulophia nuda* Lindl., an endangered medicinal orchid. *Proc. Natl. Acad. Sci. India Sect. B Biol. Sci.* **2014**, *84*, 837–846. [CrossRef]
36. Puspitaningtyas, D.M.; Handini, E. Ex-situ conservation of *Cymbidium finlaysonianum* by seed storage. *Biodiversitas J. Biol. Divers.* **2020**, *21*. [CrossRef]
37. Gantait, S.; Mitra, M. Applications of synthetic seed technology for propagation, storage, and conservation of orchid germplasms. In *Synthetic Seeds: Germplasm Regeneration, Preservation and Prospects*; Springer: Cham, Switzerland, 2019; pp. 301–321.
38. Garg, R.; Maheshwari, S. Synthetic seed technology, application and future trends. *EPH-Int. J. Agric. Environ. Res.* **2023**, *9*, 1–10. [CrossRef]

39. Patavardhan, S.S.; Ignatius, S.; Thiyam, R.; Lasrado, Q.; Karkala, S.; D'Souza, L.; Nivas, S.K. Asymbiotic seed germination and in vitro development of orchid Papilionanthe Miss Joaquim. *Ornam. Hortic.* **2022**, *28*, 246–255. [CrossRef]
40. Seaton, P.; Kendon, J.P.; Pritchard, H.W.; Puspitaningtyas, D.M.; Marks, T.R. Orchid conservation: The next ten years. *Lankesteriana* **2013**, *13*, 93–101. [CrossRef]
41. Whigham, D.F.; O'Neill, J.P.; Rasmunssen, H.N.; Caldwell, B.A.; McCormick, M.K. Seed longevity in terrestrial orchids-Potential for persistent in-situ seed banks. *Biol. Conserv.* **2006**, *129*, 2–30. [CrossRef]
42. Suzuki, R.M.; Moreira, V.C.; Pescador, R.; de Melo Ferreira, W. Asymbiotic seed germination and in vitro seedling development of the threatened orchid *Hoffmannseggella cinnabarina*. *In vitro Cell. Dev. Biol. -Plant* **2012**, *48*, 500–511. [CrossRef]
43. Paudel, M.; Pradhan, S.; Pant, B. In vitro seed germination and seedling development of *Esmeralda clarkei* Rchb. f. (Orchidaceae). *Plant Tissue Cult. Biotechnol.* **2012**, *22*, 107–111. [CrossRef]
44. Cardoso, J.C.; Zanello, C.A.; Chen, J.-T. An Overview of Orchid Protocorm-Like Bodies: Mass Propagation, Biotechnology, Molecular Aspects, and Breeding. *Int. J. Mol. Sci.* **2020**, *21*, 985. [CrossRef]
45. Udomdee, W.; Wen, P.J.; Lee, C.Y.; Chin, S.W.; Chen, F.C. Effect of sucrose concentration and seed maturity on in vitro germination of *Dendrobium nobile* hybrids. *Plant Growth Regul.* **2014**, *72*, 249–255. [CrossRef]
46. Hossain, M.M.; Dey, R. Multiple regeneration pathways in *Spathoglottis plicata* Blume—A study in vitro. *S. Afr. J. Bot.* **2013**, *85*, 56–62. [CrossRef]
47. Vudala, S.M.; Ribas, L.L.F. Seed storage and asymbiotic germination of *Hadrolaelia grandis* (Orchidaceae). *S. Afr. J. Bot.* **2017**, *108*, 1–7. [CrossRef]
48. Huh, Y.S.; Lee, J.K.; Nam, S.Y.; Hong, E.Y.; Paek, K.Y.; Son, S.W. Effects of altering medium strength and sucrose concentration on in vitro germination and seedling growth of *Cypripedium macranthos* Sw. *J. Plant Biotechnol.* **2016**, *43*, 132–137. [CrossRef]
49. Zeng, S.; Zhang, Y.; Teixeira da Silva, J.A.; Wu, K.; Zhang, J.; Duan, J. Seed biology and in vitro seed germination of *Cypripedium*. *Crit. Rev. Biotechnol.* **2014**, *34*, 358–371. [CrossRef] [PubMed]
50. Zhang, Y.F.; Yan, S.; Zhang, Y. Factors affecting germination and propagators of artificial seeds of *Dendrobium candidum*. In Proceedings of the International Conference on Agricultural and Biosystems Engineering, Amsterdam, Netherlands, 13–15 July 2011; pp. 1–2.
51. Lu, Y.X.; Li, C.J.; Zhang, F.S. Transpiration, potassium uptake and flow in tobacco as affected by nitrogen forms and nutrient levels. *Ann. Bot.* **2005**, *95*, 991–998. [CrossRef]
52. Xiang, M.A.; Zheng, F.W.; Li, Y.; Liu, L.; Wu, J. Symbiotic seed germination and seedling growth promoted by Rhizoctonia fungi in *Cymbidium mastersii*, an endangered orchid species endemic to Southwest of China. In Proceedings of the 18th EOCCE-What Future for Orchids, Paris, France, 24 March 2018.
53. Teixeira da Silva, J.A.; Norikane, A.; Tanaka, M. *Cymbidium*: Successful in vitro growth and subsequent acclimatization. *Acta Hort.* **2015**, *748*, 207–214. [CrossRef]
54. Kim, Y.J.; Lee, H.J.; Kim, K.S. Night interruption promotes vegetative growth and flowering of *Cymbidium*. *Sci. Hortic.* **2011**, *130*, 887–893. [CrossRef]
55. Teixeira da Silva, J.A.; Yam, T.; Fukai, S.; Nayak, N.; Tanaka, M. Establishment of optimum nutrient media for in vitro propagation of *Cymbidium* Sw. (Orchidaceae) using protocorm-like body segments. *Propag. Ornam. Plants* **2016**, *5*, 129–136.
56. Chand, K.; Shah, S.; Sharma, J.; Paudel, M.R.; Pant, B. Isolation, characterization, and plant growth-promoting activities of endophytic fungi from a wild orchid *Vanda cristata*. *Plant Signal. Behav.* **2020**, *15*, 174–294. [CrossRef] [PubMed]
57. Acemi, A.; Özen, F. Optimization of in vitro asymbiotic seed germination protocol for *Serapias vomeracea*. *EuroBiotech J.* **2019**, *3*, 143–151. [CrossRef]
58. Rasmussen, H.N.; Rasmussen, F.N. Seedling mycorrhiza: A discussion of origin and evolution in Orchidaceae. *Bot. J. Linn. Soc.* **2014**, *175*, 313–327. [CrossRef]
59. Yam, T.W.; Arditti, J. History of orchid propagation: A mirror of the history of biotechnology. *Plant Biotechnol. Rep.* **2009**, *3*, 1–56. [CrossRef]
60. Valadares, R.B.S.; Perotto, S.; Santos, E.C.; Lambais, M.R. Proteome changes in *Oncidium sphacelatum* (Orchidaceae) at different trophic stages of symbiotic germination. *Mycorrhiza* **2014**, *24*, 349–360. [CrossRef]
61. Rafter, M.; Yokoya, K.; Schofield, E.J.; Zettler, L.W.; Sarasan, V. Non-specific symbiotic germination of *Cynorkis purpurea* (Thouars) Kraezl a habitat-specific terrestrial orchid from the Central Highlands of Madagascar. *Mycorrhiza* **2016**, *26*, 541–552. [CrossRef] [PubMed]
62. Nikabadi, S.; Bunn, E.; Stevens, J.; Newman, B.; Turner, S.R.; Dixon, K.W. Germination responses of four native terrestrial orchids from south-west Western Australia to temperature and light treatments. *Plant Cell Tissue Organ Cult.* **2014**, *118*, 559–569. [CrossRef]
63. Mala, B.; Kuegkong, K.; Sa-Ngiaemsri, N.; Nontachaiyapoom, S. Effect of germination media on in vitro symbiotic seed germination of three Dendrobium orchids. *S. Afr. J. Bot.* **2017**, *112*, 521–526. [CrossRef]
64. Yang, Q.; Xu, L.; Xia, W.; Liang, L.; Bai, X.; Li, L.; Xu, L.; Liu, L. Mycorrhizal compatibility and germination-Promoting activity of Tulasnella species in two species of orchid (Cymbidium mannii and Epidendrum radicans). *Horticulturae* **2021**, *7*, 472. [CrossRef]
65. Shao, S.C.; Burgess, K.S.; Cruse-Sanders, J.M.; Liu, Q.; Fan, X.L.; Huang, H.; Gao, J.Y. Using in situ symbiotic seed germination to restore over-collected medicinal orchids in Southwest China. *Front. Plant Sci.* **2017**, *8*, 888. [CrossRef]

66. Smith, Z.F.; James, E.A.; Mclean, C.B. Experimental reintroduction of the threatened terrestrial orchid *Diuris fragrantissima*. *Lankesteriana Int. J. Orchid.* **2007**, *7*, 377–380. [CrossRef]
67. Rasmussen, H.N.; Dixon, K.W.; Jersáková, J.; Těšitelová, T. Germination and seedling establishment in orchids: A complex of requirements. *Ann. Bot.* **2015**, *116*, 391–402. [CrossRef] [PubMed]
68. Dong, F.; Liu, H.X.; Jin, H.; Luo, Y.B. Symbiosis between fungi and the hybrid *Cymbidium* and its mycorrhizal microstructures. *For. Stud. China* **2008**, *10*, 41–44. [CrossRef]
69. Liu, S.; Liu, M.; Liao, Q.G.; Lü, F.B.; Zhao, X.L. Effects of inoculated mycorrhizal fungi and non-mycorrhizal beneficial microorganisms on plant traits, nutrient uptake and root associated fungal community composition of the *Cymbidium* hybrid in greenhouse. *J. Appl. Microbiol.* **2021**, *131*, 413–424. [CrossRef]
70. Gogoi, K.; Kumaria, S.; Tandon, P. Ex situ conservation of *Cymbidium eburneum* Lindl.: A threatened and vulnerable orchid, by asymbiotic seed germination. *3 Biotech.* **2012**, *2*, 337–343. [CrossRef]
71. Herrera, H.; Valadares, R.; Contreras, D.; Bashan, Y.; Arriagada, C. Mycorrhizal compatibility and symbiotic seed germination of orchids from the Coastal Range and Andes in south central Chile. *Mycorrhiza* **2017**, *27*, 175–188. [CrossRef]
72. Zi, X.M.; Sheng, C.L.; Goodale, U.M.; Shao, S.C.; Gao, J.Y. In situ seed baiting to isolate germination-enhancing fungi for an epiphytic orchid, *Dendrobium aphyllum* (Orchidaceae). *Mycorrhiza* **2014**, *24*, 487–499. [CrossRef] [PubMed]
73. Sheng, C.L.; Lee, Y.; Gao, J.Y. Ex situ symbiotic seed germination, isolation and identification of effective symbiotic fungus in *Cymbidium mannii* (Orchidaceae). *Chin. J. Plant Ecol.* **2012**, *36*, 859. [CrossRef]
74. Prutsch, J.; Schardt, A.; Schill, R. Adaptations of an orchid seed to water uptake and-storage. *Plant Syst. Evol.* **2000**, *220*, 69–75. [CrossRef]
75. Chang, C.; Shiu, Q.J. January. Pollination, Seed Development and In vitro Germination of *Cymbidium* sinense. *I Int. Orchid Symp.* **2010**, *878*, 251–259.
76. Tandon, P.; Kumaria, S. Prospects of plant conservation biotechnology in India with special reference to Northeastern region. In *Biodiversity: Status and Prospects*; Tandon, P., Sharma, M., Swarup, R., Eds.; Narosa Publshing House: New Delhi, India, 2005; pp. 79–91.
77. Mohanty, P.; Paul, S.; Das, M.C.; Kumaria, S.; Tandon, P. A simple and efficient protocol for the mass propagation of *Cymbidium mastersii*: An ornamental orchid of Northeast India. *AoB Plants* **2012**, *2012*, pls023. [CrossRef] [PubMed]
78. Deb, C.R.; Pongener, A. Asymbiotic seed germination and in vitro seedling development of *Cymbidium aloifolium* (L.) Sw: A multipurpose orchid. *J. Plant Biochem. Biotechnol.* **2011**, *20*, 90–95. [CrossRef]
79. Murashige, T.; Skoog, F. A revised medium for rapid growth and bio assays with tobacco tissue cultures. *Physiol. Plant.* **1962**, *15*, 473–497. [CrossRef]
80. Dohling, S.; Kumaria, S.; Tandon, P. Optimization of nutrient requirements for asymbiotic seed germination of *D. longicornu* and *D. formosumi* Roxb. *Proc. Indian Natl. Sci. Acad.* **2008**, *74*, 167–171.
81. Utami, E.S.W.; Hariyanto, S. Organic compounds: Contents and their role in improving seed germination and protocorm development in orchids. *Int. J. Agron.* **2020**, *2020*, 2795108. [CrossRef]
82. Dutra, D.; Johnson, T.R.; Kauth, P.J.; Stewart, S.L.; Kane, M.E.; Richardson, L. Asymbiotic seed germination, in vitro seedling development, and greenhouse acclimatization of the threatened terrestrial orchid *Bletia purpurea*. *Plant Cell Tissue Organ Cult.* **2008**, *94*, 11–21. [CrossRef]
83. Hajong, S.; Kumaria, S.; Tandon, P. In vitro propagation of medicinal orchid *Dendrobium chrysanthum*. *Proc. Indian Natl. Sci. Acad.* **2010**, *76*, 67–70.
84. Bhowmik, T.K.; Rahman, M.M. Effect of different basal media and PGRs on in vitro seed germination and seedling development of medicinally important orchid *Cymbidium aloifolium* (L.) Sw. *J. Pharmacogn. Phytochem.* **2017**, *6*, 167–172.
85. Bhattacharjee, D.K.; Hossain, M.M. Effect of plant growth regulators and explants on propagation on a monopodial and sympodial orchid: A study in vitro. *J. Orchid Soc. India* **2015**, *29*, 91–102.
86. Teixeira da Silva, J.A.; Tanaka, M. Multiple regeneration pathways via Thin Cell Layers in hybrid Cymbidium (Orchidaceae). *J. Plant Growth Regul.* **2006**, *25*, 203. [CrossRef]
87. Morel, G.M. A new means of clonal propagation of orchids. *Am. Orchid Soc. Bull.* **1964**, *31*, 437–477.
88. Chang, C.; Chang, W.C. Micropropagation of *Cymbidium ensifolium* var. Misericors through Callus-Derived Rhizomes. In vitro Cellular & Developmental Biology. *Plant* **2000**, *36*, 517–520.
89. Wimber, D.E. Clonal multiplication of *Cymbidiums* through tissue culture of the shoot meristem. *Am. Orchid Soc. Bull.* **1963**, *32*, 105–107.
90. Malabadi, R.B.; Teixeira da Silva, J.A.; Nataraja, K.; Mulgund, G.S. Shoot tip transverse thin cell layers and 2,4-epibrassinolide in the micropropagation of *Cymbidium bicolor* Lindl. *Floric. Ornam. Biotech.* **2008**, *2*, 44–48.
91. Begum, A.A.; Tamaki, M.; Kako, S. Formation of protocorm-like bodies (PLBs) and shoot development through in vitro culture of outer tissue of *Cymbidium* PLB. *J. Jpn. Soc. Hort. Sci.* **1994**, *63*, 663–673. [CrossRef]
92. Hossain, M.M.; Sharma, M.; Pathak, P. In vitro propagation of *Dendrobium aphyllum* (Orchidaceae)-seed germination to flowering. *J. Plant Biochem. Biotechnol.* **2013**, *22*, 157–167. [CrossRef]
93. Teixeira da Silva, J.A.; Singh, N.; Tanaka, M. Priming biotic factors for optimal protocorm-like body and callus induction in hybrid *Cymbidium* (Orchidaceae), and assessment of cytogenetic stability in regenerated plantlets. *Plant Cell Tissue Organ Cult.* **2006**, *84*, 135–144. [CrossRef]

94. Shimasaki, K.; Uemoto, S. Micropropagation of a terrestrial *Cymbidium* species using rhizomes developed from seeds and pseudobulbs. *Plant Cell Tissue Organ Cult.* **1990**, *22*, 237–244. [CrossRef]
95. Chugh, S.; Guha, S.; Rao, I.U. Micropropagation of orchids: A review on the potential of different explants. *Sci. Hortic.* **2009**, *122*, 507–520. [CrossRef]
96. Hossain, M.M.; Sharma, M.; Pathak, P. In vitro mass propagation of an economically important orchid, *Cymbidium aloifolium* (L.) Sw. *J. Orchid Soc.* **2008**, *22*, 91–95.
97. Yasugi, S.; Sakamoto, K.; Onodera, K.; Tamashiro, M. Plantlet regeneration in root segment culture of *Cymbidium* Kenny 'Wine Color'. *Plant Tissue Cult. Lett.* **1994**, *11*, 150–152. [CrossRef]
98. Begum, A.A.; Tamaki, M.; Tahara, M.; Kako, S. Somatic embryogenesis in *Cymbidium* through in vitro culture of inner tissue of protocorm-like bodies. *J. Jpn. Soc. Hort. Sci.* **1994**, *63*, 419–427. [CrossRef]
99. Huan, L.V.T.; Takamura, T.; Tanaka, M. Callus formation and plant regeneration from callus through somatic embryo structures in *Cymbidium* orchid. *Plant Sci.* **2004**, *166*, 1443–1449. [CrossRef]
100. Das, M.C.; Kumeria, S.; Tandon, P. Protocorm regeneration, multiple shoot induction and *ex vitro* establishment of *Cymbidium devonianum* Paxt. *Asian J. Plant Sci.* **2007**, *6*, 349–353. [CrossRef]
101. Pradhan, S.; Tiruwa, B.; Subedee, B.R.; Pant, B. In vitro germination and propagation of a threatened medicinal orchid, *Cymbidium aloifolium* (L.) Sw. through artificial seed. *Asian Pac. J. Trop. Biomed.* **2014**, *4*, 971–976. [CrossRef]
102. Rihan, H.Z.; Kareem, F.; El-Mahrouk, M.E.; Fuller, M.P. Artificial seeds (principle, aspects and applications). *Agronomy* **2017**, *7*, 71. [CrossRef]
103. da Silva, J.A.T. Production of synseed for hybrid *Cymbidium* using protocorm-like bodies. *J. Fruit Ornam. Plant Res.* **2012**, *20*, 135–146. [CrossRef]
104. Deb, C.R.; Pongener, A. Studies on the in vitro regenerative competence of aerial roots of two horticultural important Cymbidium species. *J. Plant Biochem. Biotechnol.* **2012**, *21*, 235–241. [CrossRef]
105. Hossain, M.M.; Sharma, M.; Pathak, P. Cost effective protocol for in vitro mass propagation of Cymbidium aloifolium (L.) Sw.—A medicinally important orchid. *Eng. Life Sci.* **2009**, *9*, 444–453. [CrossRef]
106. Hossain, M.M.; Sharma, M.; da Silva, J.A.T.; Pathak, P. Seed germination and tissue culture of *Cymbidium giganteum* Wall. ex Lindl. *Sci. Hortic.* **2010**, *123*, 479–487. [CrossRef]
107. Nahar, S.J.; Shimasaki, K.; Haque, S.M. Effect of different light and two polysaccharides on the proliferation of protocorm-like bodies of Cymbidium cultured in vitro. *Acta Hortic.* **2012**, 307–314. [CrossRef]
108. Parmar, G.; Pant, B. In vitro seed germination and seedling development of the orchid *Coelogyne stricta* (D. Don) Schltr. *Afr. J. Biotechnol.* **2016**, *15*, 105–109.
109. Pradhan, S.; Tiruwa, B.L.; Subedee, B.R.; Pant, B. Efficient plant regeneration of *Cymbidium aloifolium* (L.) Sw., a threatened orchid of Nepal through artificial seed technology. *Am. J. Plant Sci.* **2016**, *7*, 1964–1974. [CrossRef]
110. Philip Robinson, J.; Jyoti, P.K.; Sebastinraj, J.; Suriya, K. In vitro seed germination of *Cymbidium aloifolium* (L.) Sw., a potential medicinal orchid from Eastern Ghats of Tamil Nadu, India. *J. Plant Biotechnol.* **2017**, *44*, 343–348. [CrossRef]
111. Paul, M.; Islam, T.; Sarker, R.H.; Hoque, M.I. In vitro mass propagation of *Cymbidium aloifolium* (L.) Sw. *Plant Tissue Cult. Biotechnol.* **2019**, *29*, 73–79. [CrossRef]


**Disclaimer/Publisher's Note:** The statements, opinions and data contained in all publications are solely those of the individual author(s) and contributor(s) and not of MDPI and/or the editor(s). MDPI and/or the editor(s) disclaim responsibility for any injury to people or property resulting from any ideas, methods, instructions or products referred to in the content.

*horticulturae*

*Article*

# Discrimination of *Syzygium samarangense* cv. 'Giant Green' Leaves at Different Maturity Stages by FTIR and GCMS Fingerprinting

**Nuruljannah Suhaida Idris [1], Mohammad Moneruzzaman Khandaker [1,*], Zalilawati Mat Rashid [2], Ali Majrashi [3], Mekhled Mutiran Alenazi [4], Ahmad Faris Mohd Adnan [5], Khairil Mahmud [6] and Nashriyah Mat [1]**

1 School of Agriculture Science & Biotechnology, Faculty of Bioresources and Food Industry, Universiti Sultan Zainal Abidin, Besut Campus, Besut 22200, Terengganu, Malaysia; nuruljannahsuhaida@gmail.com (N.S.I.); nashriyahbintimat@gmail.com (N.M.)
2 School of Food Industry, Faculty of Bioresources and Food Industry, Universiti Sultan Zainal Abidin, Besut Campus, Besut 22200, Terengganu, Malaysia; zalilawati@unisza.edu.my
3 Department of Biology, College of Science, Taif University, Taif 21944, Saudi Arabia; aa.majrashi@tu.edu.sa
4 Plant Production Department, College of Food and Agricultural Sciences, King Saud University, Riyadh 11451, Saudi Arabia; amekhled@ksu.edu.sa
5 Institute of Biological Sciences, Faculty of Science, Universiti Malaya, Kuala Lumpur 50603, Selangor, Malaysia; ahmad_farisz@um.edu.my
6 Department of Crop Science, Faculty of Agriculture, Universiti Putra Malaysia, Seri Kembangan 43400, Selangor, Malaysia; khairilmahmud@upm.edu.my
* Correspondence: moneruzzaman@unisza.edu.my; Tel.: +60-9699-3450

**Abstract:** 'Giant Green' is one of the *Syzygium samarangense* cultivars planted throughout Malaysia because it has great potential for benefitting human health. However, its variation in chemical compounds, especially in the leaves at different maturity stages, cannot be systematically discriminated. Hence, Fourier transform infrared spectroscopy (FTIR) and gas chromatography–mass spectrometry (GCMS) coupled with chemometric tools were applied to discriminate between the different stages of leaves, namely, young, mature, and old leaves. The chemical variability among the samples was evaluated by using principal component analysis (PCA) and hierarchical clustering analysis (HCA) techniques. For discrimination, partial least squares discrimination analysis (PLS-DA) was applied, and then partial least squares (PLS) was used to determine the correlation between biological activities (antioxidant and alpha-glucosidase inhibitory assay) and maturity stages of 'Giant Green' leaves. As a result, the PCA, HCA, and PLS-DA of the FTIR and GC-MS data showed the separation between clusters for the different maturity stages of the leaves. Additionally, the PLS result demonstrated that the young leaves showed a strong correlation between metabolite quantities and biological activities. The findings of this study revealed that FTIR and GC-MS coupled with chemometric analyses can be used as a rapid method for the discrimination of bioactive structural functions in relation to their biological activity.

**Keywords:** Giant Green cultivar; antioxidant; alpha-glucosidase inhibition; FTIR; GC-MS; PLS; HCA; PLS-DA; PLS

check for updates

**Citation:** Idris, N.S.; Khandaker, M.M.; Rashid, Z.M.; Majrashi, A.; Alenazi, M.M.; Adnan, A.F.M.; Mahmud, K.; Mat, N. Discrimination of *Syzygium samarangense* cv. 'Giant Green' Leaves at Different Maturity Stages by FTIR and GCMS Fingerprinting. *Horticulturae* **2023**, *9*, 609. https://doi.org/10.3390/horticulturae9050609

Academic Editor: Wajid Zaman

Received: 31 March 2023
Revised: 4 May 2023
Accepted: 4 May 2023
Published: 22 May 2023

**Copyright:** © 2023 by the authors. Licensee MDPI, Basel, Switzerland. This article is an open access article distributed under the terms and conditions of the Creative Commons Attribution (CC BY) license (https://creativecommons.org/licenses/by/4.0/).

## 1. Introduction

*Syzygium samarangense*, commonly known as wax apple, jambu air, water apple, or bell fruit, is a nonclimacteric tropical fruit plant that has been cultivated in Malaysia and other neighboring countries such as Thailand, the Philippines, Vietnam, and Taiwan [1]. The three major *S. samarangense* cultivars are Giant Green, Masam Manis Pink, and Jambu Madu Red [2]. Traditionally, it is a medicinal plant: various parts are used to treat some health problems such as edema, cracked tongue, asthma, diarrhea, bronchitis, fever, ulcer, sore throat, and to reduce blood pressure [3,4]. Additionally, the Giant Green cultivar contains an abundance of valuable phytochemicals such as phenolic acids, flavonoids, anthocyanins, and carotenes [5], which show antioxidant, antibacterial, antidiabetic, anticancer, and anti-inflammatory activities [6–8].

Metabolomics is the comprehensive analysis of a metabolite profile, either as a targeted or global application in drug discovery, phytomedicine, toxicology, and disease development [9]. Various spectroscopy and chromatography techniques are applied to detect and characterize the presence of metabolites in experimental samples [10]. Two analytical techniques for this task are Fourier transform infrared spectroscopy (FTIR) and gas chromatography–mass spectrometry (GC-MS). FTIR is an important technique used to identify the type of functional groups present in a compound. It also is a useful spectroscopic tool for profiling and fingerprinting molecular structures because it is non-destructive, simple to use, quick, and accurate [11]. The common absorption range used in plant studies is the mid-infrared (mid-IR) range. In the mid-IR range, infrared radiation is passed through a sample with a range of absorbance from 4000 $cm^{-1}$ to 400 $cm^{-1}$ [12]. Not all of the infrared radiation is absorbed by the sample: some of it passes through the sample and is transmitted to a detector. The resulting spectrum represents the molecular absorption and transmission, creating a molecular fingerprint of the sample. GC-MS is the most commonly used instrument for the separation and identification of compounds, especially in the drug discovery, pharmacology, and food industry fields [13]. The advantages of GC-MS are its low viscosity, higher sensitivity, rapid mass transfer velocity, and high resistance, so it has been widely used in chemical fingerprinting [14].

However, the abundance of metabolites present in plants poses challenges: analyzing them precisely without using a comprehensive method is difficult. Therefore, for several decades, many researchers have applied chemometric analyses coupled with spectroscopy and chromatography techniques to analyze the metabolites present in medicinal plants. Unsupervised multivariate analysis (MVDA) including principal component analysis (PCA); hierarchical clustering analysis (HCA); and supervised MVDA, including partial least squares discrimination analysis (PLS-DA) and partial least squares (PLS), are required to handle the huge dataset of the whole spectra recorded from plant samples. For example, Wijayanti et al. [15] successfully classified and discriminated the *Curcuma xanthorrhiza* from different regions using PCA and PLS-DA tools. Basyirah et al. [16] used PCA and HCA to classify and discriminate *Heterotrigona itama* propolis using different extraction methods (maceration, sonication, and Soxhlet). Additionally, PLS correlated the antioxidant activity and chemical contents of five varieties of Pegaga (Centella) extract [17]. From these studies, it can be concluded that chemometric analysis coupled with spectroscopy or chromatography techniques is a reliable tool that can be used in the metabolomics field.

Judging from the literature, it can be concluded that the metabolites in plants can be identified using spectroscopy and chromatography techniques. Additionally, the discrimination between experimental samples and the relationship between metabolites and biological activity can also be determined using multivariate data analysis (MVDA). Hence, this work aimed to discriminate the leaves of *Syzygium samarangense* cv. Giant Green at different stages of maturity and to correlate these maturity stages with their antioxidant and alpha-glucosidase inhibitory activities using FTIR- and GCMS-based metabolomics coupled with chemometrics. The findings help with identifying the most promising stages of Giant Green leaves to be used in pharmaceuticals.

## 2. Materials and Methods

### *2.1. Collection and Preparation of Plant Materials*

The Giant Green cultivar of wax apple leaves, namely, young (YL), mature (ML), and old (OL) leaves, at three maturity stages were collected several times from an orchard located at Kampung Olak Lempit, Banting, Selangor, Malaysia (1028° N, 1110°20′ E), at an elevation of about 45 m above sea level. Five biological replicates of each sample were used in this study. The leaves were selected carefully based on the below picture (Figure 1) physical examination reported by our previous study [18].

**Figure 1.** Different types of Giant Green leaves were used in this study. (**A**,**A1**)—adaxial and abaxial surfaces of the young leaf (YL), (**B**,**B1**)—adaxial and abaxial surfaces of the mature leaf (ML), and (**C**,**C1**)—adaxial and abaxial surfaces of the old leaf (OL).

### *2.2. Extraction Procedure*

The fresh samples (5 g) were crushed using a mortar and pestle. The samples were soaked in methanol (25 mL) and kept for three days. Then, the extracts were heated in a water bath (70 °C for 15 min), followed by being centrifuged (1789× $g$ for 15 min). The supernatants were collected and put under a fumehood until methanol was removed. The extracts were completely dried by being freeze-dried (24 h) and were stored at 4 °C before being used.

### *2.3. Attenuated Total Reflectance Fourier Transform Infrared (ATR-FTIR)*

#### 2.3.1. Sample Analysis using ATR-FTIR Machine

The diamond crystal stand was cleaned with ethanol and non-abrasive tissue. The small quantities of extracts were located on the surface stand and screwed in tightly before being analyzed by using a Shimadzu Prestige-21 Spectrophotometer (Shimadzu Brand, Kyoto, Japan) equipped with an air-cooled Deutrated Triglycin Sulphate (DTGS) detector (Shimadzu Brand, Kyoto, Japan) and scanned with a Golden Gate Single Reflection Diamond ATR accessory with an incident angle of 45° (Shimadzu Brand, Kyoto, Japan). The IR spectra of the extracts were measured with absorbance at 4000–400 $cm^{-1}$ using 4 $cm^{-1}$ and 16 scans of resolution. For each sample, three repetition measurements were collected.

#### 2.3.2. Data Pre-Processing

The spectra were normalized and smoothed to reduce the error during data analysis. The data were saved in .txt format and then copied to Microsoft Excel.

### 2.4. Gas Chromatography-Mass Spectrometry (GC-MS)

#### 2.4.1. Sample Preparation

The stock solution was prepared by diluting 2.5 mg of extract with 100% methanol (1 mL). For phytochemical screening, 200 µL of each leaf sample with a similar maturity stage was transferred out from the stock solution and put into the same vial for producing the final concentration of 500 µg/mL. However, for multivariate data analysis, individual samples (30 samples) were prepared by diluting 200 µL stock solution with 1 mL of 100% methanol. The solutions were vortexed for one minute and ready for analysis by using a GC-MS Agilent (19091S-433UI system) machine (Agilent Brand, Santa Clara, CA, USA).

#### 2.4.2. GC-MS Condition

The condition system of GC-MS and Oven temperature parameters were used as below (Table 1).

**Table 1.** Condition system of GC-MS and Oven temperature parameter.

| **Item** | **Description** | | | |
|---|---|---|---|---|
| Column type | HP5MS (30 m × 250 µm) | | | |
| Film thickness | 0.25 µm | | | |
| Carrier gas | Helium | | | |
| Flow rate and pressure | 1.0 mL/min; 9.3825 Psi | | | |
| Volume of injection | 1.0 µL | | | |
| Temperature of detector | 250 °C | | | |
| Temperature of injector | 250 °C | | | |
| Temperature of oven | 80 °C | | | |
| Temperature of transfer line | 150 °C | | | |
| Mode | Splitless | | | |
| Mass scan mode | 50–55 $m/z$ | | | |
| **Oven temperature parameter** | | | | |
| **Item** | **Rate (°C/min)** | **Value (°C)** | **Hold Time (min)** | **Run Time (min)** |
| Initial | - | 80 | 4 | 4 |
| Ramp 1 | 7 | 105 | 1 | 9 |
| Ramp 2 | 7 | 180 | 1 | 20 |
| Ramp 3 | 5 | 235 | 1 | 32 |
| Ramp 4 | 5 | 275 | 2 | 42 |

#### 2.4.3. Data Pre-Processing

Each of the GC-MS spectra that contain the peak with a percentage of probability score of 80% and above was accepted as a particular compound and used in this analysis. For chemometric analysis purposes, the data of the percentage relative area (RA) of the compound detected in the spectrum was used. RA (%) was calculated based on the formula below [19].

$$\text{Percentage of relative area} = (\text{area of particular compound}/\text{total area of all compound detected}) \times 100$$

### 2.5. Chemometric Analysis

FTIR and GC-MS spectra were subjected to four chemometric tools which are Principal Component Analysis (PCA), Hierarchical Cluster Analysis (HCA), Partial Least Squares Discriminant Analysis (PLS-DA), and Partial Least Squares (PLS). All of the data were analyzed using XLSTAT Pro 2014 software (Addinsoft, Paris, France), add-in Microsoft Excel.

#### 2.5.1. Principal Component Analysis (PCA)

PCA is unsupervised multivariate data analysis (MVDA) which is used to find a relationship between two or more groups of 'Giant Green' leaves regarding the most variation of those variables. In the PCA technique, the new variables formed and are equal to the number of original variables. The new variables are known as principal components (PCs) and the values of new variables are known as principal component score (PCS). These variables are not correlated with each other. The first new variable, PC1, explains the most information among the samples. Then, PC2 carries the residual information, and so on [20].

2.5.2. Hierarchical Cluster Analysis (HCA)

Hierarchical Cluster Analysis (HCA) is a technique combination of the same characteristic among the samples into one group or cluster. The technique of Ward's method and Euclidean distance was used for grouping 'Giant Green' leaves at different maturity stages into certain classes (clusters).

2.5.3. Partial Least Squares Discriminant Analysis (PLS-DA)

PLS-DA is a supervised method for classifying each sample into predefined classes. PLS-DA is complementary to PCA analysis whereas the separations between the groups of samples have been well improved. In this analysis, the dummy Y-axis (maturity stages) was responsible for separating 'Giant Green' leaves into different clusters in the score plot. The variables such as wavenumber or peak area (x-axis) that contributed to the discrimination among samples were identified from the loading plot. The global goodness of fit and quality model was confirmed by the cumulative $Q^2$, $R^2Y$, and $R^2X$ values. The accuracy and preciseness of the model were detected by using the confusion matrix. The confusion matrix represented the classifying of the observation (in percentage). The value closest to 100% shows a well-classified observation [15]. Besides, the variable importance to projections (VIP) was used to validate the variable contributed to discriminating of samples. A VIP value greater than 0.85 is known as a strong variable. The highest VIP value indicated the most relevant variable that influences the separation between the samples [21].

2.5.4. Partial Least Squares (PLS)

PLS is a supervised multivariate data analysis and is used when complex data with a lot of explanatory variables are involved [22]. Two variables, the dependent variable (Y) and explanatory variable (X) are used. In this study, PLS was applied to find a correlation between the FTIR fingerprint (for spectroscopy) and metabolite (for chromatography), (X) contribution in biological activities (antioxidant and alpha-glucosidase), (Y). The cumulative value of $Q^2$ is >0.5 and $R^2$ is close to 1, indicating a good model [23]. The variable (X) responsible in biological activity was identified by the variable importance in the projection (VIP). Only a VIP value greater than 0.85 indicated that a strong impact on the model was chosen.

## 3. Results

### *3.1. ATR-FTIR Fingerprint and Chemometric Analysis*

3.1.1. Assignment and Comparison of ATR-FTIR Spectra

Similar patterns of IR spectra were shown in old, mature and young leaves (Figure 2). It was observed that the broad peak at 3300 $cm^{-1}$ was assigned to intermolecular hydrogen bond (O-H) of alcohol, phenol, or carboxylic acid groups [24]. Two strong signals of C=O stretching and C-N stretching were present at 1610 $cm^{-1}$ [25] and C-O stretching at 1040 $cm^{-1}$ [26]. The C-O stretching and C-C stretching was detected at 1440 $cm^{-1}$ [27], C-N stretching at 1340 $cm^{-1}$ [21] and C-O stretching or O-H bending at 1204 $cm^{-1}$ [27]. Besides, there was methylene ($CH_2$) stretching presence at 2928 $cm^{-1}$ and 2857 $cm^{-1}$ [24], C=O stretching at 1710 $cm^{-1}$ and [26,27], and C-H out-of-plane bending at 924 $cm^{-1}$ [24]. Two peaks of aromatic group presence at 824 $cm^{-1}$ and 765 $cm^{-1}$ attributed to C-H out-of-plane bending [24]. The peak at around 586 $cm^{-1}$ was assigned to the vibration of O-H out-of-plane bending [28]. The assignment of each peak is summarized in Table 2.

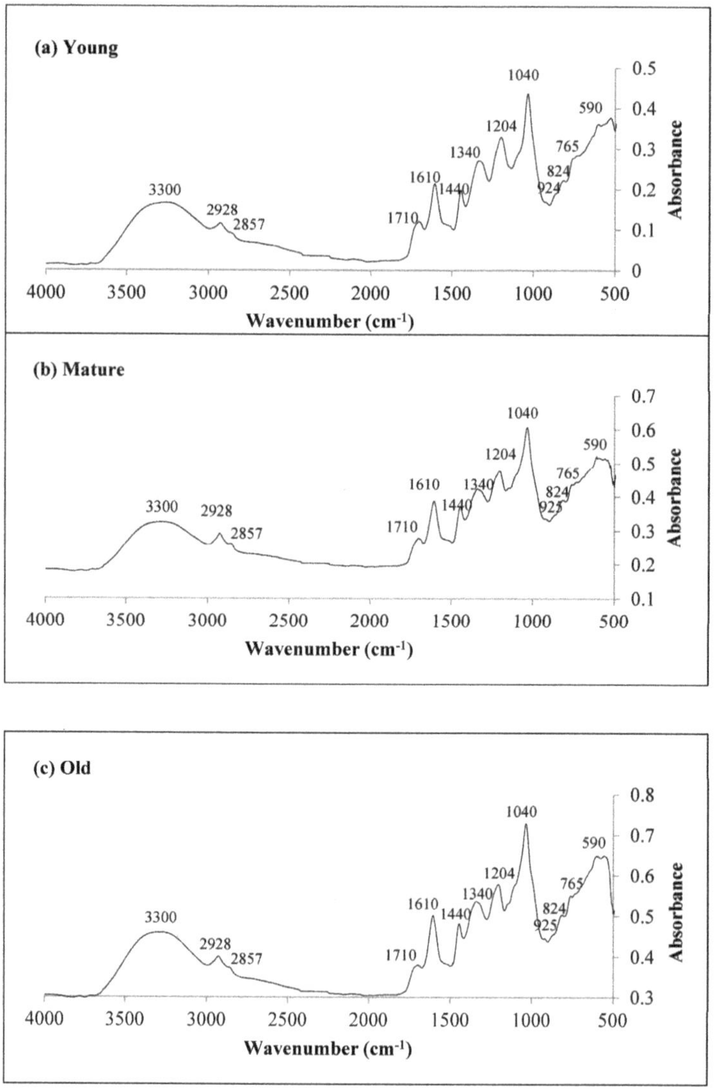


**Figure 2.** FTIR Spectra of 'Giant Green' cultivar of *S. samarangense* leaves at three maturity stages (**a**) Young leaves (**b**) Mature leaves (**c**) Old leaves.

**Table 2.** List of assignments of FTIR for 'Giant Green' cultivar of *S. samarangense* leaves at three maturity stages.

| Frequency Range ($cm^{-1}$) | Assignment of FTIR | Specific Frequency ($cm^{-1}$) | Leaves | | |
|---|---|---|---|---|---|
| | | | YL | ML | OL |
| 3500–3200 | O-H stretching | 3300 | P | P | P |
| 2942–2904 | C-H stretching asymmetric | 2928, 2933 | P | P | P |
| 2863–2846 | C-H stretching symmetric | 2857 | P | P | P |
| 1715–1710 | C=O stretching | 1710, 1715 | P | P | P |
| 1700–1600 | C=O stretching and C-N stretching (amide I) | 1610, 1642 | P | P | P |
| 1580–1510 | N-H bending and C-N stretching (amide II) | 1535 | A | A | A |
| 1450–1380 | C-O stretching and C-C stretching | 1440, 1443 | P | P | P |

**Table 2.** *Cont.*

| Frequency Range ($cm^{-1}$) | Assignment of FTIR | Specific Frequency ($cm^{-1}$) | Leaves | | |
|---|---|---|---|---|---|
| | | | YL | ML | OL |
| 1365–1343 | C-N stretching | 1340, 1345 | P | P | P |
| 1270–1150 | C-O stretching or O-H bending | 1204, 1217 | P | P | P |
| 1052–1035 | C-O stretching | 1040, 1045 | P | P | P |
| 925–910 | C-H out-of-plane bending (alkene) | 925 | P | P | P |
| 826–824 | C-H out-of-plane bending of the aromatic ring (meta) | 824 | P | P | P |
| 773–743 | C-H out-of-plane bending of the aromatic ring (para) | 765, 773 | P | P | P |
| 590–586 | O-H out-of-plane bending of alcohol | 590 | P | P | P |

P = present, A = absent, YL = young leaves, ML = mature leaves, and OL = old leaves.

#### 3.1.2. Chemometric Analysis

##### Principal Component Analysis (PCA)

PCA was performed in this study for unsupervised classification of leaves of the 'Giant Green' cultivar of *S. samarangense* at different maturity stages. Based on the score plot (Figure 3A), the total variance accounting for the first two principal components in the leaves extract was 96.64% (PC1: 50.57%; PC2: 46.07%) (Figure 3B). The model showed a separation between the maturity stages of leaf samples. The interpretation of the score plot within the loading plot gave a clear picture of the factor influencing the clustering of leaf extracts. The loading plots with a score ≥0.75 were accepted as a strong factor. Table 3 shows the variables that contributed to leaf variation along PC1 and PC2. The loading plots of leaf extracts revealed that wavenumbers at 3300, 2928, 1710, 1610, 1440, 1340, 1204, 1040, 924, 824, 765, and 590 $cm^{-1}$ contributed to variation in PC1.

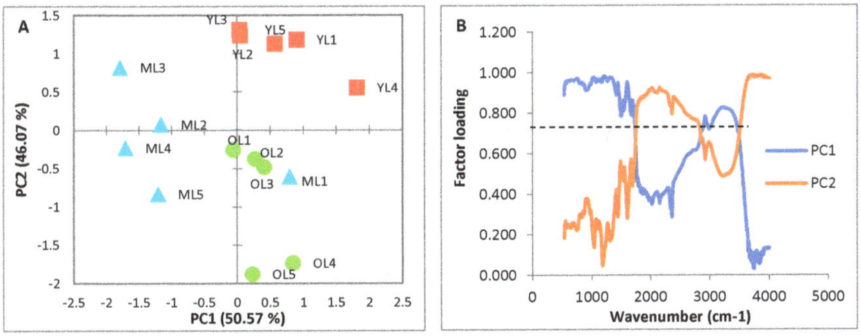


**Figure 3.** PCA-Derived of FTIR spectra representing 'Giant Green' cultivar of *S. samarangense* leaves at three maturity stages (**A**) Score Plot (**B**) Loading Plot of PC1 and PC2.

**Table 3.** Summary of strong loading variables (≥0.75) on the varimax rotation of principal component (PCs) analysis for the 'Giant Green' cultivar of *S. samarangense* leaves.

| Variable | Name of Metabolite | PCs |
|---|---|---|
| 3 | Cyclotetradecane | PC1 |
| 8 | 2,6,11,15-Tetramethylhexadecane | PC2 |
| 9 | 1-Iodododecane | PC2 |
| 10 | 9-Methyl-1-undecene | PC1 |
| 11 | 2-Butyl-1-decene | PC1 |
| 12 | (E)-9-Eicosene | PC1 |
| 14 | Phthalic acid, butyl hept-4-yl ester | PC2 |
| 21 | Phosphonofluoridic acid, methyl-, nonyl ester | PC2 |
| 23 | Diethylene glycol dibenzoate | PC1 |
| 25 | Decanol | PC1 |
| 27 | Hexadecanol | PC1 |
| 29 | Octadecanol | PC1 |
| 32 | 6,10,14-Trimethyl-2-pentadecanone | PC2 |

Hierarchical Cluster Analysis (HCA)

The similarities and differences among the three maturity stages of 'Giant Green' leaves were evaluated with HCA analysis (Figure 4). Three clusters of leaf samples were suggested. Cluster one contained all old leaf replicates and one replicate from the mature leaf sample. Then, cluster two contained the rest of the mature leaf replicates and cluster three contained all young leaf replicates. The sample formed with the same cluster tended to have a high similarity parameter.

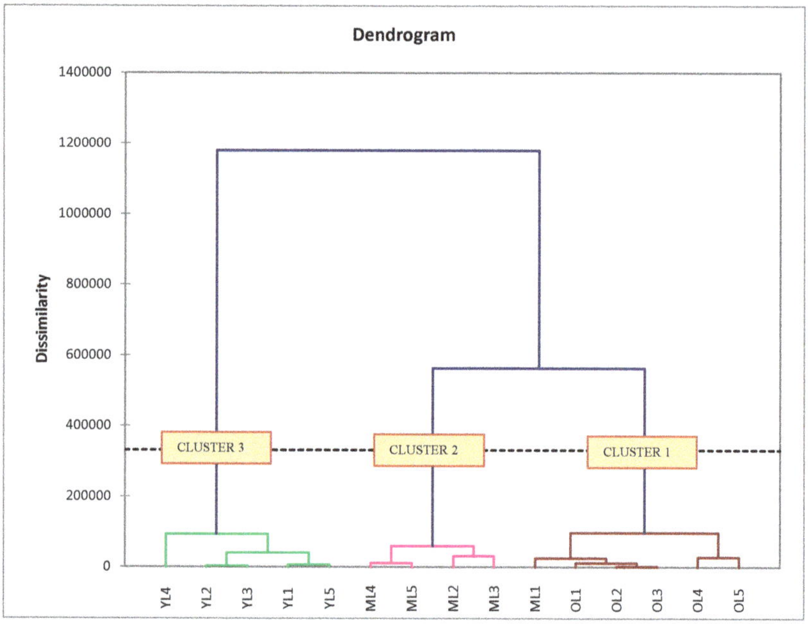


**Figure 4.** Dendrogram from Hierarchical Cluster Analysis (HCA) of 'Giant Green' cultivar of *S. samarangense* leaves at three maturity stages.

Partial Least Square–Discriminant Analysis (PLS-DA)

PLS-DA belongs to the supervised pattern recognition method. It was complementary to PCA analysis. It also had abilities to improve the separation between the groups of samples. As seen in Figure 5A,B, the separation of the three maturity stages of leaf samples was improved. This PLS-DA model had an overall $Q^2$ cumulative of 0.444, $R^2Y$ cumulative of 0.608, and $R^2X$ cumulative of 0.972. The model had a $Q^2$ cumulative value of <0.5, indicating no global goodness of fit. This suggested that the quality of the fit varies a lot depending on the maturity stage of the leaves. Besides, the efficiency of PLS-DA in classifying and discriminating the samples can be accessed through the confusion matrix. The confusion matrix result showed that all of the leaf extracts have been classified with 93.33% of correction. Besides, the variable importance to projections (VIP) was used to validate the variable contributed to the discriminating of samples. Most of the peaks have VIP values greater than 0.85 except 1204 and 1040 $cm^{-1}$ which contributed the highest in discrimination between young, mature, and old leaves. The overall VIP values are shown in Table 4.

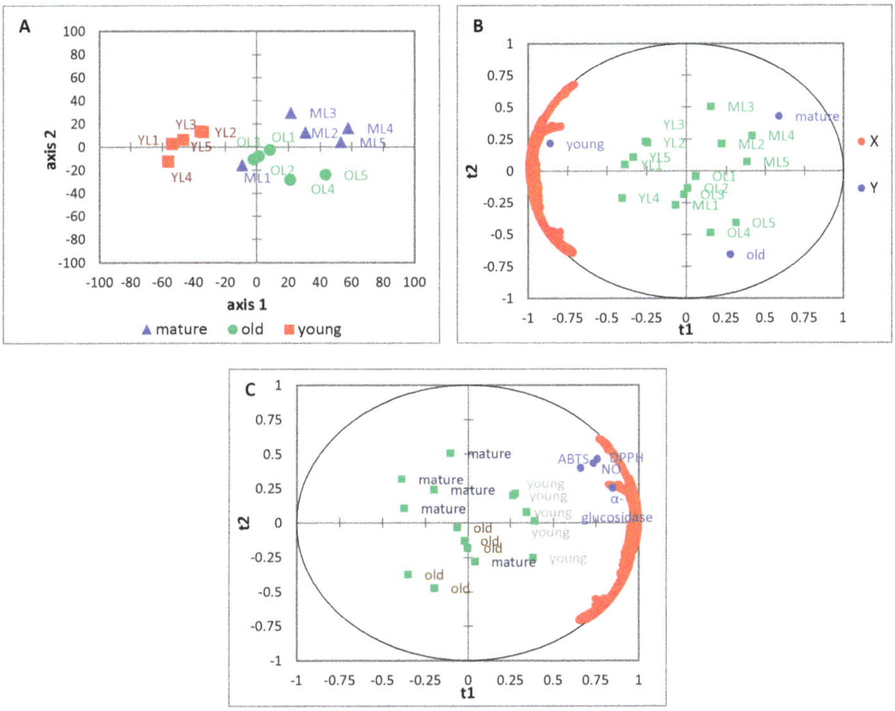


**Figure 5.** The PC1 and PC2 of FTIR results of the 'Giant Green' cultivar of *S. samarangense* leaves at three maturity stages of (**A**) PLS-DA score plot (**B**) PLS-DA bi-plot (X = FTIR wavenumber; Y = leaf maturity stages) (**C**) PLS bi-plot.

**Table 4.** Summary of strong variable importance to projections (VIP) scores (≥0.85) correspond to the partial-least square-discrimination analysis (PLS-DA) and partial-least square analysis (PLS) of 'Giant Green' cultivar of *S. samarangense* leaves at three maturity stages.

| Variables ($cm^{-1}$) | Assignment of FTIR (PLS-DA) | VIP Score | |
|---|---|---|---|
| 2857 | C-H stretching symmetric | 1.11 | |
| 2928 | C-H stretching asymmetric | 1.10 | |
| 1709 | C=O stretching | 1.02 | |
| 924 | C-H out-of-plane bending (alkene) | 1.00 | |
| 3300 | O-H stretching | 0.98 | |
| 1439 | C-O stretching and C-C stretching | 0.96 | |
| 824 | C-H out-of-plane bending of the aromatic ring (meta) | 0.96 | |
| 1609 | C=O stretching and C-N stretching (amide I) | 0.94 | |
| 764 | C-H out-of-plane bending of the aromatic ring (para) | 0.90 | |
| 590 | O-H out-of-plane bending of alcohol | 0.89 | |
| 1339 | C-N stretching | 0.85 | |
| **Variables ($cm^{-1}$)** | **Assignment of FTIR (PLS)** | **VIP Score** | **PCs** |
| 2857 | C-H stretching symmetric | 1.0 | PC1 |
| 2928 | C-H stretching asymmetric | 0.89 | PC1 |
| 3300 | O-H stretching | 0.87 | PC1 |

Partial Least Square (PLS)

PLS was established to investigate the relationship between bioactivities of antioxidants and alpha-glucosidase (Y variable) with FTIR fingerprint (x variable). The details about the antioxidant and alpha-glucosidase inhibitory activities of 'Giant Green' leaves at three maturity stages were reported in our previous study [18]. As seen in the bi-plot (Figure 5C), most young leaf samples were located at the upper right-hand quadrant on the t1 axis. This revealed that the young leaves possessed the strongest bioactivities as compared to mature and old leaves. This analysis showed a good PLS prediction model with the cumulative values of $Q^2$ at 0.591, $R^2Y$ at 0.724, and $R^2X$ at 0.972. The peaks related to this relationship were evaluated based on the loading plot (w*c). The influencer peaks were 2857, 2928, and 3300 $cm^{-1}$ which were detected to have the highest w*c [1] values. This result also aligned with data from VIP coefficients where the peak at 2857 $cm^{-1}$ had the highest VIP value than other peaks. The overall VIP values are shown in Table 4.

### *3.2. Gas Chromatography-Mass Spectrometry (GC-MS) and Chemometric Analysis*

#### 3.2.1. Assignment and Comparison of GC-MS Spectra

Based on the chromatogram shown in Figure 6a–c, the total number of metabolites identified in leaf extracts was 37. The quantities of metabolites that were found in each of the samples were not the same, whereby in young leaves it was 25 (Figure 6a), in mature leaves it was 29 (Figure 6b) and in old leaves it was 27 (Figure 6c). Six major metabolites were detected in all of the maturity stages of leaf samples which were aligned in similar retention times. These major metabolites were cyclotetradecane; octadecanol; 5(2,4-di-tert-butylphenoxy)-5-oxopentanoic acid; methoxyl; hexadecanol; and demethoxymatteucinol. Moreover, some metabolites were detected only in one or two leaf samples, while some of them were also present in all the samples but differed in peak intensities. For example, the metabolites of methyl (9Z,15Z)-9,15-octadecadienoate and stearic acid, butyl ester were only detected in young leaves but 2,6,11,15-tetramethylhexadecane, 1-iodododecane and phytol were detected in mature and old leaves. The detailed information on these metabolites is tabulated in Table 5.

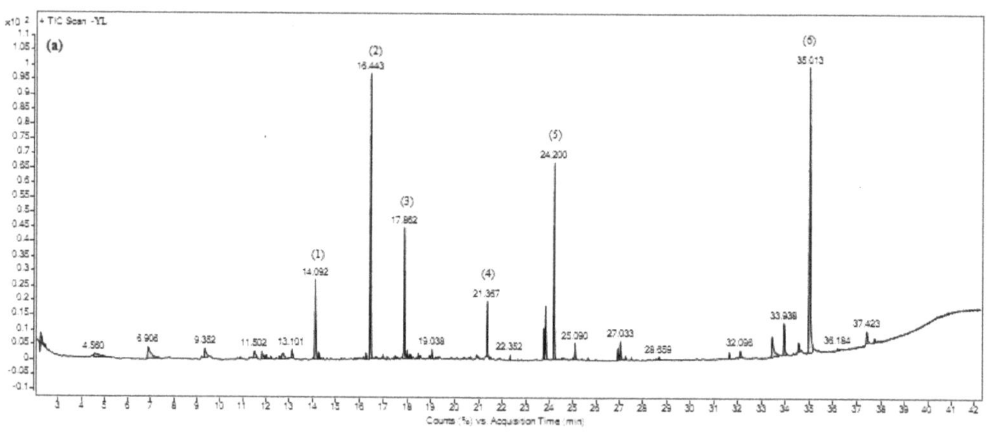


**Figure 6.** *Cont.*

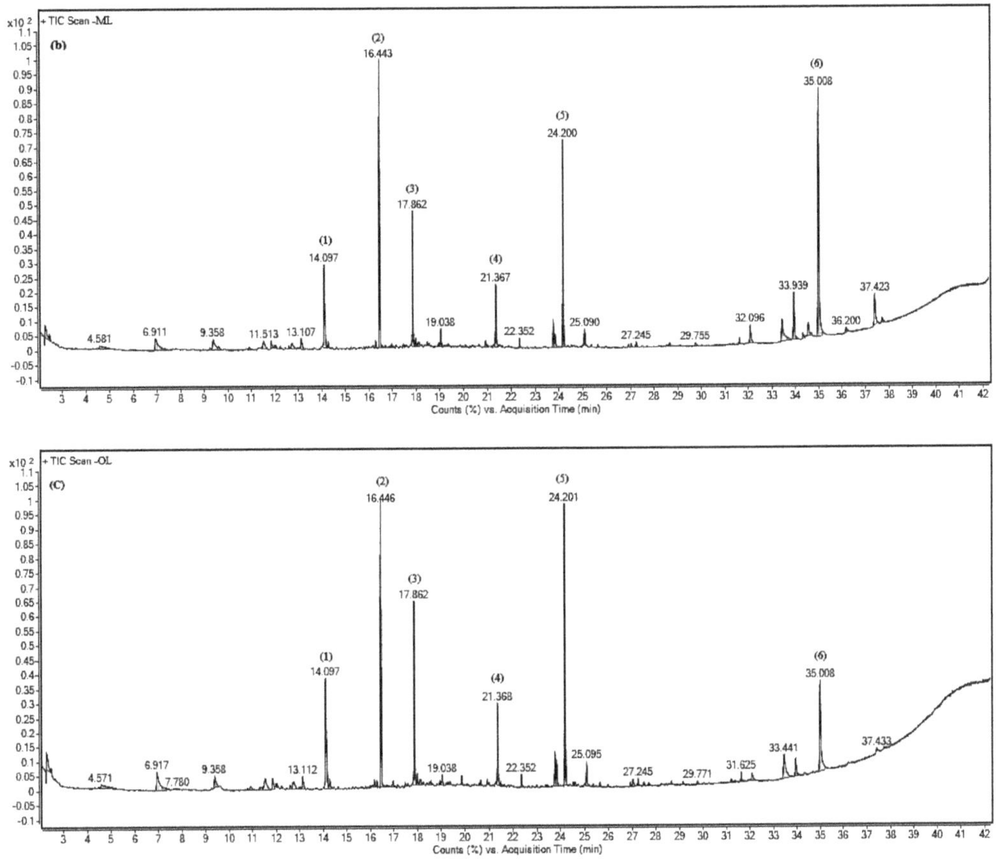


**Figure 6.** GC-MS chromatogram of major volatile metabolites present in 'Giant Green' cultivar of *S. samarangense* leaves at three maturity stages (**a**) young leaves (**b**) mature leave (**c**) old leaves; (1): cyclotetradecane; (2): 5(2,4-di-tert-butylphenoxy)-5-oxopentanoic acid; (3): hexadecanol; (4): octadecanol; (5): methoxyl; and (6): demethoxymatteucinol.

**Table 5.** List of metabolites present in the 'Giant Green' cultivar of *S. samarangense* leaves at three maturity stages with a percentage of probability score of 80% and above.

| Group | Variable | Name of Metabolite | RT (Min) | Molecular Formula | Relative Area (%) | | |
|---|---|---|---|---|---|---|---|
| | | | | | YL | ML | OL |
| Alkane | 1 | Decane | 4.57 | $C_{10}H_{22}$ | 0.94 ± 0.16 [a] | 0.73 ± 0.25 [a] | 0.39 ± 0.53 [a] |
| | 2 | 2,3-Dimethyloctane | 12.73 | $C_{10}H_{22}$ | 0.09 ± 0.21 [a] | 0.09 ± 0.20 [a] | 0.10 ± 0.22 [a] |
| | 3 | Cyclotetradecane | 14.1 | $C_{14}H_{28}$ | 5.79 ± 0.82 [a] | 4.78 ± 0.54 [a] | 4.57 ± 1.21 [a] |
| | 4 | Tetradecane | 14.26 | $C_{14}H_{30}$ | 0.40 ± 0.07 [a] | 0.26 ± 0.15 [ab] | 0.14 ± 0.91 [b] |
| | 5 | 3,5-Dimethylundecane | 16.97 | $C_{13}H_{28}$ | 0.09 ± 0.13 [a] | 0.05 ± 0.11 [a] | 0.24 ± 0.31 [a] |
| | 6 | 4,6-Dimethyldodecane | 16.15 | $C_{14}H_{30}$ | ND | 0.06 ± 0.14 [a] | ND |
| | 7 | 6-Ethyl-2-methyldecane | 17.99 | $C_{13}H_{28}$ | 0.06 ± 0.14 [a] | ND | ND |
| | 8 | 2,6,11,15-Tetramethylhexadecane | 19.86 | $C_{20}H_{42}$ | ND | 0.12 ± 0.27 [a] | 0.22 ± 0.34 [a] |
| | 9 | 1-Iodododecane | 20.66 | $C_{12}H_{25}I$ | ND | 0.05 ± 0.12 [a] | 0.05 ± 0.11 [a] |
| Alkene | 10 | 9-Methyl-1-undecene | 11.52 | $C_{12}H_{24}$ | 0.91 ± 0.16 [a] | 0.78 ± 0.08 [a] | 0.64 ± 0.30 [a] |
| | 11 | 2-Butyl-1-decene | 13.11 | $C_{14}H_{28}$ | 0.74 ± 0.10 [a] | 0.61 ± 0.08 [a] | 0.51 ± 0.29 [a] |
| | 12 | (E)-9-Eicosene | 25.09 | $C_{20}H_{40}$ | 0.90 ± 0.17 [a] | 0.79 ± 0.09 [a] | 0.66 ± 0.38 [a] |
| Ether | 13 | Decyl octyl ether | 11.83 | $C_{18}H_{38}O$ | 0.12 ± 0.28 [a] | 0.10 ± 0.22 [a] | 0.20 ± 0.27 [a] |
| Ester | 14 | Phthalic acid, butyl hept-4-yl ester | 24.54 | $C_{19}H_{28}O_4$ | ND | 0.03 ± 0.07 [a] | 0.04 ± 0.08 [a] |
| | 15 | Methyl benzoate | 6.92 | $C_8H_8O_2$ | 2.00 ± 0.22 [a] | 1.76 ± 0.49 [a] | 1.65 ± 0.15 [a] |
| | 16 | Bis(2-ethylhexyl) carbonate | 10.9 | $C_{17}H_{34}O_3$ | 0.37 ± 0.71 [a] | ND | 0.07 ± 0.10 [b] |
| | 17 | 5(2,4-Di-tert-butylphenoxy)-5-oxopentanoic acid | 16.44 | $C_{19}H_{28}O_4$ | 10.43 ± 1.20 [a] | 8.86 ± 1.17 [a] | 8.89 ± 1.92 [a] |
| | 18 | Methyl palmitate | 23.85 | $C_{17}H_{34}O_2$ | 2.00 ± 1.04 [a] | 0.49 ± 0.30 [b] | 0.50 ± 0.46 [b] |
| | 19 | Methylox | 24.20 | $C_{18}H_{28}O_3$ | 8.80 ± 1.23 [a] | 7.40 ± 0.62 [a] | 7.95 ± 0.73 [a] |
| | 20 | Methyl (9Z,15Z)-9,15-octadecadienoate | 26.92 | $C_{19}H_{36}O_2$ | 0.07 ± 0.16 [a] | ND | ND |

**Table 5.** *Cont.*

| Group | Variable | Name of Metabolite | RT (Min) | Molecular Formula | Relative Area (%) | | |
|---|---|---|---|---|---|---|---|
| | | | | | YL | ML | OL |
| | 21 | Phosphonofluoridic acid, methyl-, nonyl ester | 27.03 | $C_{12}H_{26}FO_2P$ | ND | ND | $0.04 \pm 0.08^{a}$ |
| | 22 | Stearic acid, butyl ester | 31.95 | $C_{22}H_{44}O_2$ | $0.21 \pm 0.46^{a}$ | ND | ND |
| | 23 | Diethylene glycol dibenzoate | 33.44 | $C_{18}H_{18}O_5$ | $2.38 \pm 0.53^{a}$ | $2.23 \pm 0.33^{a}$ | $1.84 \pm 1.07^{a}$ |
| | 24 | 4-Methylhexanol | 12.02 | $C_7H_{16}O$ | $0.09 \pm 0.13^{a}$ | $0.04 \pm 0.10^{a}$ | ND |
| Alcohol | 25 | Decanol | 9.36 | $C_{10}H_{22}O$ | $1.35 \pm 0.23^{a}$ | $1.13 \pm 0.08^{a}$ | $0.99 \pm 0.56^{a}$ |
| | 26 | Dodecanol | 14.12 | $C_{12}H_{26}O$ | $7.55 \pm 6.68^{a}$ | $10.16 \pm 6.84^{b}$ | $4.83 \pm 5.08^{c}$ |
| | 27 | Hexadecanol | 17.86 | $C_{16}H_{34}O$ | $6.07 \pm 0.89^{a}$ | $5.04 \pm 0.54^{a}$ | $4.92 \pm 1.09^{a}$ |
| | 28 | Intermedeol | 19.04 | $C_{15}H_{26}O$ | ND | $0.33 \pm 0.32^{a}$ | $0.11 \pm 0.24^{a}$ |
| | 29 | Octadecanol | 21.37 | $C_{18}H_{38}O$ | $3.06 \pm 0.48^{a}$ | $2.61 \pm 0.28^{a}$ | $2.47 \pm 0.69^{a}$ |
| | 30 | Phytol | 27.25 | $C_{20}H_{40}O$ | ND | $0.15 \pm 0.33^{a}$ | $0.05 \pm 0.12^{b}$ |
| Ketone | 31 | 2-(1,1-Dimethylethyl)-cyclobutanone | 18.1 | $C_8H_{14}O$ | ND | ND | $0.04 \pm 0.10^{a}$ |
| | 32 | 6,10,14-Trimethyl-2-pentadecanone | 22.35 | $C_{18}H_{36}O$ | ND | $0.13 \pm 0.18^{a}$ | $0.23 \pm 0.34^{a}$ |
| | 33 | 7,9-Di-tert-butyl-1-oxaspiro(4,5)deca-6,9-diene-2,8-dione | 23.77 | $C_{17}H_{24}O_3$ | $1.32 \pm 0.30^{a}$ | $0.96 \pm 0.21^{ab}$ | $0.69 \pm 0.47^{b}$ |
| | 34 | Pinostrobin chalcone | 32.1 | $C_{16}H_{14}O_4$ | ND | $0.54 \pm 0.53^{a}$ | $0.28 \pm 0.38^{a}$ |
| | 35 | Pinocembrin | 33.94 | $C_{17}H_{16}O_4$ | $0.72 \pm 0.72^{a}$ | $1.66 \pm 0.98^{b}$ | $0.78 \pm 0.67^{a}$ |
| | 36 | Demethoxymatteucinol | 35.02 | $C_{17}H_{16}O_4$ | $0.03 \pm 0.07^{a}$ | ND | ND |
| Amine | 37 | 5-Aminotetrazole | 9.58 | $CH_3N_5$ | $1.30 \pm 2.90^{a}$ | ND | $0.51 \pm 1.14^{b}$ |

ND = not detected. YL = young leaves; ML = mature leaves; OL = old leaves. Values are the means ± standard deviation for five biological replicates of experiments (n = 5). Data from the same horizontal row with different superscript letters refer to a significant difference ($p < 0.05$).

### 3.2.2. Chemometric Analysis

Principal Component Analysis (PCA)

The PCA analysis was established to give better information about the similarities and differences between the three maturity stages of leaves of 'Giant Green' in the context of their metabolites. It can be seen from Figure 7A, the score plot of leaf samples revealed that the total variance of the first two principal components was 43.50% with values of PC1 at 27.79% and PC2 at 15.71%. The model showed no separation between the maturity stages of leaf samples. The loading line plot (Figure 7B) of leaf samples identified the metabolites contributing to the variation in PC1 that were 3, 10, 11, 12, 23, 25, 27, and 29. Meanwhile, the metabolites 8, 9, 14, 21, and 32 contributed to PC2. The details about the metabolites are explained in Table 6.

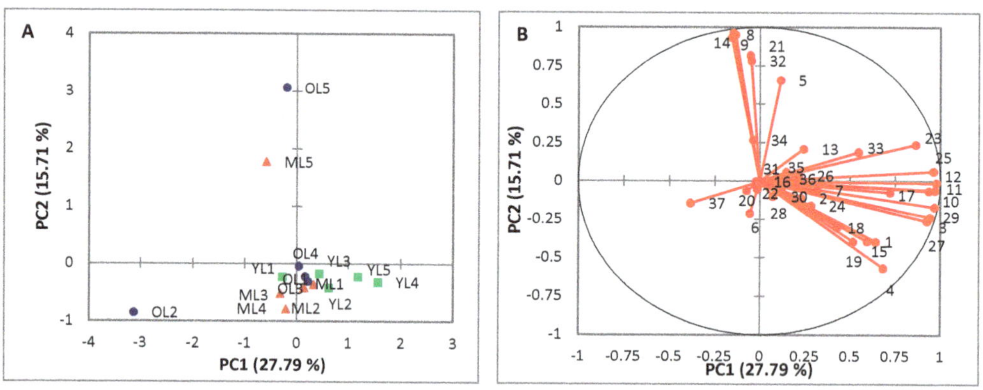


**Figure 7.** PCA-Derived of the GC-MS result representing the 'Giant Green' cultivar of *S. samarangense* leaves at three maturity stages (**A**) Score Plot (**B**) Loading Plot of PC1 and PC2.

**Table 6.** Summary of strong loading variables (≥0.75) on the varimax rotation of principal component (PCs) analysis for the 'Giant Green' cultivar of *S. samarangense* leaves.

| Variable | Name of Metabolite | PCs |
|---|---|---|
| 3 | Cyclotetradecane | PC1 |
| 8 | 2,6,11,15-Tetramethylhexadecane | PC2 |
| 9 | 1-Iodododecane | PC2 |
| 10 | 9-Methyl-1-undecene | PC1 |
| 11 | 2-Butyl-1-decene | PC1 |
| 12 | (E)-9-Eicosene | PC1 |
| 14 | Phthalic acid, butyl hept-4-yl ester | PC2 |
| 21 | Phosphonofluoridic acid, methyl-, nonyl ester | PC2 |
| 23 | Diethylene glycol dibenzoate | PC1 |
| 25 | Decanol | PC1 |
| 27 | Hexadecanol | PC1 |
| 29 | Octadecanol | PC1 |
| 32 | 6,10,14-Trimethyl-2-pentadecanone | PC2 |

Hierarchical Cluster Analysis (HCA)

The cluster analysis of 'Giant Green' leaves was illustrated clearly in the HCA dendrogram. Based on GC-MS data, the samples were grouped into three groups. The result of HCA revealed that all of the leaf samples in similar maturity stages formed a heterogeneous cluster (Figure 8). Cluster one represented three replicates from old leaves and young leaves and two replicates from mature leaves. Cluster two represented two replicates from mature and young leaves and one replicate from old leaves. Cluster three represented one replicate from mature and old leaves.

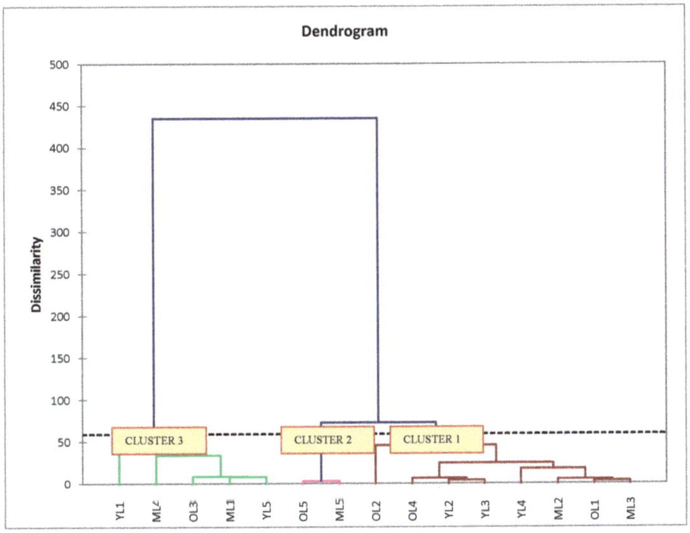


**Figure 8.** Dendrogram from Hierarchical Cluster Analysis (HCA) corresponded to GC-MS data for the 'Giant Green' cultivar of *S. samarangense* leaves at three maturity stages.

Partial Least Square-Discriminant Analysis (PLS-DA)

PLS-DA analysis was done to improve the separation between leaf samples obtained from PCA results. Pattern recognition of PLS-DA from leaves extract was carried out and is shown in Figure 9A,B. The results revealed that the separation between the maturity stages of leaf samples had improved. The model had a $Q^2$ cumulative of 0.875, $R^2Y$ cumulative of 1.000, and $R^2X$ cumulative of 0.959. This model was indicated as a good model because it had a $Q^2$ cumulative value greater than 0.5. The result of the confusion matrix also showed that young, mature, and old leaves were discriminated efficiently with a 100% correctly produced overall classification rate with no misclassified sample. Based on the results of the variable importance to projections (VIP), 24 metabolites were identified that influenced the separation of young, mature, and old leaf samples. These metabolites were 18, 33, 4, 3, 27, 1, 19, 11, 29, 17, 15, 10, 34, 25, 32, 24, 12, 16, 20, 22, 36, 7, 8, and 28. The overall result is summarized in Table 7.

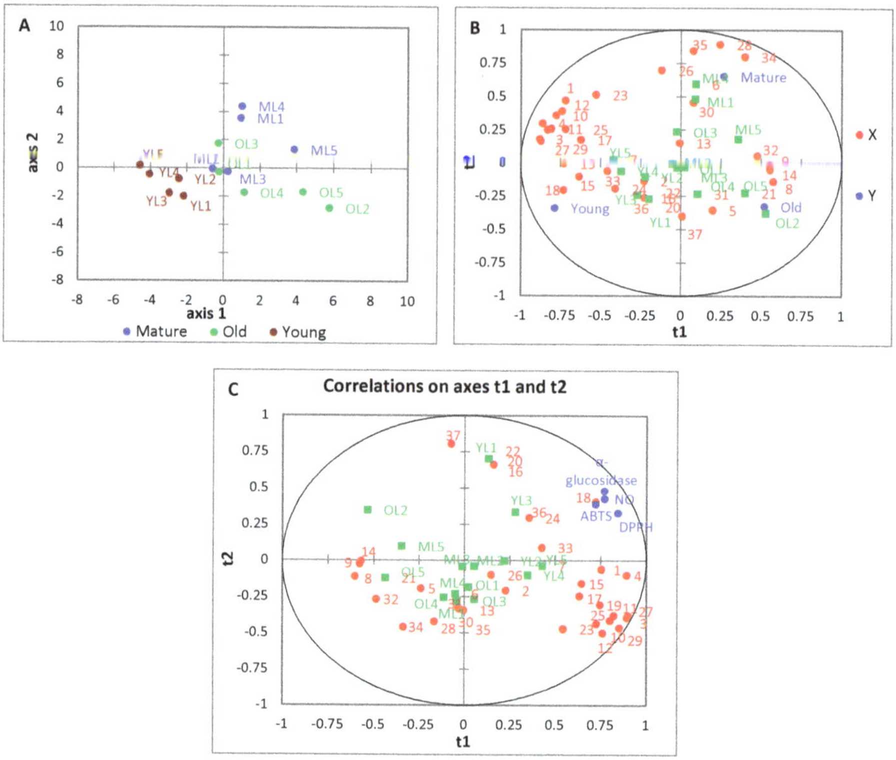


**Figure 9.** The PC1 and PC2 of GC-MS results of the 'Giant Green' cultivar of *S. samarangense* leaves at three maturity stages of (**A**) PLS-DA score plot (**B**) PLS-DA bi-plot (X = metabolites; Y = leaf maturity stages) (**C**) PLS bi-plot.

**Table 7.** Summary of strong variable importance to projections (VIP) scores (≥0.85) correspond to the partial-least square-discrimination analysis (PLS-DA) of the 'Giant Green' cultivar of *S. samarangense* leaves at three maturity stages.

| Variable | Name of Metabolite | VIP Score |
|---|---|---|
| 18 | Methyl palmitate | 1.87 |
| 33 | 7,9-Di-tert-butyl-1-oxaspiro(4,5)deca-6,9-diene-2,8-dione | 1.58 |
| 4 | Tetradecane | 1.51 |
| 3 | Cyclotetradecane | 1.39 |
| 27 | Hexadecanol | 1.38 |
| 1 | Decane | 1.33 |
| 19 | Methylox | 1.22 |
| 11 | 2-Butyl-1-decene | 1.21 |
| 29 | Octadecanol | 1.21 |
| 17 | 5(2,4-Di-tert-butylphenoxy)-5-oxopentanoic acid | 1.20 |

Table 7. *Cont.*

| Variable | Name of Metabolite | VIP Score |
|---|---|---|
| 15 | Methyl benzoate | 1.14 |
| 10 | 9-Methyl-1-undecene | 1.09 |
| 34 | Pinostrobin chalcone | 1.07 |
| 25 | Decanol | 1.05 |
| 32 | 6,10,14-Trimethyl-2-pentadecanone | 1.04 |
| 24 | 4-Methylhexanol | 1.00 |
| 12 | (E)-9-Eicosene | 0.96 |
| 16 | Bis(2-ethylhexyl) carbonate | 0.95 |
| 20 | Dodecanol | 0.94 |
| 22 | Stearic acid, butyl ester | 0.94 |
| 36 | Demethoxymatteucinol | 0.94 |
| 7 | 6-Ethyl-2-methyldecane | 0.94 |
| 8 | 2,6,11,15-Tetramethylhexadecane | 0.89 |
| 28 | Intermedeol | 0.86 |

Partial Least Square (PLS)

Variation of metabolites between the leaves of 'Giant Green' at different maturity stages and their correlation with bioactivities (antioxidant and alpha-glucosidase inhibitory activities) was evaluated using PLS analysis. The details about these bioactivities were reported in our previous study [18]. As seen in the biplot (Figure 9C), the young leaves were located at the upper right-hand quadrant on the t1 axis which is influencing a strong relationship with antioxidant and alpha-glucosidase activities compared to mature and old leaves samples. Besides, it also revealed that this PLS model was good. It had the values of $Q^2$ cumulative at 0.903, $R^2Y$ cumulative at 1.000, and $R^2X$ cumulative at 0.980. The metabolites' correlations with bioactivities in leaf samples were deduced from the loading plot; 17 metabolites were identified that contributed to this relationship. These metabolites were 18, 4, 1, 3, 33, 27, 11, 15, 29, 10, 36, 19, 17, 25, 12, 7, and 24. The variable importance to projections (VIP) was further examined to validate the significance variable. This result also aligned with data from VIP coefficients where the metabolite (18) (rt: 23.85) had the highest VIP value than other peaks which indicates it is the highest influencer to antioxidant and alpha-glucosidase activities. The details about the metabolites are summarized in Table 8.

**Table 8.** Summary of strong loading variables that correspond to the partial-least square analysis (PLS) of the 'Giant Green' cultivar of *S. samarangense* leaves at three maturity stages.

| Variable | Name of Metabolite | VIP Score | PCs |
|---|---|---|---|
| 18 | Methyl palmitate | 1.89 | PC1 |
| 4 | Tetradecane | 1.65 | PC1 |
| 1 | Decane | 1.50 | PC1 |
| 3 | Cyclotetradecane | 1.40 | PC1 |
| 33 | 7,9-Di-tert-butyl-1-oxaspiro(4,5)deca-6,9-diene-2,8-dione | 1.39 | PC1 |
| 27 | Hexadecanol | 1.38 | PC1 |
| 11 | 2-Butyl-1-decene | 1.27 | PC1 |
| 15 | Benzoic acid, methyl ester | 1.24 | PC1 |
| 29 | Octadecanol | 1.23 | PC1 |
| 10 | 9-Methyl-1-undecene | 1.20 | PC1 |
| 36 | Demethoxymatteucinol | 1.16 | PC1 |
| 19 | Methylox | 1.14 | PC1 |
| 17 | 5(2,4-Di-tert-butylphenoxy)-5-oxopentanoic acid | 1.09 | PC1 |
| 25 | Decanol | 1.07 | PC1 |
| 12 | (E)-9-Eicosene | 1.04 | PC1 |
| 7 | 6-Ethyl-2-methyldecane | 0.99 | PC1 |
| 24 | 4-Methylhexanol | 0.85 | PC1 |

## 4. Discussion

In the present study, the fingerprints of old, mature, and young leaves were found to be similar but slightly different in peak intensity and this might be due to the presence of the same type or quantity of metabolites. The similarities between the absorbance peaks also are related to the insignificance among leaf stages in antioxidant and antibacterial activities reported in our previous study. The peak at 1710 $cm^{-1}$ which is assigned to the C=O in the phenolic group is believed to have contributed to these bioactivities. Besides, this is in agreement with the finding by Easmin et al. [24], where they found that FTIR spectra for ethanol and water extracts of *Phaleria macrocarpa* fruit look similar because of the similarity in their chemical composition. Besides, the variation of peak detected between the leaf samples also might be related to different levels of enzyme activities for each maturity stage [21]. In addition, Kharbach et al. [29] reported that the resulting compound fingerprint is mostly related to plant maturity, variation of season, and location of geographic. However, the comparison among spectra only cannot provide the final conclusion about the specific fingerprint that contributed to variations between 'Giant Green' leaves at three maturity stages. So, the data of FTIR was further analyzed and subjected to chemometric analysis.

Principal Component Analysis (PCA) is unsupervised multivariate data analysis (MVDA) that is used to reduce the dimensional large dataset and at the same time has preserved important information. The most important information from the dataset is explained in PC1 and the second most important information is explained in PC2. The score plot was used to differentiate among the samples and the loading plot was used to determine the variable contributed to the samples cluster. In the present study, the PCA was established to find the relationship between 'Giant Green' leaves at three maturity stages and identify the functional group that contributed to the sample separation. The PCA result showed that the young leaves were clearly separated from mature and old leaves. But one biological replicate from mature leaves (ML1) is located near the old leaf samples. This might be why some metabolites in ML1 are also present in old leaf samples or their quantity is almost similar. Some of the previous studies also found no defined cluster between *Ficus deltodeia* syconia varieties [30], cabbage cultivar [31], and *Ipomoea aquatica* [32] because of the identicalness of their chemical contents. Besides, *Eugenia uniflora* leaves showed a clear distance of cluster between the different fruit color biotypes due to their distinctive volatile compounds [33].

Unsupervised Hierarchical Cluster Analysis (HCA) is complementary to PCA analysis. HCA was applied to determine the similarities and dissimilarities between the individual experimental samples. The sample with similar in investigated variable matched in the same cluster but the sample showed the highest dissimilarity was arranged in other clusters. The position of the cluster in the dendrogram also takes into account the far position among clusters that shows the highest dissimilarity between the individual samples [34]. The results obtained from this study demonstrate that the leaf extracts at three maturity stages were arranged in three different clusters and might be influenced by metabolite biosynthesis. Lee et al. [35] reported that metabolite presence varies at the young, mature, and old stages of *S. samarangense* cv. pink leaves. The HCA model also revealed that most leaf extracts present in the same maturity stages formed a homogeneous cluster. It was expected that the samples with the same maturity stage was similar because they consisted of metabolites of the same quality and quantity. However, one of the mature leaves (ML1) deviated away from other mature leaf samples, but arranged in cluster 1 belonging to old leaf samples, indicating that the metabolite can also develop differences among leaves at the same maturity stage. This phenomenon also might be affected by environmental factors, cultivar practices, plant ages, and soil factors of wax apple cultivar. The environmental factors such as temperature, light intensities, and climatic change influence the changes of metabolites in plants [36].

Partial Least Square-Discriminant Analysis (PLS-DA) is a supervised multivariate data analysis (MVDA) tool that is gaining more interest, especially in the analysis of metabolomics data. PLS-DA has the capability to improve the classification of experimental data that cannot be achieved by using PCA. Unlike PCA, PLS-DA is focused on class reductive in achieving the separation between the samples. In this study, PLS-DA is performed to discriminate the 'Giant Green' leaves at three maturity stages based on their FTIR dataset. In good accord with PCA and HCA analysis, one of the samples from the mature leaf stage that is ML1 cannot clearly be separated from the samples of the old leaf stage might be due to the similarities of metabolites in both samples. However, the other samples from young, mature, and old leaves were well improved in their separation than in PCA. The prediction ability of this PLS-DA model in the classification of the leaf samples with different maturity stages has been validated by the achievement of a 93.33% score in the confusion matrix. The maximum data among class (maturity stage) were collected in PLS-DA analysis, making the variable that discriminates in this model may be unlike those with PCA. Most of the discriminative variables detected in FTIR spectra contributed to the classification of young, mature, and old leaves according to variable importance to projection (VIP) coefficients. The highest VIP score represented the stronger variable attributed to the clustering of the sample. The peaks at 2857, 2928, 1709, and 924 $cm^{-1}$ were mostly related to describing the differences between the maturity stages of leaf samples. The previous literature reported that the asymmetrical (2970 $cm^{-1}$) and symmetrical (2856 $cm^{-1}$) of the methylene group ($CH_2$) and C=H bond (980 $cm^{-1}$) in the FTIR spectrum were commonly related to flavonoid structure [37]. Then, the peak around 1718 $cm^{-1}$ could be attributed to the presence of the ester compound [26]. This finding revealed that the flavonoid and ester compounds could be the largest influencer in separating between maturity stages in leaf samples. So, it can be concluded that the variation of metabolites could be attributed to discrimination between three maturity stages of 'Giant Green' leaves. In good accordance with the previous study, as reported by Lee et al. [35], the chemical compounds such as terpene and terpenoid of wax apple leave cv. pink changes during the maturation stages. Other than that, Gouvinhas et al. [38] reported that the oil from three stages of olive fruit was successfully discriminated by using supervised MVDA. They found that the changes in biochemicals happened with the ripening stages of the olive fruit. Considering the discrimination explained in the PLS-DA model, it is proven that the 'Giant Green' leaves were well-classified according to their maturity stages than using PCA.

Partial Least Square (PLS) belongs to supervised MVDA where the Y-axis represented a dependent variable and X-axis represented an independent variable in the PLS model. PLS is used to find the correlation between the two variables that are generated from the dataset of spectroscopic or chromatographic analysis and bioactivity. The validation and prediction of the goodness of the model are evaluated based on $R^2Y$ (variance explained in predictor variable), $R^2X$ (variance explained in response variable), and $Q^2Y$ (variance predictive of the goodness of fit according to cross-validation). A cross-validated correlation coefficient ($Q^2$) value higher than 0.5 indicates a good PLS model. In the current study, the relationship between FTIR spectra absorbance (wavenumber) with biological activities such as antioxidant (DPPH, NO, and ABTS) and alpha-glucosidase inhibitory activity were investigated. However, the information accessed from FTIR was limited because it just provided a clue about the class of metabolite but the specific metabolite that is responsible for activeness in biological activities is still unknown. The relationship between biological activity (Y-axis) and wavenumber (X-axis) of leaf samples at three maturity stages was illustrated in the bi-plot. Bi-plot was the combination of a score plot and a loading plot. Based on the present results, the bi-plot of leaf extracts revealed that the Y-variables (DPPH, NO, ABTS, and alpha-glucosidase) were located near the sample of the young leaf stage. It revealed that young leaf samples were highly correlated with biological activities. The strongest peaks were obtained at 2857, 2928, and 3300 $cm^{-1}$ which possessed the highest value in the loading plot and VIP score and may be responsible for antioxidant and alpha-glucosidase inhibitory activities of young leaves. The peak at 2857 $cm^{-1}$ and 2928 $cm^{-1}$ may be due to methylene stretching of asymmetrical and symmetrical vibration in methoxyl derivative and aldehyde group, and at 3300 $cm^{-1}$ may be assigned to intermolecular hydrogen bond in alcohol, phenol or carboxylic acid. These peaks showed that the possibility of primary metabolites such as carbohydrates, proteins, lipids, and polysaccharides, and secondary metabolites such as phenolic acids, flavonoid, terpenes, and terpenoids were present in the leaf sample. In good accordance with a previous study as reported by Christou et al. [39] where they found that the most important peaks in the FTIR spectrum were at the 4000–2500 $cm^{-1}$ which indicates the presence of carbohydrate, protein, lipid, and polysaccharide groups. Saidan et al. [40] also revealed that the sharp peak in the range of 1760–1600 $cm^{-1}$ may be characterized by the presence of flavonoid and terpenoid groups. In addition, the leaves of *S. samarangense* have been reported with an abundance of valuable metabolites such as quercetin, ellagic acid, myricetin, lupeol, sitosterol, triterpenes, betulin, *p*-cymene α-pinene, β-pinene and limonene [41]. These metabolites have been proven to have a significant effect on bioactivities such as antioxidant and alpha-glucosidase inhibitory activities [41–43].

Gas chromatography (GC) is the most intensive instrument used for separation of compounds in a mixture [44]. It becomes the crucial tool in identification of compounds especially in drug discovery or pharmacology and food industry fields. In this study, 37 compounds were detected in three maturity stages of 'Giant Green' leaf extracts. However, only six major compounds were identified and present at the same retention time in all of the leaf extracts. The variation of metabolites in the samples may influent their potency in biological activities. Thus, the strongest antioxidant, antibacterial and alpha-glucosidase activities of 'Giant Green' leaves in our previous study [18] could be related to the greater number of metabolites present in each of the leaf extracts. Some of the metabolites from classes of phenolic, triterpenes, ester, alkane, and carbohydrate have been given more attention by researchers because these metabolites can exhibit various pharmacological activities [45–47]. Previous literature had reported that alkane-based compounds such as tetradecane, hexadecane [48], and nonadecane [49] showed antibacterial and antifungal effects. The presence of metabolites such as methyl benzoate; methyl (9Z,15Z)-9,15-octadecadienoate [50,51], diethylene glycol dibenzoate [52] and 9-Eicosene [53] also have potent antibacterial activity. Other than that, Saleh et al. [46] reported that the metabolites based on fatty acid, organic acid, phenolic acid, carbohydrate, alkane, and sterol may possess alpha-glucosidase inhibitory activity. Fatty acids such as palmitic acid and stearic acid were known to exhibit potent alpha-glucosidase inhibitory activity [54,55] as well as possess strong antioxidant and antibacterial, antitumor, anticholesteremic, immunostimulant properties and anti-inflammatory activities [56,57]. Another metabolite that had the strongest alpha-glucosidase inhibitory activity is phytol [54,58]. Phytol is an acyclic diterpene alcohol and is commonly produced through the degradation process of the plant cell wall [54]. The same metabolite also was reported by other researchers to inhibit the strongest antimicrobial, antioxidant, antinociceptive, and anticancer activities [59,60]. However, the other metabolites found in this analysis might not yet be described in detail by previous literature. Hence, this study revealed that GC-MS is an efficient tool to profile the untargeted peak of the 'Giant Green' cultivar of wax apple leaf samples. However, the huge dataset which was obtained from hundreds of peaks of GC-MS analysis provided a barrier to providing a significant conclusion in terms of specific metabolites that contribute to discrimination between 'Giant Green' leaves at three different maturity stages. Thus, a more manageable size of GC-MS data was obtained by chemometric analysis that applied multivariate data analysis (MVDA).

PCA is performed to reduce the dataset aiming at the structuring of data and clustering of experimental samples. PCA detected the similarities between the samples and classified them into similar clusters. In this study, the 'Giant Green' cultivar of *S. samarangense* leaves did not provide good separation between their maturity stages. The grouping in PCA is based on the strength of variables in the loading plot on PC1 and PC2 axis. Similarly, our findings agreed with the work of Steingass et al. [21]. In their study, one of the green-ripe pineapple fruit did not match with other samples with the same maturity stage and the authors ascribed the variation due to the development of metabolites among the individual fruits that were different even at the same maturity stages. However, our result contradicted a previous study as reported by Maamoun et al. [61] in which there was clear discrimination between two stages of the ripening stage of *Luffa egyptiaca* Mill fruit. They noted that young fruit exhibits a negative score along PC1 and old mature fruit exhibits a positive score along PC2. Zhang et al. [62] also found a good separation between the three stages of tobacco leaves. The rosette and vigorous growth stages are located along PC1 whereas the mature leaves are located along PC2. The accumulation of certain compounds such as nicotine, sucrose, D-glucose, L-proline, D-fructose, quinic acid, glyceric acid, L-threonic acid, inositol, and DL-malic acid at various quantities in tobacco growth stages were indicated may contribute to this separation. Hence, it can be concluded that the variation and quantity of metabolites in each of the experimental samples played a significant role in the discrimination between them.

Complex chemical reactions occur at each of the maturity stages of plants suitable for their growth and cell development process. So, this process automatically changes the composition of metabolite in the plant. Hierarchical cluster analysis (HCA) is an unsupervised MVDA used to identify the natural grouping between the plant samples characterized by the values of a set of measured properties [63]. The similarity and dissimilarity of the entire set of samples are displayed in the HCA dendrogram. The results revealed that 'Giant Green' leaves were discriminated into three clusters, similar to the results in the PCA. However, each of the clusters did not represent the different maturity stages of leaf samples as expected. It showed that the data as accessed from GC-MS analysis was not able to well-discriminate between 'Giant Green' leaves at three maturity stages. Many factors could be influencing this result such as the similarities of metabolites in each of the samples, location of sampling, and biological replication of samples [30,64,65]. Despite this fact, all of the samples and their biological replicates were collected at a similar location, which has been attributed to the slight differences between metabolites as compared to those samples collected from other locations.

Partial Least Square-Discriminant Analysis (PLS-DA) is further adapted from the unsupervised classification of PCA. The supervised PLS-DA model was applied to investigate the metabolites that contributed to discrimination between 'Giant Green' leaves at three maturity stages. Its results were not in accordance with previous PCA and HCA results where all stages of leaves improved their separation from each other. The validation of the model was also proven with 100% of the confusion matrix result. The metabolites involved in this separation were confirmed with variable importance in the projection (VIP) values. The twenty-four metabolites were identified that consisted of VIP values greater than 0.85 in leaf extracts. From the results, it showed that the variation of metabolites from the groups of alcohol, ester, alkene, alkane, and ketone were involved in the discrimination of leaves (young, mature, and old leaf stages) samples. However, the understanding of factors that influence the discrimination among samples was very complex. Some researchers revealed that the factors of climate, soil, temperature, maturity stage, irrigation, and fertilizer vary the composition of metabolites in plants [31,36,66]. According to Yunusa [30], two possible factors responsible for the separation between the samples are the particular metabolite presence in all samples but different in concentration, and undetected particular metabolites in certain samples. All of these factors also affected the results of PLS-DA analysis. In addition, it was also expected that the PLS-DA model showed better performance in the classification of 'Giant Green' leaves at three maturity stages than PCA since PLS-DA was most effective in discriminating the samples based on their similarities and dissimilarities of metabolite profile.

Partial least square (PLS) is applied to find the correlation between the biological activities (antioxidant and alpha-glucosidase) and metabolites in three maturity stages of 'Giant Green' leaves. From the PLS bi-plot, the young leaves were located near the bioactivities. This finding confirmed the biological activity results, which showed that the sample from young leaf stages had the highest activity compared to samples from other stages. Based on the VIP score, the metabolites contributing the highest to these activities in leaf extract along PC1 were fatty acid (methyl palmitate, 18) and alkane-based compound (tetradecane, 4; decane, 1; cyclotetradecane, 3). This result was consistent with previous literature showing that the metabolites from fatty acid and alkane derivatives possessed antioxidant and alpha-glucosidase activities [46,54,67]. Anh et al. [68] found that the methyl palmitate presence in *Clausena indica* fruit possesses potent antioxidant and antidiabetic activities. Another study as reported by Murugesu et al. [54], revealed that methyl palmitate and pentadecanoic acid presence in *Clinacanthus nutans* Lindau leaves inhibited the stronger alpha-glucosidase inhibitory activity. The alkanes, fatty acid, methyl ester, and aromatics chain in the essential oil of *Daphne mucronata* Royle leaves and stems showed good antioxidant and antibacterial activities [69]. Considering the relationship as described in the PLS model, it was proven that the metabolites and biological activities in the 'Giant Green' leaves were well-correlated. Then, young leaves also showed the strongest relationship with antioxidant and alpha-glucosidase inhibitory activities than mature and old leaves.

## 5. Conclusions

The ATR-FTIR and GC-MS-based metabolomics approach have well-determined the metabolite variation between the three maturity stages of the 'Giant Green' cultivar of *S. samarangense* leaves. Unsupervised and supervised MVDA were successfully discriminated between the leaf samples and visualized the specific metabolites correlated to the biological activities (antioxidant and alpha-glucosidase). Thus, this work concludes that spectroscopy and chromatography fingerprinting coupled with chemometrics can be applied to select the best maturity of 'Giant Green' leaves for further use in the pharmacological field.

**Author Contributions:** Data curation, N.S.I. and M.M.K.; funding acquisition, M.M.K.; investigation, M.M.K.; methodology, N.S.I., Z.M.R. and M.M.K.; supervision, M.M.K., N.M. and Z.M.R.; validation, A.M., K.M. and M.M.A.; writing—original draft, N.S.I. and M.M.K.; writing—review and editing, N.M., M.M.K., M.M.A. and A.F.M.A. All authors have read and agreed to the published version of the manuscript.

**Funding:** This research was supported by the Ministry of Higher Education, Malaysia (MOHE), project grant No: RACE/F1/SG5/UNISZA/5 with the collaboration of Universiti Sultan Zainal Abidin, Terengganu and Universiti Malaya, Kuala Lumpur, Malaysia.

**Data Availability Statement:** The data related to the findings of this research are available upon request from the corresponding author.

**Acknowledgments:** All of the authors wish to thank the Center for Research Excellence and Incubation Management (CREIM), Universiti Sultan Zainal Abidin, Campus Gong Badak, Terengganu, Malaysia for supporting this project and publication of the findings. The Researchers also acknowledge the Deanship of Scientific Research, Taif University for editing and partial publication support.

**Conflicts of Interest:** The authors declare no conflict of interest.

## References

1. Khandaker, M.M.; Saidi, A.; Badaluddin, N.A.; Yusoff, N.; Majrashi, M.; Alenazi, M.M.; Saifuddin, M.; Alam, M.A.; Mohd, K.S. Effects of Indole-3-Butyric Acid (IBA) and Rooting Media on Rooting and Survival of Air Layered Wax Apple (*Syzygium samarangense*) cv. Jambu Madu. *Braz. J. Biol.* **2022**, *82*. [CrossRef] [PubMed]
2. Idris, N.S.; Ismail, S.Z.; Mat, N.; Khandaker, M.M. Comparative Physiological Response of Three Wax Apple (*Syzygium samarangense*) Tree Cultivars at Flower Bud Development Stage. *Biosci. Res.* **2018**, *15*, 402–411.
3. Khandaker, M.M.; Awang, I.; Ismail, S.Z. Effects of Naphthalene Acetic Acid and Gibberellic Acid on Plant Physiological Characteristics of Wax Apple (Var. Jambu Madu). *Bulg. J. Agric. Sci.* **2017**, *23*, 396–404.
4. Adodo, A.; Iwu, M.M. *Healing Plants of Nigeria. Ethnomedicine and Therapeutic Applications*; CRC Press: Boca Raton, FL, USA, 2020.
5. Banadka, A.; Wudali, N.S.; Al-Khayri, J.M.; Nagella, P. The Role of *Syzygium samarangense* in Nutrition and Economy: An Overview. *S. Afr. J. Bot.* **2022**, *145*, 481–492. [CrossRef]
6. Negreanu-Pirjol, B.-S.; Oprea, O.C.; Negreanu-Pirjol, T.; Roncea, F.N.; Prelipcean, A.-M.; Craciunescu, O.; Iosageanu, A.; Artem, V.; Ranca, A.; Motelica, L.; et al. Health Benefits of Antioxidant Bioactive Compounds in the Fruits and Leaves of *Lonicera caerulea* L. *and Aronia melanocarpa (Michx.) Elliot. Antioxidants* **2023**, *12*, 951. [PubMed]

7. Hossain, R.; Rahman, M.A.; Rafi, M.K.J.; Siddique, T.A.; Noman, A.A.; Makki, A.; Alelwani, W.; Hajjar, D.; Tangpong, J. Pharmacological and ADMET-based Pharmacokinetic Properties of *Syzygium samarangense* var. Parviflorum Leaf Extract In Vitro, In Vivo and In Silico Models. *Not. Bot. Horti. Agrobo.* **2020**, *48*, 1155–1175. [CrossRef]
8. Moneruzzaman, M.K.; Sarwar, J.; Mat, N.; Boyce, A.N. Bioactive Constituents, Antioxidant and Antimicrobial Activities of Three Cultivars of Wax Apple (*Syzygium samarangense* L.) Fruits. *Res. J. Biotechnol.* **2015**, *10*, 7–16.
9. Aborode, A.T.; Awuah, W.A.; Mikhailova, T.; Abdul-Rahman, T.; Pavlock, S.; Nansubuga, E.P.; Kundu, M.; Yarlagadda, R.; Pustake, M.; Correia, I.F.S.; et al. OMICs Technologies for Natural Compounds-based Drug Development. *Curr. Top Med. Chem.* **2022**, *22*, 1751–1765.
10. Kim, T.J.; Park, J.G.; Kim, H.Y.; Ha, S.; Lee, B.; Park, S.U.; Seo, W.D.; Kim, J.K. Metabolite Profiling and Chemometric Study for the Discrimination Analyses of Geographic Origin of Perilla (*Perilla frutescens*) and Sesame (*Sesamum indicum*) Seeds. *Foods* **2020**, *9*, 989. [CrossRef]
11. Weber, A.; Hoplight, B.; Ogilvie, R.; Muro, C.; Khandasammy, S.R.; Perez-Almodovar, L.; Sears, S.; Lednev, I.K. Innovative Vibrational Spectroscopy Research for Forensic Application. *Anal. Chem.* **2023**, *95*, 167–205. [CrossRef]
12. Azemin, A.; Dharmaraj, S.; Hamdan, M.R.; Mat, N.; Ismail, Z.; Mohd, S. Discriminating *Ficus deltoidea* var. bornensis from Different Localities by HPTLC and FTIR Fingerprinting. *J. App. Pharm. Sci.* **2014**, *4*, 69–75.
13. Jwaili, M. Pharmaceutical Applications of Gas Chromatography. *Open J. App. Sci.* **2019**, *9*, 683–690. [CrossRef]
14. El-deen, A.K.; Shimizu, K. Modified μ -QuEChERS Coupled to Diethyl Carbonate-Based Liquid Microextraction for PAHs Determination in Coffee, Tea, and Water prior to GC–MS Analysis: An Insight to Reducing the Impact of Caffeine on the GC–MS Measurement. *J. Chromatograp. B.* **2021**, *1171*, 122555. [CrossRef] [PubMed]
15. Wijayanti, T.; Riyanto, S.; Lukitaningsih, E.; Rohman, A. Application of $^1$H-NMR Specta and Multivariate Analysis for the Authentication of *Curcuma Xanthorrhiza* from *Zingiber Cassumunar*. *Int. J. App. Pharm.* **2019**, *11*, 258–263. [CrossRef]
16. Basyirah, N.; Azemin, A.; Muslim, M.; Rodi, M.; Mat, Z. Application of FTIR Fingerprints Coupled with Chemometric for Comparison of Stingless Bee Propolis from Different Extraction Methods. *Malaysian J. Fund. App. Sci.* **2019**, 350–355.
17. Abas, F.; Shitan, M.; Shaari, K.; Lajis, N.H. Comparison of Partial Least Squares and Artificial Neural Network for the Prediction of Antioxidant Activity in Extract of *Pegaga (Centella)* Varieties from $^1$H Nuclear Magnetic Resonance spectroscopy. *Food Res. Int.* **2013**, *54*, 852–860.
18. Idris, N.S.; Khandaker, M.M.; Rashid, Z.M.; Majrashi, A.; Alenazi, M.M.; Nor, Z.M.; Mohd Adnan, A.F.; Mat, N. Polyphenolic Compounds and Biological Activities of Leaves and Fruits of *Syzygium samarangense* cv. 'Giant Green' at Three Different Maturities. *Horticulturae* **2023**, *9*, 326. [CrossRef]
19. Ezhilan, B.P.; Neelamegam, R. GC-MS Analysis of Pytocomponents in the Ethanol Extract of *Polygonum chinensis* L. *Pharmacol. Res.* **2012**, *4*, 11–14.
20. Zin, M.N.B.; Azemin, A.; Rodi, M.M.M.; Mohd, K.S. Chemical Composition and Antioxidant Activity of Stingless Bee Propolis from Different Extraction Methods. *Int. J. Eng. Technol.* **2018**, *7*, 90–95.
21. Steingass, C.B.; Jutzi, M.; Muller, J.; Carle, R.; Schmarr, H. Ripening-Dependent Metabolic Changes in the Volatiles of Pineapple (*Ananas comosus* (L.) Merr.) Fruit: II. Multivariate Statistical Profiling of Pineapple Aroma Compounds Based on Comprehensive Two-Dimensional Gas Chromatography-Mass Spectrometry. *Anal. Bioanal. Chem.* **2015**, *407*, 2609–2624. [CrossRef]
22. Yunusa, A.K. Chemical Analysis and Bioactivities of Raw and Process Syconia of Seven Variaties of *Ficus deltoidea* Jack. Ph.D. Thesis, University of Sultan Zainal Abidin, Terengganu, Malaysia, 2019.
23. Park, Y.J.; Baek, S.; Choi, Y.; Kim, J.K.; Park, S.U. Metabolic Profiling of Nine Mentha Species and Prediction of Their Antioxidant Properties Using Chemometrics. *Molecules* **2019**, *24*, 258. [CrossRef] [PubMed]
24. Easmin, S.; Islam, Z.; Ghafoor, K.; Ferdosh, S.; Jaffri, J.; Ali, E.; Mirhosseini, H.; Al-Juhaimi, F.Y.; Perumal, V.; Khatib, A. Rapid Investigation of α-Glucosidase Inhibitory Activity of *Phaleria macrocarpa* Extracts using FTIR-ATR Based Fingerprinting. *J. Food Drug Anal.* **2016**, *25*, 306–315. [CrossRef] [PubMed]
25. Rana, R.; Herz, K.; Bruelheide, H.; Dietz, S.; Haider, S.; Jandt, U.; Pena, R. Leaf Attenuated Total Reflection Fourier Transform Infrared (ATR-FTIR) Biochemical Profile of Grassland Plant Species Related to Land-Use Intensity. *Ecol. Indicat.* **2018**, *84*, 803–810. [CrossRef]
26. Agatonovic-Kustrin, S.; Ramenskaya, G.; Kustrin, E.; Ortakand, D.B.; Morton, D.W.A. New Integrated HPTLC-ATR/FTIR Approach in Marine Algae Bioprofiling. *J. Pharm. Biomed. Anal.* **2020**, *189*, 113488. [CrossRef]
27. Vijayalakshmi, R.; Ravindhran, R. Comparative Fingerprint and Extraction Yield of *Diospyrus Ferrea (Willd.) Bakh.* Root with Phenol Compounds (Gallic Acid), as Determined by Uv-Vis and Ft-Ir Spectroscopy. *Asian Pac. J. Trop. Biomed.* **2012**, *2*, S1367–S1371. [CrossRef]
28. Bhattacharyya, A.S.; Kumar, S.; Sharma, A.; Kumar, D.; Patel, S.B.; Paul, D.; Dutta, P.P.; Bhattacharjee, G. Metallization and APPJ treatment of bismaleimide. *High Perform. Polym.* **2015**, *29*, 1–11. [CrossRef]
29. Kharbach, M.; Marmouzi, I.; Jemli, M.E.; Bouklouze, A.; Heyden, Y.V. Recent Advances in Untargeted and Targeted Approaches Applied in Herbal-Extracts and Essential-Oils Fingerprinting-A Review. *J. Pharm. Biomed. Anal.* **2019**, *177*, 112849. [CrossRef]
30. Yunusa, A.K.; Rashid, Z.M.; Mat, N.; Abdullah, A.B.C.; Ali, A.M. Bioactive fingerprints of aqueous extracts of *Ficus deltoidea* syconia via FTIR spectroscopy coupled with chemometrics. *Biosci. Res.* **2019**, *16*, 157–167.
31. Kim, J.; Jung., Y.; Song, B.; Bong., Y.; Hyun, D.; Lee, K.; Hwang, G. Discrimination of Cabbage (*Brassica rapa* ssp. *pekinensis*) Cultivars Grown in Different Geographical Areas using $^1$H NMR-Based Metabolomics. *Food Chem.* **2013**, *137*, 68–75. [CrossRef]

32. Lawal, U.; Maulidiani, M.; Shaari, K.; Ismail, I.S.; Khatib, A.; Abas, F. Discrimination of *Ipomoea aquatica* Cultivars and Bioactivity Correlations using NMR-Based Metabolomics Approach. *Plant Biosyst.* **2016**, *3504*, 833–843. [CrossRef]
33. Mesquita, P.R.R.; Dos Santos, F.N.; Bastos, L.P.; Costa, M.A.P.C.; Rodrigues, F.M.; Andrade, J.B. Discrimination of *Eugenia uniflora* L. Biotypes Based on Volatile Compounds in Leaves using HS-SPME/GC–MS and Chemometric Analysis. *Microchem. J.* **2016**, *130*, 79–87. [CrossRef]
34. Bougrini, M.; Tahri, K.; Saidi, T.; Hassani, N.E.A.; Bouchikhi, B.; Bari, N.E. Classification of Honey According to Geographical and Botanical Origins and Detection of Its Adulteration Using Voltammetric Electronic Tongue. *Food Anal. Methods* **2016**, *9*, 2161–2173. [CrossRef]
35. Lee, P.C.; Guo, H.Y.; Huang, C.C.; Chan, C.F. Chemical Composition of Leaf Essential Oils of *Syzygium samarangense* (BL.) Merr. et Perry cv. Pink at Three Maturity Stages. *Int. J. App. Res. Nat. Prod.* **2016**, *9*, 9–13.
36. Tiwari, U.; Cummins, E. Factors Influencing Levels of Phytochemicals in Selected Fruit and Vegetables During Pre- and Post-Harvest Food Processing Operations. *Food Res. Int.* **2013**, *50*, 497–506. [CrossRef]
37. Noh, C.H.C.; Azmin, N.F.M.; Amid, A. Principal Component Analysis Application on Flavonoids Characterization. *Ad. Sci. Technol. Eng. Syst. J.* **2017**, *2*, 435–440. [CrossRef]
38. Gouvinhas, I.; Almeida, J.M.M.M.; De Carvalho, T.; Machado, N.; Barros, A.I.R.N.A. Discrimination and Characterisation of Extra Virgin Olive Oils from Three Cultivars in Different Maturation Stages using Fourier Transform Infrared Spectroscopy in Tandem with Chemometrics. *Food Chem.* **2015**, *174*, 226–232. [CrossRef]
39. Christou, C.; Agapiou, A.; Kokkinofta, R. Use of FTIR Spectroscopy and Chemometrics for the Classification of Carobs Origin. *J. Advan. Res.* **2017**, *10*, 1–8. [CrossRef]
40. Saidan, N.H.; Hamil, M.S.R.; Memon, A.H.; Abdelbari, M.M.; Hamdan, M.R.; Mohd, K.S.; Majid, A.M.S.A.; Ismail, Z. Selected Metabolites Profiling of *Orthosiphon Stamineus* Benth Leaves Extracts Combined with Chemometrics Analysis and Correlation with Biological Activities. *BMC Complement Altern. Med.* **2015**, *15*, 350. [CrossRef]
41. Mukaromah, A.S. Wax apple (*Syzygium samarangense* (Blume) Merr. & L.M. Perry): A Comprehensive Review in Phytochemical and Physiological Perspectives. *J. Biol. App. Biol.* **2020**, *3*, 40–58.
42. Alam, S.; Dhar, A.; Hasan, M.; Richi, F.T.; Emon, N.U.; Aziz, M.; Mamun, A.A.; Chowdhury, M.N.R.; Hossain, M.J.; Kim, J.K.; et al. Antidiabetic Potential of Commonly Available Fruit Plants in Bangladesh: Updates on Prospective Phytochemicals and Their Reported MoAs. *Molecules* **2022**, *27*, 8709. [CrossRef]
43. Ragasa, C.Y.; Franco, J.F.C.; Raga, D.D.; Shen, C.C. Chemical Constituents of *Syzygium samarangense*. *Der Pharm. Chem.* **2014**, *6*, 256–260.
44. Ribeiro, C.; Gonçalves, R.; Tiritan, M.E.; Gonc, R. Separation of Enantiomers Using Gas Chromatography: Application in Forensic Toxicology, Food and Environmental Analysis. *Crit. Rev. Anal. Chem.* **2020**, *51*, 787–811. [CrossRef] [PubMed]
45. Javadi, N.; Abas, Q.; Hamid, A.A.; Simoh, A.; Shaari, K.; Ismail, I.S.; Mediani, A.; Khatib, A. GC-MS Based Metabolite Profiling of *Cosmos caudate* Leaves Possessing Alpha-Glucosidase Inhibitory Activity. *J. Food Sci.* **2014**, *79*, 1130–1136. [CrossRef]
46. Saleh, M.S.M.; Awanis, D.; Bukhari, M.; Jamshed, M.; Siddiqui, A.; Kasmuri, A.R.; Murugesu, S.; Khatib, A. GC-MS Analysis of Metabolites from Soxhlet Extraction, Ultrasound-Assisted Extraction and Supercritical Fluid Extraction of *Salacca zalacca* flesh and its Alpha-Glucosidase Inhibitory Activity. *Nat. Product Res.* **2019**, *34*, 1341–1344. [CrossRef] [PubMed]
47. Bourhia, M.; Bouothmany, K.; Bakrim, H.; Hadrach, S.; Salamatullah, A.M.; Alzahrani, A.; Alyahya, H.K.; Albadr, N.A.; Gmouh, S.; Laglaoui, A.; et al. Antibacterial Potentials of Chemically Characterized Extract of *Citrullus colocynthis* L. Seeds. *Separations* **2021**, *8*, 114. [CrossRef]
48. Yogeswari, S.; Ramalakshmi, S.; Neelavathy, R.; Muthumary, J. Identification and Comparative Studies of Different Volatile Fractions from *Monochaetia Kansensis* by GCMS. *Glob. J. Pharmacol.* **2012**, *6*, 65–71.
49. Rouis-Soussi, L.S.; Ayeb-Zakhama, A.E.; Aouni, M.; Guido, F.; Hichem, B.J.; Harzallah-Skhiri, F. Chemical Composition and Antibacterial Activity of Essential Oils from the Tunisian *Allium nigrum* L. *Exc. J.* **2014**, *13*, 526–535.
50. Kim, J.; Seob, J.; Baec, M.; Baed, C.; Yooe, J.; Bangf, M.; Choa, S.; Parkb, D. Antimicrobial Constituents from *Allium hookeri* Root. *Nat. Prod. Comm.* **2016**, *11*, 237–238. [CrossRef]
51. Mohamad, O.A.A.; Li, L.; Ma, J.; Hatab, S.; Xu, L.; Guo, J.; Rasulov, B.A.; Liu, Y.; Hedlund, B.P.; Li, W. Evaluation of the Antimicrobial Activity of Endophytic Bacterial Populations From Chinese Traditional Medicinal Plant Licorice and Characterization of the Bioactive Secondary Metabolites Produced by *Bacillus atrophaeus* Against *Verticillium dahliae*. *Front. Microbiol.* **2018**, *9*, 924. [CrossRef]
52. Noh, I.A.; Yi, V.; Jong, M. Phytochemicals, Antimicrobials and Antioxidants Studies of the Stem Bark Extract from *Calophyllum ferrugineum*. *Sci. Res. J.* **2020**, *17*, 1–12. [CrossRef]
53. Sumithra, D.; Purushothaman, D.S. Phytochemical Profiling of Ethanolic Leaves Extract of *Commelina benghalensis* L. World. *J. Pharm. Res.* **2016**, *6*, 1101–1107.
54. Murugesu, S.; Ibrahim, Z.; Ahmed, Q.; Yusoff, N.N.; Uzir, B.; Perumal, V.; Abas, F.; Saari, K.; El-Seedi, H.; Khatib, A. Characterization of α-Glucosidase Inhibitors from *Clinacanthus nutans* Lindau Leaves by Gas Metabolomics and Molecular Docking Simulation. *Molecules* **2018**, *23*, 2402. [CrossRef] [PubMed]
55. Alqahtani, S.S.; Makeen, H.A.; Menachery, S.J.; Moni, S.S. Documentation of Bioactive Principles of The Flower from *Caralluma retrospiciens* (*Ehrenb*) and In Vitro Antibacterial Activity—Part B. *Arabian J. Chem.* **2020**, *13*, 7370–7377. [CrossRef]

56. Maghdu, N.A.; Palaniyappan, K.D. In Vitro Antifungal Potentials of Bioactive Compounds Heptadecane, 9- Hexyl and Ethyl Isoallocholate Isolated from *Lepidagathis cristata* Willd. (*Acanthaceae*) leaf. *British Biomed. Bul.* **2015**, *3*, 336–343.
57. Al-Rubaye, A.F.; Ashwak, F.K.; Imad, H.H. Phytochemical Screening of Methanolic Leaves Extract of *Malva sylvestris. Int. J. Pharm. Phytochem. Res.* **2017**, *9*, 537–552. [CrossRef]
58. Gazali, M.; Jolanda, O.; Husni, A.; Nurjanah; Majid, F.A.A.; Syafitri, R. In Vitro α-Amylase and α-Glucosidase Inhibitory Activity of Green Seaweed *Halimeda tuna* Extract from the Coast of Lhok Bubon, Aceh. *Plants* **2023**, *12*, 393. [CrossRef]
59. De Moraes, J.; De Oliveira, R.N.; Costa, J.P.; Junior, A.L.G.; De Sousa, D.P.; Freitas, R.M. Phytol, a Diterpene Alcohol from Chlorophyll, as a Drug Against Neglected Tropical Disease *Schistosomiasis mansoni. PLOS Neg. Trop. Dis.* **2014**, *8*, 51. [CrossRef]
60. Swamy, M.K.; Sinniah, U.R. A Comprehensive Review on the Phytochemical Constituents and Pharmacological Activities of *Pogostemon cablin* Benth: An Aromatic Medicinal Plant of Industrial Importance. *Molecules* **2015**, *20*, 8521–8547. [CrossRef]
61. Maamoun, A.A.; El-akkad, R.H.; Farag, M.A. Mapping Metabolome Changes in *Luffa aegyptiaca* Mill Fruits at Different Maturation Stages via MS-Based Metabolomics and Chemometrics. *J. Advan. Res.* **2019**, *29*, 179–189. [CrossRef]
62. Zhang, Q.W.; Lin, L.G.; Ye, W.C. Techniques for Extraction and Isolation of Natural Products: A Comprehensive Review. *Chinese Med.* **2018**, *13*, 20. [CrossRef]
63. Braga, M.C.; Antonio, A.; Zielinski, F.; Marques, K.; Koch, F.; Souza, F.D.; Pietrowski, G.D.A.M.; Couto, M.; Granato, D.; Wosiacki, G.; et al. Classification of Juices and Fermented Beverages Made from Unripe, Ripe and Senescent Apples Based on the Aromatic Profile Using Chemometrics. *Food Chem.* **2013**, *141*, 967–974. [CrossRef] [PubMed]
64. Rohaeti, E.; Rafi, M.; Dyah, U.; Heryanto, R. Fourier Transform Infrared Spectroscopy Combined with Chemometrics for Discrimination of *Curcuma longa, Curcuma xanthorrhiza* and *Zingiber cassumunar*. *Spectrochim. Acta Part A Mol. Biomol. Spectros.* **2015**, *137*, 1244–1249. [CrossRef]
65. Deetae, P.; Parichanon, P.; Trakunleewatthana, P.; Chanseetis, C. Antioxidant and Anti-Glycation Properties of Thai Herbal Teas in Comparison with Conventional Teas. *Food Chem.* **2012**, *133*, 953–959. [CrossRef]
66. Zou, Z.; Xi, W.; Hu, Y.; Nie, C.; Zhou, Z. Antioxidant Activity of *Citrus* Fruits. *Food Chem.* **2015**, *196*, 885–896. [CrossRef]
67. Karimi, E.; Jaafar, H.Z.; Ghasemzadeh, A.; Ebrahimi, M. Fatty acid composition, antioxidant and antibacterial properties of the microwave aqueous extract of three varieties of Labisia pumila Benth. *Biol Res.* **2015**, *48*, 9. [CrossRef] [PubMed]
68. Anh, L.H.; Xuan, T.D.; Thi, N.; Thuy, D.; Quan, N.V. Antioxidant and α-amylase Inhibitory Activities and Phytocompounds of *Clausena indica* Fruits. *Medicines* **2020**, *7*, 10.
69. Ashraf, I.; Zubair, M.; Rizwan, K.; Rasool, N.; Jamil, M.; Khan, S.A.; Tareen, R.B.; Ahmad, V.U.; Mahmood, A.; Riaz, M.; et al. Chemical Composition, Antioxidant and Antimicrobial Potential of Essential Oils from Different Parts of *Daphne mucronata* Royle. *Chem. Cen. J.* **2018**, *12*, 135. [CrossRef]


**Disclaimer/Publisher's Note:** The statements, opinions and data contained in all publications are solely those of the individual author(s) and contributor(s) and not of MDPI and/or the editor(s). MDPI and/or the editor(s) disclaim responsibility for any injury to people or property resulting from any ideas, methods, instructions or products referred to in the content.

horticulturae

MDPI

*Article*

# Comparison of Growth Patterns and Metabolite Composition of Different Ginseng Cultivars (Yunpoong and K-1) Grown in a Vertical Farm

**Ga Oun Lee [1], Seong-Nam Jang [1], Min Ju Kim [1], Du Yong Cho [1], Kye Man Cho [1,2], Ji Hyun Lee [3] and Ki-Ho Son [1,4,*]**

1 Department of GreenBio Science, Gyeongsang National University, Jinju 52725, Republic of Korea; riyung@naver.com (G.O.L.); tjdska1346@naver.com (S.-N.J.); minju4492@naver.com (M.J.K.); endyd6098@naver.com (D.Y.C.); kmcho@gnu.ac.kr (K.M.C.)
2 Department of Food Science, Gyeongsang National University, Jinju 52725, Republic of Korea
3 National Institute of Horticultural and Herbal Science, Rural Development Administration, Wanju-gun 55365, Republic of Korea; leejh80@korea.kr
4 Division of Horticultural Science, Gyeongsang National University, Jinju 52725, Republic of Korea
* Correspondence: sonkh@gnu.ac.kr; Tel.: +82-55-772-3253

**Abstract:** This study analyzed growth patterns, biological compounds, antioxidant properties, ginsenoside contents, metabolites, and the annual net production of 'Yunpoong' and 'K-1' to find the optimal harvesting time of ginseng sprouts. One-year-old ginseng seedlings were cultivated in a container-type vertical farm under a temperature of 20 °C, a humidity of 60%, and average light intensity of 46.4 μmol $m^{-2}$ $s^{-1}$ (16 h photoperiod). Growth patterns at 2, 3, 4, and 5 weeks after transplanting (WAT) differed between cultivars. Regarding biological compounds and antioxidant properties, 'Yunpoong' took 5 WAT (43.59%; 2,2-diphenyl-1-picryl-hydrazine-hydrate radical scavenging activity, 1.47 $OD_{593nm}$; ferric reducing antioxidant power assay, 78.01%; 2,2′-azino-bis (3-ethylbenzothiazoline-6-sulfonic acid) radical scavenging activity), and 'K-1' took 4 WAT (0.98 Re mg $g^{-1}$; total flavonoid contents, 35.93%; DPPH) to show a high content. Two cultivars showed the highest total ginsenoside contents at 5 WAT. Most of the analyzed metabolites had a higher content in 'Yunpoong' than in 'K-1'. In both cultivars, it was confirmed that the longer the growth period (3 − > 5 WAT), the lower the yield and the annual ginsenoside net production. Therefore, 'Yunpoong' and 'K-1' cultivars should be grown as ginseng sprouts in the vertical farms for approximately 3 WAT and 4 WAT, respectively.

**Keywords:** antioxidant properties; biological compounds; ginsenoside contents; *Panax ginseng*; plant factory

check for updates

**Citation:** Lee, G.O.; Jang, S.-N.; Kim, M.J.; Cho, D.Y.; Cho, K.M.; Lee, J.H.; Son, K.-H. Comparison of Growth Patterns and Metabolite Composition of Different Ginseng Cultivars (Yunpoong and K-1) Grown in a Vertical Farm. *Horticulturae* **2023**, *9*, 583. https://doi.org/10.3390/horticulturae9050583

Academic Editor: Wajid Zaman

Received: 23 April 2023
Revised: 9 May 2023
Accepted: 12 May 2023
Published: 14 May 2023

**Copyright:** © 2023 by the authors. Licensee MDPI, Basel, Switzerland. This article is an open access article distributed under the terms and conditions of the Creative Commons Attribution (CC BY) license (https://creativecommons.org/licenses/by/4.0/).

## 1. Introduction

Ginseng (*Panax ginseng* C.A. Meyer, National Institute of Horticultural and Herbal Science, Wanju-gun, Republic of Korea), a perennial plant belonging to the family Araliaceae and genus *Panax*, is an important medicinal crop cultivated in Asian countries, such as the Republic of Korea and China. The area of ginseng grown in the Republic of Korea was 14,729 hectares, with a production of 20,772 tons in 2021 [1]. The shape and growth features of ginseng differ depending on the cultivar [2]. The ginseng cultivars have various morphological and physiological properties [3], seed yield, ginsenoside content, and pest resistance [4,5]. 'Yunpoong' is a Jakyung species with a short plant length, multiple purple stems, one red fruit, and stipule-shaped leaves [6]. Their body of the root is cylindrical, with a medium length and large thickness, and has a high yield of fresh ginseng quality. Another Jakyung species, 'K-1', mostly has one dark-purple-colored stem with red fruits and long, triple-tipped oval leaves [7]. The structure of the K-1 root, a developed tap root with 2–3 lateral and numerous fine roots, is cylindrical, suitable for red ginseng production, and has stronger disease resistance.

Ginsenosides are saponins and constitute the main components in ginseng with various physiological and pharmacological properties, such as anticancer [8], anti-stress [9], neuroprotective [10], and anti-diabetic [11]. The saponin content in ginseng differs based on the harvest time of ginseng leaves during general cultivation [12]. It can vary depending on the cultivar [3], age [13], planting location [14], and growth stage [15]. Ginseng leaves contain small amounts of ginsenoside-F1, ginsenoside-F2, and ginsenoside-F3 that are not found in the roots [15].

The ginseng leaves possess excellent physiological properties [16]. One-year-old ginseng leaves had higher ginsenoside content than roots [17,18]. A vertical farm is a multi-tier indoor plant production system that precisely controls all growth factors, including light, temperature, humidity, $CO_2$ concentration, water, and nutrients, to produce high volumes of plants [19]. By controlling the environment, such as light, temperature, and humidity, in vertical farms, it is possible to produce high-value-added crops year-round [20]. In particular, vertical farms are emerging as a new method for cultivating medicinal plants for medicines and functional foods, providing customized environmental conditions suitable for high yield and high-quality plants [21]. For example, ginseng grown in vertical farms has a shorter growth period [22] and higher saponin content than field-grown ginseng [23]. Vertical farms can increase the yield [24,25] of various crops and enable year-round production [19]. Additionally, ginseng sprouts have a short growing period and can be grown in a vertical farm where it is not affected by seasonal variations [26].

Many studies are being conducted on the comparison of ginseng varieties. Ahn et al. [3] examined the variation in ginsenoside content among different parts of ginseng roots based on the ginseng variety. The growth characteristics were compared to the 'Gopoong', 'Chunpoong', and 'Yunpoong' ginseng varieties [6]. However, there has yet to be a study on the changes in growth patterns and ginsenoside content for each ginseng variety. Therefore, this study was conducted to determine the optimal harvest time of the 'Yunpoong' and 'K-1' ginseng varieties grown in a vertical farm. Additionally, we compared the metabolite content and differences between 'Yunpoong' and 'K-1' through metabolite analysis.

## 2. Materials and Methods

### 2.1. Plant Materials and Growth Conditions

One-year-old ginseng seedlings ('Yunpoong' and 'K-1') were provided to the National Institute of Horticultural and Herbal Science, and approximately 7–10 cm ginseng seedlings were used as experimental material. Ginseng seedlings were immersed in 100 ppm of gibberellin aqueous solution (IAP, JahngRyu Industries Co., Ltd., Cheongju, Republic of Korea) for 30 min and then transplanted into a plastic tray (51.5 × 36 × 8.5 cm, L × W × H) filled with an artificial ginseng soil mix (Myeongpum-Insamsangto, Shinsung Mineral Co., Ltd., Goesan, Republic of Korea). For each cultivar, 42 ginseng sprouts for 3 repetitions were planted in a plastic container and cultivated in a smart farm cube (Dream PF Corp., Sacheon, Republic of Korea). Overhead sprays were performed with tap water 3–4 times weekly. The growth environment was maintained at 20 ± 3 °C, 60 ± 10% relative humidity, 46.4 ± 0.3 $\mu mol\ m^{-2}\ s^{-1}$ photosynthetic photon flux density (PPFD; Red and Blue LEDs), and 16 h photoperiod for 5 weeks after transplanting (WAT).

### 2.2. Growth Characteristics

The growth characteristics of ginseng sprouts were measured at 2, 3, 4, and 5 WAT by dividing the shoot and root, such as shoot and root fresh and dry weights—this is the method of Nguyen et al. [27]. The fresh weight of shoots and roots was measured with an electronic scale (PAG214C, Ohaus Corp., Parsippany, NJ, USA). Then, the samples were dried in an oven (WOF-155, Daihan Scientific, Seoul, Republic of Korea) at 70 °C for 3 days, and the dried weight was measured. The top/root ratio was divided as the fresh weight ratio of the top (shoot) to the root. The total yield was expressed by totaling the shoot and root fresh weights.

### 2.3. Analysis

#### 2.3.1. Biological Compounds and Antioxidant Properties

Ginseng sprouts from 42 plants for 3 repetitions were harvested at 3, 4, and 5 WAT and prepared as follows. Dry powder samples (1 g) were extracted in 20 mL of 50% ethanol using a flask shaker (SH-800, SH Scientific Co., Ltd., Sejong, Republic of Korea) at about 25 °C for 12 h. The extract was centrifuged for 5 min, then the supernatant of 5 mL was filtered through a 0.45 µm membrane filter, and the supernatant was used for analysis [28].

For total phenolic contents, a slightly modified Folin–Denis method [29] was used. The ginseng extract (0.5 mL) was added to a 2 mL microcentrifuge tube, and 0.5 mL of 25% $Na_2CO_3$ solution was added. The mixture was allowed to stand for 3 min, whereafter 0.25 mL of 2N Folin–Ciocalteu reagent (Sigma-Aldrich Co., Ltd., St. Louis, MO, USA) was added. The mixture was left for 1 h at 30 °C for color development to occur and then centrifuged for 1 min. The absorbance of the sample was measured at 750 nm using a spectrophotometer (UV-1800 240 V, Shimadzu Corp., Kyoto, Japan). A standard calibration curve to estimate the total phenol content was drawn using gallic acid. Results are expressed in milligrams of gallic acid equivalents per gram on a dry weight basis (mg GAE $g^{-1}$ DW).

For total flavonoid contents, the method of Kim et al. [29] was used. Briefly, 0.5 mL of the ginseng extract was placed in a 2 mL microcentrifuge tube containing 1.0 mL of diethylene glycol (Sigma-Aldrich Co., Ltd., St. Louis, MO, USA), followed by the addition of 0.01 mL of 1N NaOH. The mixture was placed in a constant temperature water bath at 37 °C for 1 h. Absorbance was measured at 420 nm using the spectrophotometer. A standard calibration curve was obtained to calculate the total flavonoid content. The contents were expressed as the average of three experiments.

For DPPH (2,2-diphenyl-1-picryl-hydrazine-hydrate; Sigma-Aldrich Co., Ltd., St. Louis, MO, USA) radical scavenging activity, the method of Cho et al. [30] was used. DPPH solution (0.8 mL; $1.5 \times 10^{-4}$ M) and the ginseng extract (0.2 mL) were dispensed into a 2 mL microcentrifuge tube, incubated in the dark for 30 min, and the absorbance value was measured at 525 nm using the spectrophotometer. The extraction solvent was used as the negative control. The radical scavenging activity was determined as follows:

$$\text{Radical scavenging activity (\%)} = [1 - (\text{absorbance value of experimental group/absorbance value of negative control group})] \times 100 \quad (1)$$

For FRAP (ferric reducing antioxidant power assay) determination, the method of Lee et al. [31] was used. The FRAP solution was prepared by mixing 300 mM of acetate buffer (pH 3.6), 10 mM of TPTZ in 40 mM of HCl, and 20 mM of $FeCl_3$ in a ratio of 10:1:1. The prepared solution was preliminarily incubated at 37 °C for 15 min. Thereafter, 0.05 mL of diluted ginseng extract samples and 0.95 mL of the FRAP solution were mixed and incubated at 37 °C for 15 min. The reaction sample was measured at 593 nm using the spectrophotometer.

For ABTS [2,2′-azino-bis (3-ethylbenzothiazoline-6-sulfonic acid); Sigma-Aldrich Co., Ltd., St. Louis, MO, USA] radical scavenging activity, the method of Cho et al. [30] was used. The ABTS (7 mM) solution and potassium persulphate (2.45 mM) were mixed in a 1:1 ratio and incubated in a dark room for 12–16 h to form ABTS cations. This ABTS+ solution was diluted with methanol to an absorbance value of 0.7 ± 0.02 at 732 nm to prepare the $ABTS^+$ solution used in the experiment. After adding 0.9 mL of $ATBS^+$ solution and 0.1 mL of ginseng extract to a 2 mL microcentrifuge tube and incubating for 3 min, the absorbance value was measured at 732 nm using the spectrophotometer. The extraction solvent was used as the negative control. The radical scavenging activity was determined by the same formula as for DPPH calculation.

#### 2.3.2. Ginsenoside Analysis

For the analysis of pre-treated ginsenosides, ginseng sprouts from 42 plants for 3 repetitions were harvested at 3, 4, and 5 WAT and prepared as follows. The preprocessing involved drying the sprouts at 70 °C for 72 h, grinding them with a mixer, and storing them at 4 °C. The dried powder samples (5 g) were extracted with 70% HPLC (high-performance liquid chromatography) methanol (20 mL) in a constant temperature bath (70 °C) for 1 h. After centrifugation at 3000 rpm for 10 min using a centrifuge (1730R, GYROZEN Co., Ltd., Gimpo, Republic of Korea), the supernatant was filtered through a 0.45 μm membrane filter. This process was repeated twice, and a total of 40 mL of ginseng sprout extract was obtained. The extraction samples were concentrated under reduced pressure at 60 °C, and the resulting extract was dissolved in 2 mL of HPLC water, filtered through a 0.45μm membrane filter, and used for HPLC analysis [32].

The quantification of ginsenosides in ginseng extract samples was carried out using HPLC analysis (HPLC Agilent 1260 series Co., Forest Hill, VIC, Australia) with a diode array detector, as described by Lee et al. [33] with a slight modification. Ginsenoside standards (Rg1, Re, Ro, Rf, F5, F3, Rg2, Rh1, Rb1, Rc, F1, Rb2, Rb3, Rd, Rd2, F2, Rg3, PPT, compound K, Rh2, and PPD) were purchased from KOC Biotech (Daejeon, Republic of Korea). HPLC-grade organic solvents used for analysis were purchased from J.T. Baker (Philipsburg, NJ, USA). Other reagents used were purchased at special grade or HPLC grades. The sample (20 μL) and the 50% ethanol were injected onto an analytical reversed-phase C18 column (TSK-ODS100Z, 4.6 × 250 nm, 5 μm; Tosoh Corp., Tokyo, Japan). The mobile phase was water (elution A) and acetonitrile (elution B), and the following gradient program was used: 0–10 min, 19% B; 15 min, 20% B; 30 min, 23% B; 42 min, 30% B; 75 min, 35% B; 80 min, 60% B; and 100 min, 90% B. The column was reconditioned with 100% B for 3 min. Other HPLC conditions were as follows: detection, 203 nm; flow rate, 1.0 mL/min; column temperature, 25 °C.

### *2.4. Analysis of Metabolites*

#### 2.4.1. Quantification Methods and Gas Chromatography–Mass Spectrometry (GC/MS) Conditions for Primary Metabolite

Extracts of sprouted ginseng ('Yunpoong' and 'K-1') were analyzed using GC/MS (Figure S1 and Table S1). The dry sample (25 mg) was homogenized with 80% methanol (800 uL) using a blender to extract metabolites. After centrifuging the extract for 10 min, 10 uL of the supernatant was dried using a speed vac (Labconco Co., Kansas, MO, USA). The dried residues were dissolved in 70 μL of hydroxymethoxy amine and incubated for 90 min at 37 °C. After, samples were then derivatized by adding 70 μL of *N*,*O*-bis (trimethylsilyl) trifluoroacetamide for 30 min at 70 °C. Then, the reaction product was centrifuged for 10 min, and the supernatant of 1 mL was used in GC/MS analysis. The derivatized metabolites were analyzed using a GC-2010 Plus (Shimadzu Corp., Kyoto, Japan) equipped with a DB-5MS column (30 m × 0.25 mm, 0.25 μm film thickness; Agilent Technologies, Santa Clara, CA, USA). The injector temperature was set at 200 °C, and helium was used as the carrier gas at a flow rate of 1 mL per min. The oven temperature program was set as holding at 70 °C for 2 min, increasing from 210 °C at 10 °C min, and holding at 320 °C for 7 min. The effluent was detected using a GCMS-TQ 8030 MS (Shimadzu Corp., Kyoto, Japan) system during 0.03 s with electron ionization at 15 eV. The ion source and interface temperatures were 200 and 280 °C, respectively. The primary metabolite analysis was performed for 5 repetitions.

#### 2.4.2. Quantification Methods and Ultra-Performance Liquid Chromatography/Time-of-Flight Mass Spectrometry (UPLC/Q-TOF-MS) Conditions for Secondary Metabolites

Extracts of sprouted ginseng ('Yunpoong' and 'K-1') were analyzed using UPLC/Q-TOF-MS (Figure S2 and Table S2). The dry sample (50 mg) was homogenized with 80% methanol (800 uL) using a blender to extract the metabolites. Following centrifugation, the supernatants were analyzed using UPLC-Q-TOF MS (Waters Corp., Milford, MA, USA). The metabolites were separated using an Acquity BEH C18 column (2.1 mm × 100 mm, 1.7 μm; Waters, Milford, MA, USA) equilibrated with water containing 0.1% formic acid (A) and acetonitrile containing 0.1% formic acid (B). The eluted metabolites were detected using Q-TOF MS with negative electrospray ionization mode. The desolation temperature and flow rates were 400 °C and 900 L $h^{-1}$, respectively, and the source temperature was 100 °C. The sampling cone and capillary voltages were 20 V and 2.5 kV, respectively. Leucine-enkephalin (554.2615 Da) was used as the lock mass with an infusion flow rate of 0.35 mL per min. MS data were obtained with a scan range of 50 to 1500 m $z^{-1}$. The secondary metabolite analysis was performed for 5 repetitions.

#### 2.4.3. Metabolite Data Processing

The spectra of substance peaks on GC/MS chromatograms were identified using the NIST 11 and Wiley 9 mass spectral libraries and retention indices, and calculated using n-alkanes. The skeptical peaks on UPLC/Q-TOF-MS chromatograms were evaluated using the Chemspider, Metlin (metlin.scripps.edu; accessed on 8 August 2022), human metabolome (www.hmdb.ca; accessed on 8 August 2022), and EZmass databases and authentic standards. The peaks were identified using the UNIFI software (version 1.9.2, Waters, Milford, MA, USA) connected to various online databases, version 3.9.

### *2.5. Statistical Analysis*

The growth characteristics, biological compounds, antioxidant properties, and ginsenoside content were measured. Ginseng sprouts of 42 plants for 3 repetitions were harvested according to the WAT. Data were analyzed using the SAS 9.4 program (SAS Institute Inc., Cary, NC, USA) with variance analysis. Duncan's multiple range test was used to verify the significant differences ($p < 0.05$) in all treatments. All graphs were created using SigmaPlot 8.0 (Systat Software Inc., San Jose, CA, USA).

Multivariate statistical analysis was carried out using SIMCA-P+ v.16.0.2 (Umetrics, Umea, Sweden) and all variables were automatically transformed and scaled to unit variance. PCA, PLS-DA, VIP, permutation test, and $p$-value were used to visualize the graphs. Statistical differences between the experimental data were analyzed using one-way analysis of variance (ANOVA) with Duncan's test using SPSS 27.0 (SPSS Inc., Chicago, IL, USA). Boxplots and heatmaps of metabolite were calculated and created using the R software.

## 3. Results and Discussion

### *3.1. Growth Characteristics*

Two ginseng cultivars ('Yunpoong' and 'K-1') showed different growth characteristic patterns at 2, 3, 4, and 5 weeks after transplanting (WAT) (Table 1). The shoot fresh weight of 'Yunpoong' showed the greatest value (1.0442 g) at 3 WAT, while 'K-1' showed the greatest value (0.8077 g) at 4 WAT. Overall, the shoot fresh weight was greater in 'Yunpoong' than in 'K-1'. The shoot dry weight showed a similar pattern to the shoot fresh weight. In the shoot dry weight, 'Yunpoong' showed the greatest value in the 3 WAT (0.1678 g) and 'K-1' in the 4 WAT (0.1271 g). The root fresh weights of 'Yunpoong' and 'K-1' showed the greatest values (0.7924 g, and 0.5502 g, respectively) at 3 WAT. Overall, the root fresh weight of 'Yunpoong' was greater than that of 'K-1'. Unlike the root fresh weight, the root dry weight showed a greater value in 'Yunpoong' at 3 WAT and 'K-1' at 5 WAT. The root dry weight of 'K-1' showed a similar trend to that of the root fresh weight of 'K-1'. Overall, the values of root dry weight were greater for 'Yunpoong' than for 'K-1'.

Table 1 presents the top/root ratio of 'Yunpoong' and 'K-1' at 2, 3, 4, and 5 WAT. A significantly higher ($p < 0.05$) top/root ratio was observed in 'K-1' than in 'Yunpoong' at 3 WAT. The top/root ratios of 'Yunpoong' and 'K-1' were the greatest at 4 WAT at 1.4058 and 1.5843, respectively. Similar to shoot fresh weight, there was a significantly difference regarding total yield between cultivars. The greatest total yield was 1.8367 g for 'Yunpoong' and 1.3315 g for 'K-1' at 4 WAT. Overall, the total yield value was greater in 'Yunpoong' than in 'K-1'.

**Table 1.** Growth patterns of 'Yunpoong' and 'K-1' at 3, 4, and 5 weeks after transplanting.

| Cultivar | WAT | Shoot (g) | | Root (g) | | Top/Root Ratio | Total Yield |
|---|---|---|---|---|---|---|---|
| | | Fresh Weight | Dry Weight | Fresh Weight | Dry Weight | | |
| Yunpoong | 2 | 0.70 ± 0.03 b | 0.09 ± 0.00 b | 0.71 ± 0.03 a | 0.10 ± 0.00 c | 0.99 ± 0.04 b | 1.42 ± 0.05 b |
| | 3 | 1.04 ± 0.04 a | 0.17 ± 0.01 a | 0.79 ± 0.04 a | 0.13 ± 0.01 ab | 1.34 ± 0.08 a | 1.84 ± 0.07 a |
| | 4 | 1.01 ± 0.07 a | 0.16 ± 0.01 a | 0.72 ± 0.05 a | 0.10 ± 0.01 bc | 1.41 ± 0.06 a | 1.73 ± 0.11 a |
| | 5 | 0.95 ± 0.05 a | 0.17 ± 0.01 a | 0.74 ± 0.04 a | 0.14 ± 0.01 a | 1.29 ± 0.04 a | 1.70 ± 0.09 a |
| K-1 | 2 | 0.40 ± 0.02 c | 0.04 ± 0.00 c | 0.42 ± 0.02 b | 0.05 ± 0.00 b | 0.96 ± 0.03 b | 0.83 ± 0.03 c |
| | 3 | 0.55 ± 0.03 b | 0.08 ± 0.01 b | 0.55 ± 0.05 a | 0.07 ± 0.01 a | 1.03 ± 0.06 b | 1.10 ± 0.07 b |
| | 4 | 0.81 ± 0.04 a | 0.13 ± 0.01 a | 0.52 ± 0.04 a | 0.08 ± 0.01 a | 1.58 ± 0.09 a | 1.33 ± 0.07 a |
| | 5 | 0.74 ± 0.03 a | 0.12 ± 0.01 a | 0.53 ± 0.04 a | 0.08 ± 0.01 a | 1.43 ± 0.08 a | 1.26 ± 0.06 a |

WAT means the weeks after transplanting. Different letters within columns indicate significant differences ($p < 0.05$), based on Duncan's test.

The growth characteristic results at 3 WAT for 'Yunpoong' and 4 WAT for 'K-1' are similar to previous studies. In previous studies, Jang et al. [34] reported that the shoot dry weight increased up to 4 WAT and then decreased. Developing the seedling and mature ginseng canopies takes approximately 4 weeks [35]. The root dry weight decreased for the first 2 WAT and then gradually increased [34]. In our study, 'K-1' showed the same trend, but 'Yunpoong' decreased at 4 WAT. These inconsistent results may be due to the different environmental conditions and cultivars.

Well-grown plants have more shoot biomass than root biomass, which increases the top/root ratio [36]. The total yield was generally identical to the shoot fresh weight because the change in shoot fresh weight over time was much more significant than that of the root fresh weight. The total yield showed the same trend as the shoot fresh weight [37], consistent with our results. In two ginseng cultivars used in a previous study ('Chenpoong' and 'Geumpoong'), 'Chenpoong' had higher significant differences than 'Geumpoong' in the fresh and dry weight by cultivar and growth period [38]. Our results also showed that the growth characteristics of ginseng were affected by cultivar and growth period.

### *3.2. Biological Compounds and Antioxidant Properties*

#### 3.2.1. Biological Compounds (Total Phenolic and Flavonoid Contents)

The total phenolic and flavonoid contents differed between the two ginseng cultivars ('Yunpoong' and 'K-1') at 2, 3, 4, and 5 WAT (Table 2). The total phenolic contents of the shoot showed a higher range in 'Yunpoong' than in 'K-1'. The total phenolic contents in the roots of 'Yunpoong' gradually increased and showed the highest content at 5 WAT (2.10 GAE mg $g^{-1}$); however, 'K-1' showed the highest content at 3 WAT (1.60 GAE mg $g^{-1}$). During 3–5 WAT, the total phenolic contents of the 'K-1' roots did not change significantly. Overall, the total flavonoid contents were higher in 'Yunpoong' than 'K-1' in both the shoots and roots. For 'Yunpoong' at 3 WAT (1.17 RE mg $g^{-1}$), the total flavonoid contents of the shoots had a higher content than at other WATs, while no significant differences occurred at 4 (1.08 RE mg $g^{-1}$) and 5 WAT (1.06 RE mg $g^{-1}$). 'K-1' showed the highest total flavonoid contents of the shoot at 4 WAT (0.98 RE mg $g^{-1}$). The total flavonoid contents in the roots of 'Yunpoong' and 'K-1' showed the highest values of 0.26 RE mg $g^{-1}$ at 4 and 5 WAT and 0.24 RE mg $g^{-1}$ at 5 WAT, respectively, showing a similar trend to the total flavonoid contents of the shoot.

Phytochemicals such as phenolic acids and flavonoids are widely present in various plants and are known as representative antioxidants derived from natural products [39,40]. Kim et al. [29] reported that the total phenolic and flavonoid contents of the shoot of ginseng sprouts were higher than those of the root; our results were similar to those of previous studies. The potential antioxidant activity was positively correlated with total phenolic and flavonoid contents [29,33,41,42]. However, our study showed different results depending on the growth period and cultivars. When comparing the total phenolic and flavonoid contents of the three cultivars in broccoli sprouts, the total phenolic contents differed significantly among the three cultivars. However, the total flavonoid contents did not differ significantly among the three cultivars [43]. Our results suggested that ginseng's phenolic and flavonoid contents differed depending on the cultivars and periods.

**Table 2.** Biological compounds of 'Yunpoong' and 'K-1' at 3, 4, and 5 weeks after transplanting.

| Cultivar | WAT | Total Phenolic Contents (GAE mg $g^{-1}$) | | Total Flavonoid Contents (RE mg $g^{-1}$) | |
|---|---|---|---|---|---|
| | | Shoot | Root | Shoot | Root |
| Yunpoong | 3 | 2.72 ± 0.04 b | 1.43 ± 0.02 b | 1.17 ± 0.01 a | 0.25 ± 0.00 a |
| | 4 | 3.19 ± 0.02 a | 1.86 ± 0.04 a | 1.08 ± 0.01 b | 0.26 ± 0.01 b |
| | 5 | 3.06 ± 0.04 a | 2.10 ± 0.09 a | 1.06 ± 0.01 b | 0.26 ± 0.00 b |
| K-1 | 3 | 2.57 ± 0.01 a | 1.60 ± 0.02 a | 0.90 ± 0.01 b | 0.18 ± 0.00 b |
| | 4 | 2.46 ± 0.01 b | 1.49 ± 0.04 a | 0.98 ± 0.02 a | 0.21 ± 0.00 a |
| | 5 | 2.32 ± 0.02 c | 1.51 ± 0.01 a | 0.79 ± 0.01 c | 0.24 ± 0.00 c |

WAT means the weeks after transplanting. Different letters within columns indicate significant differences ($p < 0.05$), based on Duncan's test.

### 3.2.2. Antioxidant Properties (DPPH, FRAP, and ABTS)

The DPPH of the shoot increased from 3 WAT (33.59%) to 5 WAT (43.59%) for 'Yunpoong' (Table 3), while that of 'K-1' was the highest at 4 WAT (35.93%). During the 3–5 WAT, DPPH of the shoots exhibited higher activity in 'Yunpoong' than 'K-1'. DPPH of the roots increased from 3 WAT to 5 WAT in both cultivars, and 'Yunpoong' showed 12.21% and 'K-1' 14.26% at 5 WAT.

**Table 3.** Antioxidant properties of 'Yunpoong' and 'K-1' at 3, 4, and 5 weeks after transplanting.

| Cultivar | WAT | DPPH Radical Scavenging Activity (%) | | Ferric Reducing/Antioxidant Power ($OD_{593nm}$) | | ABTS Radical Scavenging Activity (%) | |
|---|---|---|---|---|---|---|---|
| | | Shoot | Root | Shoot | Root | Shoot | Root |
| Yunpoong | 3 | 33.59 ± 0.22 c | 10.01 ± 0.38 b | 1.30 ± 0.01 c | 0.34 ± 0.00 b | 75.22 ± 0.32 b | 25.02 ± 0.72 c |
| | 4 | 37.70 ± 0.53 b | 11.43 ± 0.11 ab | 1.40 ± 0.01 b | 0.41 ± 0.01 a | 73.95 ± 0.56 c | 35.02 ± 0.47 b |
| | 5 | 43.59 ± 0.05 a | 12.21 ± 0.23 a | 1.47 ± 0.02 a | 0.40 ± 0.01 a | 78.01 ± 0.42 a | 39.17 ± 1.00 a |
| K-1 | 3 | 32.73 ± 0.31 b | 11.16 ± 0.23 c | 1.12 ± 0.01 b | 0.42 ± 0.00 c | 71.03 ± 0.30 a | 36.13 ± 0.28 b |
| | 4 | 35.93 ± 0.76 a | 12.99 ± 0.16 b | 1.25 ± 0.01 a | 0.48 ± 0.00 b | 61.35 ± 0.40 b | 35.44 ± 0.35 b |
| | 5 | 33.35 ± 0.29 ab | 14.26 ± 0.42 a | 1.26 ± 0.01 a | 0.53 ± 0.00 a | 62.09 ± 0.10 b | 39.74 ± 0.61 a |

WAT means the weeks after transplanting. Different letters within columns indicate significant differences ($p < 0.05$), based on Duncan's test.

The FRAP of the shoot content was higher in 'Yunpoong' than 'K-1', but the FRAP of the root content was higher in 'K-1' than in 'Yunpoong'. FRAP of the shoots increased from 3 to 5 WAT for 'Yunpoong' (1.30 and 1.47 $OD_{593nm}$, respectively) and 'K-1' (1.12 and 1.26 $OD_{593nm}$, respectively). FRAP of the roots showed a higher content (0.51 $OD_{593nm}$) at 4 WAT in 'Yunpoong' than in 'K-1'. FRAP of the roots increased from 3 to 5 WAT in 'K-1' (0.42 and 0.53 $OD_{593nm}$, respectively).

The ABTS of the shoots and roots in 'K-1' was higher than in 'Yunpoong' during the WAT. The ABTS of the shoots was the highest at 5 WAT and 3 WAT for 'Yunpoong' (78.01%) and 'K-1' (62.09%), respectively. The highest root activity for 'Yunpoong' (39.17%) and 'K-1' (39.74%) was observed at 5 WAT.

Antioxidant properties are used to determine the antioxidant capacity of fruits [44] and vegetables [45]. DPPH is a method to confirm the activity of scavenging anion radicals [46]. It is a relatively stable free radical with a deep purple color and is reduced by sulfur-containing amino acids and L-ascorbic acid to confirm antioxidant power. FRAP is a method that uses the fact that $Fe^{3+}$ can donate a hydrogen ion, and the radical is stabilized and reduced to $Fe^{2+}$ [47]. ABTS is a method for confirming the activity of scavenging cation radicals. It reacts with hydroxyl, peroxyl, alkoxyl, and inorganic radicals to form stable cation radicals, enabling the measurement of the anti-oxidation of hydrophilic and hydrophobic substances [48]. Kang et al. [49] concluded that DPPH radical scavenging activity is an indicator of antioxidant activity for the total phenolic contents; the higher the reducing power, the higher the activity. Our results showed no correlation between biological compounds and antioxidant properties. In grains and fruits, the correlation between total phenol content and antioxidant activity is high. However, it has been reported that there is little correlation between total phenol content and antioxidant activity in medicinal plants and fruits containing a large number of anthocyanins [50]. Judging from the report that phenolic compounds contained in ginseng have structural characteristics that make it difficult to access the DPPH radical center, the types and contents of phenolic compounds differed [51]. Presumably, this is because trace components other than phenolic compounds have a combined effect. There was no significant difference in the DPPH, FRAP, and ABTS contents between the shoots and roots [47]. Our results showed that the shoot (leaves) had higher antioxidant properties than the roots in all WAT.

### 3.3. Ginsenoside

The ginsenoside contents showed significant differences between the ginseng cultivars ('Yunpoong' and 'K-1') at 3, 4, and 5 WAT (Tables 4 and 5). Regarding the protopanaxatriol-type (PPT-type) compounds of the shoot, the contents of 'Yunpoong' and 'K-1' showed an increase in Rg1 and Re. The ginsenoside Rh1 was not detected in 'Yunpoong' at 4 WAT. The total PPT-type content of the shoot was 25.26 and 34.28 mg $g^{-1}$ for 'Yunpoong' and 'K-1', respectively, showing approximately 1.36 times higher content in 'K-1' at 5 WAT. Among the protopanaxadiol (PPD-type) compounds in the shoot, Rb2, Rb3, and F2 tended to increase over time. The compound K (C.K) was not detected in 'Yunpoong' at 3 WAT, and C.K and ginsenoside Rd2 did not differ significantly between 'Yunpoong' and 'K-1' at 3–5 WAT. The total PPD-type of the shoot was approximately 1.45 times higher for 'K-1' (69.44 mg $g^{-1}$) than for 'Yunpoong' (47.75 mg $g^{-1}$). Oleanane types (OA-type) in the 'Yunpoong' shoots decreased from 3 WAT (0.38 mg $g^{-1}$) to 5 WAT (0.24 mg $g^{-1}$). OA-type in the 'K-1' shoots increased from 3 WAT (0.32 mg $g^{-1}$) to 5 WAT (0.49 mg $g^{-1}$). The total ginsenoside contents of the shoots were 47.75 mg $g^{-1}$ and 69.44 mg $g^{-1}$ for 'Yunpoong' and 'K-1', respectively, at 5 WAT. The major ginsenoside derivatives of the shoots (at concentrations > 9 mg $g^{-1}$) were Re and Rd in 'Yunpoong' and 'K-1'at 3–5 WAT. The individual ginsenoside content distributions of the 'Yunpoong' shoots at 5 WAT were PPT-type (25.26 mg $g^{-1}$) >PPD-type (22.25 mg $g^{-1}$) >OA-type (0.24 mg $g^{-1}$), while those of 'K-1' were PPD-type (34.68 mg $g^{-1}$) > PPT-type (34.28 mg $g^{-1}$) > OA-type (0.49 mg $g^{-1}$).

The ginsenoside contents of the roots of the two ginseng cultivars ('Yunpoong' and 'K-1') were not significantly affected by WAT. PPT-type compounds of the 'Yunpoong' roots contained ginsenoside Re (3, 4, and 5 WAT; 1.09, 1.08, and 1.52 mg $g^{-1}$, respectively), F5 (5 WAT; 4.45 mg $g^{-1}$) and F3 (3 WAT; 2.75 mg $g^{-1}$) and the remaining components were detected at low levels (<2.0 mg $g^{-1}$). Ginsenoside Re was detected in the 'K-1' roots at 3, 4, and 5 WAT (3.78, 4.32, and 3.33 mg $g^{-1}$, respectively), and the remaining PPT-type compounds were detected at low levels. The PPD-type of the 'Yunpoong' roots was higher at 5 WAT (10.61 mg $g^{-1}$) than at 3 WAT (6.47 mg $g^{-1}$), while 'K-1' showed a high value at 3 WAT (9.89 mg $g^{-1}$). The OA-type of the roots was higher at 3 WAT (0.92 and 1.48 mg $g^{-1}$, respectively) than at 5 WAT (0.39 and 0.58 mg $g^{-1}$, respectively) for 'Yunpoong' and 'K-1'. Total ginsenoside contents of the 'Yunpoong' roots increased from 3 WAT (18.51 mg $g^{-1}$) to 5 WAT (25.52 mg $g^{-1}$), while that of the 'K-1' roots declined from 3 WAT (16.89 mg $g^{-1}$) to 5 WAT (15.29 mg $g^{-1}$). The individual ginsenoside content distributions of the 'Yunpoong' roots at 5 WAT were in the following order: PPT-type (13.44 mg $g^{-1}$) > PPD-type (10.61 mg $g^{-1}$) > OA-type (1.48 mg $g^{-1}$), while 'K-1' showed as the following order: PPD-type (9.41 mg $g^{-1}$) > PPT-type (5.30 mg $g^{-1}$) > OA-type (0.58 mg $g^{-1}$).

**Table 4.** Protopanaxatriol types in the 'Yunpoong' and 'K-1'shoots at 3, 4, and 5 weeks after transplanting.

| | Cultivar | WAT | Protopanaxatriol Types (mg g$^{-1}$) | | | | | | | | | |
|---|---|---|---|---|---|---|---|---|---|---|---|---|
| | | | Rg1 | Re | Rf | F5 | F3 | Rg2 | Rh1 | F1 | PPT | Total PPT-Type |
| Shoot | Yunpoong | 3 | 2.71 ± 0.03 c | 12.62 ± 0.02 b | 0.19 ± 0.00 a | 1.10 ± 0.01 b | 1.61 ± 0.01 b | 0.59 ± 0.01 b | 0.05 ± 0.03 a | 1.38 ± 0.00 c | 0.13 ± 0.00 b | 20.38 ± 0.01 c |
| | | 4 | 2.85 ± 0.01 b | 15.25 ± 0.08 a | 0.54 ± 0.08 a | 1.26 ± 0.04 ab | 1.88 ± 0.02 a | 0.69 ± 0.01 a | ND | 1.52 ± 0.01 a | 0.16 ± 0.00 a | 24.16 ± 0.23 b |
| | | 5 | 3.67 ± 0.02 a | 15.31 ± 0.09 a | 0.58 ± 0.20 a | 1.45 ± 0.09 a | 1.91 ± 0.00 a | 0.70 ± 0.01 a | 0.05 ± 0.02 a | 1.47 ± 0.00 b | 0.12 ± 0.00 b | 25.26 ± 0.27 a |
| | K-1 | 3 | 4.50 ± 0.01 c | 14.87 ± 0.02 c | 1.08 ± 0.02 a | 1.77 ± 0.03 a | 1.83 ± 0.01 c | 0.75 ± 0.02 b | 0.08 ± 0.00 b | 1.69 ± 0.01 b | 0.15 ± 0.01 b | 26.73 ± 0.02 c |
| | | 4 | 5.28 ± 0.01 b | 17.79 ± 0.03 b | 0.31 ± 0.21 b | 1.72 ± 0.07 a | 2.12 ± 0.02 b | 0.90 ± 0.01 a | 0.13 ± 0.01 a | 1.72 ± 0.02 b | 0.14 ± 0.01 b | 30.10 ± 0.20 b |
| | | 5 | 6.67 ± 0.31 a | 19.67 ± 0.29 a | 0.40 ± 0.15 b | 1.95 ± 0.10 a | 2.41 ± 0.02 a | 0.93 ± 0.04 a | 0.13 ± 0.00 a | 1.94 ± 0.04 a | 0.19 ± 0.00 a | 34.28 ± 0.19 a |
| Root | Yunpoong | 3 | 1.09 ± 0.01 b | 4.95 ± 0.01 b | 0.92 ± 0.10 a | 0.47 ± 0.03 b | 2.75 ± 0.04 a | 0.82 ± 0.08 a | ND | 0.05 ± 0.02 a | 0.06 ± 0.00 a | 11.12 ± 0.21 b |
| | | 4 | 1.08 ± 0.06 b | 4.68 ± 0.06 c | 0.89 ± 0.03 a | 0.51 ± 0.02 b | 0.19 ± 0.10 c | 0.56 ± 0.03 b | ND | 0.05 ± 0.00 a | 0.03 ± 0.00 b | 7.99 ± 0.13 c |
| | | 5 | 1.52 ± 0.05 a | 5.71 ± 0.03 a | 0.94 ± 0.03 a | 4.45 ± 0.06 a | 0.66 ± 0.07 b | ND | ND | 0.09 ± 0.01 a | 0.06 ± 0.00 a | 13.44 ± 0.13 a |
| | K-1 | 3 | 0.99 ± 0.01 a | 3.78 ± 0.01 b | 0.61 ± 0.03 a | 0.43 ± 0.00 b | 0.27 ± 0.01 a | 0.44 ± 0.00 b | ND | 0.05 ± 0.00 a | 0.03 ± 0.00 b | 6.61 ± 0.04 b |
| | | 4 | 0.86 ± 0.01 b | 4.32 ± 0.05 a | 0.59 ± 0.01 a | 0.53 ± 0.01 a | 0.19 ± 0.02 b | 0.55 ± 0.01 a | ND | 0.04 ± 0.00 b | 0.07 ± 0.00 a | 7.14 ± 0.09 a |
| | | 5 | 0.77 ± 0.02 c | 3.33 ± 0.01 c | 0.47 ± 0.02 b | 0.35 ± 0.00 c | ND | 0.31 ± 0.00 c | ND | 0.04 ± 0.00 b | 0.03 ± 0.00 b | 5.30 ± 0.05 c |

WAT means the weeks after transplanting. Different letters within columns indicate significant differences ($p < 0.05$), based on Duncan's test. ND: Not detected.

**Table 5.** Protopanaxadiol types, oleanane types, and total ginsenoside in the 'Yunpoong' and 'K-1' shoots at 3, 4, and 5 weeks after transplanting.

| | Cultivar | WAT | Protopanaxadiol Type (mg g$^{-1}$) | | | | | | | | | | | Oleanane Type | Total Ginsenoside |
|---|---|---|---|---|---|---|---|---|---|---|---|---|---|---|---|
| | | | Rb1 | Rc | Rb2 | Rb3 | Rd | Rd2 | F2 | Rg3 | C.K | Rh2 | PPD | Total PPD-Type | Ro | |
| Shoot | Yunpoong | 3 | 0.55 ± 0.01 c | 2.48 ± 0.08 b | 3.49 ± 0.03 b | 0.40 ± 0.04 a | 11.17 ± 0.22 b | 0.22 ± 0.02 a | 0.51 ± 0.09 a | 0.18 ± 0.03 a | ND | 0.10 ± 0.00 a | 0.92 ± 0.01 b | 20.03 ± 0.44 b | 0.38 ± 0.01 a | 40.80 ± 0.45 b |
| | | 4 | 0.63 ± 0.02 b | 2.84 ± 0.01 a | 3.95 ± 0.03 a | 0.45 ± 0.02 a | 12.50 ± 0.09 a | 0.28 ± 0.01 a | 0.58 ± 0.01 a | 0.17 ± 0.01 a | 0.01 ± 0.00 a | 0.07 ± 0.00 b | 1.10 ± 0.01 a | 22.59 ± 0.15 a | 0.37 ± 0.00 a | 47.12 ± 0.38 a |
| | | 5 | 0.68 ± 0.01 a | 2.71 ± 0.05 a | 3.99 ± 0.04 a | 0.49 ± 0.05 a | 12.20 ± 0.28 a | 0.27 ± 0.08 a | 0.70 ± 0.01 a | 0.15 ± 0.00 a | 0.01 ± 0.01 a | 0.06 ± 0.00 b | 0.99 ± 0.04 b | 22.25 ± 0.39 a | 0.24 ± 0.01 b | 47.75 ± 0.50 a |
| | K-1 | 3 | 0.95 ± 0.02 a | 3.23 ± 0.04 c | 4.98 ± 0.03 c | 0.52 ± 0.01 b | 13.46 ± 0.08 c | 0.19 ± 0.02 a | 0.54 ± 0.01 b | 0.28 ± 0.11 a | 0.01 ± 0.00 a | 0.13 ± 0.00 a | 0.84 ± 0.01 b | 25.14 ± 0.17 c | 0.32 ± 0.01 b | 52.19 ± 0.18 c |
| | | 4 | 0.94 ± 0.03 a | 4.20 ± 0.03 b | 5.89 ± 0.01 b | 0.68 ± 0.06 a | 14.63 ± 0.40 b | 0.26 ± 0.01 a | 0.69 ± 0.06 b | 0.19 ± 0.02 a | 0.01 ± 0.01 a | 0.12 ± 0.00 b | 1.38 ± 0.05 a | 28.99 ± 0.49 b | 0.46 ± 0.02 a | 59.55 ± 0.47 b |
| | | 5 | 1.04 ± 0.04 a | 4.77 ± 0.10 a | 6.54 ± 0.03 a | 0.75 ± 0.01 a | 18.62 ± 0.22 a | 0.40 ± 0.08 a | 0.87 ± 0.03 a | 0.26 ± 0.01 a | 0.01 ± 0.00 a | 0.11 ± 0.00 b | 1.30 ± 0.03 a | 34.68 ± 0.07 a | 0.49 ± 0.06 a | 69.44 ± 0.20 a |
| Root | Yunpoong | 3 | 0.01 ± 0.00 b | 2.82 ± 0.02 b | 1.73 ± 0.02 b | 0.27 ± 0.00 ab | 0.80 ± 0.01 b | 0.04 ± 0.00 a | 0.01 ± 0.00 b | 0.06 ± 0.01 a | 0.03 ± 0.00 c | 0.25 ± 0.00 b | 0.45 ± 0.01 a | 6.47 ± 0.03 b | 0.92 ± 0.01 a | 18.51 ± 0.25 b |
| | | 4 | 3.92 ± 0.17 a | 2.61 ± 0.03 c | 1.36 ± 0.01 c | 0.17 ± 0.08 b | 0.77 ± 0.01 b | 0.05 ± 0.02 a | ND | 0.04 ± 0.01 b | 0.05 ± 0.00 b | 0.23 ± 0.00 c | 0.48 ± 0.16 a | 9.66 ± 0.38 a | 1.24 ± 0.15 a | 18.90 ± 0.34 b |
| | | 5 | 0.37 ± 0.01 b | 4.37 ± 0.03 a | 2.72 ± 0.03 a | 0.40 ± 0.00 a | 1.60 ± 0.01 a | 0.07 ± 0.01 a | 0.04 ± 0.01 a | 0.05 ± 0.01 ab | 0.11 ± 0.00 a | 0.67 ± 0.00 a | 0.22 ± 0.00 a | 10.61 ± 0.08 a | 1.48 ± 0.39 a | 25.52 ± 0.18 a |
| | K-1 | 3 | 3.59 ± 0.01 a | 2.85 ± 0.01 a | 1.63 ± 0.02 a | 0.27 ± 0.01 a | 0.74 ± 0.02 a | 0.04 ± 0.00 b | 0.01 ± 0.00 a | 0.03 ± 0.00 a | 0.06 ± 0.00 b | 0.36 ± 0.00 a | 0.31 ± 0.01 b | 9.89 ± 0.02 a | 0.39 ± 0.05 b | 16.89 ± 0.05 a |
| | | 4 | 3.42 ± 0.05 b | 2.51 ± 0.04 b | 1.16 ± 0.02 c | 0.23 ± 0.01 b | 0.57 ± 0.01 b | 0.04 ± 0.00 ab | 0.02 ± 0.01 a | 0.03 ± 0.00 a | 0.03 ± 0.00 c | 0.31 ± 0.01 b | 0.42 ± 0.01 a | 8.74 ± 0.15 c | 0.43 ± 0.03 ab | 16.31 ± 0.27 a |
| | | 5 | 3.59 ± 0.01 a | 2.51 ± 0.01 b | 1.51 ± 0.01 b | 0.27 ± 0.01 a | 0.70 ± 0.01 a | 0.06 ± 0.01 a | 0.03 ± 0.00 a | 0.04 ± 0.00 a | 0.09 ± 0.00 a | 0.31 ± 0.01 b | 0.30 ± 0.00 b | 9.41 ± 0.03 b | 0.58 ± 0.01 a | 15.29 ± 0.07 b |

WAT means the weeks after transplanting. Different letters within columns indicate significant differences ($p < 0.05$), based on Duncan's test. ND: Not detected.

Ginsenosides are generally classified into two groups based on their chemical structure: the four-cyclic structure dammarane type and the five-cyclic structure OA-type. In addition, dammarane-type ginsenosides are divided into two groups, PPD-type and PPT-type [52]. Ginseng leaves had a 6–8 times higher total ginsenoside content than the roots [18]. Kim et al. [29] and Hwang et al. [53] reported on the growth in a vertical farm; total ginsenoside contents of the shoots (61.31–110 and 101.73 mg $g^{-1}$) were higher than that of the roots (25.81–34.19 and 37.39 mg $g^{-1}$). Our results showed that the total ginsenoside contents in 'Yunpoong' and 'K-1' were higher in the shoots (40.80–47.75 mg $g^{-1}$ and 52.19–69.44 mg $g^{-1}$, respectively) than in the roots (18.51–25.52 mg $g^{-1}$ and 15.29–16.89 mg $g^{-1}$, respectively). The major ginsenoside compounds in ginseng sprout leaves were ginsenoside Re and Rd [37,54,55]. In our study, the Rd and Re contents were higher in 'K-1' than in 'Yunpoong' (Rd, 11.17–12.50 mg $g^{-1}$ and 13.46–18.62 mg $g^{-1}$, respectively; Re, 12.62–15.31 mg $g^{-1}$ and 14.87–19.67 mg $g^{-1}$, respectively). Choi et al. [12] reported that the total ginsenoside contents were 97.29, 66.42, 67.61, and 36.24 mg $g^{-1}$ at 5, 7, 9, and 15 weeks, showing a gradually decreasing content.

### *3.4. Metabolites*

The primary metabolites measured at 7 WAT differed for the two ginseng cultivars ('Yunpoong' and 'K-1') (Figure 1). The identification of 'Yunpoong' and 'K-1' metabolites at 7 WAT using GC/MS was visualized using a partial least-squares discriminant analysis (PLS-DA) score plot (Figure S3). The goodness of fit ($R_2X = 0.721$; $R_2Y = 0.993$), predictability ($Q_2 = 0.974$), *p*-value (0.001), and cross-validation determined using the permutation test ($n = 5$) indicated that the PLS-DA model was statistically acceptable. In the PLS-DA score plot, 'Yunpoong' and 'K-1' were separated by t1 and t2. The VIP and *p*-values of individual metabolites were analyzed to find primary metabolites that contributed to the difference between ginseng cultivars (Figure S1 and Table S1). In total, 18 metabolites were detected: 7 amino acids, 1 inorganic acid, 4 organic acids, and 6 sugars were detected; their VIP was VIP $\geq 0.86$ and their *p*-values were $p < 0.05$. The primary metabolomic differences between 'Yunpoong' and 'K-1' are shown using boxplots and a heatmap (Figure 1). It showed that the content of primary metabolites differed between ginseng cultivars. 'Yunpoong' had a higher metabolite content than 'K-1', except for malic acid (an organic acid) and maltose (a sugar).

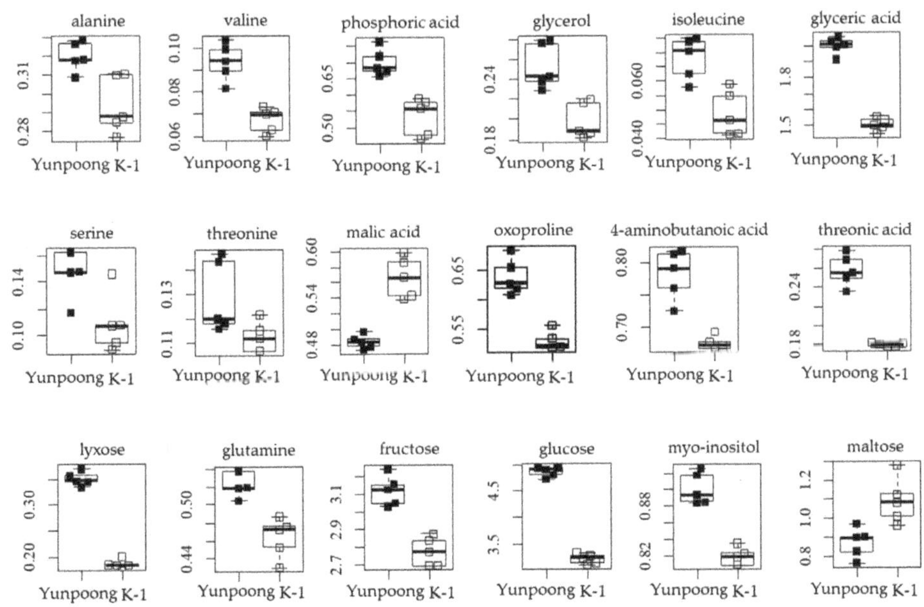


**Figure 1.** Box plots for primary metabolites in 'Yunpoong' and 'K-1' cultivars.

At week 7, the secondary metabolite patterns differed between the two ginseng cultivars ('Yunpoong' and 'K-1'), as detected using UPLC/Q-TOF-MS (Figure 2). The UPLC/Q-TOF-MS result was visualized using a PLS-DA to discriminate between the two ginseng cultivars ('Yunpoong' and 'K-1') (Figure S4). The goodness of fit ($R_2X = 0.736$; $R_2Y = 0.999$), predictability ($Q_2 = 0.995$), *p*-value ($6.19 \times 10^{-4}$), and cross-validation determined using the permutation test ($n = 5$) indicated that the PLS-DA model was statistically acceptable. In the PLS-DA score plot, 'Yunpoong' and 'K-1' were separated by t1 and t2. The VIP and *p*-values of individual metabolites were analyzed to find secondary metabolites that contributed to the difference between ginseng cultivars (Figure S2 and Table S2). In total, 32 metabolites were detected. The boxplots and heatmap showed differences in secondary metabolites between 'Yunpoong' and 'K-1' cultivars (Figure 2). It shows that the content of secondary metabolites is different for each cultivar of ginseng and that 'Yunpoong' had higher contents of all secondary metabolites than 'K-1'.

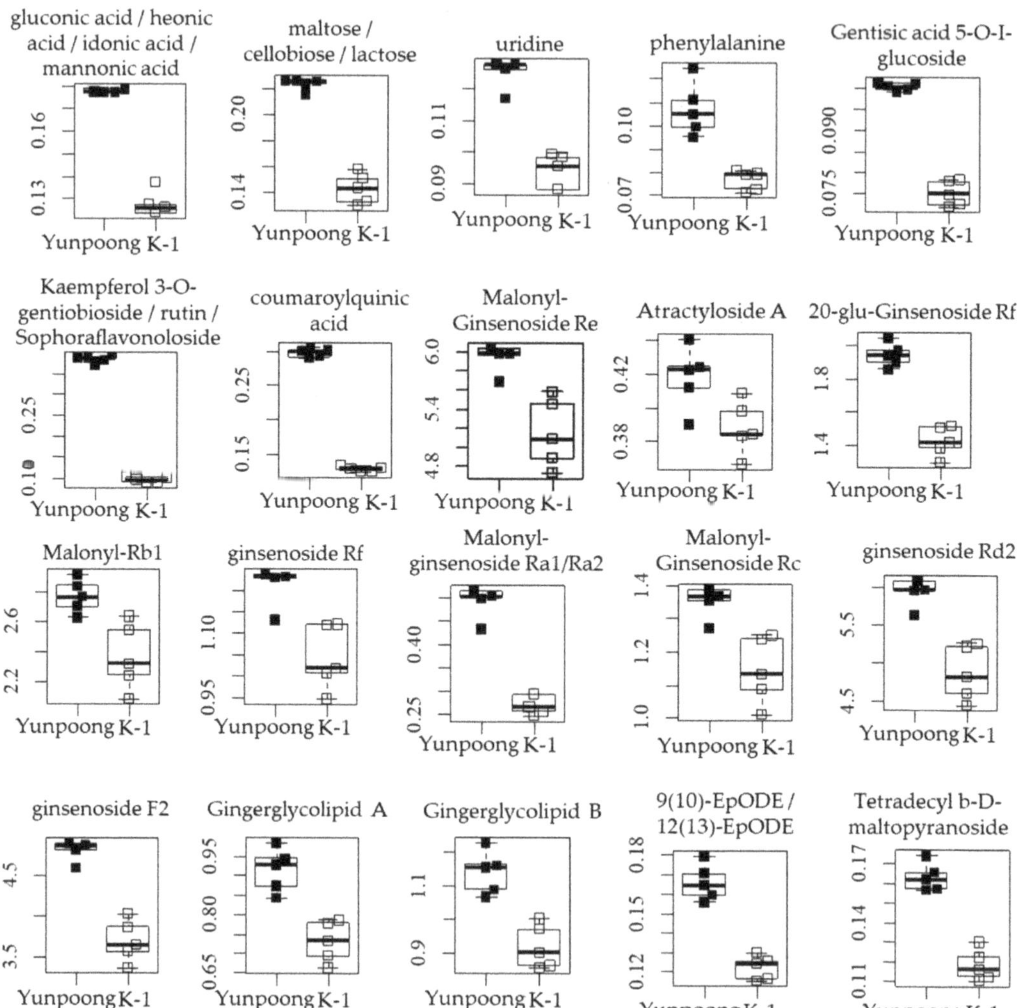


**Figure 2.** Box plots for secondary metabolites in 'Yunpoong' and 'K-1' cultivars.

Eom et al. [40] reported that ginseng metabolites are related to properties such as anti-aging, anti-diabetic, anti-cancer, and anti-inflammatory. Previously, 21 and 34 metabolites detected using GC/MS were identified and reported in Asian and Western ginseng, respectively [56]. However, in our results, only 18 metabolites were detected through GC/MS.

### 3.5. Annual Net Production

Since year-round production is possible in plant factories, it is necessary to compare net production with actual production for efficiency. The annual net production of 'Yunpoong' showed a higher total yield and ginsenoside production per year than 'K-1' (Table 6). Through analyzing the growth of ginseng and the net production of ginsenosides at 3, 4, and 5 WAT, it was confirmed that the annual net production and the content of ginsenosides decreased as the cultivation period increased for both cultivars.

**Table 6.** Annual net production of 'Yunpoong' and 'K-1' at 3, 4, and 5 weeks after transplanting.

| Cultivar | WAT | Annual Net Production | |
|---|---|---|---|
| | | **Yield (g m$^2$ Year)** | **Ginsenoside (mg m$^2$ Year)** |
| Yunpoong | 3 | 1.72 ± 0.06 a | 102.12 ± 1.01 a |
| | 4 | 1.22 ± 0.08 b | 80.28 ± 0.22 b |
| | 5 | 0.96 ± 0.05 c | 69.99 ± 0.64 c |
| K-1 | 3 | 1.03 ± 0.06 a | 71.26 ± 0.16 a |
| | 4 | 0.94 ± 0.05 a | 71.02 ± 0.68 a |
| | 5 | 0.71 ± 0.03 b | 60.18 ± 0.12 b |
| Significance | | | |
| Cultivar (C) | | <0.0001 | <0.0001 |
| Week (W) | | <0.0001 | <0.0001 |
| C × W | | 0.0005 | <0.0001 |

WAT means the weeks after transplanting. Different letters within columns indicate significant differences ($p < 0.05$), based on Duncan's test.

Stable production of medicinal crops can be obtained in vertical farms [25]. Park et al. [57] showed that the yield of ginseng grown in vertical farms was higher than in open fields. In this analysis of the efficiency for the net production in plant factories, the efficient cultivation period may be 3 and 4 WAT for 'Yunpoong' and 'K-1', respectively.

## 4. Conclusions

Ginseng sprouts are widely used as a medicinal crop. According to ginseng cultivars, growth and quality (for example, ginsenosides) differ. This study compared two ginseng cultivars grown in a plant factory. Growth characteristics showed higher values for growth parameters at 3 and 4 WAT in 'Yunpoong' and 'K-1', respectively. Moreover, the 'Yunpoong' cultivar yield was higher than that of the 'K-1' cultivar. Biological compounds and antioxidant properties were the highest at 5 and 4 WAT in 'Yunpoong' and 'K-1', respectively. The highest ginsenoside content was detected at 5 WAT in both 'Yunpoong' and 'K-1' cultivars. 'Yunpoong' had a higher content than 'K-1' for most primary and secondary metabolites. Yunpoong at 3 WAT and K-1 at 3 and 4 WAT showed the highest yield and ginsenoside content in the annual net production. In conclusion, the cultivar and harvesting period might be important factors for the production, yield, and quality of ginseng. Moreover, it is essential to determine a vertical farm's annual net production efficiency for the optimum cultivation time.

**Supplementary Materials:** The following supporting information can be downloaded at: https://www.mdpi.com/article/10.3390/horticulturae9050583/s1, Figure S1: Standard chromatograms analyzed using GC/MS; Figure S2: Standard chromatogram analyzed using UPLC/Q-TOF-MS; Figure S3: Partial least-squares discriminant analysis and Permutation in 'Yunpoong' and 'K-1' of primary metabolites; Figure S4: Partial least-squares discriminant analysis and Permutation in 'Yunpoong' and 'K-1' of secondary metabolites. Table S1: Identification of significant primary metabolites contributing to the separation among sample groups; Table S2: Identification of major secondary metabolites contributing to the separation among sample groups.

**Author Contributions:** K.-H.S., J.H.L. and K.M.C. conceived the project. G.O.L., S.-N.J., M.J.K. and D.Y.C. performed experiments. G.O.L. wrote the manuscript. All authors have read and agreed to the published version of the manuscript.

**Funding:** This work was carried out with the support of "Cooperative Research Program for Agriculture Science and Technology Development (project no. PJ015655)" Rural Development Administration, Republic of Korea.

**Data Availability Statement:** All data are available in the paper.

**Acknowledgments:** We are grateful for helpful co-working and discussions with members of the Cho's laboratory and National Institute of Horticultural and Herbal Science.

**Conflicts of Interest:** The authors declare no conflict of interest.

## References

1. Ministry of Agriculture, Food and Rural Affairs (MAFRA). *2021 Ginseng Statistics*; Ministry of Agriculture, Food and Rural Affairs: Sejong, Republic of Korea, 2022; pp. 1–36.
2. Kim, Y.C.; Kim, Y.B.; Park, H.W.; Bang, K.Y.; Kim, J.U.; Jo, I.H.; Kim, K.H.; Song, B.H.; Kim, D.H. Optimal harvesting time of ginseng seeds and effect of gibberellic acid ($GA_3$) treatment for improving stratification rate of ginseng (*Panax ginseng* C. A. Meyer) seeds. *Korean J. Med. Crop Sci.* **2014**, *22*, 423–428. [CrossRef]
3. Ahn, I.O.; Lee, S.S.; Lee, J.H.; Lee, M.J.; Jo, B.G. Comparison of ginsenoside contents and pattern similarity between root parts of new cultivars in *Panax ginseng* C. A. Meyer. *J. Ginseng Res.* **2008**, *32*, 15–18. [CrossRef]
4. Kwon, W.S.; Lee, M.G.; Lee, J.H. Characteristics of flowering and fruiting in new varieties and lines of *Panax ginseng* C. A. Meyer. *J. Ginseng Res.* **2001**, *25*, 41–44.
5. Lee, S.O.; Kim, M.J.; Kim, D.G.; Choi, H.J. Antioxidative activities of temperature-stepwise water extracts from Inonotus obliquus. *J. Korean Soc. Food Sci. Nutr.* **2005**, *34*, 139–147. [CrossRef]
6. Kwon, W.S.; Lee, M.G.; Choi, K.T. Breeding process and characteristics of Yunpoong; a new variety of *Panax ginseng* C. A. Meyer. *J. Ginseng Res.* **2000**, *24*, 1–7.
7. Wang, H.; Xu, F.; Wang, X.; Kwon, W.S.; Yang, D.C. Molecular discrimination of *Panax ginseng* cultivar K-1 using pathogenesis-related protein 5 gene. *J. Ginseng Res.* **2019**, *43*, 482–487. [CrossRef]
8. Kim, J.H.; Yi, Y.S.; Kim, M.Y.; Cho, J.Y. Role of ginsenosides; the main active components of *Panax ginseng*; in inflammatory responses and diseases. *J. Ginseng Res.* **2017**, *41*, 435–443. [CrossRef]
9. Kim, D.H.; Moon, Y.S.; Lee, T.H.; Jung, J.S.; Suh, H.W.; Song, D.K. The inhibitory effect of ginseng saponins on the stress-induced plasma interleukin-6 level in mice. *Neurosci. Lett.* **2003**, *353*, 13–16. [CrossRef]
10. López, M.V.N.; Cuadrado, M.P.G.-S.; RuizPoveda, O.M.P.; Del Fresno, A.M.V.; Accame, M.E.C. Neuroprotective effect of individual ginsenosides on astrocytes primary culture. *Biochim. Biophys. Acta* **2007**, *1770*, 1308–1316. [CrossRef]
11. Vuksan, V.; Sung, M.K.; Sievenpiper, J.L.; Stavro, P.M.; Jenkins, A.L.; Di Buono, M.; Lee, K.S.; Leiter, L.A.; Nam, K.Y.; Arnason, J.T. Korean red ginseng (*Panax ginseng*) improves glucose and insulin regulation in well-controlled; type 2 diabetes, Results of a randomized; double-blind; placebo-controlled study of efficacy and safety. *Nutr. Metabol. Cardiovasc. Dis.* **2008**, *18*, 46–56. [CrossRef]
12. Choi, J.E.; Li, X.; Han, Y.H.; Lee, K.T. Changes of saponin contents of leaves; stems and flower buds of *Panax ginseng* C. A. Meyer by harvesting days. *Korean J. Med. Crop Sci.* **2009**, *17*, 251–256.
13. Lim, W.S. Effects of interactions among age; cultivation method (location) and population on ginsenoside content of wild *Panax quinquefolium* L. one year after transplanting from wild. *Korean J. Med. Crop Sci.* **2005**, *13*, 254–261.
14. Li, X.G.; Nam, K.Y.; Choi, J.E. Difference of the ginsenosides contents according to the planting location in *Panax ginseng* C. A. Meyer. *Korean J. Crop Sci.* **2009**, *54*, 159–164.
15. Yahara, S.; Tanaka, O.; Komori, T. Saponins of the leaves of *Panax ginseng* C. A. Meyer. *Chem. Pharm. Bull.* **1976**, *24*, 2204–2208. [CrossRef]
16. Xie, J.T.; Lin, E.; Wang, C.Z.; Yuan, C.S. Constituents and effects of ginseng leaf. *Orient. Pharm. Exp. Med.* **2004**, *4*, 1–8. [CrossRef]
17. Wang, Y.; Pan, J.Y.; Xiao, X.Y.; Lin, R.C.; Cheng, Y.Y. Simultaneous determination of ginsenosides in *Panax ginseng* with different growth ages using high-performance liquid chromatography-mass spectrometry. *Phytochem. Anal.* **2006**, *17*, 424–430. [CrossRef]

18. Kim, G.S.; Hyun, D.Y.; Kim, Y.O.; Lee, S.E.; Kwon, H.; Cha, S.W.; Park, C.B.; Kim, Y.B. Investigation of ginsenosides in different parts of *Panax ginseng* cultured by hydroponics. *Korean J. Hortic. Sci. Technol.* **2010**, *28*, 216–226.
19. Kozai, T. Why LED Lighting for Urban Agriculture? In *LED Lighting for Urban Agriculture*, 1st ed.; Kosai, T., Fujiwara, K., Runkle, E., Eds.; Springer: Singapore, 2016; pp. 3–18.
20. Kim, J.H.; Chang, S.D. Industrialization condition and possibility of plant factory. *Korean J. Agric. Manag. Policy* **2009**, *36*, 918–948.
21. Goto, E. Chapter 15—Production of Pharmaceuticals in a Specially Designed Plant Factory A2—Kozai, Toyoki. In *Plant Factory*; Niu, G., Takagaki, M., Eds.; Academic Press: San Diego, CA, USA, 2016; pp. 193–200. [CrossRef]
22. Cheon, S.K.; Mok, S.K.; Lee, S.S.; Shin, D.Y. Effects of light intensity and quality on the growth and quality of Korean ginseng (*Panax ginseng* C. A. Meyer) I. Effects of light intensity on the growth and yield of ginseng plants. *J. Ginseng Res.* **1991**, *15*, 21–30.
23. Kim, M.J.; Li, X.; Han, J.S.; Lee, S.E.; Choi, J.E. Effect of blue and red LED irradiation on growth characteristics and saponin contents in *Panax ginseng* C. A. Meyer. *Korean J. Med. Crop Sci.* **2009**, *17*, 187–191. [CrossRef]
24. Johkan, M.; Shoji, K.; Goto, F.; Hahida, S.; Yoshihara, T. Effect of green light wavelength and intensity on photomorphogenesis and photosynthesis in Lactuca sativa. *Environ. Exp. Bot.* **2012**, *75*, 128–133. [CrossRef]
25. Urrestarazu, M.; Nájera, C.; del Mar Gea, M. Effect of the spectral quality and intensity of light-emitting diodes on several horticultural crops. *HortScience* **2016**, *51*, 268–271. [CrossRef]
26. Lee, D.U.; Ku, H.B.; Lee, Y.J.; Kim, G.N.; Lee, S.C. Antioxidant and antimelanogenic activities of *Panax ginseng* sprout extract. *J. Korean Soc. Food Sci. Nutr.* **2019**, *48*, 699–703. [CrossRef]
27. Nguyen, T.K.L.; Lee, J.-H.; Lee, G.O.; Cho, K.M.; Cho, D.Y.; Son, K.-H. Optimization of Cultivation Type and Temperature for the Production of Balloon Flower (Platycodongrandiflorum A. DC) Sprouts in a Plant Factory with Artificial Lighting. *Horticulturae* **2022**, *8*, 315. [CrossRef]
28. Cho, K.M.; Lee, H.Y.; Lee, Y.M.; Seo, E.Y.; Kim, D.H.; Son, K.-H.; Lee, J.; Cho, D.Y.; Lee, J.H. Comparative assessment of compositional constituents and antioxidant effects in ginseng sprouts (*Panax ginseng*) through aging and fermentation processes. *LWT* **2022**, *164*, 113644. [CrossRef]
29. Kim, S.C.; Kang, Y.M.; Seong, J.A.; Lee, H.Y.; Cho, D.Y.; Joo, O.S.; Lee, J.H.; Cho, K.M. Comprehensive changes of nutritional constituents and antioxidant activities of ginseng sprouts according to the roasting process. *Korean J. Food Preserv.* **2021**, *28*, 72–87. [CrossRef]
30. Cho, K.M.; Hwang, C.E.; Joo, O.S. Change of physicochemical properties; phytochemical contents and biological activities during the vinegar fermentation of Elaeagnus multiflora fruit. *Korean J. Food Preserv.* **2017**, *24*, 125–133. [CrossRef]
31. Lee, H.Y.; Lee, D.H.; Kim, S.C.; Cho, D.Y.; Cho, K.M. Changes in nutritional components and antioxidant activities from soybean leaves containing high isoflavone contents according to different storage temperatures and periods. *J. Appl. Biol. Chem.* **2020**, *63*, 305–317. [CrossRef]
32. Jang, S.-N.; Lee, G.O.; Sim, H.-S.; Bae, J.-S.; Lee, A.-R.; Cho, D.-Y.; Cho, K.-M.; Son, K.-H. Effect of Pre-Harvest Irradiation of UV-A and UV-B LED in Ginsenosides Content of Ginseng Sprouts. *J. Bio-Environ. Control.* **2022**, *31*, 28–34. [CrossRef]
33. Lee, J.H.; Kim, S.C.; Lee, H.Y.; Cho, D.Y.; Jung, J.G.; Kang, D.; Kang, S.S.; Cho, K.M. Changes in nutritional compositions of processed mountain-cultivated ginseng sprouts (*Panax ginseng*) and screening for their antioxidant and anti-inflammatory properties. *J. Funct. Foods* **2021**, *86*, 104668. [CrossRef]
34. Jang, I.B.; Yu, J.; Suh, S.J.; Jang, I.B.; Kwon, K.B. Growth and Ginsenoside Content in Different Parts of Ginseng Sprouts Depending on Harvest Time. *Korean J. Med. Crop Sci.* **2018**, *26*, 205–213. [CrossRef]
35. Proctor, J.T.A.; Dorais, M.; Bleiholder, H.; Willis, A.; Hack, H.; Meier, V. Phenological growth stages of North American ginseng (*Panax quinquefolius*). *Ann. Appl. Biol.* **2003**, *143*, 311–317. [CrossRef]
36. Minghui, J.; Xiaotong, J.; Yunlong, Z. Effects of long-term nitrogen addition on community aboveground and belowground biomass and their ratio in a typical steppe of Inner Mongolia. *Chin. J. Ecol.* **2020**, *39*, 3185–3193. [CrossRef]
37. Kim, Y.J.; Nguyen, T.K.L.; Oh, M.M. Growth and Ginsenosides Content of Ginseng Sprouts According to LED-Based Light Quality Changes. *Agronomy* **2020**, *10*, 1979. [CrossRef]
38. Song, B.H.; Chang, Y.G.; Lee, K.A.; Lee, S.W.; Kang, S.W.; Cha, S.W. Studies on Analysis of growth characteristics; ability of dry matter production; and yield of *Panax ginseng* C. A. Meyer at different growth stages with different cultivars and shading nets in paddy field. *Korean J. Med. Crop Sci.* **2011**, *19*, 90–96. [CrossRef]
39. Haralampidis, K.; Trojanowska, M.; Osbourn, A.E. Biosynthesis of triterpenoid saponin in plants. *Adv. Biochem. Eng. Biotechnol.* **2002**, *75*, 31–49. [CrossRef]
40. Eom, S.J.; Hwang, J.E.; Kim, H.S.; Kim, K.T.; Paik, H.D. Anti-inflammatory and cytotoxic effects of ginseng extract bioconverted by Leuconostoc mesenteroides KCCM 12010P isolated from kimchi. *Int. J. Food Sci. Technol.* **2018**, *53*, 1331–1337. [CrossRef]
41. Lee, K.S.; Seong, B.J.; Kim, G.H.; Kim, S.I.; Han, S.H.; Kim, H.H.; Baik, N.D. Ginsenoside; phenolic acid composition and physiological significances of fermented ginseng leaf. *J. Korean Soc. Food Sci. Nutr.* **2010**, *39*, 1194–1200. [CrossRef]
42. Chung, I.M.; Lim, J.J.; Ahn, M.S.; Jeong, H.N.; An, T.J.; Kim, S.H. Comparative phenolic compound profiles and antioxidant activity of the fruit; leaves; and roots of Korean ginseng (*Panax ginseng* Meyer) according to cultivation years. *J. Ginseng Res.* **2016**, *40*, 68–75. [CrossRef]
43. Tian, M.; Xu, X.; Liu, Y.; Xie, L.; Pan, S. Effect of Se treatment on glucosinolate metabolism and health-promoting compounds in the broccoli sprouts of three cultivars. *Food Chem.* **2016**, *190*, 374–380. [CrossRef]

44. Hashemi, S.M.B.; Khaneghah, A.M.; Barba, F.J.; Nemati, Z.; Shokofti, S.S.; Alizadeh, F. Fermented sweet lemon juice (*Citrus limetta*) using Lactobacillus plantarum LS5, Chemical composition; antioxidant and antibacterial activities. *J. Funct. Foods* **2017**, *38*, 409–414. [CrossRef]
45. Cho, K.M.; Ha, T.J.; Lee, Y.B.; Seo, W.D.; Kim, J.Y.; Ryu, H.W.; Jeong, S.H.; Kang, Y.M.; Lee, J.H. Soluble phenolics and antioxidant properties of soybean (*Glycine max* L.) cultivars with varying seed coat colours. *J. Funct. Foods* **2013**, *5*, 1065–1076. [CrossRef]
46. Lee, S.S.; Lee, J.H.; Ahn, I.O. Characteristics of new cultivars in *Panax ginseng* C.A. Meyer. In *Proceedings of the Ginseng Society Conference*; Korean Society Ginseng: Seoul, Republic of Korea, 2005; Volume 18, pp. 3–18.
47. Park, S.J. Antioxidant activities and whitening effects of ethanol extract from *Panax ginseng* sprout powder. *J. Korean Soc. Food Sci. Nutr.* **2019**, *48*, 276–281. [CrossRef]
48. Re, R.; Pellegrini, N.; Proteggente, A.; Pannala, A.; Yang, M.; Rice-Evans, C. Antioxidant activity applying an improved ABTS radical cation decolorization assay. *Free Radic. Biol. Med.* **1999**, *26*, 1231–1237. [CrossRef] [PubMed]
49. Kang, K.S.; Kim, H.Y.; Yamabe, N.; Yokozawa, T. Stereospecificity in hydroxyl radical scavenging activities of four ginsenosides produced by heat processing. *Bioorg. Med. Chem. Lett.* **2006**, *16*, 5028–5031. [CrossRef] [PubMed]
50. Velioglu, Y.S.; Mazza, G.; Gao, L.; Oomah, B.D. Antioxidant activity and total phenolics in selected fruits; vegetables; and grain products. *J. Agric. Food Chem.* **1998**, *46*, 4113–4117. [CrossRef]
51. Yoshida, T.; Mori, K.; Hatano, T.; Okumura, T.; Uehara, I.; Komagoe, K.; Fujita, Y.; Okuda, T. Studies on inhibition mechanism of autoxidation by tannins and flavonoids. V. Radical-scavenging effects of tannins and related polyphenols on 1;1-diphenyl-2-picrylhydrazyl radical. *Chem. Pharm. Bull.* **1989**, *37*, 1919–1921. [CrossRef]
52. Kim, K.H.; Lee, D.; Lee, H.L.; Kim, C.E.; Jung, K.; Kang, K.S. Beneficial effects of *Panax ginseng* for the treatment and prevention of neurodegenerative diseases, Past findings and future directions. *J. Ginseng Res.* **2018**, *42*, 239–247. [CrossRef]
53. Hwang, S.I.; Joo, J.M.; Joo, S.Y. ICT-based smart farm factory systems through the case of hydroponic ginseng plant factory. *J. Korean Inst. Commun. Inf. Sci.* **2015**, *40*, 780–790. [CrossRef]
54. Seong, B.J.; Kim, S.I.; Jee, M.G.; Lee, H.C.; Kwon, A.R.; Kim, H.H.; Won, J.Y.; Lee, K.S. Changes in Growth; Active Ingredients; and Rheological Properties of Greenhouse-Cultivated Ginseng Sprout during Its Growth Period. *Korean J. Med. Crop Sci.* **2019**, *27*, 126–135. [CrossRef]
55. Hwang, S.H.; Kim, S.C.; Seong, J.A.; Lee, H.Y.; Cho, D.Y.; Kim, M.J.; Jung, J.G.; Jeong, E.H.; Son, K.-H.; Cho, K.M. Comparison of ginsenoside contents and antioxidant activity according to the size of ginseng sprout has produced in a plant factory. *J. Appl. Biol. Chem.* **2021**, *64*, 253–261. [CrossRef]
56. Yang, L.; Yu, Q.T.; Ge, Y.Z.; Zhang, W.S.; Fan, Y.; Ma, C.W.; Liu, Q.; Qi, L.W. Distinct urine metabolome after Asian ginseng and American ginseng intervention based on GC-MS metabolomics approach. *Sci. Rep.* **2016**, *6*, 39045. [CrossRef] [PubMed]
57. Park, J.-E.; Kim, H.; Kim, J.; Choi, S.-J.; Ham, J.; Nho, C.W.; Yoo, G. A comparative study of ginseng berry production in a vertical farm and an open field. *Ind. Crops Prod.* **2019**, *140*, 111612. [CrossRef]

**Disclaimer/Publisher's Note:** The statements, opinions and data contained in all publications are solely those of the individual author(s) and contributor(s) and not of MDPI and/or the editor(s). MDPI and/or the editor(s) disclaim responsibility for any injury to people or property resulting from any ideas, methods, instructions or products referred to in the content.

horticulturae

Article

# Comparative Analysis of Glucosinolate and Phenolic Compounds in Green and Red Kimchi Cabbage (*Brassica rapa* L. ssp. *pekinensis*) Hairy Roots after Exposure to Light and Dark Conditions

Sook Young Lee [1], Haejin Kwon [2], Jae Kwang Kim [3], Chang Ha Park [4], Ramaraj Sathasivam [2,*] and Sang Un Park [2,5,*]

1 Marine Bio Research Center, Chosun University, 61-220 Myeongsasimni, Sinji-myeon, Wando-gun 59146, Republic of Korea
2 Department of Crop Science, Chungnam National University, 99 Daehak-ro, Yuseong-gu, Daejeon 34134, Republic of Korea
3 Division of Life Sciences and Convergence Research Center for Insect Vectors, College of Life Sciences and Bioengineering, Incheon National University, Yeonsu-gu, Incheon 22012, Republic of Korea
4 Department of Biological Sciences, Keimyung University, Dalgubeol-daero 1095, Dalseo-gu, Daegu 42601, Republic of Korea
5 Department of Smart Agriculture Systems, Chungnam National University, 99 Daehak-ro, Yuseong-gu, Daejeon 34134, Republic of Korea
* Correspondence: ramarajbiotech@gmail.com (R.S.); supark@cnu.ac.kr (S.U.P.)

**Abstract:** *Brassica rapa* L. ssp. *pekinensis* (Lour.) Hanelt (kimchi cabbage) is a major vegetable cultivated in Korea, and its hairy roots (HRs) are rich in glucosinolates and phenolic compounds. This study aimed to induce HRs from cotyledon explants via the transformation of the *Agroacterium rhizogenes* strain R1000 and examine the glucosinolate and phenolic compounds present in the HRs of two kimchi cabbage (green and red) cultivars after exposure to 16 h light/8 h dark conditions (photosynthetic photon flux density of 54.6 μmol $m^{-2}$ $s^{-1}$) and continuous dark conditions. The highest HR production was achieved in the green kimchi cabbage grown under dark conditions (0.37 ± 0.01 DW g/30 mL). The highest glucosinolate and phenolic contents were neoglucobrassicin and catechin hydrate, which were highest in the green kimchi HRs grown under dark (GKHD) conditions (5268.29 ± 292.84 μg/g DW) and green HRs grown under light (GKHL) conditions (203.49 ± 4.70 μg/g DW), respectively. A heat map showed that the red kimchi HRs grown under dark conditions (RKHD) and the GKHL condition accumulated the highest glucosinolate and phenolic contents. Principal component (PCA) and partial least-squares discriminant (PLS-DA) analyses of the 13 identified metabolites showed a clear separation. According to a variable importance in projection (VIP) analysis, quercetin was the most important metabolite, leading to a clear separation. The most suitable conditions for enhancing the glucosinolate and phenolic contents were the GKHD and GKHL conditions, respectively, whereas both compounds were enhanced in the RKHD condition. HRs cultures cultivated under light and dark conditions are a promising method to enhance the production of specific health-promoting bioactive metabolites, which might be helpful in the pharmaceutical and nutraceutical industries.

**Keywords:** *Agrobacterium rhizogenes*; *Brassica rapa* L. ssp. *pekinensis*; hairy root; glucosinolate; phenolic compounds

check for updates

**Citation:** Lee, S.Y.; Kwon, H.; Kim, J.K.; Park, C.H.; Sathasivam, R.; Park, S.U. Comparative Analysis of Glucosinolate and Phenolic Compounds in Green and Red Kimchi Cabbage (*Brassica rapa* L. ssp. *pekinensis*) Hairy Roots after Exposure to Light and Dark Conditions. *Horticulturae* **2023**, *9*, 466. https://doi.org/10.3390/horticulturae9040466

Academic Editor: Wajid Zaman

Received: 10 March 2023
Revised: 4 April 2023
Accepted: 5 April 2023
Published: 7 April 2023

**Copyright:** © 2023 by the authors. Licensee MDPI, Basel, Switzerland. This article is an open access article distributed under the terms and conditions of the Creative Commons Attribution (CC BY) license (https://creativecommons.org/licenses/by/4.0/).

## 1. Introduction

*Brassica rapa* L. ssp. *pekinensis* (Lour.) Hanelt, otherwise called Kimchi cabbage, belongs to the Brassicaceae family, and it is the main ingredient of traditional Korean foods. It is prepared through a series of fermentation processes [1]. In kimchi cabbage, several phenolic compounds with antioxidant activity and flavonoids have been identified [2,3].

In addition, it contains aliphatic glucosinolates, aromatic glucosinolates, and relatively high amounts of indolic glucosinolates, a precursor to indole-3-carbinol, which chemically prevents cancer [4,5].

Different varieties of kimchi cabbage have been produced through introgression breeding techniques. Among these, Xie et al. [6] introduced the red phenotypic kimchi cabbage variety crossed with the red *Brassica juncea* using the embryo rescue procedure. In another study, reddish purple and red kimchi cabbage were obtained via interspecific crossing between red and kimchi cabbage [7]. The color of red kimchi cabbage is reddish purple, and it has an abundant anthocyanin content. Due to its attractive color and high antioxidant properties, it is popularly used in salads [7].

Glucosinolates (GSLs) are secondary metabolites (SMs) containing nitrogen and sulfur that play vital roles in human health and plant defense mechanisms [8]. To date, more than 200 GSLs have been detected in plants, mainly in the Brassica family [9]. GSLs are derivatives of amino acids—phenylalanine, alanine, methionine, isoleucine, tryptophan, tyrosine, and valine—and are separated into three major groups corresponding to their amino acid precursors, namely aliphatic, aromatic, and indolic GSLs [9,10].

Phenolic compounds are SMs composed of aromatic rings containing one or more hydroxyl groups, and they are very common in plants [11]. They can be classified into flavonoids (anthocyanidins, flavanones, flavan-3-ols, flavonols, flavones, isoflavones, and others) and non-flavonoids (hydroxycinnamates, stilbenes, phenolic acids, and others) based on their arrangement and number of carbon atoms [12]. In addition, phenolic compounds play a vital role in human health, with anti-inflammatory, antibacterial, and antioxidant actions [13–15].

*Agrobacterium rhizogenes* infects higher plants and is responsible for the growth of hairy root (HR) disease, mainly in dicotyledonous plants and a few monocotyledonous plants [16]. HR cultures have been developed from several plant species through *A. rhizogenes* transformation to enhance SM production [17–19]. The phenotypic pattern of HRs is defined by genetic stability, a lack of geotropism, hormone-independent growth, lateral branching, and the ability to produce specific SMs. The main benefit of HRs is that they often show a better biosynthetic capacity for SM production than their parent plants [20].

Light is a major factor in the growth and production of SMs in HR cultures [21]. Several studies have shown that light affects the biosynthesis of SMs in HR cultures [22–24]. In this respect, the purpose of this study was to develop HRs of green and red Kimchi cabbage using an *A. rhizogenes*-mediated HR transformation and analyze their glucosinolates and phenolic compounds after exposure to light and dark conditions. These results will provide basic information for further bio-engineering research to increase the glucosinolates and phenylpropanoid contents in the HRs of kimchi cabbage varieties.

## 2. Materials and Methods

### 2.1. Seed Sterilization and Germination

Seeds of green (cv. CR carotene) and red (cv. Ppalgang 3) Kimchi cabbage were purchased from the Kwonnong Seed Company (Cheongju, Republic of Korea). The seeds were surface-sterilized with 70% (*v/v*) ethanol for 30 s and then soaked with a 4% (*v/v*) NaClO solution containing one drop of a Tween 20 solution for 10 min. The seeds were then washed 5–6 times with sterilized distilled water. The sterilized seeds were taken to a clean bench and dried with sterilized tissue paper. Seven seeds were placed on a Petri dish containing solidified hormone-free 1/2 Murashige and Skoog Basal Medium (MS) (Kisan Bio, Seoul, Republic of Korea) [25] containing 0.8% (*w/v*) plant agar. The Petri dishes were kept in a growth chamber under 16 h light/8 h dark photoperiod cycles at 25 °C until seedlings were established. The seedlings were then moved into solidified hormone-free 1/2 MS medium after two to three weeks.

*2.2. Growth of A. rhizogenes*

Wild-type *A. rhizogenes* strain R1000 bacterial cells were cultured in a flask containing 30 mL of Luria–Bertani (LB) liquid medium (Ambrothia, Daejeon, Republic of Korea) and incubated in a continuous rotary shaker (Hanbaek scientific Co., HB-201SF, Republic of Korea) at 28 °C for 20 h at 200 rpm in dark conditions. The bacterial cell suspension was centrifuged (Mega 21R, Hanil, Incheon, Republic of Korea) at 3000× *g* for 10 min at 4 °C to collect the cells and resuspended with $\frac{1}{2}$ MS liquid medium to maintain a final cell density of $OD_{600}$ = 1.0 for plant inoculation.

*2.3. Establishment of HR Cultures*

From the 10-day-old kimchi cabbage cultivar seedlings, the cotyledons were excised and cut into tiny pieces (about 0.5 × 0.5 cm) with a sterile blade. The sliced explants were co-cultured with the bacterial suspension for 10–15 min. After drying with sterile tissue paper, the co-cultured explants were placed in Petri dishes containing hormone-free 1/2 MS medium with 0.8% (*w/v*) agar and were incubated for two days at 25 °C in the dark. The explants were removed after 2 days of co-cultivation, washed 5–6 times in sterile distilled water, and then moved to a hormone-free 1/2 MS medium containing 500 mg/L cefotaxime (GoldBio, St. Louis, MO, USA) for HR induction [26]. From the wounded site, several HRs emerged within 3–4 weeks, and the HRs were excised, transferred to the hormone-free 1/2 MS medium containing 500 mg/L cefotaxime, and incubated at 25 °C in dark conditions. This procedure was repeated 3–4 times.

*2.4. HR Culture under Light and Dark Conditions*

Approximately 4 g of HRs of both green and red kimchi cabbage were transferred to 30 mL of the hormone-free $\frac{1}{2}$ MS liquid medium and incubated on a shaker (Hanbaek scientific Co., HB-201SLI, Republic of Korea) at 110 rpm for 2 weeks at 25 °C. Each type of HR was cultured in six flasks; three flasks were placed under 16 h light/8 h dark conditions (photosynthetic photon flux density of 54.6 $\mu$mol $m^{-2}$ $s^{-1}$), and the other three flasks were placed under continuous dark conditions [27]. After 2 weeks, HRs were harvested and dried using a freeze dryer (HyperCool HC3055, Hanil Scientific Inc., Gimpo, Republic of Korea) for at least 3 days to measure the dry weight (DW) of the HRs and to analyze the secondary metabolites.

*2.5. Extraction and Analysis of Glucosinolate Compounds*

Glucosinolates were extracted and analyzed as described by Sathasivam et al. [10,28], with slight modifications. Approximately 100 mg of freeze-dried green and red kimchi was dissolved with 4.5 mL of 70% methanol (*v/v*), thoroughly mixed by vortexing (KMC-1300 V, Vision Scientific Co., Ltd., Daejeon-Si, Republic of Korea), incubated at 70 °C for 5 min, and centrifuged at 14,000 rpm for 10 min at 4 °C. A diethylaminoethanol Sephadex A-25 column (GE Healthcare, Uppsala, Sweden) was used to separate and filter the upper layer, and the column was washed with 3 mL of autoclaved distilled water. For the desulfation of the eluted mixture, 75 µL of purified arylsulfatase was added, and it was kept at room temperature overnight. After the overnight incubation, the mixture was eluted with 0.5 mL of ultrapure water, filtered, and sterilized using a 0.22 µm PTFE syringe filter (Sterlitech Corp., Kent, WA, USA) before being injected for high-performance liquid chromatography (HPLC) analysis. The HPLC model, column conditions, operating conditions, and gradient program are shown in Table S1. The glucosinolate content was analyzed, measured, and quantified according to the method described by Sathasivam et al. [28].

*2.6. Extraction and Analysis of Phenolic Compounds*

Phenolic compounds were extracted and analyzed following the protocol described by Sathasivam et al. [10,28]. To 100 mg of freeze-dried green and red kimchi samples, 3 mL of an 80% MeOH (*v/v*) solution was added. The samples were then vortexed for 1 min and immediately sonicated at 37 °C for 1 h. At 4 °C, the mixture was centrifuged for 15 min at

10,000 rpm. The resulting supernatants were filtered through a 0.45 μm PTFE syringe filter (Millipore, Bedford, MA, USA) before being injected for HPLC analysis. The HPLC model, column conditions, operating conditions, and gradient program are shown in Table S1. The phenolic content was measured and quantified with reference to a corresponding standard calibration curve [28].

### 2.7. Statistical Analysis

The statistical analysis of the data was performed using SPSS version 26.0 (IBM Corp., Armonk, NY, USA). The data were calculated as the means ± standard deviations of three replicates, and Duncan's multiple range test (DMRT) was used to determine whether there was a significant difference ($p < 0.05$) among the means. PCA, PLS-DA, a heat map analysis, a Pearson correlation analysis, and a VIP analysis of 13 metabolites identified in the GKHL, GKHD, RKHL, and RKHD conditions were performed using MetaboAnalyst 5.0. Autoscaling was carried out as a preprocessing approach for the data set.

## 3. Results

### 3.1. Growth Patterns of Kimchi Cabbage HRs under Light and Dark Conditions

The growth of the green kimchi cabbage HRs grown under light (GKHL) conditions, green kimchi cabbage HRs grown under dark (GKHD) conditions, red kimchi cabbage HRs grown under light (RKHL) conditions, and red kimchi cabbage HRs grown under dark (RKHD) conditions ranged from 0.37 to 0.27 g/30 mL (Figure 1). The highest HR growth was obtained in the GKHD condition (0.37 ± 0.01 DW g/30 mL), followed by the RKHD condition (0.33 ± 0.01 g/30 mL), whereas the GKHL condition (0.31 ± 0.02 DW g/30 mL) and the RKHL condition (0.27 ± 0.01 g/30 mL) showed the lowest growth patterns. The growth was reduced by 1.12-, 1.19-, and 1.37-fold by the RKHD condition, GKHL condition, and RKHL condition, respectively, when compared to the GKHD condition. These findings suggest that HRs grown under dark conditions are more suitable for the growth of green and red kimchi cabbage compared to light conditions.

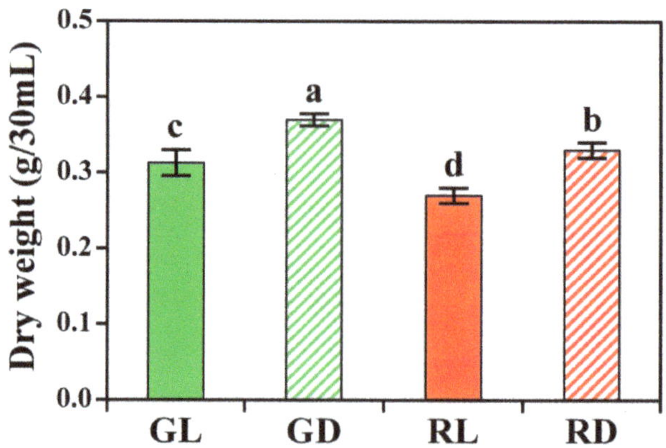


**Figure 1.** Growth of green and red kimchi cabbage HRs grown under light and dark conditions. Samples were harvested after 10 days of HR growth. For light and dark conditions, HRs were placed under 16 h light/8 h dark conditions and continuous dark conditions, respectively. Significant differences exist between mean values with distinct letters ($p < 0.05$, ANOVA and DMRT). GL—green kimchi cabbage grown under light conditions; GD—green kimchi cabbage grown under dark conditions; RL—red kimchi cabbage grown under light conditions; RD—red kimchi cabbage grown under dark conditions.

### 3.2. Glucosinolate Accumulation in Response to Light and Dark Conditions

The glucosinolate accumulation in green and red kimchi cabbage HRs was analyzed using HPLC. Five glucosinolate compounds, i.e., glucobrassicin, glucoerucin, 4-methoxyglucobrassicin, gluconasturtiin, and neoglucobrassicin, were detected in the GKHL, GKHD, RKHL, and RKHD conditions (Figure 2). Their contents were significantly different under light and dark conditions. In the GKHL, GKHD, RKHL, and RKHD conditions, neoglucobrassicin comprised the largest portion of the quantified glucosinolate content. Neoglucobrassicin was more highly accumulated in the GKHD condition than in the other kimchi cabbage HRs grown under light and dark conditions. Similarly, the GKHD condition also enhanced gluconasturtiin production. The glucobrassicin content of the kimchi cabbage HRs was more affected by the light and dark conditions when compared to the other four glucosinolate compounds. Compared with the RKHD condition, the amount of glucobrassicin was 2.72-, 2.61-, and 2.16-fold lower in GKHL, RKHL, and GKHD conditions, respectively. In addition, the levels of glucoerucin and 4-methoxyglucobrassicin were also highest in the RKHD condition. The glucoerucin contents were 1.27-, 1.22-, and 1.27-fold lower in the GKHL, GKHD, and RKHL conditions, respectively, than in the RKHD condition. Moreover, the level of 4-methoxyglucobrassicin was highest in the RKHD condition, which was 1.19-, 1.14-, and 1.09-fold higher compared to the GKHL, RKHL, and GKHD conditions, respectively. The total glucosinolate contents ranged from 6884.55 to 8864.58 μg/DW in response to different HRs of kimchi cabbages grown under light and dark conditions. In particular, the total glucosinolate content of the RKHD condition was the highest, which was 1.29-fold higher than that of the RKHL condition. The total amounts of glucosinolate in the GKHL and GKHD conditions were 1.29- and 1.06-fold lower than in the RKHD condition, respectively. This overall result showed that dark conditions enhance glucosinolate accumulation in the HRs of kimchi cabbage.

### 3.3. Phenolic Accumulation in Response to Light and Dark Conditions

The phenolic accumulation in green and red kimchi cabbage HRs was analyzed using HPLC. Seven different phenolic compounds, i.e., caffeic acid, catechin hydrate, chlorogenic acid, *p*-coumaric acid, epicatechin, quercetin, trans-cinnamic acid, and 4-hydroxybenzoic acid, were detected in the GKHL, GKHD, RKHL, and RKHD conditions (Figure 3). All phenolic compounds showed differential accumulation in the GKHL, GKHD, RKHL, and RKHD conditions. The total phenolic contents ranged from 250.53 to 289.22 μg/g DW in response to different HRs of kimchi cabbages grown under light and dark conditions. The total phenolic contents in the GKHL, RKHL, and RKHD conditions did not show any significant differences, whereas they were slightly decreased in the GKHD condition. The total phenolic content in the GKHD condition was 1.15-, 1.14-, and 1.13-fold lower than those of the GKHL, RKHD, and RKHL conditions, respectively. In particular, catechin hydrate comprised the largest portion of the quantified phenolic content in the GKHL, GKHD, RKHL, and RKHD conditions. Among the two HRs grown under different conditions, most of the individual phenolic compounds, such as trans-cinnamic acid, caffeic acid, catechin hydrate, chlorogenic acid, and *p*-coumaric acid, were highest in the GKHL condition. The quercetin contents were significantly higher in the RKHL and RKHD conditions compared to those of the GKHL and GKHD conditions. In contrast, the 4-hydroxybenzoic acid contents were significantly higher in the GKHL and GKHD conditions than in the RKHL and RKHD conditions. Interestingly, the trans-cinnamic acid content was significantly higher in the GKHL condition, whereas it was not detected in the RKHD condition. The trans-cinnamic acid contents were 1.93- and 2.32-fold lower in the GKHD and RKHL conditions, respectively, than in green kimchi cabbage HRs grown under light conditions. This result shows that most of the individual and total phenolic contents were highest in the GKHL condition.

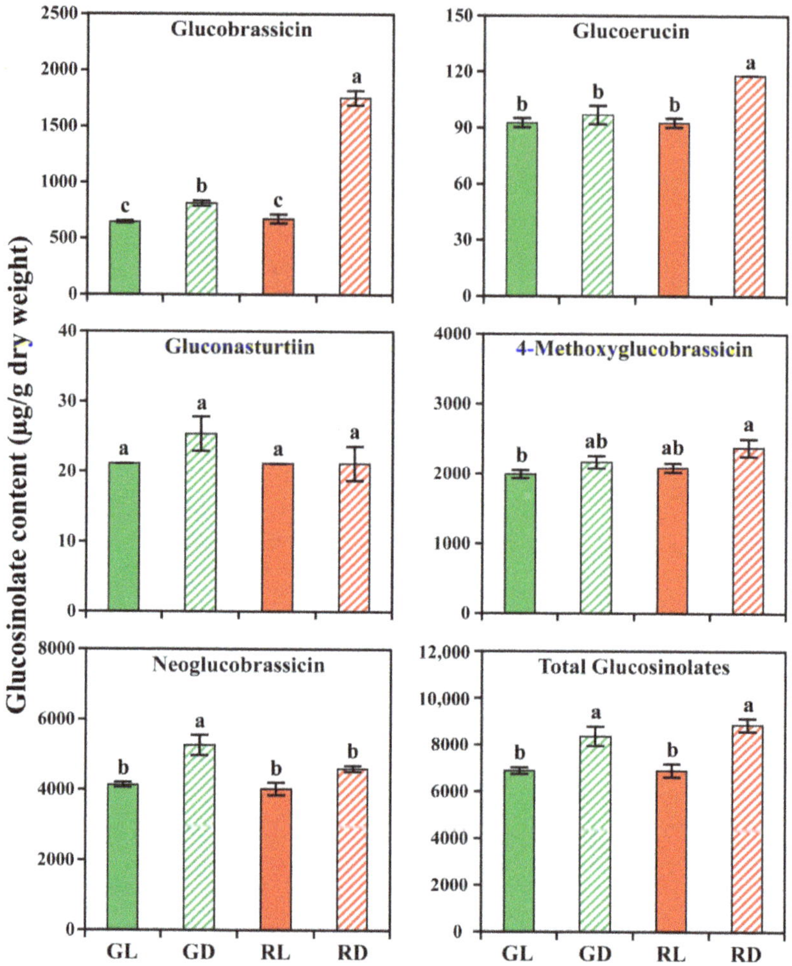


**Figure 2.** Effect of light and dark conditions on glucosinolate compound (µg/g DW of the HR cultures) accumulation in HR cultures of green and red kimchi cabbage. Samples were harvested after 10 days of HR growth. Significant differences exist between mean values with distinct letters ($p < 0.05$, ANOVA and DMRT). GL—green kimchi cabbage grown under light conditions; GD—green kimchi cabbage grown under dark conditions; RL—red kimchi cabbage grown under light conditions; RD—red kimchi cabbage grown under dark conditions.

### *3.4. Metabolic Profiling of Identified Metabolites in Response to Light and Dark Conditions*

In total, five individual glucosinolate compounds and eight phenolic compounds were identified and quantified in samples from the GKHL, GKHD, RKHL, and RKHD conditions using HPLC. A heat map showed that most of the individual glucosinolate and phenolic compounds were significantly higher in the RKHD and the GKHL conditions, respectively (Figure 4). The heat map was divided into two major clusters, namely cluster 1 and cluster 2, but they were further subdivided into subclusters 1-1, 1-2, 2-1, and 2-2. Clusters 1-1, 2-1, and 2-2 were further subgrouped into two subclusters, namely 1-1a and 1-1b, 2-1a and 2-1b, and 2-2a and 2-2b, respectively. Cluster 1 contained the metabolites abundantly present in the GKHL condition, whereas cluster 2 was separated based on the lowest metabolites present in the GKHL condition. Cluster 1-1 was further separated into two clusters, namely 1-1a and 1-1b. Cluster 1-1a was separated based on the highest amounts of phenolic compounds present in the HRs exposed to the light conditions (GKHL and RKHL conditions), whereas cluster 1-1b formed a group based on the phenolic compounds rich in specific cultivars (GKHL and GKHD conditions). In addition, most of the metabolites identified in the GKHL, GKHD, RKHL, and RKHD conditions were separated into distinct groups (e.g., caffeic acid was significantly abundant in the GKHL, RKHL, and RKHD conditions (cluster 1-2)). Cluster 2-1 was also separated into a distinct group. Cluster 2-1a formed a cluster based on the abundant metabolites present in specific cultivars (GKHL and GKHD conditions), whereas 2-1b was separated based on the metabolites that were abundant in specific conditions (RKHD and GKHD conditions). Cluster 2-2 was separated based on the metabolites that were significantly abundant in the RKHD conditions. This cluster was divided into two subgroups, namely 2-2a and 2-2b. Subgroups 2-2a and 2-2b formed a group based on the metabolites that were significantly abundant in different cultivars grown under different conditions (RKHD and GKHD dark conditions). Regarding the glucosinolates, the metabolites present in the HRs of two cultivars grown under light and dark conditions were grouped in clusters 2-1b and 2-2a (gluconasturtiin, neoglucobrassicin, 4-methoxyglucobrassicin, glucoerucin, and glucobrassicin), whereas the phenolic compounds and metabolites from the HRs of two cultivars grown under light and dark conditions were separated into three groups (Figure 4). Cluster 1 consisted of catechin hydrate, *p*-coumaric acid, chlorogenic acid, caffeic acid, and trans-cinnamic acid. In group 2-la, 4-hydroxybenzoic acid formed a separate cluster, whereas in group 2-2b, epicatechin and quercetin were clustered. Significantly higher ($p \leq 0.05$) levels of glucosinolate and phenolic compounds in the GKHL, GKHD, RKHL, and RKHD conditions are shown in Figure S1, which demonstrates that unique metabolite signature characteristics were found in the GKHL, GKHD, RKHL, and RKHD conditions.

According to the PCA findings, PC1 and PC2 showed 47.5% and 23.6% of the variance, respectively (Figure 5). The PCA showed a clear separation among the different HRs grown under light and dark conditions. This clear separation among the different cultivars grown under light and dark conditions might be due to 4-methoxyglucobrassicin, glucoerucin, glucobrassicin, epicatechin, and quercetin, and their associated eigenvector values were −0.32494, −0.29928, −0.29511, −0.26775, and −0.25909, respectively. Those of trans-cinnamic acid, *p*-coumaric acid, catechin hydrate, chlorogenic acid, and 4-hydroxybenzoic acid were 0.34956, 0.34201, 0.27291, 0.25167, and 0.19125, respectively (Figure 5A). In addition, a PLS-DA was performed to maximize the separation between different kimchi cabbage HRs grown under light and dark conditions. The PLS-DA model showed a clear separation between the different HRs grown under light and dark conditions, which were 24.9% (PC1) and 39.5% (PC2). The PLS-DA also showed a clear separation between the HRs grown under light and dark conditions. This supports the PCA results. This clear separation might be due to trans-cinnamic acid, 4-hydroxybenzoic acid, neoglucobrassicin, gluconasturtiin, and chlorogenic acid, and their associated eigenvector values were −0.49375, −0.34139, −0.3139, −0.30284, and −0.26393, respectively. Those of quercetin, glucobrassicin, glucoerucin, epicatechin, and 4-methoxyglucobrassicin were 0.46467, 0.27146, 0.21669, 0.21551, and 0.12372, respectively (Figure 5B). In addition, quercetin, 4-hydroxybenzoic acid, and neoglucobrassicin were the most significant metabolites in the prediction, according to the VIP analysis, which led to a clearer separation between the green and red kimchi cabbage than the different conditions (Figure 6). The green and red kimchi cabbage HRs grown under dark conditions showed a clear separation in both the PCA and PLS-DA compared to the GKHL and RKHL conditions. This might have been because the glucosinolate contents were significantly higher in the HRs grown under dark conditions than under light conditions. However, the GKHL and RKHL conditions showed a slightly closer group than the GKHD and RKHD conditions. This was because the total glucosinolate and total phenolic contents were similar in both cultivars grown under light conditions. From these results, for the enhancement of the glucosinolate and phenolic contents in the HRs of both kimchi cabbage cultivars, the dark and light conditions were the most suitable, respectively.

This supports the heat map results, which showed that most of the individual glucosinolate and phenolic compounds were highest in kimchi cabbage HRs grown under dark and light conditions, respectively.

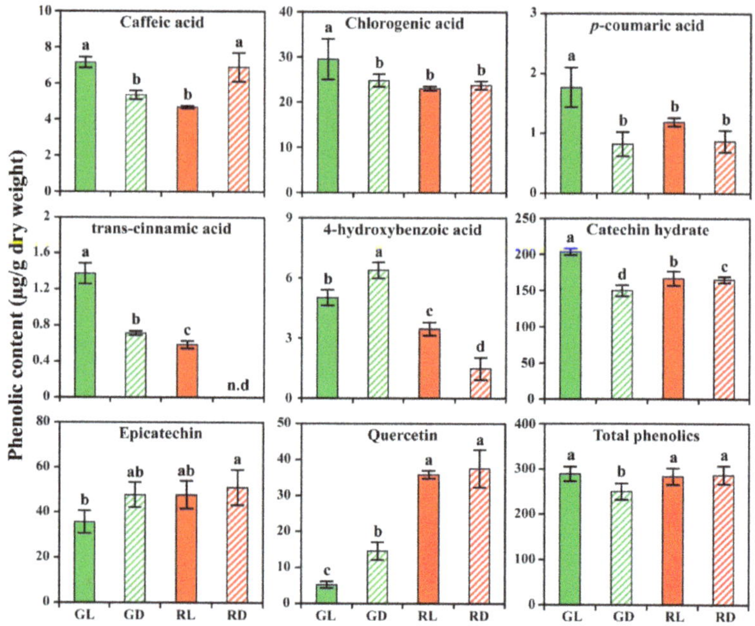


**Figure 3.** Effect of light and dark treatments on the phenolic compound (µg/g DW of the HR cultures) accumulation in HR cultures of green and red kimchi cabbage. Samples were harvested after 10 days of HR growth. Significant differences exist between mean values with distinct letters ($p < 0.05$, ANOVA and DMRT). n.d.—not detected. GL—green kimchi cabbage grown under light conditions; GD—green kimchi cabbage grown under dark conditions; RL—red kimchi cabbage grown under light conditions; RD—red kimchi cabbage grown under dark conditions.

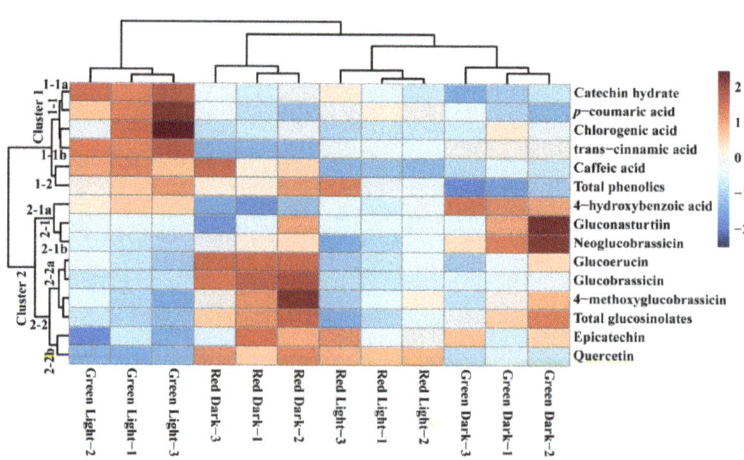


**Figure 4.** Heatmap showing the differences in the relative metabolite contents in the HRs of green and red kimchi cabbage after exposure to light and dark conditions. Blue denotes a reduction, whereas red indicates an increase in metabolite concentration.

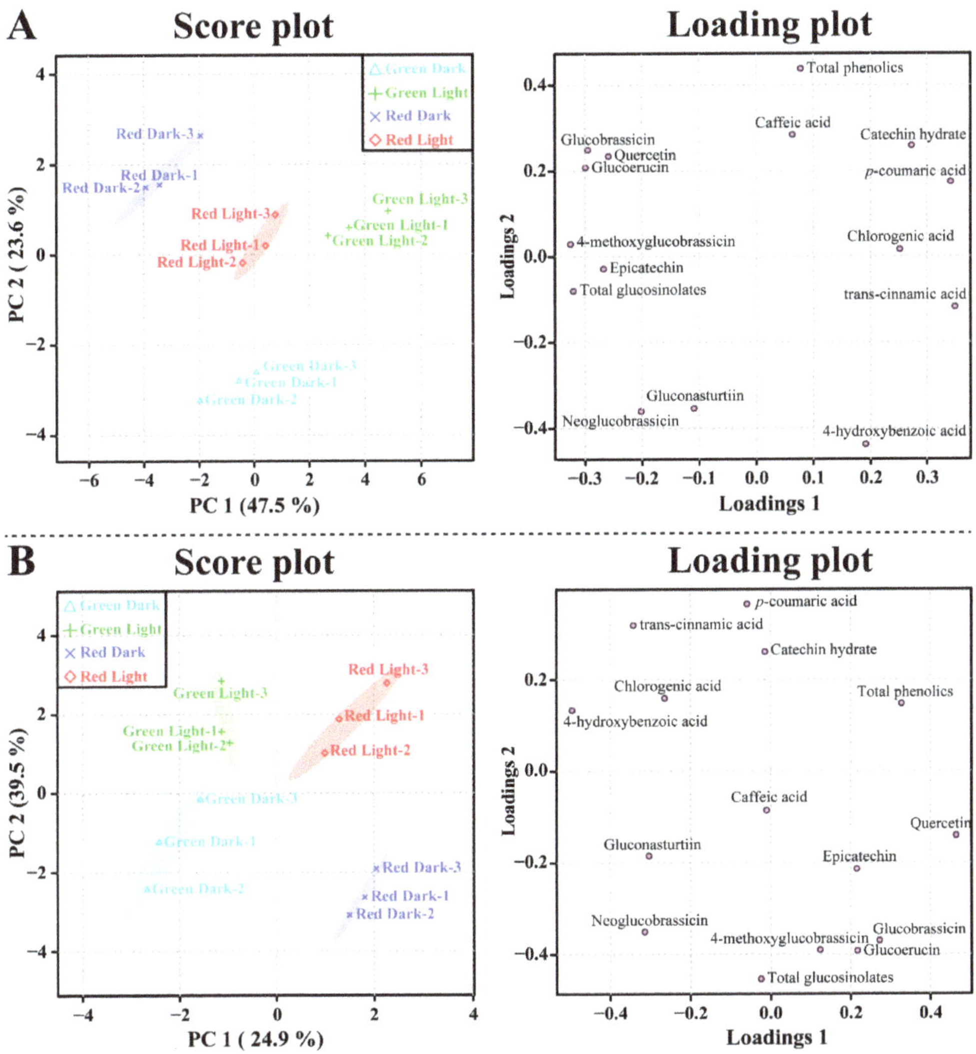


**Figure 5.** Score and loading plots of the PCA (**A**) and PLS-DA (**B**) models of the metabolites found in the HRs of green and red kimchi cabbage after exposure to light and dark conditions.

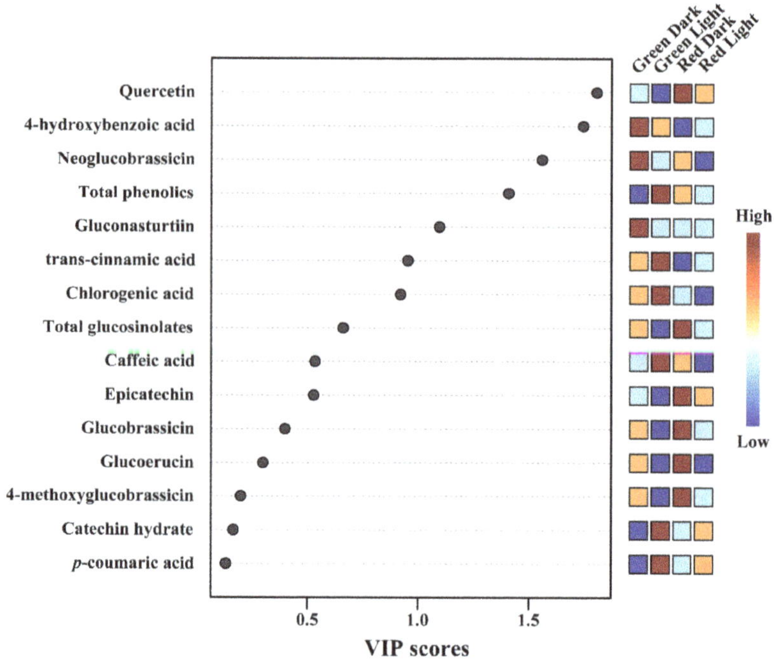


**Figure 6.** The main components separating the HRs of green and red kimchi cabbage after exposure to light and dark conditions, based on the VIP scores obtained using the PLS-DA model.

A correlation matrix analysis of 13 metabolites identified after exposing green and red HRs to light and dark conditions (Figure 7) was clustered into two main groups that consisted of both individual glucosinolates and phenolic compounds based on positive correlations. The most significant metabolite, according to VIP analysis, was quercetin, which showed positive correlations with epicatechin ($r$ = 0.55481, $p$ = 0.061166), glucoerucin ($r$ = 0.51565, $p$ = 0.086173), glucobrassicin ($r$ = 0.57067, $p$ = 0.052656), and 4-methoxyglucobrassicin ($r$ = 0.49907, $p$ = 0.098581), whereas it showed negative correlations with most of the individual glucosinolate and phenolic compounds, namely chlorogenic acid, gluconasturtiin, neoglucobrassicin, caffeic acid, catechin hydrate, *p*-coumaric acid, 4-hydroxybenzoic acid, and trans-cinnamic acid. Among the identified individual glucosinolate compounds, neoglucobrassicin showed the highest content and showed strong positive correlations with gluconasturtiin, epicatechin, glucoerucin, glucobrassicin, 4-methoxyglucobrassicin, and 4-hydroxybenzoic acid, whereas catechin hydrate showed the highest content when compared to the other individual phenolic compounds and showed positive correlations with caffeic acid ($r$ = 0.57526, $p$ = 0.050358), *p*-coumaric acid ($r$ = 0.80613, $p$ = 0.0015418), chlorogenic acid ($r$ = 0.65078, $p$ = 0.021916), and trans-cinnamic acid ($r$ = 0.70098, $p$ = 0.011095). However, catechin hydrate showed a negative correlation with several compounds, namely gluconasturtiin, neoglucobrassicin, epicatechin, quercetin, glucoerucin, glucobrassicin, 4-methoxyglucobrassicin, and 4-hydroxybenzoic acid.

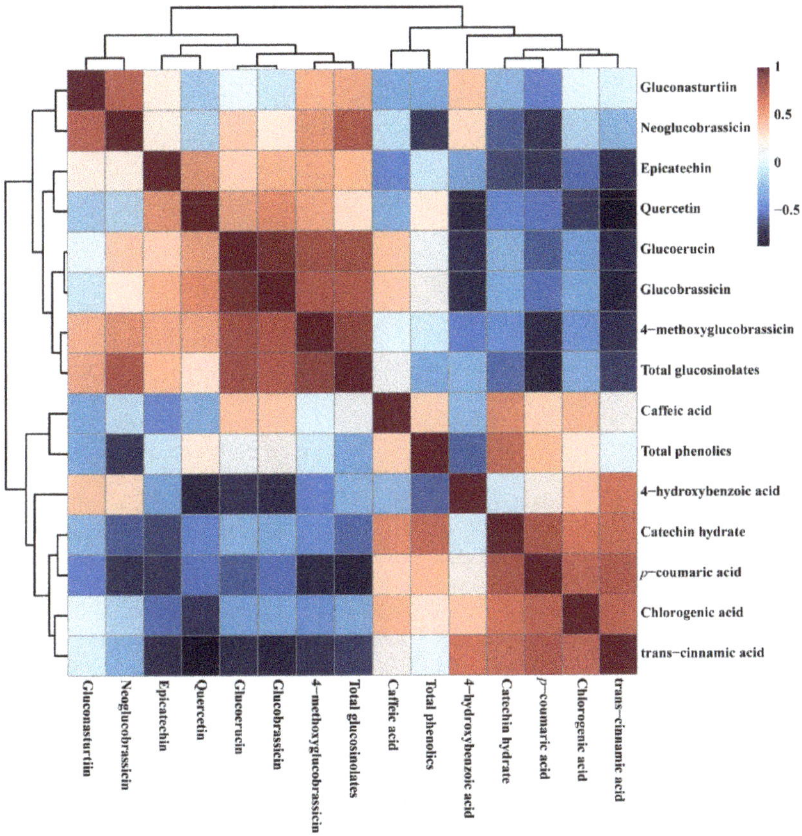


**Figure 7.** Correlation matrix showing the relationships between the metabolites found in the HRs of green and red kimchi cabbage grown under light and dark conditions. The intensity of the blue or red color, as shown on the color scale, denotes the correlation coefficient value for each colored box, which indicates Pearson's correlation coefficient for a pair of metabolites.

## 4. Discussion

HRs can grow vigorously on a medium without adding hormones to produce a significant amount of valuable SMs [29,30]. In this study, dark conditions enhanced the DW of HRs in both green and red kimchi cabbage. A similar result was obtained in a previous study in which the HRs of green and purple *Ocimum basilicum* were grown under light and dark conditions; the results showed that the highest growth rate was achieved under dark conditions [31]. Therefore, to enhance HR growth in plant species, dark conditions are the most appropriate.

Previously, several studies reported that HRs accumulate significant phenolic and flavonoid contents [32–34]. For instance, HR cultures of *Dracocephalum moldavica* and *Momordica charantia* showed higher accumulations of total flavonoid and phenolic contents than non-transformed roots [33,34]. Similarly, an HR culture of *Beta vulgaris* accumulated 20-fold higher levels of phenolic compounds than a control culture [35]. A previous study demonstrated that higher glucosinolate concentrations were achieved in the roots than in the shoots at an identical developmental stage in cabbage and radish [36]. In transgenic HRs, 4-hydroxyglucobrassicin, gluconasturtiin, 4-methoxyglucobrassicin, glucobrassicin, and neoglucobrassicin were highly present in watercress [37]. A similar result was obtained in this study; these compounds were also rich in the HRs of green and red kimchi cabbage. Kim et al. [38] reported that *Fagopyrum tataricum* transgenic HRs produced higher levels of phenolic compounds, such as caffeic acid, chlorogenic acid, catechin, gallic acid, ferulic acid, quercetin, and rutin. In another study, the chlorogenic acid, protocatechuic acid, and ferulic acid contents were higher in the transgenic HRs of tomatoes [39]. In addition, the glucosinolates and phenolic compounds such as 4-methoxyglucobrassicin, 4-hydroxyglucobrassicin, glucoallysin, glucobrassicanapin, glucobrassicin, gluconapin, gluconasturtiin, neoglucobrassicin, progoitrin, sinigrin, catechin, ferulic acid, cinnamic acid, chlorogenic acid, *p*-hydroxybenzoic acid, *p*-coumaric acid, quercetin, and rutin were high in the transgenic HRs of *Brassica rapa* spp. *rapa* [32]. In this study, we identified most of the glucosinolate and phenolic compounds in the green and red kimchi cabbage HRs. Therefore, most of the HRs shared common glucosinolates and phenolic compounds.

Light plays a vital role in the growth, biosynthesis, and production of SMs, such as anthocyanins, flavonoids, and terpenoids [40–45]. The stimulatory effects of light on SM production have been reported in several plants, such as *Artimisia annua* [46], *O. basilicum* [31], *Perilla frutiescnens* [47], *Peteroselinum hortense*, and *Rudbeckia hirata* [48]. In contrast, light also has an inhibitory effect on SM accumulation, such as for shikonin and nicotine [49]. In this study, most of the individual glucosinolate and phenolic metabolites were significantly higher in the dark and light conditions, respectively. The accumulation of rosmarinic acid content varied in *O. basilicum* HRs when cultured under light and dark conditions [31]. In a previous study, the amino acid level of *B. juncea* increased when grown under dark conditions [50]. In contrast, higher anthocyanin accumulation was achieved in Tartary buckwheat T10 sprouts grown under light conditions compared to those grown under dark conditions [50]. In another study, the accumulation of compounds varied based on light or dark conditions. For example, free amino acids and $\gamma$-aminobutyric acid were enhanced in *B. juncea* seedlings when they were grown under light and dark conditions, respectively [51]. These studies have shown that a specific condition might be efficient for the induction of specific compounds in a particular species.

Different cultivars (Tartary buckwheat Hokkai T8 and T10) grown under similar conditions showed differential phenolic content accumulation [52]. This is consistent with our study results because we observed that the green and red kimchi cultivars showed varied production and accumulation of glucosinolate and phenolic contents in the HRs of kimchi cabbage. In addition, neoglucobrassicin, followed by 4-methoxyglucobrassicin and glucobrassicin, comprised the highest portion of the glucosinolates in the HRs of green and red kimchi. A similar result was obtained in green and red kale HRs, where the highest to lowest glucosinolate contents occurred in the following order: neoglucobrassicin, 4-methoxyglucobrassicin, and glucobrassicin [53]. In addition, the HRs of broccoli cultured in MS medium showed the highest accumulation of neoglucobrassicin content [54]. In this study, the neoglucobrassicin, 4-methoxyglucobrassicin, and glucobrassicin contents were significantly higher in green kimchi HRs, red kimchi HRs, and red kimchi HRs, respectively. In contrast to this study, in the HRs of green kale, the neoglucobrassicin accumulation was lowest, whereas the 4-methoxyglucobrassicin and glucobrassicin accumulation were significantly higher when cultured using MS medium [55]. Therefore, most HRs consist of common glucosinolates. However, the levels might differ based on the species.

## 5. Conclusions

HR cultures are an important alternative strategy for the mass production of valuable health-promoting SMs, such as glucosinolates and phenolic compounds, in kimchi cabbage. The HR growth in the GKHD condition showed the highest DW compared to that of the light conditions. HRs produced more glucosinolates in the GKHD condition, whereas the highest phenolic compound content was achieved in the RKHL condition. These findings shed light on the underlying mechanisms of the abiotic elicitors that enhance glucosinolate and phenolic compounds in kimchi cabbage HRs and can potentially be used as "biological factories" for the large-scale production of bioactive substances, such as glucosinolates and phenolic compounds. However, the antioxidant activity of HRs is currently under examination. In the future, further studies will be required to fully comprehend the influences of various light intensities and light quality on glucosinolate and phenylpropanoid pathway gene expression and the accumulation of associated compounds in the HR cultures of different kimchi cabbage cultivars.

**Supplementary Materials:** The following supporting information can be downloaded at https://www.mdpi.com/article/10.3390/horticulturae9040466/s1, Figure S1: Individual metabolites that are significantly ($p \leq 0.05$) higher in hairy roots of green and red kimchi cabbage grown under light and dark conditions; Table S1: HPLC conditions for phytochemical analysis of metabolites.

**Author Contributions:** Conceptualization, S.Y.L., R.S. and S.U.P.; methodology, H.K., J.K.K., C.H.P. and R.S.; software, R.S.; validation, S.Y.L., R.S. and S.U.P.; formal analysis, H.K., J.K.K., C.H.P. and R.S.; investigation, S.Y.L. and S.U.P.; resources, S.Y.L. and S.U.P.; data curation, H.K., J.K.K., C.H.P. and R.S.; writing—original draft preparation, S.Y.L. and R.S.; writing—review and editing, S.Y.L., R.S. and S.U.P.; visualization, S.Y.L. and S.U.P.; supervision, S.Y.L. and S.U.P.; project administration, S.Y.L. and S.U.P.; funding acquisition, S.U.P. All authors have read and agreed to the published version of the manuscript.

**Funding:** This study was supported by a research fund from Chosun University in 2022.

**Institutional Review Board Statement:** Not applicable.

**Informed Consent Statement:** Not applicable.

**Data Availability Statement:** Data reported are available in the Supplementary Materials.

**Conflicts of Interest:** The authors declare no conflict of interest.

## References

1. Warwick, S.I. Brassicaceae in agriculture. In *Genetics and Genomics of the Brassicaceae*; Schmidt, R., Bancroft, I., Eds.; Springer: New York, NY, USA, 2011; pp. 33–65, ISBN 978-1-4419-7118-0.
2. Li, Z.; Lee, H.W.; Liang, X.; Liang, D.; Wang, Q.; Huang, D.; Ong, C.N. Profiling of phenolic compounds and antioxidant activity of 12 cruciferous vegetables. *Molecules* **2018**, *23*, 1139. [CrossRef] [PubMed]
3. Seong, G.-U.; Hwang, I.-W.; Chung, S.-K. Antioxidant capacities and polyphenolics of Chinese cabbage (*Brassica rapa* L. ssp. *pekinensis*) leaves. *Food Chem.* **2016**, *199*, 612–618. [CrossRef] [PubMed]
4. Kim, J.K.; Chu, S.M.; Kim, S.J.; Lee, D.J.; Lee, S.Y.; Lim, S.H.; Ha, S.-H.; Kweon, S.J.; Cho, H.S. Variation of glucosinolates in vegetable crops of *Brassica rapa* L. ssp. *pekinensis*. *Food Chem.* **2010**, *119*, 423–428. [CrossRef]
5. Lee, M.-K.; Chun, J.-H.; Byeon, D.H.; Chung, S.-O.; Park, S.U.; Park, S.; Arasu, M.V.; Al-Dhabi, N.A.; Lim, Y.-P.; Kim, S.-J. Variation of glucosinolates in 62 varieties of Chinese cabbage (*Brassica rapa* L. ssp. *pekinensis*) and their antioxidant activity. *LWT-Food Sci. Technol.* **2014**, *58*, 93–101. [CrossRef]
6. Xie, L.; Li, F.; Zhang, S.; Zhang, H.; Qian, W.; Li, P.; Zhang, S.; Sun, R. Mining for candidate genes in an introgression line by using RNA sequencing: The anthocyanin overaccumulation phenotype in Brassica. *Front. Plant Sci.* **2016**, *7*, 1245. [CrossRef]
7. Lee, H.; Oh, I.-N.; Kim, J.; Jung, D.; Cuong, N.P.; Kim, Y.; Lee, J.; Kwon, O.; Park, S.U.; Lim, Y. Phenolic compound profiles and their seasonal variations in new red-phenotype head-forming Chinese cabbages. *LWT* **2018**, *90*, 433–439. [CrossRef]
8. Sønderby, I.E.; Geu-Flores, F.; Halkier, B.A. Biosynthesis of glucosinolates–gene discovery and beyond. *Trends Plant Sci.* **2010**, *15*, 283–290. [CrossRef]
9. Clarke, D.B. Glucosinolates, structures and analysis in food. *Anal. Methods* **2010**, *2*, 310–325. [CrossRef]
10. Sathasivam, R.; Park, S.U.; Kim, J.K.; Park, Y.J.; Kim, M.C.; Nguyen, B.V.; Lee, S.Y. Metabolic profiling of primary and secondary metabolites in kohlrabi (*Brassica oleracea* var. *gongylodes*) sprouts exposed to different light-emitting diodes. *Plants* **2023**, *12*, 1296. [CrossRef]
11. Fernandez-Panchon, M.; Villano, D.; Troncoso, A.; Garcia-Parrilla, M. Antioxidant activity of phenolic compounds: From in vitro results to in vivo evidence. *Crit. Rev. Food Sci. Nutr.* **2008**, *48*, 649–671. [CrossRef]
12. Crozier, A.; Jaganath, I.B.; Clifford, M.N. Phenols, polyphenols and tannins: An overview. *Plant Second. Metab. Occur. Struct. Role Hum. Diet* **2006**, *1*, 1–25.
13. Ambriz-Pérez, D.L.; Leyva-López, N.; Gutierrez-Grijalva, E.P.; Heredia, J.B. Phenolic compounds: Natural alternative in inflammation treatment. A Review. *Cogent Food Agric.* **2016**, *2*, 1131412.
14. Cushnie, T.T.; Lamb, A.J. Antimicrobial activity of flavonoids. *Int. J. Antimicrob. Agents* **2005**, *26*, 343–356. [CrossRef] [PubMed]
15. Rice-Evans, C.; Miller, N.; Paganga, G. Antioxidant properties of phenolic compounds. *Trends Plant Sci.* **1997**, *2*, 152–159. [CrossRef]
16. Mauro, M.L.; Bettini, P.P. *Agrobacterium rhizogenes* rolB oncogene: An intriguing player for many roles. *Plant Physiol. Biochem.* **2021**, *165*, 10–18. [CrossRef]

17. Kim, Y.-K.; Sathasivam, R.; Kim, Y.B.; Kim, J.K.; Park, S.U. Transcriptomic analysis, cloning, characterization, and expression analysis of triterpene biosynthetic genes and triterpene accumulation in the hairy roots of *Platycodon grandiflorum* exposed to methyl jasmonate. *ACS Omega* **2021**, *6*, 12820–12830. [CrossRef]
18. Morey, K.J.; Peebles, C.A. Hairy roots: An untapped potential for production of plant products. *Front. Plant Sci.* **2022**, *13*, 937095. [CrossRef]
19. Roy, A. Hairy root culture an alternative for bioactive compound production from medicinal plants. *Curr. Pharm. Biotechnol.* **2021**, *22*, 136–149. [CrossRef]
20. Kim, Y.; Wyslouzil, B.E.; Weathers, P.J. Secondary metabolism of hairy root cultures in bioreactors. *Vitr. Cell. Dev. Biol.-Plant* **2002**, *38*, 1–10. [CrossRef]
21. Hussain, M.J.; Abbas, Y.; Nazli, N.; Fatima, S.; Drouet, S.; Hano, C.; Abbasi, B.H. Root cultures, a boon for the production of valuable compounds: A comparative review. *Plants* **2022**, *11*, 439. [CrossRef] [PubMed]
22. Abbasi, B.H.; Tian, C.-L.; Murch, S.J.; Saxena, P.K.; Liu, C.-Z. Light-enhanced caffeic acid derivatives biosynthesis in hairy root cultures of *Echinacea purpurea*. *Plant Cell Rep.* **2007**, *26*, 1367–1372. [CrossRef]
23. Mukherjee, C.; Samanta, T.; Mitra, A. Redirection of metabolite biosynthesis from hydroxybenzoates to volatile terpenoids in green hairy roots of *Daucus carota*. *Planta* **2016**, *243*, 305–320. [CrossRef]
24. Yu, K.-W.; Murthy, H.N.; Hahn, E.-J.; Paek, K.-Y. Ginsenoside production by hairy root cultures of *Panax ginseng*: Influence of temperature and light quality. *Biochem. Eng. J.* **2005**, *23*, 53–56. [CrossRef]
25. Murashige, T.; Skoog, F. A revised medium for rapid growth and bioassays with tobacco tissue cultures. *Physiol. Plant* **1962**, *15*, 473–497. [CrossRef]
26. Sathasivam, R.; Choi, M.; Radhakrishnan, R.; Kwon, H.; Yoon, J.; Yang, S.H.; Kim, J.K.; Chung, Y.-S.; Park, S.U. Effects of various strains of *Agrobacterium rhizogenes* strains on hairy root induction and analyses of primary and secondary metabolites in *Ocimum basilicum*. *Front. Plant Sci.* **2022**, *13*, 983776. [CrossRef]
27. Park, C.H.; Park, Y.E.; Yeo, H.J.; Park, N.I.; Park, S.U. Effect of light and dark on the phenolic compound accumulation in Tartary buckwheat hairy roots overexpressing *ZmLC*. *Int. J. Mol. Sci.* **2021**, *22*, 4702. [CrossRef] [PubMed]
28. Sathasivam, R.; Kim, M.C.; Yeo, H.J.; Nguyen, B.V.; Sohn, S.I.; Park, S.U.; Kim, J. Accumulation of phenolic compounds and glucosinolates in sprouts of pale green and purple kohlrabi (*Brassica oleracea* var. *gongylodes*) under light and dark conditions. *Agronomy* **2021**, *11*, 1939. [CrossRef]
29. Choi, M.; Sathasivam, R.; Nguyen, B.V.; Park, N.I.; Woo, S.-H.; Park, S.U. Expression analysis of phenylpropanoid pathway genes and metabolomic analysis of phenylpropanoid compounds in adventitious, hairy, and seedling roots of Tartary buckwheat. *Plants* **2021**, *11*, 90. [CrossRef] [PubMed]
30. Su, W.W.; Lee, K.-T. Plant cell and hairy root cultures–Process characteristics, products, and applications. In *Bioprocessing for Value-Added Products from Renewable Resources*; Su, W.W., Lee, K.-T., Eds.; Elsevier: Amsterdam, The Netherlands, 2007; pp. 263–292.
31. Kwon, D.Y.; Kim, Y.B.; Kim, J.K.; Park, S.U. Production of rosmarinic acid and correlated gene expression in hairy root cultures of green and purple basil (*Ocimum basilicum* L.). *Prep. Biochem. Biotechnol.* **2021**, *51*, 35–43. [CrossRef] [PubMed]
32. Chung, I.-M.; Rekha, K.; Rajakumar, G.; Thiruvengadam, M. Production of glucosinolates, phenolic compounds and associated gene expression profiles of hairy root cultures in turnip (*Brassica rapa* ssp. *rapa*). *3 Biotech* **2016**, *6*, 1–16. [CrossRef] [PubMed]
33. Weremczuk-Jeżyna, I.; Grzegorczyk-Karolak, I.; Frydrych, B.; Królicka, A.; Wysokińska, H. Hairy roots of *Dracocephalum moldavica*: Rosmarinic acid content and antioxidant potential. *Acta Physiol. Plant* **2013**, *35*, 2095–2103. [CrossRef]
34. Thiruvengadam, M.; Praveen, N.; Maria John, K.; Yang, Y.-S.; Kim, S.-H.; Chung, I.-M. Establishment of *Momordica charantia* hairy root cultures for the production of phenolic compounds and determination of their biological activities. *Plant Cell Tiss. Org. Cult.* **2014**, *118*, 545–557. [CrossRef]
35. Georgiev, V.G.; Weber, J.; Kneschke, E.-M.; Denev, P.N.; Bley, T.; Pavlov, A.I. Antioxidant activity and phenolic content of betalain extracts from intact plants and hairy root cultures of the red beetroot *Beta vulgaris* cv. Detroit dark red. *Plant Foods Hum. Nutr.* **2010**, *65*, 105–111. [CrossRef] [PubMed]
36. Bhandari, S.R.; Jo, J.S.; Lee, J.G. Comparison of glucosinolate profiles in different tissues of nine Brassica crops. *Molecules* **2015**, *20*, 15827–15841. [CrossRef] [PubMed]
37. Park, N.I.; Kim, J.K.; Park, W.T.; Cho, J.W.; Lim, Y.P.; Park, S.U. An efficient protocol for genetic transformation of watercress (*Nasturtium officinale*) using *Agrobacterium rhizogenes*. *Mol. Biol. Rep.* **2011**, *38*, 4947–4953. [CrossRef]
38. Kim, Y.K.; Li, X.; Xu, H.; Il Park, N.; Uddin, M.R.; Pyon, J.Y.; Park, S.U. Production of phenolic compounds in hairy root culture of tartary buckwheat (*Fagopyrum tataricum* Gaertn). *J. Crop Sci. Biotechnol.* **2009**, *12*, 53–57. [CrossRef]
39. Singh, H.; Dixit, S.; Verma, P.C.; Singh, P.K. Evaluation of total phenolic compounds and insecticidal and antioxidant activities of tomato hairy root extract. *J. Agric. Food Chem.* **2014**, *62*, 2588–2594. [CrossRef]
40. Gerhardt, K.E.; Lampi, M.A.; Greenberg, B.M. The effects of far-red light on plant growth and flavonoid accumulation in Brassica napus in the presence of ultraviolet B radiation. *Photochem. Photobiol.* **2008**, *84*, 1445–1454. [CrossRef]
41. Kawoosa, T.; Singh, H.; Kumar, A.; Sharma, S.K.; Devi, K.; Dutt, S.; Vats, S.K.; Sharma, M.; Ahuja, P.S.; Kumar, S. Light and temperature regulated terpene biosynthesis: Hepatoprotective monoterpene picroside accumulation in *Picrorhiza kurrooa*. *Funct. Integr. Genom.* **2010**, *10*, 393–404. [CrossRef]

42. Koyama, K.; Ikeda, H.; Poudel, P.R.; Goto-Yamamoto, N. Light quality affects flavonoid biosynthesis in young berries of Cabernet Sauvignon grape. *Phytochemistry* **2012**, *78*, 54–64. [CrossRef] [PubMed]
43. Lei, H.; Dong, B.-C.; Yang, X.-J.; Huang, C.-B.; Wang, X.-D.; Wu, X.-J. Effect of light on flavonoids biosynthesis in red rice Rdh. *Agric. Sci. China* **2009**, *8*, 746–752.
44. Vogt, T.; Ibdah, M.; Schmidt, J.; Wray, V.; Nimtz, M.; Strack, D. Light-induced betacyanin and flavonol accumulation in bladder cells of *Mesembryanthemum crystallinum*. *Phytochemistry* **1999**, *52*, 583–592. [CrossRef] [PubMed]
45. Wang, Y.; Gao, L.; Shan, Y.; Liu, Y.; Tian, Y.; Xia, T. Influence of shade on flavonoid biosynthesis in tea (*Camellia sinensis* (L.) O. Kuntze). *Sci. Hortic.-Amst.* **2012**, *141*, 7–16. [CrossRef]
46. Liu, C.-Z.; Guo, C.; Wang, Y.-C.; Ouyang, F. Effect of light irradiation on hairy root growth and artemisinin biosynthesis of *Artemisia annua* L. *Process Biochem.* **2002**, *38*, 581–585. [CrossRef]
47. Zhong, J.J.; Seki, T.; Kinoshita, S.I.; Yoshida, T. Effect of light irradiation on anthocyanin production by suspended culture of *Perilla frutescens*. *Biotechnol. Bioeng.* **1991**, *38*, 653–658. [CrossRef]
48. Łuczkiewicz, M.; Zárate, R.; Dembińska-Migas, W.; Migas, P.; Verpoorte, R. Production of pulchelin E in hairy roots, callus and suspension cultures of *Rudbeckia hirta* L. *Plant Sci.* **2002**, *163*, 91–100. [CrossRef]
49. Tabata, M.; Mizukami, H.; Hiraoka, N.; Konoshima, M. Pigment formation in callus cultures of *Lithospermum erythrorhizon*. *Phytochemistry* **1974**, *13*, 927–932. [CrossRef]
50. Park, N.I.; Li, X.; Suzuki, T.; Kim, S.-J.; Woo, S.-H.; Park, C.H.; Park, S.U. Differential expression of anthocyanin biosynthetic genes and anthocyanin accumulation in tartary buckwheat cultivars 'Hokkai T8'and 'Hokkai T10'. *J. Agric. Food Chem.* **2011**, *59*, 2356–2361. [CrossRef]
51. Li, X.; Kim, Y.B.; Uddin, M.R.; Lee, S.; Kim, S.-J.; Park, S.U. Influence of light on the free amino acid content and γ-aminobutyric acid synthesis in *Brassica juncea* seedlings. *J. Agric. Food Chem.* **2013**, *61*, 8624–8631. [CrossRef]
52. Thwe, A.A.; Kim, J.K.; Li, X.; Bok Kim, Y.; Romij Uddin, M.; Kim, S.J.; Suzuki, T.; Park, N.I.; Park, S.U. Metabolomic analysis and phenylpropanoid biosynthesis in hairy root culture of tartary buckwheat cultivars. *PLoS ONE* **2013**, *8*, e65349. [CrossRef]
53. Cuong, D.M.; Park, S.U.; Park, C.H.; Kim, N.S.; Bong, S.J.; Lee, S.Y. Comparative analysis of glucosinolate production in hairy roots of green and red kale (*Brassica oleracea* var. *acephala*). *Prep. Biochem. Biotechnol.* **2019**, *49*, 775–782. [CrossRef] [PubMed]
54. Kim, H.H.; Kwon, D.Y.; Bae, H.; Kim, S.-J.; Kim, Y.B.; Uddin, M.R.; Park, S.U. Influence of auxins on glucosinolate biosynthesis in hairy root cultures of broccoli (*Brassica oleracea* var. *italica*). *Asian J. Chem.* **2013**, *25*, 6099. [CrossRef]
55. Lee, S.Y.; Bong, S.J.; Kim, J.K.; Park, S.U. Glucosinolate biosynthesis as influenced by growth media and auxin in hairy root cultures of kale (*Brassica oleracea* var. *acephala*). *Emir. J. Food Agric.* **2016**, 277–282. [CrossRef]


**Disclaimer/Publisher's Note:** The statements, opinions and data contained in all publications are solely those of the individual author(s) and contributor(s) and not of MDPI and/or the editor(s). MDPI and/or the editor(s) disclaim responsibility for any injury to people or property resulting from any ideas, methods, instructions or products referred to in the content.

*horticulturae*

MDPI

Article

# Polyphenolic Compounds and Biological Activities of Leaves and Fruits of *Syzygium samarangense* cv. 'Giant Green' at Three Different Maturities

**Nuruljannah Suhaida Idris [1], Mohammad Moneruzzaman Khandaker [1,*], Zalilawati Mat Rashid [2], Ali Majrashi [3], Mekhled Mutiran Alenazi [4], Zanariah Mohd Nor [1], Ahmad Faris Mohd Adnan [5] and Nashriyah Mat [1]**

1 School of Agriculture Science & Biotechnology, Faculty of Bioresources and Food Industry, Universiti Sultan Zainal Abidin, Besut Campus, Besut 22200, Terengganu, Malaysia
2 School of Food Industry, Faculty of Bioresources and Food Industry, Universiti Sultan Zainal Abidin, Besut Campus, Besut 22200, Terengganu, Malaysia
3 Department of Biology, College of Science, Taif University, Taif 21944, Saudi Arabia
4 Plant Production Department, College of Food and Agricultural Sciences, King Saud University, Riyadh 11451, Saudi Arabia
5 Institute of Biological Sciences, Faculty of Science, Universiti Malaya, Kuala Lumpur 50603, Selangor, Malaysia
* Correspondence: moneruzzaman@unisza.edu.my; Tel.: +60-9699-3450

**Abstract:** *Syzygium samarangense* cv. 'Giant Green' is an underutilised fruit that can be found in Malaysia and other Asian countries. Since this fruit is not fully commercialised, the information about its potential health benefits is limited. Thus, this study was carried out to determine the polyphenolic contents (total phenolic and total flavonoid) and biological activities (antioxidant, alpha-glucosidase and antibacterial assay) of 'Giant Green' leaves and fruits at different maturity stages. The young, mature and old leaves, and unripe, half-ripened and ripened fruits were analysed. The results showed that the young leaves increased the TPC and TFC by 35% and 41%, over the old leaves. Similarly, TPC and TFC contents were 37% and 54% higher in unripe fruits compared to the ripened fruits. In addition, young leaves exhibited the strongest scavenging activity towards DPPH, NO and ABTS radicals with $IC_{50}$ values increasing 1.6-fold, 1.7-fold and 2.3-fold, respectively, over the old leaves. However, in fruit samples, only unripe fruits were able to inhibit more than 50% of radicals. A comparable trend was observed in alpha-glucosidase inhibitory assay whereas young leaves and unripe fruits recorded 81% and 99% increases in $IC_{50}$ values, respectively, from young leaves to old leaves and unripe fruits to ripened fruits. Identically, young leaves also showed a significant effect in antibacterial assay with an inhibition zone increase of 19%, 36%, 32%, and 31% in *S. aureus*, *E. faecalis*, *S. typhimurium* and *E. coli*, respectively, over the old leaves. However, only unripe fruits were most effective against all tested bacteria while half-ripened fruits were only effective against *E. faecalis* with a 1.1-fold increase in the inhibition zone compared to unripe fruits. Ripened fruits were resistant to all of the bacteria. These results suggest that the young leaves and unripe fruits of 'Giant Green' cultivar of *S. samarangense* could be a potential candidate for the management of some diseases coming from harmful free radicals or bacterial infection.

**Keywords:** *Syzygium samarangense*; 'Giant Green' cultivar; phenolic; flavonoid; antioxidant; alpha-glucosidase inhibitory; antibacterial

check for updates

**Citation:** Idris, N.S.; Khandaker, M.M.; Rashid, Z.M.; Majrashi, A.; Alenazi, M.M.; Nor, Z.M.; Mohd Adnan, A.F.; Mat, N. Polyphenolic Compounds and Biological Activities of Leaves and Fruits of *Syzygium samarangense* cv. 'Giant Green' at Three Different Maturities. *Horticulturae* **2023**, *9*, 326. https://doi.org/10.3390/horticulturae9030326

Academic Editor: Wajid Zaman

Received: 23 December 2022
Revised: 18 February 2023
Accepted: 22 February 2023
Published: 1 March 2023

**Copyright:** © 2023 by the authors. Licensee MDPI, Basel, Switzerland. This article is an open access article distributed under the terms and conditions of the Creative Commons Attribution (CC BY) license (https://creativecommons.org/licenses/by/4.0/).

## 1. Introduction

Plant parts such as leaf, flower, fruit, root and stem barks have been used for thousands of years as a traditional medicine which is important in health promotion and wellness. The various plant parts can be used as effective therapeutic agents in reducing the development of certain chronic diseases such as cardiovascular, diabetes, cancer, arthritis, atherosclerosis, rejuvenation of aged skin and diseases associated with cerebral aging [1,2]. Eleven (11) percent of the 252 basic and essential drugs originated from plant parts while the rest come from synthetic drugs derived from natural precursors [3]. Human bodies are always exposed to harmful reactive oxygen species (ROS) such as superoxide anion, hydrogen peroxide ($H_2O_2$), peroxyl (ROO) radicals, peroxynitrite anion (ONOO) and reactive hydroxyl (OH) radicals which are formed by pollution, smoking or pesticides used in crops [4]. ROS is an essential part of our body at the normal rate in aerobic metabolism and is involved in host defence. Nevertheless, the excess production of ROS is positively correlated to various chronic diseases such as inflammation, diabetes mellitus, cancer, atherosclerosis and hypertension [5,6]. Plant parts are an ultimate source of phenols, flavonoids, tannin and volatile compounds that act as bioactive compounds, natural antioxidants against ROS, reduce oxidative stress and protect humans from chronic diseases [7].

The wax apple (*Syzygium samarangense*) is a non-climacteric tropical fruit that belongs to the genus of *Syzygium* in the Myrtaceae family [8]. The wax apple fruit is also known as jambu air, water apple, wax jambu, bell fruit, makopa, samarang rose apple and java apple [9]. The fruit tree is usually cultivated and grown in Malaysia, Thailand, Taiwan, the Philippines, Vietnam, Laos, China, India, Bangladesh and Indonesia [10]. The fruits are bell or pear-shaped, usually red or light red, pink, green, greenish-white or cream coloured. They are crispy and have an aromatic flavour with a subtle sweet taste. The 'Jambu Madu Red' and 'Masam Manis Pink' are popular cultivars in Malaysia and other Southeast Asian Countries [11,12]. Furthermore, the 'Giant Green' cultivar of wax apple is more aromatic but less popular and underutilised among the cultivars. The cultivar has a spreading canopy with a 3 m canopy width. The density of light green leaves in the canopy of this cultivar is lower than the other two cultivars. The cultivar produces 8 to 10 creamy white flowers in a cluster at the branch tips or in the axils of leaves or any points on the trunk. The flowers bloom within 1 month after bud development and the fruits can be harvested around 50 days after anthesis [8]. The fruit production of this cultivar is non-seasonal and can be harvested three times per year in tropical areas. The fruit length of 'Giant Green' is around 6.3 cm and the diameter is around 5.2 cm [11]. The cultivar has the largest fruit weight (90 g) and produces more seeds than the other two cultivars. The fruit has a rough surface and almost all the fruit is edible. The 'Giant Green' fruits are eaten after ripening, with salt or cooked as a sauce. The leaves of the wax apple contain flavanones, ellagitannins, flavonol glycosides, triterpenoids, anthocyanidins, proanthocyanidins, chalcones and terpenoids [6]. The fruits of the wax apple are a rich source of phenols, flavonoids and several antioxidant compounds and as a result, have potential benefits for human health and are used in traditional medicine.

The plant parts contain valuable chemical content that was believed to benefit human health. In folk medicine, wax apple root is used as a medicine to treat oedema, reduce the itching of the skin and release menstrual cycle pain [8]. The bark and stem of the wax apple have antifungal properties which have been used in wound treatment [6]. The leaves have traditionally been used to treat a cracked tongue, asthma, bronchitis, fever, for bathing purposes, itches and waist pain [13]. The fruit is used to cure mouth ulcers, as a stimulant to increase the urine level, to improve blood circulation in the pelvis and uterus, for treatment of fever, sore throat and to reduce blood pressure. The flower is used to relieve fever and halt diarrhoea [8]. The phenolic and flavonoid contents of wax apple fruits perform antioxidant, antibacterial, antidiabetic and positive cytotoxic activity against the SW-480 human colon cancer cell line [14,15]. The flavonoids, tannins, alkaloids, terpenoids and essential oils of wax apple leaves are also active in anti-inflammatory, spasmolytic, antioxidant, antidiabetic, anticancer and analgesic activities [16,17].

The 'Giant Green' cultivar is underutilised and is not fully commercialised among the three cultivars of wax apple fruits in Malaysia. Previous studies have shown antioxidant and antibacterial activities of this cultivar as well as the presence of valuable phytochemicals such as phenolic, flavonoid, carotenoid and vitamin C in their leaf, fruit and bark [18,19]. However, studies evaluating the biological activity and the changes of its phytochemical in leaves and fruits of 'Giant Green' at different maturity stages have not been reported in the literature. In addition, the leaves and fruits undergo various morpho-physiological, biochemical and biological changes during maturation and ripening. Hence, the proper maturity stage is crucial for the harvesting of leaves and fruits of the 'Giant Green' cultivar to obtain maximum benefit from them for pharmaceutical uses. Keeping this in consideration, the present study was designed to determine the effects of maturity level on polyphenolic content and biological activities such as antioxidants, alpha-glucosidase inhibitory and antibacterial activities of 'Giant Green' leaves and fruits (i.e., young leaf, mature leaf, old leaf, unripe fruit, half-ripened fruit and ripened fruit). This study proposes that the maturity level of leaves and fruits of the 'Giant Green' cultivar can regulate the accumulation of polyphenolic contents, antioxidant activity, α-glucosidase inhibitory activity and antibacterial activity.

## 2. Materials and Methods

### 2.1. Collection and Preparation of Plant Materials

The leaves and fruits of the 'Giant Green' cultivar of wax apple were collected several times from the experimental trees of wax apple from an orchard located at Kampung Olek Lempit, Banting, Selangor, Malaysia (1°28 N, 111°20 E), at an elevation of about 45 m above sea level in a hot and humid tropical climate. The type of soil in the orchard was peat and the mean pH was around 4.6 [18]. Three maturity stages of leaves were examined, namely young (YL), mature (ML) and old (OL) leaves whereas for fruits the stages were unripe (UF), half-ripened (HF) and ripened (RF) fruit which were collected and analysed. Five biological replicates of each sample were used in this study. The leaves and fruits were selected based on the physical examination reported by the previous studies. The leaves were collected based on the method reported by Lee et al. [20] (Figure 1). The old leaves were dark green in colour, with a very hard surface. The position of the leaves was lower on the branch, showing the first sign of epiphylly or senescence, with a length of 18–18.5 cm and a diameter of 8.0–8.5 cm. The mature leaves were green in colour, had a hard surface, had a fully developed structure and had a position in the middle of the branch, with a length of 16–16.5 cm and a diameter of 6.0–6.5 cm. The young leaves were light green in colour, had a short growth phase, a soft surface, were fully sized but still lacking in their structure, the position on the branch was higher than the previous stage and the length was 11–11.5 cm and the diameter 3.5–4.0 cm.

The maturity of fruits was selected based on their physical parameters such as colour, texture and size [21] (Figure 1). The ripened fruit was reddish-green in skin colour, greenish-white in pulp, more soft and juicy in texture, had a length of 6.0–7.5 cm and a diameter of 5.5–6.0 cm. The half-ripened fruit was pale green in skin colour, was less soft and juicy in texture, had a length of 5.0–5.5 cm and a diameter 4.5–5.0 cm. Unripe fruit was green in skin colour, was more tough and not juicy in texture, had a length of 3.5–4.0 cm and a diameter of 3.5–4.0 cm.

### 2.2. Extraction Procedure

The sample was wiped with wet tissue and crushed using a mortar and pestle. The crushed sample (5 g) was soaked in methanol (25 mL) and allowed to stand for three days. Then, it was heated in a water bath at 70 °C for 15 min, followed by being centrifuged at 4000 rpm for 15 min. The supernatant was collected and put in the fumehood to remove all of the methanol solvent. Then, the sample was further lyophilised with the freeze-dried method (24 h). The extract was kept in an air-tight container and stored at 4 °C until analysis.

(A)

(B)

**Figure 1.** Leaves (**A**) and Fruits (**B**) of 'Giant Green' cultivar of *Syzygium. samarangense* at three maturity stages.

## 2.3. Percentage of Extraction Yield

The percentage of extraction yield of 'Giant Green' leaves and fruit extracts was calculated according to Zin et al. [22] using the formula below:

$$Percentage\ of\ extraction\ yield = \frac{Weight\ of\ extracts\ (g)}{Weight\ of\ \ raw\ sample\ (g)} \times 100 \quad (1)$$

## 2.4. Determination of Total Phenolic Assay

The total phenolic content (TPC) assay was measured according to Zin et al. [22]. Approximately 1 mg of standard (gallic acid) and 5 mg of plant crude extracts were dissolved in 1 mL of dimethylsulfoxide (DMSO) which gave 1 mg/mL and 5 mg/mL, respectively, of stock solutions. Then, the standard stock solution was diluted with DMSO by a serial dilution method to produce the final concentrations of 200, 180, 160, 120, 100, 80 and 60 μg/mL. The plant sample was evaluated at a final concentration of 1 mg/mL. So, 60 μL of stock solution from the plant sample was transferred out from the microtube and DMSO was made up to 100 μL. Next, 200 μL of Folin–Ciocalteau was added into standard and plant sample plates, then, the mixture was vortexed vigorously. The reaction was terminated by adding 800 μL of sodium carbonate (7.5%). Then, the mixture was vortexed again and incubated for two hours in the dark at room temperature for blue colour development. The absorbance of the mixture was read at 765 nm using a microplate reader. TPC value of the plant extract was calculated by using the formula below and expressed as milligrams of gallic acid equivalents per gram dry weight (mg GAE/g DW).

$$C = \frac{cV}{m} \quad (2)$$

where C is total phenolic content, *c* is the concentration of gallic acid (μg/mL) obtained from the calibration curve, *V* is the final volume of plant extract and *m* is the weight of dried extract.

*2.5. Determination of Total Flavonoid Assay*

The total flavonoid content (TFC) was quantified using a calorimetric assay according to Cunha et al. [23] with some modifications. Quercetin was used as a standard for the calibration curve. Briefly, 0.5 mg of quercetin was dissolved in 1 mL of dimethyl sulfoxide (DMSO) to complete the concentration of stock solution to 1 mg/mL. Exactly 5 mg of plant sample was weighted and diluted in 1 mg/mL of DMSO. The stock solution (quercetin) was prepared by serial dilutions to final concentrations of 100, 50, 25, 12.50, 6.25, 3.125, 1.563 and 0.781 μg/mL. Then, 140 μL of plant extract (5 mg/mL) was pipetted out into a 96-well plate. Standard quercetin solution and plant extract were separately mixed with 150 μL of aluminium chloride (10%) and 150 μL of potassium acetate (1 M). After mixing, 700 μL was made up with distilled water and the solution was let to stand at room temperature for 30 min in a dark condition. The absorbance was measured at 415 nm wavelength using a microplate reader. The TFC value of the plant samples was obtained from the linear regression equation of quercetin and calculated using the equation below. The results were expressed as milligrams of quercetin equivalents per gram dry weight (mg QE/g DW).

$$C = \frac{cV}{m} \tag{3}$$

where *C* is total flavonoid content, *c* is the concentration of quercetin (μg/mL) obtained from the calibration curve, *V* is the final volume of plant extract and *m* is the weight of the dried extract.

*2.6. Antioxidant Assay*

2.6.1. DPPH Radical Scavenging Assay

The antioxidant capacity using 1,1-phenyl-2-picrylhydrazyl (DPPH) radical was determined using the method reported by Zin et al. [22]. Exactly 25 μL of extract with different concentrations (0–250 μg/mL) was added to a 96-well plate and then, mixed with 200 μL of methanolic solution of DPPH (0.1 mM). The control was prepared by substitution of crude extract with DMSO (25 μL) plus DPPH solution (200 μL). The plate was wrapped with aluminium foil and incubated for 30 min in the dark at room temperature. After that, the mixture was measured at 517 nm wavelength using a microplate reader. Quercetin was used as a positive control. For negative control, the equivalent of DMSO without extract was prepared in a similar manner. The percentage inhibition of DPPH radical was calculated as below.

$$Percentage\ inhibition\ (\%) = \frac{Abs.\ C - Abs.\ S}{Abs.\ C} \times 100 \tag{4}$$

where

*Abs. C* = Absorbance of control

*Abs. S* = Absorbance of sample in the presence of extract or positive control

2.6.2. Nitric Oxide (NO) Radical Scavenging Assay

Nitric oxide radical scavenging assay was obtained from the reaction of sodium nitroprusside and measured by Griess reagent [24]. Sodium nitroprusside solution (10 mM) was prepared by diluting 0.3 g of sodium nitroprusside into 100 mL of phosphate buffer saline. A volume of 20 μL of extract, as well as a positive control (quercetin) with various concentrations (0–250 μg/mL), was added with 80 μL of sodium nitroprusside solution in a 96-well plate. Then, the plate was incubated under light for 150 min at room temperature. After that, 100 μL of Griess reagent was added to the plate. The Griess reagent was prepared freshly before being used by mixing 1% sulphanilamide solution and 0.1% naphthyl ethylenediamine ethylene diamine dihydrochloride solution in a similar volume. The sulphanilamide solution was prepared by dilution of 0.5 μg of sulphanilamide in 50 μL of 20% glacial acetic acid, and naphthyl ethylene diamine dihydrochloride solution was prepared by dilution of 0.05 μg of naphthyl ethylene diamine dihydrochloride in 50 μL of distilled water. Then, the plate was kept for 10 min before absorbance of the solution was taken using a microplate reader at 540 nm. The equivalent of DMSO without extract was prepared in the same manner and is known as negative control. The $IC_{50}$ values of extract and positive control were determined. The percentage of inhibition nitrite scavenging was measured using a formula as described in the DPPH assay.

2.6.3. ABTS Radical Scavenging Activities

The ability of the extract to scavenge 2,2,-azinobis-(3-ethylbenzothiazoline-6-sulfonic acid) ($ABTS^+$) was measured as described by Shalaby and Shanab [25] with small modifications. ABTS salt (0.0768 g) and potassium persulfate (0.0132 g) were dissolved in distilled water (20 mL) to form monocation radical ABTS ($ABTS^+$). $ABTS^+$ is very sensitive to light. Then, the tube containing the $ABTS^+$ solution was wrapped with aluminium foil and allowed to stand in a dark condition for 16 h at room temperature before being used. Then, the solution was further diluted with methanol to an absorbance of 0.700 ± 0.02. Methanolic $ABTS^+$ solution (189 µL) was mixed with 21 µL of various concentrations of all of the extracts and positive control, quercetin (0–250 µg/mL). A similar manner was used to prepare the negative control without extract but substitute with an equal volume of DMSO. Thereafter, the mixture was measured at an absorbance of 734 nm wavelength. $ABTS^+$ radical inhibition was estimated using the equation as described in the DPPH assay.

2.6.4. Reducing Power

The reducing power assay was carried out according to the method of Mayur et al. [26] with slight modifications. Aliquots (20 µL) of plant extract with different concentrations (0–250 µg/mL) was mixed with 30 µL of 0.1 M phosphate buffer (pH 6.6) and 30 µL of potassium ferracyanide (1%, *w/v* in distilled water). The solution was incubated at 50 °C for 20 min. Then, the reaction of the mixture was stopped by the addition of 30 µL of 10% trichloroacetic acid. The solution was mixed well. Next, 20 µL of ferric chloride (0.1% *w/v*) and 70 µL of distilled water were added. The capacity-reducing power of extract was compared to the quercetin as a positive control. The absorbance was recorded at 700 nm by using a microplate reader. The higher absorbance value indicates a higher reducing power of the extract.

*2.7. Alpha-Glucosidase Inhibitory Assay*

The alpha-glucosidase inhibitory assay was carried out using the method by Misbah et al. [27]. In this assay, the enzyme solution contained 0.2 unit/mL of α-glucosidase enzyme (*Bacillus stearothermophilus*, Sigma Aldrich Brand, Gillingham, UK) and 0.1 M phosphate buffer (pH 7); and the substrate solution consisted of ϱ-Nitrophenyl-α-D-glucopyranoside (0.5 mM) and a similar phosphate buffer was used. For the stock solution, 5 mg of plant extract was dissolved in 1 mL of 98% DMSO. Then, the working solution was prepared by dissolving stock solution with 10% DMSO at certain concentrations (0–45 µg/mL). A total of 10 µL of the extract had been put into a 96-well plate and added with 25 µL of enzyme solution. The mixture was incubated at 37 °C. After 10 min, 25 µL of substrate solution was added and incubated again for 30 min at 37 °C. To terminate the reaction, 100 µL of 0.2 M sodium carbonate was added to the mixture. The absorbance was measured at a 410 nm wavelength. The solution was used as a control in the absence of plant extract, while, without enzyme and substrate it was used as a blank. The percentage inhibition was calculated as in the formula below and the $IC_{50}$ value was determined.

$$Percentage\ inhibition\ (\%) = \frac{(Abs.\ C - Abs.\ CB) - (Abs.\ S - Abs.\ SB)}{1!Abs.C - Abs.CB} \times 100 \quad (5)$$

where,

*Abs. C* = Absorbance of control (10% DMSO + enzyme + substrate)
*Abs. CB* = Absorbance of control blank (10% DMSO without enzyme and substrate)
*Abs. S* = Absorbance of sample (sample solution + enzyme + substrate)

*Abs. SB* = Absorbance of sample blank (sample solution without enzyme and substrate)

### 2.8. Antibacterial Assay

The antibacterial activity was evaluated by using agar well diffusion according to the methods of Shanmugam et al. [28]. Two Gram-positive bacteria, (*Staphylococcus aureus* (ATCC 33591) and *Enterococcus faecalis* (ATCC 29212)) and two Gram-negative bacteria (*Escherichia coli* (ATCC 35218) and *Salmonella Typhimurium* (ATCC 14028)), were obtained from the Microbiology Laboratory, Universiti Sultan Zainal Abidin, Besut Campus, Besut, Terengganu. The isolated colony of bacteria was selected with a sterile wire loop and grew in nutrient agar at 37 °C for 24 h. The pure colony was picked and transferred into a Muller–Hilton Broth (MHB) and was incubated at 37 °C until it reached the turbidity of 0.5 McFarland standards ($1.5 \times 10^8$ CFU/mL). MHA medium was inoculated with 100 µL volume of standardised inoculums of bacterial strains (0.5 McFarland). The bacterial suspension was spread over MHA using a sterile cotton bud and then allowed to stand for a minute to dry the medium. Five of the wells with a 6 mm diameter were made by using a sterile cork borer on the plate. A total of 10 mg of extracts were dissolved in 1 mL DMSO as a stock solution. Using a micropipette, 100, 80, 40 and 20 µL of extracts were added to the respective wells. A total of 100 µL of DMSO was used as a negative control. For a comparative study of antibacterial properties of the wax apple plant, eight commercial standard antibiotics (Gentamycin, Tetracycline, Vancomycin, Amoxycillin, Cephazolin, Neomycin, Metronidazole and Cefaclor) were also analysed. All of the plates were incubated in an upright position at 37 °C for 24 h. The diameter of zone inhibition (mm) was measured. Every experiment was conducted in triplicate.

### 2.9. Statistical Analysis

Statistical analysis was performed on the data using XLSTAT software version 2014. The significant difference between the parameters studied was analysed by using one-way ANOVA followed by post hoc Tukey's Honestly Significant Difference (HSD) test. Values were expressed as mean ± standard deviation (SD). The confidence levels of all analyses were performed at 95% with $p < 0.05$ representing a significant difference.

## 3. Results

### 3.1. Percentage of Extraction Yields

The percentage of extraction yields of leaves and fruits of the 'Giant Green' cultivar of wax apple are shown in Table 1. The results showed that the leaves and fruit extracts yield varied from 3.57 to 4.46% in descending order of old leaves > young leaves > mature leaves. The maturity stages of the leaf did not produce a significant effect on extract yield; however, fruit maturity stages yielded significant effects on the percentage of extract yield. The highest percentage of extraction yield of fruit extracts was recorded in ripened fruit with a value of 4.87%. The lowest percentage of extract yield, 2.98%, was found in half-ripened fruits (Table 1).

### 3.2. Total Phenolic Content (TPC)

The total phenolic content (TPC) of the 'Giant Green' cultivar of wax apple leaves and fruits was determined to see the effects of maturity stages on the accumulation of phenols in plant parts. The TPC values were calculated using the gallic acid standard curve (Figure 2A). The results indicated that total phenolic content decreased significantly with the advancement of leaf and fruit maturity. The young leaves had the highest TPC followed by mature leaves with a value of 66.56 mg GAE/g DW and 50.37 mg GAE/g DW. The old leaves contained the lowest amount of phenolic content 43.16 mg GAE/g DW. The TPC value of fruit extracts ranged from 34.08 to 54.11 mg GAE/g DW (Table 1). The unripe fruit had a 1.59-fold TPC compared to the ripened fruits.

**Table 1.** Percentage yield, total phenolic content (TPC) and total flavonoid content (TFC) of leaves and fruit extracts of the 'Giant Green' cultivar of *Syzygium samarangense* at three maturity stages.

| Samples | Percentage of Extraction Yield (%) | TPC (mg GAE/g DW) | TFC (mg QE/g DW) |
|---|---|---|---|
| **Leaves** | | | |
| Young | 3.87 ± 0.03 $^{A}$ | 66.56 ± 0.86 $^{A}$ | 17.25 ± 0.44 $^{A}$ |
| Mature | 3.57 ± 0.03 $^{A}$ | 50.37 ± 1.05 $^{B}$ | 11.75 ± 0.36 $^{B}$ |
| Old | 4.46 ± 0.05 $^{A}$ | 43.16 ± 0.83 $^{C}$ | 10.20 ± 0.29 $^{C}$ |
| **Fruit** | | | |
| Unripe | 3.13 ± 0.02 $^{a}$ | 54.11 ± 0.42 $^{a}$ | 2.34 ± 0.06 $^{a}$ |
| Half-ripen | 2.97 ± 0.01 $^{a}$ | 39.97 ± 0.31 $^{b}$ | 1.65 ± 0.06 $^{b}$ |
| Ripen | 4.87 ± 0.02 $^{b}$ | 34.08 ± 0.44 $^{c}$ | 1.07 ± 0.04 $^{c}$ |

Values are expressed as the means ± standard deviation. The different superscript letter refers to significant difference ($p < 0.05$) by comparing the three maturity stage of leaves and fruit samples.

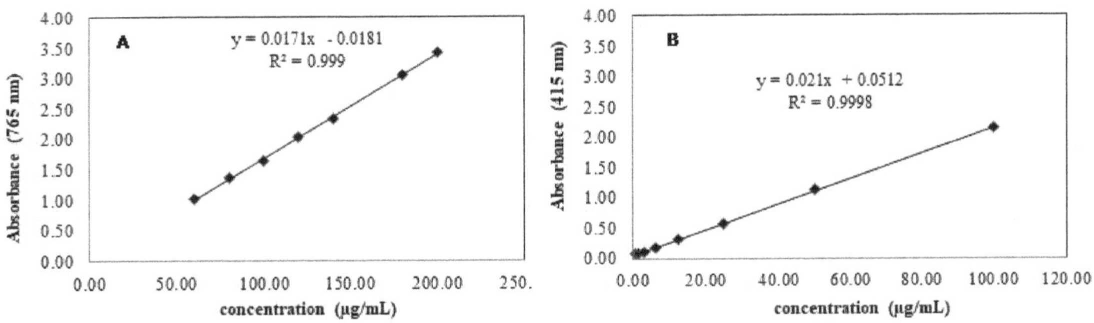


**Figure 2.** Calibration curve of quercetin and gallic acid standards for determination of total phenolic content (TPC) (**A**) and total flavonoid content (TFC) (**B**).

### 3.3. Total Flavonoid Content

This result showed that the amount of flavonoid for 'Giant Green' leaves and fruits have a significant effect among three maturity stages (Figure 2B). TFC values of leaf extracts ranged from 10.20 to 17.25 mg QE/g DW with descending order of young leaves > mature leaves > old leaves. The TFC value depicted a 1.7-fold decrease from young leaves to old leaves. For fruit extracts, the flavonoid contents ranged from 1.07 to 2.34 mg QE/g DW (Table 1) in descending order of unripe fruit > half-ripened fruit > ripened fruit. Unripe fruits have flavonoid content two times higher than ripened fruits.

### 3.4. Antioxidant Assays

#### 3.4.1. DPPH Free Radical Scavenging Activity

Figure 3a illustrates the abilities of 'Giant Green' leaves to scavenge the DPPH radical. The result showed that young leaves exhibited the strongest antioxidant activity, and the weakest was in the old leaves extract. The $IC_{50}$ values of leaf extracts ranged from 13.66 to 21.79 µg/mL (Table 2). Young leaves had 1.6-fold lower $IC_{50}$ values compared to old leaf extract and 1.2-fold lower than mature leaf extract. Meanwhile, the percentage inhibitions of DPPH radicals in fruit extracts are shown in Figure 4a. The unripe fruit showed the strongest antioxidant activity followed by half-ripened while the least one was ripened fruit extract. Between the samples, only unripe fruit showed the ability to scavenge DPPH radicals at more than 50% with $IC_{50}$: 13.79 µg/mL (Table 2).

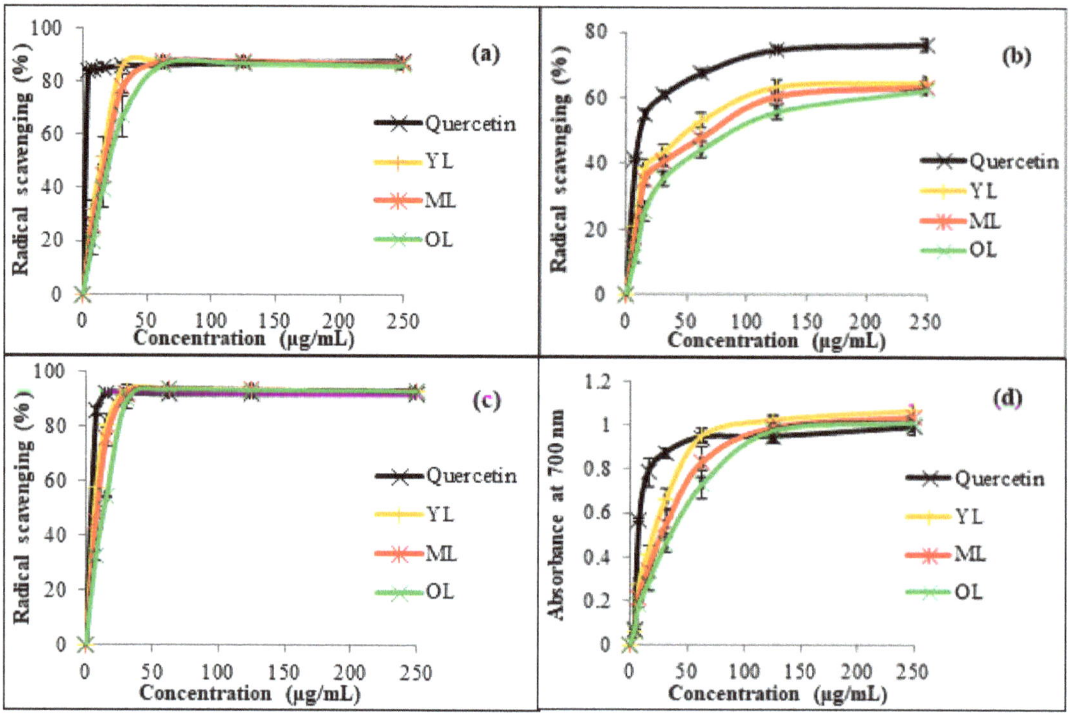


**Figure 3.** Antioxidant activities of leaves of the 'Giant Green' cultivar of *Syzygium samarangense* at three different maturity stages: (**a**) DPPH; (**b**) NO; (**c**) ABTS; (**d**) Reducing power assay.

**Table 2.** $IC_{50}$ values of antioxidant assays (DPPH, ABTS and NO) for leaves and fruit extracts of 'Giant Green' cultivar of *Syzygium samarangense* at three maturity stages.

| Sample | $IC_{50}$ Value (µg/mL) | | |
|---|---|---|---|
| | DPPH | NO | ABTS |
| Quercetin | $10.20 \pm 0.26$ $^{Aa}$ | $11.23 \pm 0.55$ $^{Aa}$ | $4.10 \pm 0.10$ $^{Aa}$ |
| **Leaves** | | | |
| Young | $13.66 \pm 0.19$ $^{B}$ | $51.57 \pm 1.71$ $^{B}$ | $6.31 \pm 0.06$ $^{B}$ |
| Mature | $16.57 \pm 0.19$ $^{C}$ | $69.29 \pm 1.03$ $^{C}$ | $8.87 \pm 0.31$ $^{C}$ |
| Old | $21.79 \pm 0.35$ $^{D}$ | $88.13 \pm 0.83$ $^{D}$ | $14.62 \pm 0.43$ $^{D}$ |
| **Fruit** | | | |
| Unripe | $13.79 \pm 0.22$ $^{b}$ | $87.80 \pm 3.50$ $^{b}$ | $46.77 \pm 0.67$ $^{b}$ |
| Half-ripened | ND | ND | $237.33 \pm 1.61$ $^{c}$ |
| Ripened | ND | ND | ND |

ND = Not detected. Values are expressed as means ± standard deviation. The different superscript letter refers to significant difference ($p < 0.05$) by comparing the three maturity stages of leaves and fruit samples. DPPH: 2,2-diphenyl-1-picrylhydrazyl; ABTS: 2,2′-azinobis-(3-ethylbenzothiazoline-6-sulphonic acid; NO: nitric oxide. Capital letter denotes leaves and small letter fruits.

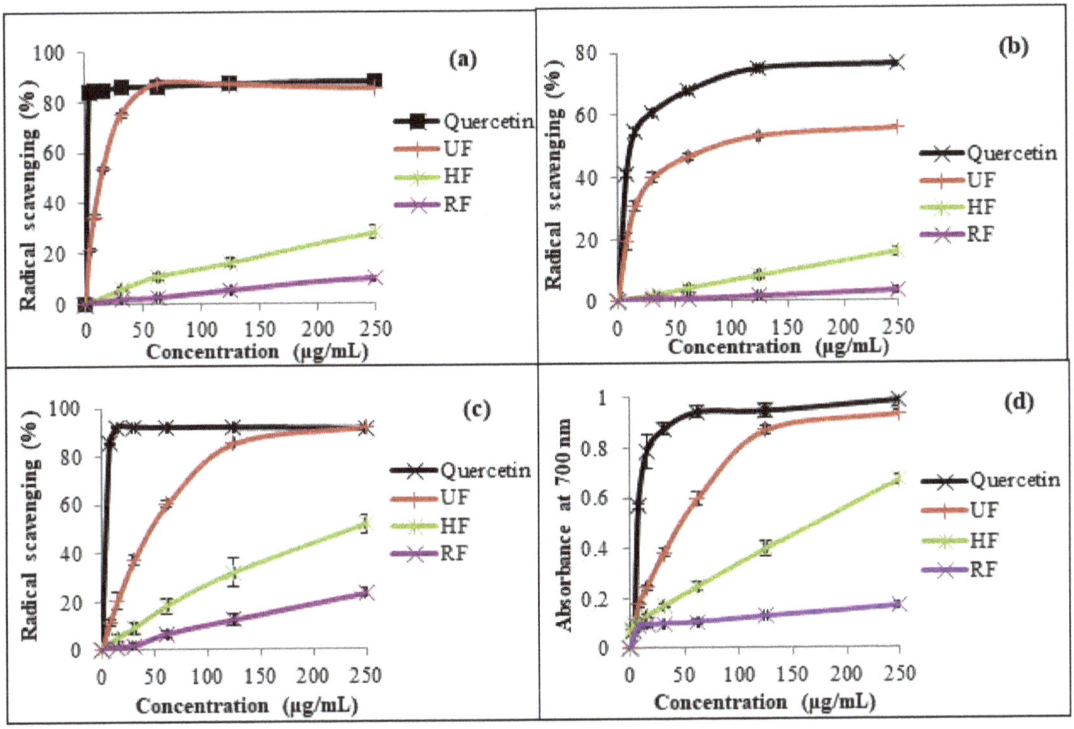


**Figure 4.** Antioxidant activities of fruits of 'Giant Green' cultivar of *Syzygium samarangense* at three different maturity stages: (**a**) DPPH; (**b**) NO; (**c**) ABTS; (**d**) Reducing power assay.

#### 3.4.2. Nitric Oxide (NO) Radical Scavenging Activity

The capabilities of 'Giant Green' leaf extracts to inhibit the nitric oxide radical is shown in Figure 3b. The results revealed that young leaves had greater NO scavenging capacity followed by mature and old leaves. The $IC_{50}$ values of NO radicals depicted a 41% increase from young leaves to old leaves (Table 2). For the 'Giant Green' fruit, unripe fruit exhibited the strongest scavenging activity towards NO radicals followed by half-ripened and ripened fruit extracts (Figure 4b). The result indicates that only unripe fruit extract had a significant effect to inhibit NO radicals with an $IC_{50}$ value of 87.80 μg/mL (Table 2).

#### 3.4.3. ABTS Radical Scavenging Activity

The percentage inhibition of the ABTS radical scavenging assay of 'Giant Green' leaves is shown in Figure 3c. The results revealed that young leaves had the strongest ability to inhibit ABTS radicals followed by mature leaves and the weakest one was old leaves. The $IC_{50}$ values of leaf extracts ranged from 6.31 to 14.62 μg/mL (Table 2). $IC_{50}$ values recorded a 2.3-fold increase in old leaves compared to young leaves. The abilities of 'Giant Green' fruit extracts in the scavenging of ABTS are shown in Figure 4c in descending order as follows: unripe fruit > half- ripened > ripened fruit. $IC_{50}$ values of fruit extracts depicted an 80% increase from unripe fruit to half-ripened fruit but were not detected in ripened fruit (Table 2).

#### 3.4.4. Reducing Power Activity

The absorbance results of reducing power for 'Giant Green' leaves are shown in Figure 3d. The reducing power of leaf extracts was in a concentration-dependent manner. The reducing power of quercetin, young, mature and old leaves was recorded with values of 0.99 ± 0.03, 1.07 ± 0.02, 1.03 ± 0.03 and 1.01 ± 0.04, respectively, at a concentration of 250 μg/mL. The young leaves showed higher reducing power followed by mature and old leaves. Furthermore, as seen in Figure 4d, the unripe fruits had a better capability to donate electrons and reduce the ferricyanide complex, while the half-ripened fruits' capability was moderate and the weakest was ripe fruits. The reducing power of all fruit samples also increases with an increase in concentration. The maximum values (at a concentration of 250 μg/mL) of reducing power of unripe, half-ripened and ripened fruits were 0.94 ± 0.01, 0.67 ± 0.02 and 0.17 ± 0.02, respectively. This result indicates that the reducing power in fruit extracts had an 82% decline from unripe fruit to ripened fruits.

### 3.4.5. Correlation Analysis

*Correlation of TPC, TFC and antioxidant assays between leaves and fruits of 'Giant Green' at different maturity stages.*

The TP content (Table 3) showed significant positive strong correlations between young leaves with half-ripened fruits, whereas mature leaves with old leaves and ripened fruits and half-ripened fruits with ripened fruits showed significant negative strong correlations. No correlations were detected between young leaves and mature leaves; mature leaves with half-ripened fruits; old leaves with unripe fruits and half-ripened fruits; unripe fruits with half-ripened and ripened fruits and half-ripened with ripened fruits.

**Table 3.** The correlation coefficients (R value) among leaves and fruits of the 'Giant Green' cultivar of *S. samarangense* in TPC, TFC and antioxidant assays (DPPH, NO and ABTS).

| Sample | Correlation Coefficient of TPC | | | | | |
|---|---|---|---|---|---|---|
| | YL | ML | OL | UF | HF | RF |
| YL | 1.00 | | | | | |
| ML | 0.16 | 1.00 | | | | |
| OL | −0.25 | −0.65 | 1.00 | | | |
| UF | −0.46 | −0.37 | −0.14 | 1.00 | | |
| HF | 0.64 | 0.15 | −0.22 | 0.17 | 1.00 | |
| RF | −0.51 | −0.78 | 0.43 | 0.28 | −0.70 | 1.00 |
| Sample | Correlation coefficient of TFC | | | | | |
| | YL | ML | OL | UF | HF | RF |
| YL | 1.00 | | | | | |
| ML | 0.80 | 1.00 | | | | |
| OL | 0.46 | 0.46 | 1.00 | | | |
| UF | 0.51 | 0.21 | −0.29 | 1.00 | | |
| HF | −0.30 | −0.40 | 0.00 | 0.33 | 1.00 | |
| RF | 0.09 | 0.50 | −0.33 | 0.34 | 0.05 | 1.00 |
| Sample | Correlation coefficient of DPPH Assay | | | | | |
| | YL | ML | OL | UF | HF | RF |
| YL | 1.00 | | | | | |
| ML | 0.92 | 1.00 | | | | |
| OL | 0.92 | 1.00 | 1.00 | | | |
| UF | −1.00 | −0.92 | −0.91 | 1.00 | | |
| HF | 0.09 | −0.31 | −0.32 | −0.10 | 1.00 | |
| RF | −0.96 | −0.78 | −0.77 | 0.97 | −0.36 | 1.00 |

**Table 3.** *Cont.*

| Sample | Correlation coefficient of NO Assay | | | | | |
|---|---|---|---|---|---|---|
| | YL | ML | OL | UF | HF | RF |
| YL | 1.00 | | | | | |
| ML | −0.35 | 1.00 | | | | |
| OL | 0.94 | −0.64 | 1.00 | | | |
| UF | 0.73 | −0.90 | 0.92 | 1.00 | | |
| HF | −0.69 | 0.92 | −0.89 | −1.00 | 1.00 | |
| RF | 0.96 | −0.60 | 1.00 | 0.89 | −0.87 | 1.00 |
| Sample | Correlation coefficient of ABTS Assay | | | | | |
| | YL | ML | OL | UF | HF | RF |
| YL | 1.00 | | | | | |
| ML | 0.49 | 1.00 | | | | |
| OL | 0.80 | 0.91 | 1.00 | | | |
| UF | −0.98 | −0.67 | −0.92 | 1.00 | | |
| HF | −0.69 | −0.97 | −0.99 | 0.83 | 1.00 | |
| RF | 0.48 | 1.00 | 0.91 | −0.66 | −0.97 | 1.00 |

Significant at $p < 0.05$; The value >0.9 a strong correlation, >0.8 a fair strong correlation and >0.6 a moderately strong correlation at $p$ value 0.05. YL: young leaves; ML: mature leaves; OL: old leaves; UF: unripe fruits: HF: half-ripened fruits; RF: ripened fruits; TPC: total phenolic content; TFC: total flavonoid content; DPPH:2,2-diphenyl-1-picrylhydrazyl; ABTS: 2,2′-azinobis-(3-ethylbenzothiazoline-6-sulphonic acid; NO: nitric oxide.

In the case of TF content (Table 3), only the mature leaves showed significant, positive and very strong correlations with young leaves. Most of the samples showed a moderate and weak correlation with each other. However, no correlation was observed between young leaves and ripened fruits, old leaves with half-ripened fruits and half-ripened fruits with ripened fruits.

Furthermore, in the DPPH assay (Table 3), a significant perfect positive correlation was shown between mature leaves and old leaves, while young leaves with unripe fruits showed a significant perfect negative correlation. Very strong correlations were found between young leaves and mature leaves; old leaves and ripened fruits; mature leaves with unripe fruits; old leaves with unripe fruits; and unripe fruits with ripened fruits. No correlation was shown between young leaves and half-ripened fruits and unripe fruit with half-ripened fruits.

For NO assay (Table 3), there was a significant perfect positive correlation between old leaves and ripened fruits, while a significant perfect negative correlation with unripe fruits and half-ripened fruits was found. In addition, young leaves with old leaves and ripened fruits; mature leaves with unripe fruits and half-ripened fruits; old leaves with unripe fruits and half-ripened fruits; unripe fruits with ripened fruits and lastly, half-ripened fruits with ripened fruits showed very strong correlations.

Moreover, in the ABTS assay (Table 3), only mature leaves with ripened fruits recorded a significant perfect positive correlation. Young leaves with mature leaves and ripened fruits showed a significant positive and weak correlation, while mature leaves with unripe fruits and unripe fruits with ripened fruits showed a significant negative and moderate correlation. However, the rest of the samples correlated very strongly with each other.

*Correlation between TPC, TFC and Antioxidant Assays.*

The correlations of TPC and TFC with antioxidant assays are as shown in Table 4. There were strong correlations between TPC and antioxidant assays which are ABTS, NO and DPPH in leaves but a moderate correlation in fruit extracts. However, both leaf and fruits extracts were shown to have a strong correlation between TFC and all of the antioxidant assays.

**Table 4.** The correlation coefficients (R value) of TPC and TFC with antioxidant assay ($IC_{50}$ value) in the 'Giant Green' cultivar of *S. samarangense*.

| Variables | Leaves | | Fruits | |
|---|---|---|---|---|
| | **TPC** | **TFC** | **TPC** | **TFC** |
| DPPH | 0.92 | 0.85 | 0.58 | 0.84 |
| ABTS | 0.96 | 0.91 | 0.68 | 0.98 |
| NO | 0.90 | 0.83 | 0.64 | 0.92 |

Significant at $p < 0.05$; TPC: total phenolic content; TFC: total flavonoid content; DPPH:2,2-diphenyl-1-picrylhydrazyl; ABTS:2,2′-azinobis-(3-ethylbenzothiazoline-6-sulphonic acid; NO: nitric oxide.

### 3.5. *Alpha-Glucosidase Inhibitory Assay*

The percentage of α-glucosidase inhibition of the 'Giant Green' leaves and fruit extracts were plotted as a function of concentration in comparison with standard, quercetin are shown in Figure 5. The results reveal that all of the extracts inhibited α-glucosidase enzyme in vitro. Based on Table 5, young leaves ($IC_{50}$: 0.80 μg/mL) exhibited good α-glucosidase enzyme compared to quercetin ($IC_{50}$: 1.54 μg/mL). Mature leaves and old leaf extracts showed appreciable inhibitory activity. The ranking of α-glucosidase inhibition based on the $IC_{50}$ values are as follows: young leaves > mature leaves > old leaves. The $IC_{50}$ values of young leaves depicted a 3.0-fold decrease compared to mature leaves and continuously decreased to 1.8-fold in old leaves as compared with mature leaves. Furthermore, out of the three types of fruit extracts, only unripe fruit and half-ripened fruit extracts showed percentages of more than 50% of alpha-glucosidase inhibitory activity (Figure 5). Unripe fruit exhibited a significant effect of α-glucosidase inhibitory activity followed by half-ripened fruit and ripened fruit. The $IC_{50}$ value dramatically increased to 93% from unripe fruit to half-ripened fruit and continued to increase until it was not detected in ripened fruit (Table 5).

**Table 5.** $IC_{50}$ values of alpha-glucosidase inhibitory activity for leaves and fruit extracts of the 'Giant Green' cultivar of *Syzygium samarangense* at three maturity stages.

| Sample | Alpha-Glucosidase ($IC_{50}$ Value, μg/mL) |
|---|---|
| Quercetin | $1.54 \pm 0.75^{ABa}$ |
| **Leaves** | |
| Young | $0.80 \pm 0.44^{A}$ |
| Mature | $2.39 \pm 0.06^{B}$ |
| Old | $4.26 \pm 0.37^{C}$ |
| **Fruit** | |
| Unripe | $1.98 \pm 0.02^{a}$ |
| Half-ripened | $28.14 \pm 0.34^{b}$ |
| Ripened | $133.06 \pm 3.54^{c}$ |

ND = Not detected. Values are expressed as means ± standard deviation. The different superscript letter refers to significant difference ($p < 0.05$) by comparing the three maturity stages of leaves and fruit samples. Capital letter denotes leaves and small letter for fruits.

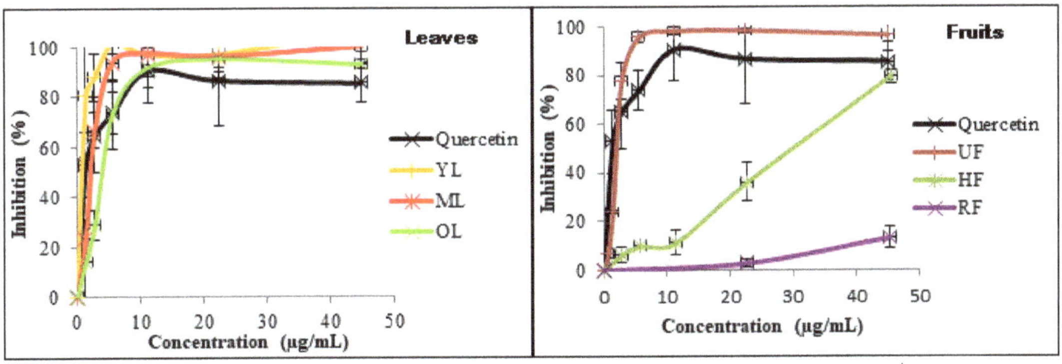


**Figure 5.** Alpha-glucosidase inhibitory assay of 'Giant Green' cultivar of *Syzygium samarangense* leaves and fruits at three different maturity stages.

Correlation Analysis

*Correlation of alpha-glucosidase inhibitory assay between leaves and fruits of 'Giant Green' at different maturity stages.*

In the present study, a significant perfect negative correlation was observed between young leaves and mature leaves, then mature leaves with half-ripened fruits (Table 6). Moreover, young leaves with old leaves; unripe fruits and half-ripened fruits; old leaves with half-ripened fruits; and unripe fruits with half-ripened fruits showed a significant positive and very strong correlation, while a significant negative and very strong correlation was recorded between mature leaves, old leaves and unripe fruits; and old leaves with ripened fruits. However, no correlation was found between unripe fruits and ripened fruits.

**Table 6.** The correlation coefficients (R value) among leaves and fruits of the 'Giant Green' cultivar of *S. samarangense* in TPC, TFC and alpha-glucosidase inhibitory assay.

| Sample | Correlation Coefficient of α-Glucosidase Inhibitory Assay | | | | | |
|---|---|---|---|---|---|---|
| | YL | ML | OL | UF | HF | RF |
| YL | 1.00 | | | | | |
| ML | −1.00 | 1.00 | | | | |
| OL | 0.83 | −0.84 | 1.00 | | | |
| UF | 0.86 | −0.85 | 0.43 | 1.00 | | |
| HF | 0.99 | −1.00 | 0.88 | 0.80 | 1.00 | |
| RF | −0.52 | 0.54 | −0.91 | −0.02 | −0.61 | 1.00 |

Significant at $p < 0.05$; YL: young leaves; ML: mature leaves; OL: old leaves; UF: unripe fruits: HF: half-ripened fruits; RF: ripened fruits.

*Correlation between TPC, TFC and Alpha-Glucosidase Inhibitory Activity*

Statistical analysis showed that the leaf extracts have a significant strong correlation between the $IC_{50}$ value of alpha-glucosidase activity and phenolic content with r = 0.938 but is moderate in fruit extracts with r = 0.577. A strong and positive correlation was also observed between alpha-glucosidase inhibitory activity ($IC_{50}$ value) and flavonoid content in leaves and fruit extracts with r = 0.930 and r = 0.850, respectively. The results demonstrated that some phenolic and flavonoid compounds were responsible for the activeness of this activity.

### 3.6. Antibacterial Assay

#### 3.6.1. Standard Drug Susceptibility/Resistance Testing

Eight standard drugs were investigated to evaluate their capacity for antibacterial activity against two Gram-positive bacteria (*Staphylococcus aureus* and *Enterococcus faecalis*) and two Gram-negative bacteria (*Salmonella typhimurium* and *Escherichia coli*). The drugs used in this assay were Gentamycin (CN), Tetracycline (TE), Vancomycin (VA), Amoxycillin/Clavulanic acid 2:1 (AMC), Cephazolin (KZ), Neomycin (N), Metronidazole (MTZ) and Cefaclor (CEC). The results of susceptibility testing are tabulated in Table 7. The data revealed that the overall susceptibility rates in Gentamycin, Tetracycline, Amoxicillin and Neomycin were the highest (100%) against the tested bacteria. Vancomycin was able to inhibit the growth of Gram-negative bacteria (*S. typhimurium* and *E. coli*) only with a susceptibility rate of 50%. Meanwhile, Cephazolin had susceptibility rates of 75% against *E. faecalis* and Cefaclor depicted 50% of susceptibility rates toward Gram-positive bacteria (*S. aureus* and *E. faecalis*).

**Table 7.** The susceptibility test of standard drug against two Gram-positive and two Gram-negative of bacteria.

| Bacterial Strains | Zone of Inhibition (mm) | | | | | | | |
|---|---|---|---|---|---|---|---|---|
| | CN (10 μg) | TE (30 μg) | VA (5 μg) | AMC (30 μg) | KZ (30 μg) | N (30 μg) | MTZ (5 μg) | CEC (30 μg) |
| *S. aureus* | $17.05 \pm 0.09^{a}$ | $7.07 \pm 0.06^{f}$ | $13.07 \pm 0.12^{c}$ | $9.10 \pm 0.10^{e}$ | $15.07 \pm 0.06^{b}$ | $10.10 \pm 0.10^{d}$ | NA | NA |
| *E. faecalis* | $20.10 \pm 0.10^{b}$ | $11.03 \pm 0.06^{d}$ | $11.07 \pm 0.12^{d}$ | $30.07 \pm 0.12^{a}$ | NA | $17.08 \pm 0.08^{c}$ | NA | NA |
| *S. typhimurium* | $23.13 \pm 0.15^{b}$ | $20.10 \pm 0.10^{e}$ | NA | $25.10 \pm 0.10^{a}$ | $22.20 \pm 0.17^{c}$ | $21.10 \pm 0.10^{d}$ | NA | $23.08 \pm 0.08^{b}$ |
| *E. coli* | $21.08 \pm 0.07^{a}$ | $18.17 \pm 0.15^{c}$ | NA | $15.17 \pm 0.15^{f}$ | $17.12 \pm 0.10^{d}$ | $19.13 \pm 0.12^{b}$ | NA | $16.10 \pm 0.10^{e}$ |

NA = not active. Values are the means ± standard deviation for three replicates of experiments (n = 3). Data from same horizontal row with different superscript letter refers to a significant difference ($p < 0.05$). CN: Gentamycin; TE: Tetracycline; VA: Vancomycin; AMC: Amoxycillin/Clavulanic acid 2:1; KZ: Cephazolin; N: Neomycin; MTZ: Metronidazole and CEC: Cefaclor.

#### 3.6.2. Plant Extract Susceptibility/Resistance Testing

The pattern inhibition of Gram-positive (*S. aureus* and *E. faecalis*) and Gram-negative bacteria (*E. coli* and *S. typhimurium*) toward 'Giant Green' leaves and fruits is shown in Figure 6. As seen, the inhibition activity of extracts increases with the increasing concentration. The data depicted that the antibacterial activity against all of the tested bacteria was in a concentration-dose dependent manner. Among the leaf extracts, the young leaves exhibited the maximum inhibition zones in all tested bacteria followed by mature leaves and the minimum inhibition was in old leaves (Table 8). Furthermore, compared to data from the standard drug susceptibility test (Table 7), the young leaf extract displayed very strong activity against all of the bacteria compared to the tested standard drugs. This study suggests that the young leaves have greater potential to be used as medicine in combating certain bacteria. The effectiveness of 'Giant Green' fruits at three maturity stages is recorded in Table 9. The best results are represented by unripe fruit extract which is active against all Gram-positive and Gram-negative bacteria. Half-ripened fruit was only inhibited by one bacterium which is *E. faecalis*. However, the ripened fruit was resistant to all of the tested bacteria. The results indicate that the unripe fruit possessed more powerful antibacterial activity than other mature fruits.

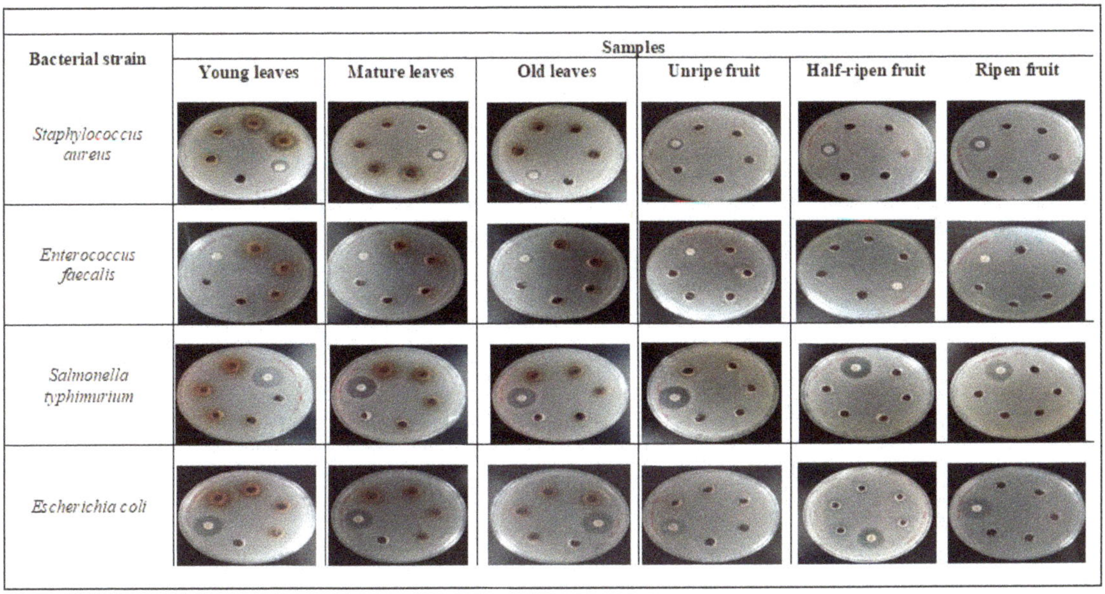


**Figure 6.** Growth inhibition of some bacteria by methanolic extract of 'Giant Green' leaves and fruits at three maturity stages.

**Table 8.** The antibacterial activity of the 'Giant Green' cultivar of Syzygium samarangense leaves at three maturity stages.

| Leaves Samples | Concentration (µg/mL) | Zone of Inhibition (mm) | | | |
|---|---|---|---|---|---|
| | | *S. aureus* | *E. faecalis* | *S. typhimurium* | *E. coli* |
| Young | 1000 | 20.47 ± 0.81 [a] | 24.13 ± 0.81 [a] | 25.40 ± 0.69 [a] | 24.33 ± 0.58 [a] |
| | 800 | 18.47 ± 0.50 [ab] | 19.47 ± 0.81 [b] | 23.07 ± 1.01 [ab] | 22.00 ± 1.00 [ab] |
| | 400 | 13.80 ± 0.72 [de] | 16.13 ± 0.81 [cd] | 19.40 ± 1.22 [c] | 18.67 ± 0.58 [cde] |
| | 200 | 10.13 ± 0.81 [g] | 13.80 ± 0.35 [def] | 15.40 ± 0.69 [def] | 17.00 ± 1.00 [def] |
| Mature | 1000 | 18.33 ± 0.58 [ab] | 19.03 ± 0.45 [bc] | 20.00 ± 0.50 [bc] | 21.13 ± 0.81 [bc] |
| | 800 | 16.33 ± 0.58 [bc] | 16.37 ± 1.42 [cd] | 18.00 ± 0.50 [cd] | 19.13 ± 0.81 [bcd] |
| | 400 | 13.33 ± 0.58 [def] | 14.70 ± 1.25 [de] | 15.00 ± 0.50 [def] | 16.13 ± 1.03 [ef] |
| | 200 | 8.33 ± 6.43 [fg] | 13.70 ± 1.25 [def] | 10.67 ± 1.61 [g] | 15.13 ± 1.03 [fg] |
| Old | 1000 | 16.67 ± 0.58 [bc] | 15.33 ± 0.58 [de] | 17.33 ± 0.58 [cde] | 16.77 ± 0.68 [def] |
| | 800 | 15.33 ± 0.58 [cd] | 12.67 ± 1.15 [efg] | 14.67 ± 0.58 [ef] | 15.43 ± 1.25 [f] |
| | 400 | 12.33 ± 0.58 [efg] | 11.67 ± 1.15 [fg] | 12.33 ± 1.53 [fg] | 12.43 ± 1.50 [gh] |
| | 200 | 11.00 ± 1.00 [g] | 10.67 ± 1.15 [g] | 10.67 ± 2.08 [g] | 11.77 ± 1.08 [h] |

Values are expressed as the means ± standard deviation. Data from the same vertical row with different superscript letters refers to significant difference ($p < 0.05$) by comparing among the three maturity stages of leaf samples.

**Table 9.** The antibacterial result of the 'Giant Green' cultivar of Syzygium samarangense fruits at three maturity stages.

| Fruit Sample | Concentration (µg/mL) | Zone of Inhibition (mm) | | | |
|---|---|---|---|---|---|
| | | *S. aureus* | *E. faecalis* | *S. typhimurium* | *E. coli* |
| Unripe | 1000 | 13.17 ± 0.29 [a] | 14.17 ± 0.29 [ab] | 15.33 ± 0.58 [a] | 14.33 ± 0.58 [a] |
| | 800 | 11.83 ± 0.76 [ab] | 13.17 ± 0.29 [bc] | 13.67 ± 0.58 [b] | 13.33 ± 0.58 [a] |
| | 400 | 10.83 ± 0.76 [bc] | 12.50 ± 0.50 [bc] | 12.67 ± 0.58 [bc] | 11.67 ± 0.58 [b] |
| | 200 | 9.83 ± 0.76 [c] | 11.50 ± 0.50 [c] | 11.67 ± 0.58 [c] | 10.33 ± 0.58 [b] |
| Half-ripened | 1000 | NA | 15.33 ± 0.58 [a] | NA | NA |
| | 800 | NA | 13.33 ± 0.58 [b] | NA | NA |
| | 400 | NA | 9.67 ± 0.58 [d] | NA | NA |
| | 200 | NA | 8.33 ± 1.15 [d] | NA | NA |
| Ripened | 1000 | NA | NA | NA | NA |
| | 800 | NA | NA | NA | NA |
| | 400 | NA | NA | NA | NA |
| | 200 | NA | NA | NA | NA |

NA = Not active. Values are expressed as the means ± standard deviation. Data from the same vertical row with different superscript letters refers to significant difference ($p < 0.05$) by comparing the three maturity stages of fruit samples.

## 4. Discussion

Our results indicate that the percentage of extraction yield of leaves and fruits of wax apple extracts varied with the maturity level. The extract yield increased with the maturity stages with the highest percentage of extract yields being recorded in old leaves and ripened fruits of wax apple. The solvent of extraction, pH, temperature, polarity, time of extraction, method of extraction and quantities of sample used can influence the values of the percentage of extraction yield [29]. Singh et al. [30] reported that the morphology of the sample matrix also influences the yield of extraction. The accumulation of extractable compounds depends on the inheritance of the plant, plant developmental stages and responsibilities at different environments. The extracts percentage yield is also affected by the growing condition, light supplied and their genetic variation which modifies the composition of components in plants through their biosynthesis [31]. The highest yield of extracts in old leaves and ripened fruits may be due to increased synthesis and accumulation of phytochemicals at late maturity. The phenolic, flavonoid, steroid, tannin and terpenoid are the phytochemicals that are present in the methanolic extract of leaves and fruits of wax apples [32,33].

Generally, the total phenolic content of all of the 'Giant Green' extracts was in the range of 30–70 mg GAE/g DW. Our results showed that the young leaves and unripe fruit contained the highest phenolic contents compared to mature and old leaves, and half-ripened and ripened fruits. The differences in phenolic between the plant parts may be due to the changes in the type and quantities of polyphenolic compounds during its growth process. Yoshioka et al. [34] revealed that phenolic compounds act as stimulants or inhibitors of enzymes and catalyse as well as control the activities of polyphenol oxidase in plants during the development and ripening stages. In addition, the distribution of phenolic compounds among the plant parts has also often been associated with defence against pathogens or herbivores [35] and function in controlling the biotic and abiotic stress in plants [36]. Similar results were previously reported by Lee et al. [20] with the young leaves of pink cultivar of wax apple containing the higher phenolics such as β-elemene, γ-terpinene and β-caryophyllene compared to the mature and old leaves. Other researchers also found that the unripe fruits of *Sonneratia caseolaris* [37], *Nypha fruticans* Wurmb [38],

*Pyracantha* [39], *Psidium guajava* [40] and *Malpighia emarginata* DC [41] displayed a high phenolic content compared to ripe fruits.

The results obtained from this study demonstrate that the flavonoid contents of leaves and fruits of 'Giant Green' decreased with the maturity process. The young leaves and unripe fruits recorded the highest flavonoid contents among the other stages. Similar results were reported by Kingne et al. [42] with the young avocado leaves and mango leaves exhibiting the highest flavonoid content compared to mature ones. Young aronia leaves also contained more flavonoid content than the old leaves with a value of 163.7 mg CE/g dry weight and 103.6 mg CE/g dry weight, respectively [43]. On other hand, some studies reported that the flavonoid content of blueberry [44], mangrove apple [39] and papaya [45] increased in the early stages and decreased at the late stages during fruit ripening. The differences in total flavonoid content obtained between the samples in this study and those reported in the previous literature can be influenced by the cultivar and variety, sunlight, growing condition and age of the plants [46].

The present study shows that the young leaves and unripe fruit exhibited the strongest DPPH, NO and ABTS radical scavenging activity as well as the highest reducing power. This may be attributed to higher phenolic and flavonoid compounds in extracts of young leaves and unripe fruits. The plants consisting of more phenolic compounds are believed to contribute to higher antioxidant activity [37]. This finding correlates with previous researchers who reported that the young leaves of *Persea americana* and *Mangifera indica* [42], *Syzygium polyanthum* [47], *Blepharocalyx salicifolius* [48] and *Aronia melanocarpa* [43] exhibited the strongest DPPH activity compared to the old and mature leaf extract. In the same way, Bakar et al. [37] demonstrated that the unripe fruits of *Sonneratica caseolaris* had good scavenging of DPPH and ABTS radical activity for both 80% methanol and aqueous extracts compared to ripe fruit. Taghizadeh et al. [49] revealed that the type of solvent used for the extraction process influences antioxidant activity. Hence, the choice of a suitable polar solvent such as methanol, ethanol or water was crucial to attract the polar compounds such as phenolic and flavonoid in the plant sample. Methanol was reported to be the best solvent to maximise the recovery of the polar compound in the plant extract [36]. This reason may be why the highest phenolic and flavonoid contents were recorded in young leaves and unripe fruit. Furthermore, the electron-donating nature of the substituent groups like -OH, -Cl and -$CH_3$ in plant compounds are able to inhibit the generation of nitrite and peroxy nitrite anions increasing antioxidant activity [50]. In addition, another factor influencing the variation of radical scavenging activity in the extracts might be influenced by pre- and post-harvest factors [51]. The pre-harvest factors such as the conditions of the environment and agronomic practices could be responsible for the changes in antioxidant levels in the plant parts [52]. The conditions of the environment such as hot or cold temperatures, dry or wet soil, period exposed to sunlight and climatic change cause a decrease or increase in valuable phytochemicals that possess antioxidant value. Zheng and Wang [53] reported that the temperature significantly influenced the antioxidant activity in citrus fruit. Furthermore, the agronomic conditions such as type and quantity of fertilizer, the effectiveness of the irrigation system, maturity stages of the plant and date of sowing influence the antioxidant activity [51]. Previous research by Rajan and Bhat [54] found that the potency of antioxidants in kundang fruit (*Bouea macrophlia* Griffith) is affected by maturity stages in which the unripe kundang fruits possess the strongest antioxidant activity compared to ripe fruit Furthermore, another factor is the postharvest storage condition [51]. The aspects of time, temperature and light intensities are very crucial to maintain the quality of antioxidant compounds in the plant. Some phenolic compounds such as flavonoid, phenolic acid, anthocyanin and ascorbic acid are sensitive or insensitive to storage temperature, thereby affecting the antioxidant activity [55].

Our study indicates that phenolic and flavonoid compounds in the leaves and fruit of the 'Giant Green' cultivar of *S. samarangense* were correlated to antiradicals. These findings indicate that phenolic and flavonoid are major compounds contributing to antioxidant activity, especially in the scavenging of harmful radicals generated from oxidative stress.

Ng et al. [56] found a correlation between TPC and TFC with DPPH and ABTS radical scavenging activity of selected medicinal plants. Furthermore, Majumder et al. [57] also revealed a good correlation between TPC and TFC with in vitro and in vivo antioxidant activities, which indicates that phenolic compounds possess strong antioxidant capacities in *S. samarangense* leaves. Zielinski et al. [58] stated that the chemical structure of the phenolic compound may influence the variation of antioxidant capacity. The compound with the highest ability to delocalise the lone electron around the aromatic ring possesses stronger antioxidant activity. However, a decrease in hydroxy and methoxy substituents and the increase in the electron-withdrawing group in the aromatic ring reduces the radical scavenging activity. Another factor contributing to the relationship of phenolic and flavonoid with antioxidant activity was the ability of these compounds to inhibit the oxidant enzyme such as nitric oxide synthase (NOS), xanthine oxidase (XO) and NADPH oxidase (NOX). Research reported by Nakao et al. [59] found that hesperetin can inhibit the production of XO, which decreases the formation of free radicals. Furthermore, phenolic compounds that directly react with reactive oxygen species (ROS) or reactive nitrogen species (RNS) also enhance antioxidant capacity. These compounds can act as a safeguard to control the accumulation of ROS and RNS in the body. In addition, the synergism among the phenolic compound or with other groups also influences the efficiency of antioxidant activity [60]. For example, the interaction of phenolic compounds with ascorbic acid and vitamin E [61], flavonoids with protein [62] and phenolic with phenolic such as naringenin with hesperidin [63] significantly increases the antioxidant capacity.

One of the properties of the wax apple plant is to have antidiabetic properties. This was proven in our study when the 'Giant Green' cultivar of *S. samarangense* showed a significant effect in alpha-glucosidase inhibitory activity, especially in young leaves and unripe fruit extracts. Phenolic and flavonoid compounds have been reported to have a major effect on alpha-glucosidase inhibitory activity [64]. Nurnaeimah et al. [65] reported that the high α-glucosidase inhibitory activities represent a potential antidiabetic agent. The highest quantity of phenolic and flavonoid compounds in young leaves and unripe fruit extracts could be the cause of why both extracts showed the strongest alpha-glucosidase inhibitory activity than other maturity stages. Some of the literature reported that the antidiabetic activity in *S. samarangense* plant parts was positively correlated with phenolic contents [64–67]. Hu et al. [68] also found that resorcinol derivatives in the *S. samarangense* leaves could inhibit the strongest alpha-glucosidase activity. Fatanah et al. [69] indicated that the youngest plants need more phenolic compounds that are believed to function in defending against ultraviolet radiation and aggression by pathogens compared to old plants. This might be one of the reasons why an active defensive mechanism by secondary metabolites in young plants exhibited significant antidiabetic activity more than in old plants. Furthermore, the hypoglycaemic effect in plant extracts also increases the efficiency of alpha-glucosidase inhibitory activity. Phenolic and flavonoid compounds have been identified to possess a hypoglycaemic effect in reducing blood glucose levels [70].

From this analysis, it is clear that the phenolic and flavonoid content of leaves and fruits of the 'Giant Green' cultivar of *S. samarangense* were correlated with alpha-glucosidase inhibitory activity. The inhibition capacity of alpha-glucosidase is closely related to the chemical structures of phenolic and flavonoid compounds such as the position and quantity of the hydrogen group attached to the aromatic ring, the methyl group substituent and the complexity of the structure [71]. The more OH group attached at the aromatic ring, the more effective the compound to inhibit alpha-glucosidase activity [72]. Then, the flavonoid compound with glucose moiety at the C-3 position and the methyl group at the position C-7 are more effective toward α-glucosidase inhibitory activity [73]. The addition of hydrophobic and bulky substituents such as methyl substituent in compounds decreases the activity of alpha-glucosidase. The synergistic effect also exhibits the greatest inhibitory activity of alpha-glucosidase [74]. The combination of phenolic compounds with other compounds such as glyceolin and luteolin significantly improved the alpha-glucosidase inhibitory activity [75].

The study found that young leaves and unripe fruit of the 'Giant Green' cultivar of *S. samarangense* possess the strongest antibacterial activity of the maturity stages. Both of the extracts also were considered the most promising for their activity against both Gram-negative bacteria, *E. coli* and *S. typhimurium* where it is more difficult to find the compounds that have the capability to penetrate the double membrane surrounding the bacterial cell wall [76]. Small hydrophilic molecules which are lipophilic macromolecules have properties to pass through the outer membrane of Gram-negative bacteria [77]. This study revealed that young leaves and unripe fruits consist of valuable chemical compounds that have the strongest ability to penetrate both the inner and outer membrane. These compounds might be accumulated in the plasma membrane resulting in the loss of cellular constituents, changes in cellular structure and function and disturbed metabolism [78]. Other than that, it is also capable of inhibiting the synthesis of the bacterial cell wall, causing the depletion of energy, mutation, cell damage and lastly leading to death [79,80]. Chemical compounds such as flavonoid, triterpenes, sterol, tannin, terpenoid and alkaloid were believed to significantly contribute to antibacterial activity [81,82]. Research by Khandaker et al. [19] showed that phenolic and flavonoid content in wax apple leaves, bark and fruit extracts are able to inhibit the growth of four bacteria, *Bacillus cereus*, *S. aureus*, *E. coli* and *Pseudomonas aeruginosa*. Moreover, the synergistic effect of the different chemical constituents, even present in small quantities, also influences the effectiveness of the extract to inhibit the growth of certain bacteria [83]. Future studies of young leaves and unripe fruit should explore the lead compound that gives greater potency in antibacterial activity, especially against Gram-negative bacteria.

## 5. Conclusions

This study has shown that young leaves and unripe fruit of 'Green Giant' cultivar of *Syzygium samarangense* had the highest total phenolic content, total flavonoid content and had strongest antioxidant, alpha-glucosidase inhibitory and antibacterial activities compared to other maturity stages. It can be concluded that young leaves and unripe fruits of the 'Giant Green' cultivar are good sources of natural antioxidants that can be used to scavenge harmful free radicals. Furthermore, they can also be used in food preservatives or the pharmaceutical industry. However, further study needs to be conducted for the identification of bioactive compounds that serve as effective antioxidant, antidiabetic and antibacterial agents in these samples.

**Author Contributions:** Data curation, N.S.I. and M.M.K.; investigation, M.M.K.; methodology, N.S.I., M.M.K. and Z.M.R.; supervision, M.M.K., N.M. and Z.M.R.; validation, A.M. and M.M.A.; writing—original draft, N.S.I. and M.M.K.; writing—review & editing, N.M., M.M.K., Z.M.N. and A.F.M.A. All authors have read and agreed to the published version of the manuscript.

**Funding:** This research was supported from the Ministry of Higher Education, Malaysia (MOHE), project grant No: RACE/F1/SG5/UNISZA/5 with collaboration of Universiti Sultan Zainal Abidin, Terengganu and Universiti Malaya, Kuala Lumpur, Malaysia.

**Data Availability Statement:** The data that support the findings of this study are available from the corresponding author upon request.

**Acknowledgments:** The authors wish to thank the Center for Research Excellence and Incubation Management (CREIM), Universiti Sultan Zainal Abidin, Campus Gong Badak, Terengganu, Malaysia for supporting this project. The Researchers also acknowledge Deanship of Scientific Research, Taif University for editing and partial publication support.

**Conflicts of Interest:** The authors declare that there is no conflict of interest in this study.

## References

1. Kempraj, V.; Bhat, S.K. Ovicidal and Larvicidal Activities of *Cyperus giganteus* Vahl and *Cyperus rotundus* Linn. Essential Oils against *Aedes albopictus* (Skuse). *Nat. Prod. Rad.* **2008**, *7*, 420–425.
2. Bhat, S.K.; Sharma, A. Therapeutic Potential of Cardiac Glycosidases of *Calotropis gigantean* for Breast Cancer. *Int. Res. J. Pharm.* **2013**, *4*, 164–167. [CrossRef]

3. WHO. *Traditional Medicine Strategy 2002–2005*; WHO Publication; WHO: Geneva, Switzerland, 2002.
4. Lee, S.; Suh, S.; Kim, S. Protective Effects of the Green Tea Polyphenol (-)-Epigallocatechin gallate against Hippocampal Neuronal Damage after Transient Global Ischemia in Gerbils. *Neurosci. Let.* **2000**, *287*, 191–194. [CrossRef] [PubMed]
5. Raga, D.D.; Cheng, C.L.C.; Lee, K.C.I.C.; Olaziman, W.Z.P.; De Guzman, V.J.A.; Shen, C.-C.; Franco Jr, F.C.; Ragasa, C.Y. Bioactivities of Triterpenes and a Sterol from *Syzygium samarangense*. *Zei. F. Nat.* **2011**, *66*, 235–244.
6. Mukaromah, A.S. Wax apple (*Syzygium samarangense* (Blume) Merr. & L.M. Perry): A Comprehensive Review in Phytochemical and Physiological Perspectives. *J. Biol. App. Biol.* **2020**, *3*, 40–58.
7. Haq, I.; Hossain, A.B.M.S.; Khandaker, M.M.; Merican, A.F.; Faruq, G.; Boyce, A.N.; Azirun, M.S. Antioxidant and antibacterialactivities of different extracts and fractions of a mangrove plant *Sonneratia alba*. *Int. J. Agric. Biol.* **2014**, *14*, 707–714.
8. Morton, J. Java apple. In *Fruits of Warm Climates*; Morton, J.F., Ed.; Wintervine, Inc.: Miami, FL, USA, 1987; pp. 381–382.
9. Moneruzzaman, K.M.; Hossain, A.B.M.S.; Normaniza, O.; Boyce, A.N. Growth, Yield and Quality Responses to GA3 of Wax Apple *Syzygium samarangense* var. *Jambu Air Madu Fruits Grown Under Field Conditions*. *Afr. J. Biotechnol.* **2011**, *10*, 11911–11918.
10. Nakasone, H.Y.; Paull, R.E. *Tropical Fruits*; CAB International Publisher: New York, NY, USA, 1998.
11. Moneruzzaman, M.K.; Sarwar, J.; Mat, N.; Boyce, A.N. Bioactive Constituents, Antioxidant and Antimicrobial Activities of Three Cultivars of Wax Apple (*Syzygium samarangense* L.) Fruits. *Res. J. Biotechnol.* **2015**, *10*, 7–16.
12. Idris, N.S.; Ismail, S.Z.; Mat, N.; Khandaker, M.M. Comparative Physiological Response of Three Wax Apple *(Syzygium Samarangense)* Tree Cultivars at Flower Bud Development Stage. *Biosci. Res.* **2018**, *15*, 402–411.
13. Shahreen, S.; Banik, J.; Hafiz, A.; Rahman, S.; Zaman, A.T.; Shoyeb, A.; Chowdhury, M.H.; Rahmatullah, M. Antihyperglycemic Activities of Leaves of Three Edible Fruit Plants (*Averrhoa carambola, Ficus hispida* and *Syzygium samarangense*) of Bangladesh. *Afr. J. Tradit. Complement. Altern. Med.* **2012**, *9*, 287–291. [CrossRef]
14. Khandaker, M.M.; Boyce, A.N.; Osman, N. The influence of hydrogen peroxide on the growth, development and quality of wax apple (Syzygium samarangense, [Blume] Merrill & L.M. Perry var. jambu madu) fruits. *Plant Physiol. Biochem.* **2012**, *53*, 101–110.
15. Simirgiotis, M.J.; Adachi, S.; To, S.; Yang, H.; Reynertson, K.A.; Basile, M.J.; Gil, R.R.; Weinstein, I.B.; Kennelly, E.J. Cytotoxic chalcones and antioxidants from the fruits of a *Syzygium samarangense* (Wax Jambu). *Food Chem.* **2008**, *107*, 813–819. [CrossRef]
16. Mollika, S.; Nesa, M.L.; Munira, M.S.; Sayem, W.; Parvin, N. Evaluation of Analgesic, Anti-Inflammatory and CNS activities of the Methanolic Extract of *Syzygium samarangense* Bark. *IOSR J. Pharm.* **2013**, *3*, 12–18. [CrossRef]
17. Hossain, R.; Rahman, M.A.; Rafi, M.K.J.; Siddique, T.A.; Noman, A.A.; Makki, A.; Alelwani, W.; Hajjar, D.; Tangpong, J. Pharmacological and ADMET-based Pharmacokinetic Properties of *Syzygium samarangense* var. Parviflorum Leaf Extract in vitro, in vivo and in silico Models. *Not. Bot. Horti. Agrobo.* **2020**, *48*, 1155–1175. [CrossRef]
18. Al-saif, A.M.; Hossain, A.B.M.S.; Taha, R.M.; Moneruzzaman, K.M. Photosynthetic Yield, Fruit Ripening and Quality Characteristics of Cultivars of *Syzygium samarangense*. *Afr. J. Agril. Res.* **2011**, *6*, 3623–3630.
19. Khandaker, M.M.; Alebidi, A.I.; Hossain, A.S.; Mat, N.; Boyce, N. Physiological and Biochemica; Properties of Three Cultivars of Wax Apple (*Syzygium samarangense* [Blume] Merrill & L. M. Perry) Fruits. *J. Sustain. Sci. Manag.* **2015**, *10*, 66–75.
20. Lee, P.C.; Guo, H.Y.; Huang, C.C.; Chan, C.F. Chemical Composition of Leaf Essential Oils of *Syzygium samarangense* (BL.) Merr. et Perry cv. Pink at Three Maturity Stages. *Int. J. App. Res. Nat. Prod.* **2016**, *9*, 9–13.
21. Adak, N.; Heybeli, N.; Ertekin, C. Infrared Drying of Strawberry. *Food Chem.* **2017**, *219*, 109–116. [CrossRef]
22. Zin, M.N.B.; Azemin, A.; Rodi, M.M.M.; Mohd, K.S. Chemical Composition and Antioxidant Activity of Stingless Bee Propolis from Different Extraction Methods. *Int. J. Eng. Technol.* **2018**, *7*, 90–95.
23. Cunha, I.; Sawaya, A.C.; Caetano, F.M.; Shimizu, M.T.; Marcucci, M.C.; Drezza, F.T.; Carvalho, P.D.O. Factors that influence the yield and composition of Brazilian propolis extracts. *J. Brazilian Chem. Soc.* **2004**, *15*, 964–970. [CrossRef]
24. Awah, F.M.; Verla, A.W. Antioxidant Activity, Nitric Oxide Scavenging Activity and Phenolic Contents of *Ocimum gratissimum* leaf extract. *J. Med. Plants Res.* **2010**, *4*, 2479–2487.
25. Shalaby, E.A.; Shanab, S.M.M. Comparison of DPPH and ABTS Assays for Determining Antioxidant Potential of Water and Methanol Extracts of *Spirulina platensis*. *Ind. J. Mar. Sci.* **2013**, *42*, 556–564.
26. Mayur, B.; Sandesh, S.; Shruti, S.; Yum, S. Antioxidant and α-glucosidase inhibitory properties of *Carpesium abrotanoides* L. *J. Med. Plants Res.* **2010**, *4*, 1547–1553.
27. Misbah, H.; Aziz, A.A.; Aminudin, N. Antidiabetic and Antioxidant Properties of *Ficus deltoidea* Fruit Extracts and Fractions. *Complement. Altern. Med.* **2013**, *13*, 118. [CrossRef]
28. Shanmugam, A.; Kathiresan, K.; Nayak, L. Preparation, Characterization and Antibacterial Activity of Chitosan and Phosphorylated Chitosan from Cuttlebone of *Sepia kobiensis* (Hoyle, 1885). *Biotechnol. Rep.* **2016**, *9*, 25–30. [CrossRef]
29. Sharma, A.; Singh, D. A Comparative Study of Effects of Extraction Solvents/Techniques on Percentage Yield, Polyhenolic Composition, and Antioxidant Potential of Various Extracts Obtained from Stems of Nepeta Leucophylla: RP-HPLC-DAD Assessment of Its Polyhenolic Constituents. *J. Food Biochem.* **2016**, *41*, 12337. [CrossRef]
30. Singh, J.; Sing, V.; Shukla, S.; Rai, A.K. Phenolic Content and Antioxidant Capacity of Selected Cucurbit Fruits Extracted with Different Solvents. *J. Nutri. Food Sci.* **2016**, *6*, 565. [CrossRef]
31. Keneni, Y.G.; Bahiru, L.A.; Marchetti, J.M. Effects of Different Extraction Solvents on Oil Extracted from Jatropha Seeds and the Potential of Seed Residues as a Heat Provider. *BioEnergy Res.* **2021**, *14*, 1207–1222. [CrossRef]
32. Madhavi, M.; Ram, M.R. Biochemical and Elemental Analysis of *Syzygium samarangense* Roots. *J. Pharmaco. Phytochem.* **2019**, *8*, 346–350.

33. Mamdouh, N.S.; Sugimoto, S.; Matsunami, K.; Otsuka, H.; Kamel, M.S. Taxiphyllin 6′-O-Gallate, Actinidioionoside 6′-O-Gallate and Myricetrin 2″-O-Sulfate from the Leaves of *Syzygium samarangense* and Their Biological Activities. *Chem. Pharma. Bull.* **2014**, *62*, 1013–1018. [CrossRef]
34. Yoshioka, T.; Inokuchi, T.; Fujioka, S.; Kimura, Y. Phenolic Compounds and Flavonoids as Plant Growth Regulators from Fruit and Leaf of *Vitex rotundifolia*. *Zeit. F. Nat.* **2004**, *59*, 509–514. [CrossRef]
35. Kennedy, D.O.; Wightman, E.E.L. Herbal Extracts and Phytochemicals: Plant Secondary Metabolites and the Enhancement of Human Brain Function. *Adv. Nutr.* **2011**, *2*, 32–50. [CrossRef]
36. Zhang, Q.W.; Lin, L.G.; Ye, W.C. Techniques For Extraction and Isolation of Natural Products: A Comprehensive Review. *Chin. Med.* **2018**, *13*, 20. [CrossRef]
37. Bakar, A.F.I.; Bakar, A.M.F.; Hassan, S.H.A.; Sanusi, S.B.; Kormin, F.; Sabran, S.F.; Fuzi, F.Z.M. Comparison of phytochemicals, antioxidant and anti-cholinesterase activity of unripe and ripe fruit of *Sonneratia caseolaris*. *Food Res.* **2020**, *4*, 507–514. [CrossRef]
38. Prasad, N.; Yang, B.; Kong, K.W.; Khoo, H.E.; Sun, J.; Azlan, A.; Ismail, A.; Romli, Z.B. Phytochemicals and antioxidant capacity from *Nypa fruticans* Wurmb. fruit. *Evid. -Based Complement. Altern. Med.* **2013**, *2013*, 154606. [CrossRef]
39. Belwal, T.; Pandey, A.; Bhatt, I.D.; Rawal, R.S.; Luo, Z. Trends of polyphenolics and anthocyanins accumulation along ripening stages of wild edible fruits of Indian Himalayan region. *Sci. Rep.* **2019**, *9*, 5894. [CrossRef]
40. Lim, Y.Y.; Lim, T.T.; Tee, J.J. Antioxidant properties of guava fruit: Comparison with some local fruits. *Sunway Acad. J.* **2006**, *3*, 9–20.
41. Righetto, A.M.; Netto, F.M.; Carraro, F. Chemical Composition and Antioxidant Activity of Juices from Mature and Immature Acerola (*Malpighia emarginata* DC). *Food Sci. Technol. Int.* **2005**, *11*, 315–321. [CrossRef]
42. Kingne, F.K.; Djikeng, F.T.; Tsafack, H.D.; Karuna, M.S.; Womeni, H.M. Phenolic content and antioxidant activity of young and mature mango (*Mangifera indica*) and avocado (*Persea americana*) leave extracts. *Int. J. Phytomedicine* **2018**, *10*, 181–190. [CrossRef]
43. Thi, N.D.; Hwang, E.-S. Bioactive Compound Contents and Antioxidant Activity in Aronia (*Aronia melanocarpa*) Leaves Collected at Different Growth Stages. *Prev. Nutr. Food Sci.* **2014**, *19*, 204–212. [CrossRef]
44. Maisarah, A.M.; Nurul Amira, B.; Asmah, R.; Fauziah, O. Antioxidant analysis of different parts of *Carica papaya*. *Int. Food Res. J.* **2013**, *20*, 1043–1048.
45. Castrejo, A.D.R.; Eichholz, I.; Rohn, S.; David, A.; Kroh, L.W.; Huyskens-keil, S. Phenolic Profile and Antioxidant Activity of Highbush Blueberry (*Vaccinium corymbosum* L.) During Fruit Maturation and Ripening. *Food Chem.* **2008**, *109*, 564–572. [CrossRef]
46. Aljuhaimi, F.; Ghafoor, K.; Adiamo, O.Q. The Effect of Harvest Time and Varieties on Total Phenolics, Antioxidant Activity and Phenolic Compounds of Olive Fruit and Leaves. *J. Food Sci. Technol.* **2019**, *56*, 2373–2385.
47. Kusuma, I.W.; Kuspradini, H.; Arung, E.T.; Aryani, F.; Min, Y.-H.; Kim, J.-S.; Kim, Y.U. Biological Activity and Phytochemical Analysis of Three Indonesian Medicinal Plants, *Murraya koenigii, Syzygium polyanthum* and *Zingiber purpurea*. *Acupunct. Meridian Stud.* **2011**, *4*, 75–79. [CrossRef]
48. Habermann, E.; Imatomi, M.; Pontes, F.C.; Gualtier, S.C.J. Antioxidant activity and phenol content of extracts of bark, stems and young and mature leaves from *Blepharocalyx salicifolius* (Kunth) O. Berg. *Brazil. J. Biol.* **2016**, *76*, 898–904. [CrossRef]
49. Taghizadeh, S.F.; Rezaee, R.; Davarynejad, G.; Karimi, G.; Nemati, S.H.; Asili, J. Phenolic Profile and Antioxidant Activity of *Pistacia vera* var. Sarakhs Hull and Kernel Extracts: The Influence of Different Solvents. *J. Food Meas. Charact.* **2018**, *12*, 2138–2144. [CrossRef]
50. Rajesh, M.P.; Natvar, J.P. In Vitro Antioxidant Activity of Coumarin Compounds by DPPH, Super Oxide and Nitric Oxide Free Radical Scavenging Methods. *J. Adv. Pharm. Educ. Res.* **2011**, *1*, 52–68.
51. Zou, Z.; Xi, W.; Hu, Y.; Nie, C.; Zhou, Z. Antioxidant Activity of *Citrus* Fruits. *Food Chem.* **2015**, *196*, 885–896. [CrossRef]
52. Tiwari, U.; Cummins, E. Factors Influencing Levels of Phytochemicals in Selected Fruit and Vegetables During Pre- and Post-Harvest Food Processing Operations. *Food Res. Int.* **2013**, *50*, 497–506. [CrossRef]
53. Zheng, W.; Wang, S.Y. Antioxidant activity and Phenolic Compounds in Selected Herbs. *J. Agril. Food Chem.* **2001**, *49*, 5165–5170. [CrossRef]
54. Rajan, N.S.; Bhat, R. Antioxidant Compounds and Antioxidant Activities in Unripe and Ripe Kundang Fruits (*Bouea macrophylla* Griffith). *Fruits* **2016**, *71*, 41–47. [CrossRef]
55. Chaudhary, P.R.; Yu, X.; Jayaprakasha, G.K.; Patil, B.S. Influence of Storage Temperature and Low-Temperature Conditioning on The Levels of Health-Promoting Compounds in Rio Red Grapefruit. *Food Sci. Nutri.* **2016**, *5*, 545–553. [CrossRef]
56. Ng, Z.X.; Tyng, Y.; Koick, T.; Yong, P.H. Comparative Analyses on Radical Scavenging and Cytotoxic Activity Of Phenolic and Flavonoid Content From Selected Medicinal Plants. *Nat. Prod. Res.* **2021**, *35*, 5271–5276. [CrossRef]
57. Majumder, R.; Alam, M.B.; Chowdhury, S.T.; Bajpai, V.K.; Shukla, S. Quantitative Measurement of Bioactive Compounds From Leaves of *Syzygium samarangense* with Antioxidant Efficacy. *J. Nat. Sci. Found.* **2017**, *45*, 169–178. [CrossRef]
58. Zielinski, A.A.F.; Granato, D.; Alberti, A.; Nogueira, A.; Demiate, I.M.; Windson, C.; Haminiuk, I. Modelling The Extraction of Phenolic Compounds and in vitro Antioxidant Activity of Mixtures of Green, White and Black Teas (*Camellia sinensis* L. Kuntze). *J. Food Sci. Technol.* **2015**, *52*, 6966–6977. [CrossRef]
59. Nakao, K.; Murata, K.; Itoh, K.; Hanamoto, Y.; Masuda, M.; Moriyama, M.; Shintani, T.; Matsuda, H. Anti-Hyperuricemia Effects of Extracts of Immature Citrus Unshiu Fruit. *J. Trad. Med.* **2011**, *28*, 10–15.
60. Nunes, M.A.; Reszczynski, F.; Pascoa, R.N.M.J.; Costa, A.S.G.; Alves, R.C.; Oliveira, M.B.P.P. Influence of Olive Pomace Blending on Antioxidant Activity: Additive, Synergistic, and Antagonistic Effects. *Molecules* **2021**, *26*, 169. [CrossRef]

61. Dasgupta, A.; Klein, K. *Antioxidants in Food, Vitamins and Supplements*; Elsevier Publisher: New York, NY, USA, 2014.
62. Arts, M.J.T.J.; Haenen, G.R.M.M.; Wilms, L.C.; Beestra, S.A.J.N.; Heijnen, C.G.M.; Voss, H.; Bast, A. Interactions between Flavonoids and Proteins: Effect on the Total Antioxidant Capacity. *J. Agril. Food Chem.* **2002**, *50*, 1184–1187. [CrossRef]
63. Freeman, B.L.; Eggett, D.L.; Parker, T.L. Synergistic and Antagonistic Interactions of Phenolic Compounds Found in Navel Oranges. *J. Food Sci.* **2010**, *75*, 570–576. [CrossRef]
64. Aung, E.E.; Kristanti, A.N.; Aminah, N.S.; Takaya, Y.; Ramadhan, R. Plant Description, Phytochemical Constituents and Bioactivities of *Syzygium* Genus: A Review. *Open Chem.* **2020**, *18*, 1256–1281. [CrossRef]
65. Nurnaeimah, N.; Mat, N.; Mohd, K.S.; Badaluddin, N.A.; Yusoff, N.; Sajili, M.H.; Mahmud, K.; Adnan, A.F.M.; Khandaker, M.M. The Effects of Hydrogen Peroxide on Plant Growth, Mineral Accumulation, as Well as Biological and Chemical Properties of *Ficus deltoidea*. *Agronomy* **2020**, *10*, 599. [CrossRef]
66. Shen, S.C.; Chang, W.C. Hypotriglyceridemic and Hypoglycemic Effects of Vescalagin From Pink Wax Apple [*Syzygium samarangense* (Blume) Merrill and Perry cv. Pink] in High-Fructose Diet-Induced Diabetic Rats. *Food Chem.* **2013**, *136*, 858–863. [CrossRef] [PubMed]
67. Wang, B.H.; Cao, J.J.; Zhang, B.; Chen, H.Q. Structural Characterization, Physicochemical Properties and β-glucosidase Inhibitory Activity of Polysaccharide from the Fruits of Wax Apple. *Carbohydr. Polym.* **2019**, *1*, 227–236. [CrossRef] [PubMed]
68. Hu, Y.K.; Wang, L.; Zhao, Y.; Liu, J.P.; Wang, J.H.; Zhao, Y. Two New Oleanane Triterpenoids from *Syzygium samarangense. Chem. Nat. Compd.* **2020**, *56*, 692–695. [CrossRef]
69. Fatanah, D.; Abdullah, N.; Hashim, N.; Hamid, A.A. Quantification of Phenolic Compounds and Sensorial Properties of *Cosmos Caudatus* Herbal Tea at Different Maturity Stages. *Adv. Environ. Biol.* **2015**, *9*, 15–20.
70. Huang, D.; Jiang, Y.; Chen, W.; Yao, F.; Huang, G.; Suna, L. Evaluation of Hypoglycemic Effects of Polyphenols and Extracts From *Penthorum chinense. J. Ethnopharmacol.* **2015**, *163*, 256–263. [CrossRef]
71. Resurreccion-Magno, M.H.; Villasenor, I.M.; Harada, N.; Monde, K. Antihyperglycaemic flavonoids from *Syzygium samarangense* (Blume) Merr. and Perry. *Phytotherapy Res.* **2005**, *19*, 246–251. [CrossRef]
72. Proenca, C.; Freitas, M.; Ribeiro, D.; Oliveira, E.F.T.; Sousa, J.L.C.; Tome, S.M.; Ramos, M.J.; Silva, A.M.S.; Fernandes, P.A.; Fernandes, E. α-Glucosidase Inhibition by Flavonoids: An *In Vitro* and *In Silico* Structure–Activity Relationship Study. *J. Enz. Inhibit. Med. Chem.* **2017**, *32*, 1216–1228. [CrossRef]
73. Manaharan, T.; Appleton, D.; Cheng, H.M.; Palanisamy, D. Flavonoid Isolated from *Syzygium aqueum* Leaf Extract as Potential Antihyperglycemic Agents. *Food Chem.* **2012**, *132*, 1802–1807. [CrossRef]
74. Yang, J.; Wang, X.; Zhang, C.; Ma, L.; Wei, T.; Zhao, Y.; Peng, X. Comparative Study of Inhibition Mechanisms of Structurally Different Flavonoid Compounds on α -Glucosidase and Synergistic Effect with Acarbose. *Food Chem.* **2021**, *347*, 129056. [CrossRef]
75. Son, H.-U.; Yoon, E.-K.; Yoo, C.-Y.; Park, C.-H.; Bae, M.-A.; Kim, T.-H.; Lee, C.L.; Lee, K.W.; Seo, H.; Kim, K.-J.; et al. Effects of Synergistic Inhibition on α-glucosidase by Phytoalexins in Soybeans. *Biomolecules* **2019**, *9*, 828. [CrossRef]
76. Norizan, N.; Ahmat, N.; Mohamad, S.A.S.; Nazri, N.A.A.M.; Ramli, S.S.A.R.; Kasim, N.M.; Zain, W.Z.W.M. Total Phenolic Content, Total Flavonoid Content, Antioxidant and Antimicrobial Activities of Malaysian *Shorea*. *Res. J. Med. Plant* **2012**, *6*, 489–499.
77. Nikaido, H.; Neidhardt, F.C. *Outer Membrane in Escherichia coli and Salmonella: Cellular and Molecular Biology*; ASM Press: Washington, DC, USA, 1996; pp. 29–47.
78. Millat, M.S.; Islam, S.; Hussain, M.S.; Moghal, M.M.R.; Islam, T. Anti-Bacterial Profiling of *Launaea sarmentosa* (Willd.) and *Bruguiera cyclindrical* (L.): Two Distinct Ethno Medicinal Plants of Bangladesh. *Eur. J. Exp. Bot.* **2017**, *7*, 6.
79. Marcucci, M.C.; Ferreres, F.; Garcia-Viguera, C.; Bankova, V.S.; De Castro, S.L.; Dantas, A.P.; Valente, P.H.; Paulino, N. Phenolic Compounds from Brazilian Propolis with Pharmacological Activities. *J. Ethnopharmacol.* **2001**, *74*, 105–112. [CrossRef]
80. Kim, D.W.; Son, K.H.; Chang, H.W.; Bae, K.; Kang, S.S.; Kim, H.P. Anti-inflammatory Activitiy of *Sedum kamtschaticum*. *J. Ethnopharmacol.* **2004**, *90*, 409–414. [CrossRef]
81. Maharaj, A.; Naidoo, Y.; Dewir, Y.H.; Rihan, H. Phytochemical Screening and Antibacterial and Antioxidant Activities of *Mangifera indica* L. Leaves. *Horticulturae* **2022**, *8*, 909. [CrossRef]
82. Rodrigues, L.A.; Almeida, A.C.; Gontijo, D.C.; Salustiano, I.V.; Almeida, A.A.; Brandao, G.C.; Ribon, A.O.B.; Leite, J.P.V. Antibacterial Screening of Plants from the Brazilian Atlantic Forest Led to the Identification of Active Compounds in *Miconia latecrenata* (DC.) Naudin. *Nat. Prod. Res.* **2021**, *35*, 5904–5908. [CrossRef] [PubMed]
83. Kamsala, R.V.; Lepakshi, B.M.; Padma, Y.; Raju, R.R.V. Studies on Antimicrobial and Antioxidant Properties of Leaf Extracts of *Syzygium alternifolium* (WT.) Walp. *Int. J. Pharm. Pharm. Sci.* **2015**, *7*, 139–143.


**Disclaimer/Publisher's Note:** The statements, opinions and data contained in all publications are solely those of the individual author(s) and contributor(s) and not of MDPI and/or the editor(s). MDPI and/or the editor(s) disclaim responsibility for any injury to people or property resulting from any ideas, methods, instructions or products referred to in the content.

*horticulturae* MDPI

*Article*

# Anatomical and Chemical Analysis of *Moringa oleifera* Stem Tissue Grown under Controlled Conditions

**Holly M. McVea and Lisa J. Wood ***

Faculty of Environment, University of Northern British Columbia, Prince George, BC V2N4Z9, Canada
* Correspondence: lisa.wood@unbc.ca

**Abstract:** *Moringa oleifera* is a relatively well-studied ethnobotanical species, but information is limited regarding its stem anatomy and the production potential of phytochemicals from bark tissue. Knowing that variation exists in the production of chemical defenses by plants with growing conditions and with developmental stages, *M. oleifera* was grown under controlled conditions to characterize stem tissues and to determine if stem bark contained the correct phytochemical compounds to be of value in medicinal treatments. We used microscopy to characterize the stem anatomy of *M. oleifera* and analyzed stem bark extracts using FTIR and GC to identify 4-(α-L-rhamnosyloxy)-benzyl isothiocyanate (moringin) and benzylamine (moringine) in tissue. We found the stems to be in transition between juvenile and mature stages of development at 4 months old under the growth conditions used. In 7-month-old stems, we found the presence of moringin in all bark samples and did not find any moringine. These results indicate that *M. oleifera* bark of 7-month-old trees grown in greenhouse conditions may be valuable for drug development.

**Keywords:** *Moringa*; stem anatomy; bark chemistry; moringin; moringine; controlled environment

check for updates

**Citation:** McVea, H.M.; Wood, L.J. Anatomical and Chemical Analysis of *Moringa oleifera* Stem Tissue Grown under Controlled Conditions. *Horticulturae* **2023**, *9*, 213. https://doi.org/10.3390/horticulturae9020213

Academic Editor: Wajid Zaman

Received: 31 December 2022
Revised: 1 February 2023
Accepted: 2 February 2023
Published: 6 February 2023

**Copyright:** © 2023 by the authors. Licensee MDPI, Basel, Switzerland. This article is an open access article distributed under the terms and conditions of the Creative Commons Attribution (CC BY) license (https://creativecommons.org/licenses/by/4.0/).

## 1. Introduction

*Moringa* species (otherwise known as miracle trees or drumstick trees) are tropical deciduous dicotyledonous trees that are distributed throughout Africa and Asia [1,2]. All parts of the plant are edible and nutritious, and they are the source of many useful compounds, making them an important famine food and source of medicinal and cleaning compounds for impoverished nations [3–5].

The leaves, seeds, and roots of *M. oleifera* are well-studied [1,6]. The leaves of *M. oleifera* are pale-green bipinnate or tripinnate and feathery with opposite ovate leaflets [1,6]. The leaves are known to have a broad array of essential nutrients in relatively large concentrations and are, therefore, a common food additive [3]. The seeds of *M. oleifera* are typically brown, roughly almond-shaped, and measure to be roughly 1.9 × 1.1 cm. The seeds are produced in large pods (the 'drumsticks') that can grow up to 50 cm in length. The seeds and pods, like the leaves, are high in nutritional quality [4]. Older roots have a vascular cambium that consists of 6–8 layers; these layers produce roundish vessel elements surrounded by xylem parenchyma cells [7]. The 3–4 layered phellogen forms rectangular or square-shaped cells, and the walls of the phellum cells are suberized. Additionally, the phelloderm is large and consists of thin-walled parenchymatous cells containing scattered groups of fibers [7]. The roots contribute to *Moringa*'s common classification as a tuber vegetable, as the roots are the most commonly eaten part of the plant [7]. Furthermore, the root bark is often harvested for various pharmaceutical and ethnobotanical uses [1,5,7].

The stem and bark of *M. oleifera* are poorly understood in relation to the other parts of the plant. To date, there has only been one anatomical analysis of mature *M. oleifera* stems [7]. There is, therefore, a gap in the literature regarding the average size and area of stem and cell tissue types across the tree's stages of growth and a further deficit of anatomical diagrams of stem sections. Each plant tissue type of *M. oleifera* has unique anatomical and phytochemical attributes, resulting in unique uses.

Originally native to the sub-Himalayan Mountains of northern India, *M. oleifera* has been cultivated for various uses in tropical and subtropical regions around the world [1]. Some uses of *M. oleifera* include biofuel production, water purification, lubrication, leather tanning, and food preparation [1,5,8]. Additionally, this species is a valuable source of phytochemicals, which assist in multiple biological activities, including oxidative DNA damage protection, promoting anti-inflammatory responses, anti-hepatoprotective processes, ulcer recovery, antibiotic immune system responses, antiperoxidative processes, and antiproliferative processes (among others) [1,5,9].

Among the many phytochemicals typically possessed by *M. oleifera* are moringine (appears chemically identical to benzylamine) and moringin (4-(α-L-rhamnosyloxy)-benzyl isothiocyanate) [5]. Moringine is an alkaloid chemical, and its presence in *M. oleifera* is the first record of a plant-produced benzylamine [10]. Moringine is considered to be a toxic compound [11], with little research elucidating the extent of its toxicity in humans. It has been used in experiments and was found to act as a potassium channel blocker, causing reduced feeding in mice [12] and resulting in a decrease in plasma-free fatty acids and water intake in rats [10]. Alternatively, moringin is a sugar derivative that has recently been shown to act medicinally against several pathologies. *M. oleifera* bark extracts containing moringin were found to effectively treat rats with aggressive breast and colorectal carcinoma [13]. Al-Asmari et al. [13] found that *M. oleifera* extracts caused a decrease in cancer cell survival and motility and an increase in malignant cell apoptosis. Moringin has also specifically been shown as promising in the treatment and prevention of ischemic stroke [14], as well as for increasing apoptosis in neuroblastoma and hepatocarcinoma cells, and as a treatment for multiple sclerosis-induced neuropathic pain [15–17].

Paikra et al. [18] reported that only the leaves of *M. oleifera* have been found to contain moringin, while seeds and roots were found to contain moringine. To date, we could not find any specific literature describing the phytochemistry of bark that included both testing for moringin (potentially beneficial) and moringine (potentially harmful).

In natural environments, quantities of secondary metabolites such as moringin and moringine vary with environmental conditions; changes in solar radiation, temperature, nutrient availability, water availability, and biotic competition may all influence chemistry [5,19,20]. In optimal environments, plants tend to invest more energy into growth and reproduction than into protective anti-herbivory measures (such as toxic secondary metabolites) [5,19]. Trees growing in optimal environments may produce fewer toxic compounds, such as moringine, as there is little need for anti-herbivory measures. No information currently exists on how levels of moringin or moringine in *M. oleifera* may be altered by the environment.

To add to the current literature, we sought to depict the distribution and size of various tissue types within *M. oleifera* stems grown in a greenhouse under specific environmental conditions (Objective 1). We also sought to evaluate the efficacy of the bark as a source of secondary plant compounds when grown under controlled conditions, considering specifically the presence of moringin and moringine (Objective 2).

## 2. Methods

### 2.1. Growing Conditions

All *M. oleifera* trees used for this study were grown at the I.K. Barber Enhanced Forestry Lab (EFL), University of Northern British Columbia (UNBC). For the *M. oleifera* trees used in anatomical analyses, germination methods were derived from *Moringa* Farms [21]. Seeds of Indian origin were imbibed in a zip-lock bag (left slightly unsealed to allow for airflow) filled with water for 24 h. After the imbibing period, the seeds were removed from the water, dried on paper towel, and placed in a closed paper bag left in a cupboard above a stove to provide a dark warm germination environment. After 14 days, most of the seeds had germinated, and the seeds were planted 2 cm deep into soil. The soil mixture consisted of 107 L peat, 20 L perlite, 20 L vermiculite, 1/3 c coir soil enhancer, 1/3 c dolomitic lime, 1/3 c MicroMax nutrients, and 2/3 L slow-release nutrients (14-14-14). The trees were watered to saturation approximately twice weekly with plain water for the first month and with water containing 14-14-14 fertilizer at every watering thereafter, as signs of nutrient deficiency, including chlorosis, were noticed after 1 month of growth. During watering, the trees were monitored for signs of diseases to ensure none were present. The trees were grown in a growth chamber at 30 °C during the day (0600 to 2200 h) under LED grow lights and cooled to 28 °C between 2200 h and 0600 h with the lights turned off. After 4 months of growth, the trees were cut down and immediately underwent anatomical analysis.

The methods used for growing the trees utilized in chemical analyses were developed by Morgan et al. [8]. Seeds were planted 2.54 cm deep into a soil mixture prepared by mixing 20 L coir, 20 L coarse sand, and 20 L peat with 60 g of slow-release nutrients (14–14–14); there was also an addition of 3 tablespoons of dolomite to the soil prior to planting. High-pressure sodium (HPS) supplemental lighting was supplied to the trees each day between 0600 to 2200 h. The trees were kept at 24 °C within the housing greenhouse bay and maintained at a relative humidity ranging from 20 to 40%. During the night (2200–0600 h), the lights were turned off and the bay was cooled to 18 °C. The trees were watered approximately three times per week to the point of saturation. These procedures were followed for 7 months, at which point the trees were cut down and frozen until the bark could be peeled from the stems. The bark was then kept frozen until the time of phytochemical analysis.

### 2.2. Anatomical Analysis

Cross-sections were obtained from five freshly harvested 4-month-old *M. oleifera* plants by hand using a razor blade. The cross-sections were stained with toluidine blue (TBO) and then photographed and examined using a DS-Ri2 Eclipse FN1 Nikon light microscope under 40× magnification. The cross-section images were analyzed using NIS-Elements Basic Research (v.5.10.01 64-bit) software. Tangential width measurements (at the widest point, from the outside of the cell wall on one side to the other) were recorded for dilated phloem rays, vessel elements, and xylem rays. Radial width measurements were recorded for the periderm, true phloem tissue, and vascular cambium (at the widest point, from the outside of the cell wall on one side to the other). Lastly, small, localized pockets of sclerenchyma tissue within the cortex were measured by tracing the perimeter of the cell walls; the imaging software then algorithmically calculated the estimated area within the tracing. Raw data measurements were collected and loaded into Microsoft Excel.

### 2.3. Chemical Analysis

The extraction methods used were derived from Oluduro et al. [22]. The bark was removed from 19 frozen samples of 7-month-old *Moringa*. Once the bark had been removed, it was ground using liquid nitrogen and an IKA A11 basic analytical mill. The mill was cleaned between samples to avoid cross-contamination. Once the bark was ground, ethanolic extracts were prepared for each of the 19 samples. A total of 5 g (±0.10 g) of plant tissue (with an average moisture content of 39.7 % (SE = 1.39 %, CV = 0.111) were mixed with 100 mL of 90% ethanol. The solutions were then left at room temperature and shaken for 30 min a day at 120 rpm for 3 days and then left undisturbed at room temperature for an additional 2 days. Once the extraction time had fully elapsed, the suspension solutions underwent gravity filtration through #1 Whatman paper until clear (at least two filtrations were necessary). The extractions were then rotovapped at room temperature to remove some of the excess ethanol and water. The concentrated extracts were then filtered through 0.45 μm microfilters via syringe. The finished extracts were then stored in air-tight 2 mL vials at 4 °C.

Extracts were analyzed using Fourier-transform infrared spectroscopy (FTIR) on a Bruker ALPHA II with a platinum ATR module to detect if moringin or moringine were present. Determining the presence of the two compounds consisted of comparisons to FTIR of standard-grade moringin (molecular weight = 311.35 g/mol) and moringine (molecular weight = 107.15 g/mol). The moringin standard was >98.0% pure, obtained from ChemFaces (CAS No. 73255-40-0, Catalog No. CFN89445), and the moringine standard, in the form of benzylamine >99.0% pure, was obtained from Sigma-Aldrich (CAS No. 100-46-9, Catalog No. 13180). Once the qualitative phytochemical presence of the two compounds of interest were determined, the extracts were quantitatively analyzed using gas chromatography (GC) on an Agilent 6890GC at Northern Analytical Laboratory Services (NALS) at the University of Northern British Columbia's Prince George campus. The 19 samples were run with a ValcoBond VB5, 30 m × 0.25 mm × 0.25 μm film thickness column, with a constant flow rate of 1.7 mL/min using helium carrier gas at a 20:1 split ratio, with an oven programmed at 120 °C for 1 min, ramped 10 °C/min to 250 °C then ramped 25 °C/min to 300 °C and then held for 4 min at 300 °C.

### 2.4. Data Analysis

Descriptive statistics were calculated for the anatomical measurements across the 4-month-old trees using Microsoft Excel 2020 and IBM SPSS version 26 statistical analytics software. When possible, all 5 samples were used in the calculation; unfortunately, in some of the samples, the phloem rays were interconnected and/or the sclerenchyma tissues ran the circumference of the stem and were not possible to measure.

A distribution report was produced to summarize the presence–absence data resulting from the FTIR analysis; sample peaks were matched to the standard outputs and to the output of pure ethanol to determine if either moringine or moringin were present in the ethanolic extracts. Using GC, compound determinations were based on retention times with standards. The peaks of the moringin standard were integrated over the whole range of peaks to acquire concentration data; standard concentration data were plotted against area to determine correlation and for inaccuracy correction ($R^2 = 0.992$).

Since this was discovery-based research and not a true experiment, we did not make any statistical comparisons aside from obtaining average measurements of anatomy and chemical concentrations.

## 3. Results

### 3.1. Moringa Stem Anatomy

At the time of harvest (4 months old), the trees were transitioning from the juvenile stage to maturity; Nielsen [23] reported that it can take up to 8 months for the tree to fully mature. This juvenile–mature transition was evident by the presence of sclerenchyma tissues, which were visible as separate regions of tissue above and beside thick pockets of phloem (indicating some immaturity), combined with relatively thick xylem tissues connected in a complete circle within the circumference of the stem (indicating a level of maturity) (Figures 1 and 2). The transitioning trees had an average of 83 xylem rays/stem, with the number of rays ranging from 32 to 104 (SD = 29.04 rays, CV = 0.3490). Each of the tissue and cell types measured in the transitioning stems can be observed in Figures 1 and 2. In addition to the diameters, the sclerenchyma tissue pockets had an average area of 296,648 μm$^2$ but ranged from 23,051 μm$^2$ to 889,283 μm$^2$ (SD = 192,290.0, CV = 834.2). The radial phloem and tangential dilated phloem ray measurements had the greatest range (SD = 237.29 μm and 112.83 μm). Individual comparisons between tissue and cell measurements can be observed in Figures 3 and 4. These measurements highlight the consistency across individuals in periderm and vascular cambial radial diameters and the variability in the radial diameter of phloem tissue (Figure 3). Figure 4 illustrates that the vessel elements are relatively consistent in tangential diameter across individuals but show a range of sizes within any given individual. Figure 4 also shows the relative consistency in xylem ray tangential diameters across and within samples.

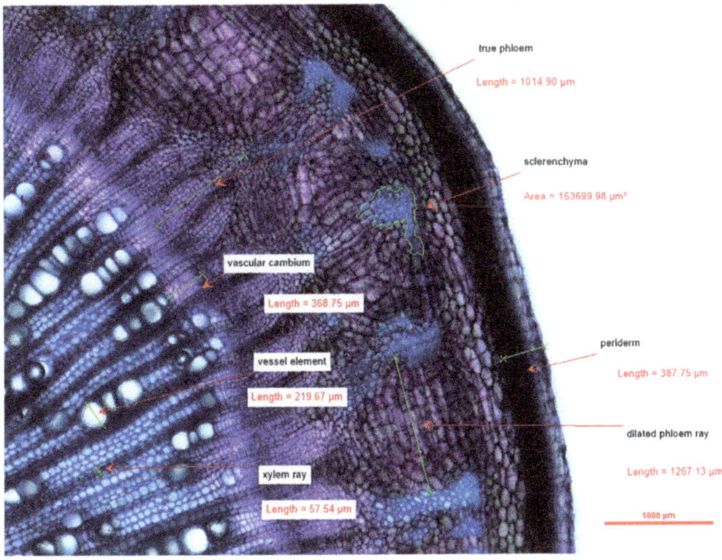


**Figure 1.** An example of a quarter-transverse section of a *Moringa oleifera* stem, stained with TBO, transitioning from the juvenile developmental phase to maturity observed under 40× magnification depicting prominent tissue and cell types and their measurements (measured at the widest point, from the outside of the cell wall on one side to the other).

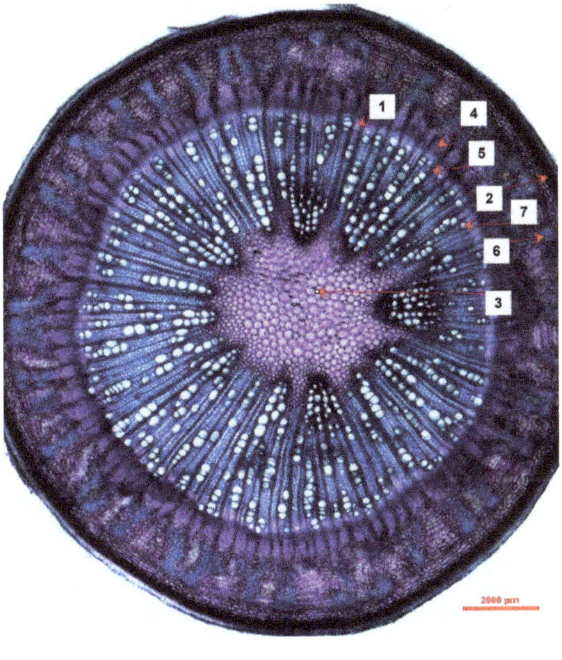


**Figure 2.** A transverse section of a *Moringa oleifera* stem transitioning from the juvenile developmental phase to maturity observed under 40× magnification depicting measured prominent tissue and cell types (1 = vascular cambium, 2 = periderm, 3 = pith, 4 = true phloem, 5 = xylem ray, 6 = sclerenchyma, 7 = vessel element).

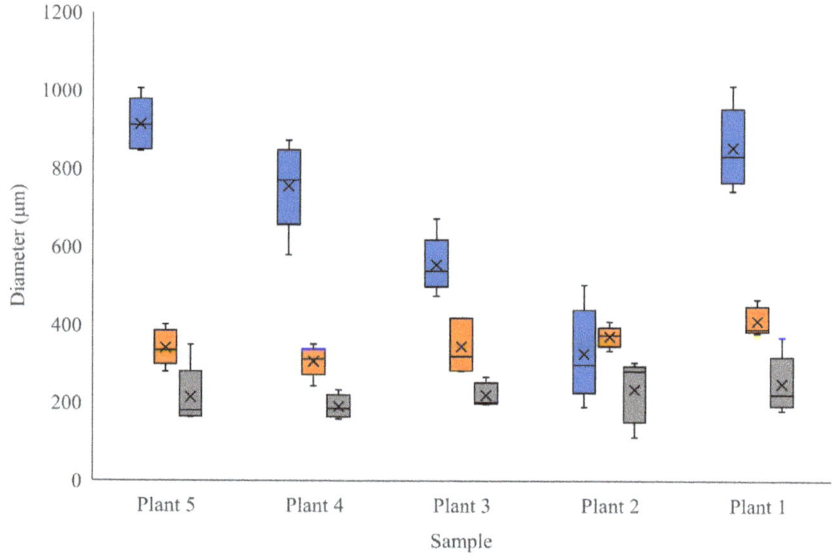


**Figure 3.** A comparison of the distributions of radially-measured tissues diameters in 5 stems of 4-month-old *Moringa oleifera* (measured at the widest point, from the outside of the cell wall on one side to the other) between tree samples (blue = true phloem, orange = periderm, grey = vascular cambium).

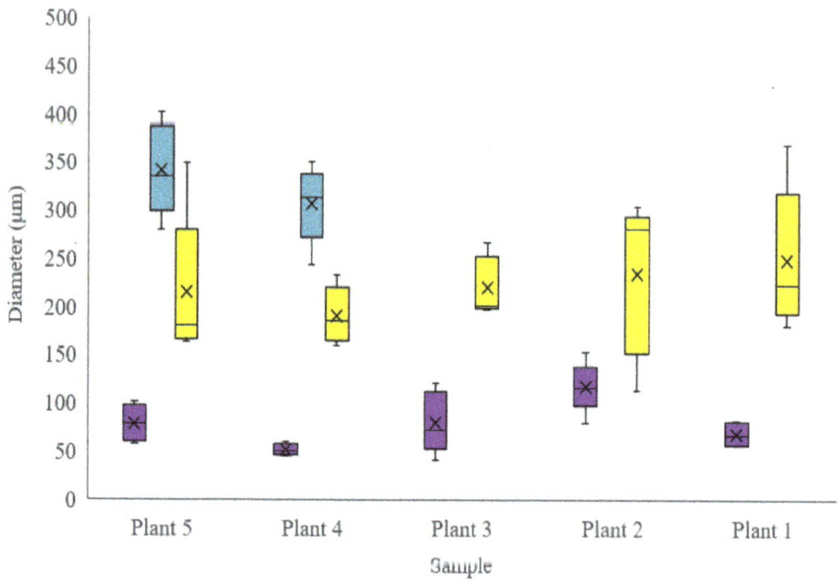


**Figure 4.** A comparison of the distributions of tangentially-measured tissues and cells diameters in 5 stems of 4-month-old *Moringa oleifera* (measured at the widest point, from the outside of the cell wall on one side to the other) between tree samples (cyan = dilated phloem ray, yellow = vessel element, purple = xylem ray).

### *3.2. Chemical Findings*

The FTIR analyses indicated that all 19 samples of stem bark from 7-month-old *Moringa* were nearly identical in their qualitative chemical composition. The FTIR indicated the presence of moringin in all 19 samples, indicated by the matching peaks in the samples and the standard (Figure 5). The peaks of interest can be observed at between 850–1500 $cm^{-1}$ and at 1920 $cm^{-1}$ (Figure 5); the remaining matching peaks are a result of the –OH group in the ethanol matching with the –OH in the moringin between 2800 $cm^{-1}$ and between 3400 $cm^{-1}$. The FTIR outputs further demonstrated the lack of moringine in all 19 samples; the lack of matching peaks can be observed in Figure 6.

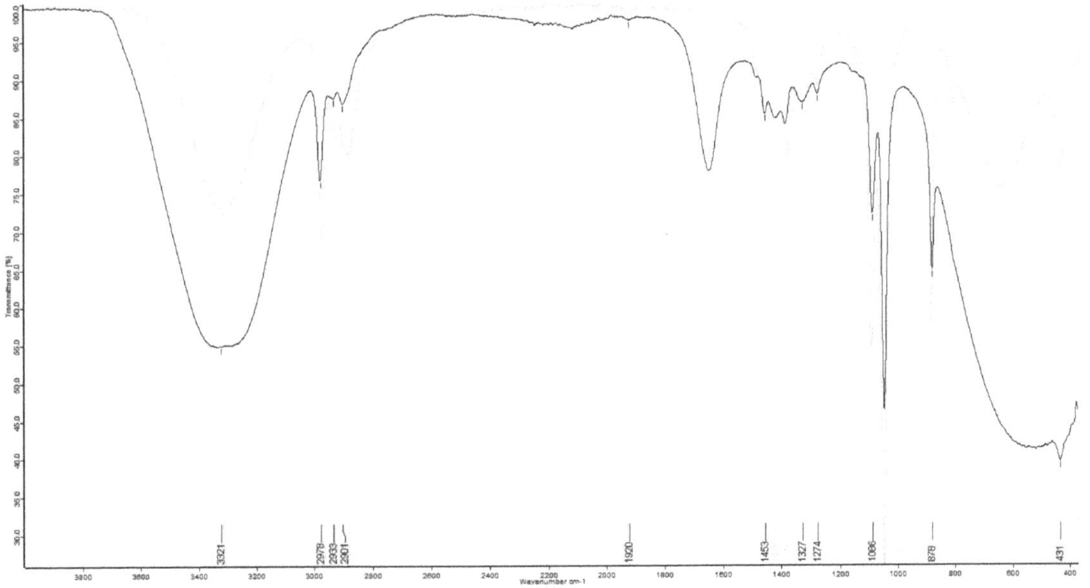


**Figure 5.** FTIR output, measuring percent transmittance over a range of wavelengths ($cm^{-1}$), comparing a representative ethanolic bark extract of *Moringa oleifera* (black) to standard-grade moringin (cyan). The many commonalities in peaks between the representative extract and the standard suggest that there is moringin present in the extracts.

The GC analyses conducted were able to quantify the amount of moringin present in the ethanolic extracts and confirm the lack of moringine (Figure 7). In ranges of peaks of interest, the chromatograms were well resolved; the moringin standard produced a cluster of peaks between 15.2 and 15.4 min. The chromatograms showed that the samples contained multiple compounds, which was expected, and is typical of plant material. A large peak was evident for the solvent (ethanol), and smaller peaks were detected at various points, including at the appropriate retention time for moringin (aligned with the standard). The moringine standard produced a large, narrow peak at 2.145 min, indicating its volatility. Samples did not show peaks corresponding to the moringine standard (Figure 7). The GC indicated that there was an average moringin concentration of 80.39 ug/mL in the extracts (SE = 2.06 ug/mL) with a range in concentrations between 8.29 ug/mL and 151.16 ug/mL (Table 1).

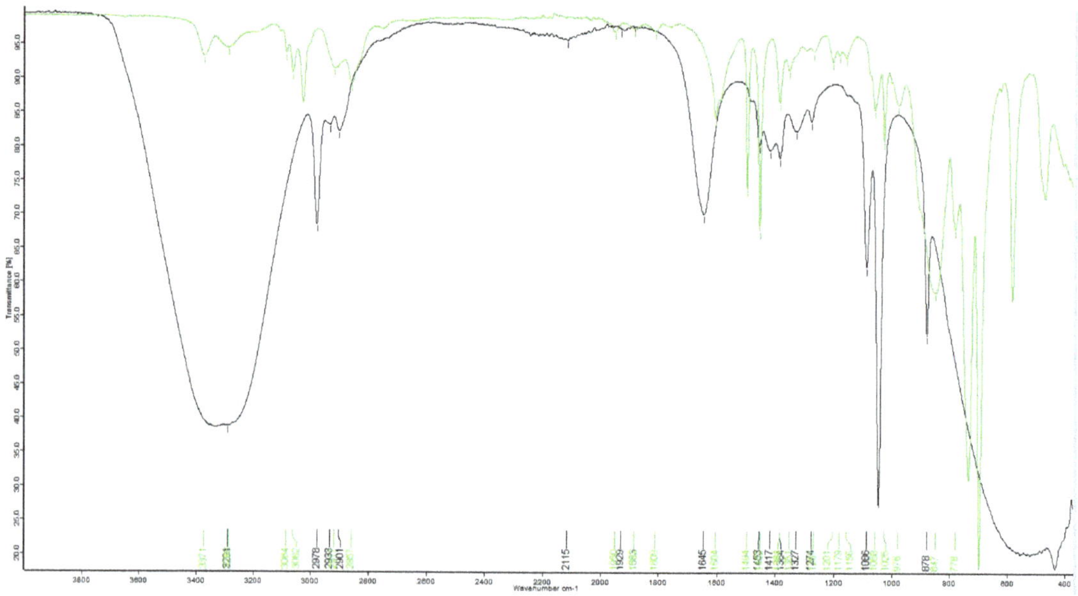


**Figure 6.** FTIR output, measuring percent transmittance over a range of wavelengths ($cm^{-1}$), comparing a representative ethanolic bark extract of *Moringa oleifera* (black) to standard-grade moringine (green). The minimal commonalities in peaks between the representative extract and the standard suggest that there is no moringine present in the extracts.

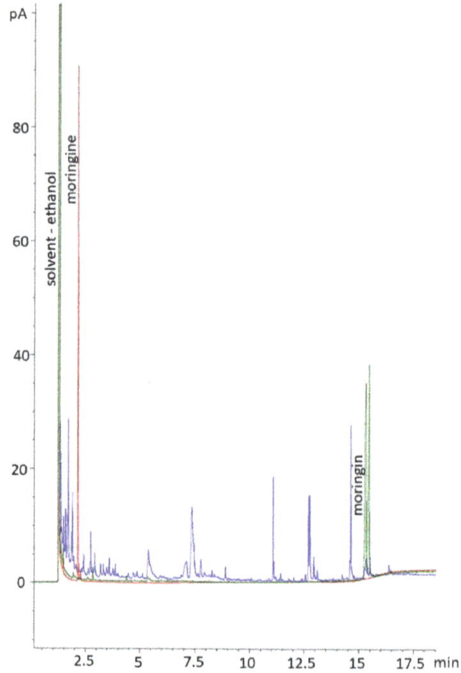


**Figure 7.** Chromatogram showing the moringin standard (green), moringine standard (red), and a representative *Moringa oleifera* sample extract, extracted from the *M. oleifera* bark (blue).

**Table 1.** Concentrations of moringine and moringin calculated using gas chromatography (forced zero) and the peak area for each of the 19 bark samples tested of *Moringa oleifera*.

| Sample ID | Benzylamine Area | Moringin Area | Benzylamine (ug/mL) | Moringin Conc (ug/mL) |
|---|---|---|---|---|
| Y18 | 0.000 | 6.814 | 0.000 | 46.693 |
| Y18-2 | 0.000 | 7.946 | 0.000 | 54.450 |
| Y63 | 0.000 | 16.786 | 0.000 | 115.031 |
| Y56 | 0.000 | 18.340 | 0.000 | 125.679 |
| Y61 | 0.000 | 6.901 | 0.000 | 47.290 |
| Y41 | 0.000 | 7.242 | 0.000 | 49.625 |
| Y29 | 0.000 | 14.365 | 0.000 | 98.440 |
| Y49 | 0.000 | 17.167 | 0.000 | 117.641 |
| Y24 | 0.000 | 5.274 | 0.000 | 36.143 |
| Y30 | 0.000 | 6.051 | 0.000 | 41.465 |
| Y48 | 0.000 | 17.477 | 0.000 | 119.764 |
| Y38 | 0.000 | 10.742 | 0.000 | 73.610 |
| Y04 | 0.000 | 16.662 | 0.000 | 114.180 |
| Y13 | 0.000 | 7.377 | 0.000 | 50.550 |
| Y07 | 0.000 | 22.058 | 0.000 | 151.162 |
| Y65 | 0.000 | 16.447 | 0.000 | 112.710 |
| Y70 | 0.000 | 10.085 | 0.000 | 69.113 |
| Y52 | 0.000 | 13.946 | 0.000 | 95.567 |
| Y15 | 0.000 | 1.210 | 0.000 | 8.292 |

## 4. Discussion

We sought to depict the distribution and size of various tissue types within *M. oleifera* stems grown in a greenhouse under specific environmental conditions (Objective 1). In a study by Vyas [7], a brief overview of *Moringa oliefera*'s stem anatomy was previously documented. They described young stems as having 16–18 vascular bundles, a large parenchymous pith, and a pericycle composed of alternate groups of parenchyma cells and fibers; these groups eventually form a circular band in mature stems [7]. Vyas [7] further describes mature stems' vascular cambia (VC) that produce large amounts of secondary xylem, which consists of uniseriate xylem rays, roundish vessel elements, and lignified thick-walled fibers; the VC also produces small amounts of secondary phloem. We add to this description by providing measurements for a variety of tissue and cell types in juvenile–mature transitional tissues of *Moringa oliefera*. This growth stage highlights the development of the ring of secondary tissues, including sclerenchyma. Interestingly, the radial diameter of the true phloem showed much more variability between individuals than did other tissue types. Phloem is the main transport tissue for photosynthates, metabolites, and other compounds and generally displays a high level of plasticity in response to environmental factors [20]. Given that these trees were all grown in the same, highly controlled environment, it is unlikely that atmospheric conditions were the reason for this variability. Seeds were also all from the same source, although they could have been derived from different parent trees. Since our plants did suffer from a nutrient deficiency at one point during their growth period, it is possible that the severity of this stress was different among individuals and, therefore, created a difference in phloem development [20], contributing to the variability shown among phloem tissues between our samples (Figure 3). The higher degree of consistency in the other tissue and cell sizes between individuals indicates that tissues other than phloem were relatively unaffected by the period of mineral deficiency (Figures 3 and 4). Our description of the stem anatomy of juvenile–mature transitioning *M. oleifera* grown under controlled conditions (Objective 1) adds to the currently available literature describing the anatomy of this species.

We qualitatively and quantitatively confirmed that 100% of the samples contained moringin using FTIR and GC, indicating that moringin is consistently produced in *Moringa oleifera* bark in trees grown under controlled greenhouse conditions. Thus far, moringin has largely been extracted from the seeds of *Moringa* for medicinal use [14,24,25]. The bark extracts we produced contained an average of 0.08 mg/mL and up to 0.15 mg/mL of moringin, but the samples consisted of an average of 39.7% water; therefore, the concentration of moringin in solution was diluted. Further refinement of the extraction process from bark could likely yield concentrations that are useful in medicine, for example, for the treatment of spinal cord injury and ischemic stroke [14,25]. Moringin was administered to rats at 3.5 mg/mL daily, producing neuroprotective properties and reducing oxidative stress and inflammation [14].

The double confirmation (through both FTIR and GC) that none of the bark samples contained moringine indicates that a greenhouse environment did not encourage the production of the potentially toxic compound [5,10–12]. In outdoor conditions, there are greater temperature, moisture, and nutrient fluctuations than under greenhouse conditions. Growing under optimal conditions and not being exposed to the dangers of herbivory, plants are less likely to produce defensive or stress-related secondary metabolites. Furthermore, plant chemical defenses vary greatly with age, and there is a particular reduction in chemical defenses during the transitional phase between being a juvenile and maturity [26]. Given that it can take up to 8 months for *M. oleifera* to mature [23], our research supports that growing *M. oleifera* in stable greenhouses for 4–7 months yields plants that do not have a risk of containing moringine. This is an important conclusion for those looking to produce *Moringa* plants for the treatment of pathologies.

We sought to evaluate the efficacy of the bark as a source of secondary plant compounds when grown under controlled conditions, considering specifically the presence of moringin and moringine (Objective 2). This research suggests that by producing fast-growing *M. oleifera* in optimal greenhouse conditions, trees can produce bark within 4–7 months, from which the extract is of good quality for use in medicinal treatments. This production can be conducted anywhere in the world, as illustrated by our location in northern Canada, and is not limited to tropical growing conditions, as would be the case in outdoor cultivation.

## 5. Conclusions

*M. oleifera* trees at 4–7 months of age contain transitional juvenile–mature anatomy. These tissues show the development of secondary tissues and relatively large and variable amounts of phloem, possibly reflective of the nutrient regime. *M. oleifera* grown under optimal growing conditions produced no moringine, but they maintained the production of moringin, likely due to the constantly available resources and lack of competition and herbivory, thus making phytotoxins superfluous. Future improvements could be made to this research through the use of another standard for GC to possibly produce a cleaner peak and through the analysis of a more concentrated extract. Overall outcomes of this study contribute to the effort to understand this 'miracle tree' in its entirety and to elucidate bark as an alternative to seed use for the production of moringin. Future studies could add to our findings by narrowing the growing conditions of *M. oleifera* to maximize moringin yields, optimizing the extraction process from bark, and using trees of different ages to determine if and when moringine is produced under alternative conditions. Furthermore, comparing trees grown in outdoor conditions to controlled conditions in a true experiment would add to current knowledge.

**Author Contributions:** H.M.M. conducted the literature review, carried out all methods, analyzed results, and wrote the initial manuscript draft as an undergraduate thesis project. L.J.W. established the methods for use, provided materials and experimental space, supervised the research, added to the literature review, and revised the writing for manuscript submission. All authors have read and agreed to the published version of the manuscript.

**Funding:** This research received no external funding. The APC was funded by the University of Northern British Columbia.

**Institutional Review Board Statement:** Not applicable.

**Informed Consent Statement:** Not applicable.

**Data Availability Statement:** Data are kept at the University of Northern British Columbia, and are available upon request. Please email the corresponding author.

**Acknowledgments:** This research received no funding; however, we would like to acknowledge Chris Opio for his expertise and guidance in the research and for providing in-kind contributions of the 7-month-old stem samples and seeds. We would also like to acknowledge Jasneek Manhas and Katie Tribe for their help in the laboratory. Lastly, we would like to thank the UNBC Chemistry Department (Beth Gentleman, Kaila Fadock, and Todd Whitcombe) and NALS (Charles Bradshaw) for the contributions of their laboratories, equipment and supplies, and expertise.

**Conflicts of Interest:** Authors declare no conflict of interest.

## References


1. Saini, R.K.; Sivanesan, I.; Keum, Y.-S. Phytochemicals of Moringa Oleifera: A Review of Their Nutritional, Therapeutic and Industrial Significance. *3 Biotech* **2016**, *6*, 203. [CrossRef]
2. Padayachee, B.; Baijnath, H. An Overview of the Medical Importance of Moringaceae. *J. Med. Plant Res.* **2012**, *6*, 5831–5839.
3. El-Taher, A.M.; Mousa, A.A.A. Comparative Anatomical Studies on Some Moringa Species Growing in Egypt. *J. Agric. Res.* **2014**, *22*, 134–145.
4. Azza, S.M. Morpho-Anatomical Variations of Leaves and Seeds among Three Moringa Species. *Life Sci. J.* **2014**, *11*, 827–832.
5. Aliyu, A.; Chukwuma, U.D.; Omoregie, E.H.; Folashade, K.O. Qualitative Phytochemical Analysis of *Moringa oleifera* Lam. From Three Climatic Zones of Nigeria. *J. Chem. Pharm. Res.* **2016**, *8*, 93–101.
6. Pandey, A.; Pradheep, K.; Gupta, R.; Nayar, E.R.; Bhandari, D.C. 'Drumstick Tree' (*Moringa oleifera* Lam.): A Multipurpose Potential Species in India. *Genet. Resour. Crop Evol.* **2011**, *58*, 453–460. [CrossRef]
7. Vyas, M.K. A Contribution of the Anatomical Characters of *Moringa oleifera* Lam. And Their Significance. *J. Pharmacogn. Phytochem.* **2019**, *8*, 576–578.
8. Morgan, C.R.; Opio, C.; Migabo, S. Chemical Composition of Moringa (*Moringa oleifera*) Root Powder Solution and Effects of Moringa Root Powder on E. Coli Growth in Contaminated Water. *S. Afr. J. Bot.* **2020**, *129*, 243–248. [CrossRef]
9. Choudhary, M.K.; Bodakhe, S.H.; Gupta, S.K. Assessment of the Antiulcer Potential of Moringa Oleifera Root-Bark Extract in Rats. *J. Acupunct. Meridian Stud.* **2013**, *6*, 214–220. [CrossRef]
10. Bour, S.; Visentin, V.; Prévot, D.; Daviaud, D.; Saulnier-Blache, J.S.; Guigne, C.; Valet, P.; Carpéné, C. Effects of Oral Administration of Benzylamine on Glucose Tolerance and Lipid Metabolism in Rats. *J. Physiol. Biochem.* **2005**, *61*, 371–379. [CrossRef]
11. World Health Organization and International Labour Organization. Icsc 1338—Benzylamine. Inchem.org. Available online: https://www.inchem.org/documents/icsc/icsc/eics1338.htm (accessed on 13 January 2023).
12. Raimondi, L.; Banchelli, G.; Ghelardini, C.; Pirisino, R. The Reduction of Food Intake Induced in Mice by Benzylamine and Its Derivatives. *Inflammopharmacology* **2003**, *11*, 189–194. [CrossRef]
13. Al-Asmari, A.K.; Albalawi, S.M.; Athar, M.T.; Khan, A.Q.; Al-Shahrani, H.; Islam, M. Moringa Oleifera as an Anti-Cancer Agent against Breast and Colorectal Cancer Cell Lines. *PLoS ONE* **2015**, *10*, e0135814. [CrossRef]
14. Galuppo, M.; Nicola, G.R.D.; Iori, R.; Dell'Utri, P.; Bramanti, P.; Mazzon, E. Antibacterial Activity of Glucomoringin Bioactivated with Myrosinase against Two Important Pathogens Affecting the Health of Long-Term Patients in Hospitals. *Molecules* **2013**, *18*, 14340–14348. [CrossRef]
15. Cirmi, S.; Ferlazzo, N.; Gugliandolo, A.; Musumeci, L.; Mazzon, E.; Bramanti, A.; Navarra, M. Moringin from Moringa Oleifera Seeds Inhibits Growth, Arrests Cell-Cycle, and Induces Apoptosis of SH-SY5Y Human Neuroblastoma Cells through the Modulation of NF-KB and Apoptotic Related Factors. *Int. J. Mol. Sci.* **2019**, *20*, 1930. [CrossRef] [PubMed]
16. Antonini, E.; Iori, R.; Ninfali, P.; Scarpa, E.S. A Combination of Moringin and Avenanthramide 2f Inhibits the Proliferation of Hep3B Liver Cancer Cells Inducing Intrinsic and Extrinsic Apoptosis. *Nutr. Cancer* **2018**, *70*, 1159–1165. [CrossRef] [PubMed]
17. Giacoppo, S.; Iori, R.; Bramanti, P.; Mazzon, E. Topical Moringin-Cream Relieves Neuropathic Pain by Suppression of Inflammatory Pathway and Voltage-Gated Ion Channels in Murine Model of Multiple Sclerosis. *Mol. Pain* **2017**, *13*, 1744806917724318. [CrossRef]
18. Paikra, B.K.; Dhongade, H.K.J.; Gidwani, B. Phytochemistry and Pharmacology of *Moringa oleifera* Lam. *J. Pharmacopunct.* **2017**, *20*, 194–200. [CrossRef]
19. Stamp, N. Out of the Quagmire of Plant Defense Hypotheses. *Q. Rev. Biol.* **2003**, *78*, 23–55. [CrossRef] [PubMed]

20. López-Salmerón, V.; Cho, H.; Tonn, N.; Greb, T. The Phloem as a Mediator of Plant Growth Plasticity. *Curr. Biol.* **2019**, *29*, R173–R181. [CrossRef]
21. Moringa Farms—The Moringa Specialists. Moringa Farms. Available online: https://moringafarms.com/ (accessed on 13 January 2023).
22. Oluduro, A.O. Evaluation of Antimicrobial Properties and Nutritional Potentials of *Moringa oleifera* Lam. Leaf in South-Western Nigeria. *Malays. J. Microbiol.* **2012**, *8*, 59–67.
23. Nielsen, L. Growing Moringa: The Majestic Drumstick Tree. Epic Gardening. Available online: https://www.epicgardening.com/growing-Moringa/ (accessed on 8 January 2023).
24. Ragasa, C.Y.; Ng, V.A.S.; Shen, C.-C. Chemical Constituents of *Moringa oleifera* Lam. Seeds. *Int. J. Pharmacogn. Phytochem. Res.* **2016**, *8*, 495–498.
25. Giacoppo, S.; Galuppo, M.; De Nicola, G.R.; Iori, R.; Bramanti, P.; Mazzon, E. 4(α-l-Rhamnosyloxy)-Benzyl Isothiocyanate, a Bioactive Phytochemical That Attenuates Secondary Damage in an Experimental Model of Spinal Cord Injury. *Bioorg. Med. Chem.* **2015**, *23*, 80–88. [CrossRef]
26. Barton, K.E.; Koricheva, J. The Ontogeny of Plant Defense and Herbivory: Characterizing General Patterns Using Meta-Analysis. *Am. Nat.* **2010**, *175*, 481–493. [CrossRef]

**Disclaimer/Publisher's Note:** The statements, opinions and data contained in all publications are solely those of the individual author(s) and contributor(s) and not of MDPI and/or the editor(s). MDPI and/or the editor(s) disclaim responsibility for any injury to people or property resulting from any ideas, methods, instructions or products referred to in the content.

horticulturae

MDPI

*Article*

# Two New *Ferula* (Apiaceae) Species from Central Anatolia: *Ferula turcica* and *Ferula latialata*

**Hüseyin Onur Tuncay [1,2,*], Emine Akalın [1,3], Aslı Doğru-Koca [4], Fatma Memnune Eruçar [5,6] and Mahmut Miski [5]**

1 Department of Pharmaceutical Botany, Faculty of Pharmacy, Istanbul University, Istanbul 34116, Türkiye
2 Department of Pharmaceutical Botany, Institute of Graduate Studies in Health Sciences, Istanbul University, Istanbul 34116, Türkiye
3 Faculty of Pharmacy, Eastern Mediterranean University, Famagusta 99628, Türkiye
4 Department of Biology, Faculty of Science, Hacettepe University, Ankara 06800, Türkiye
5 Department of Pharmacognosy, Faculty of Pharmacy, Istanbul University, Istanbul 34116, Türkiye
6 Department of Pharmacognosy, Institute of Graduate Studies in Health Sciences, Istanbul University, Istanbul 34116, Türkiye
* Correspondence: onur.tuncay@istanbul.edu.tr

**Abstract:** *Ferula turcica* and *Ferula latialata* are two novel endemic species discovered in the Konya and Kırşehir provinces of the central Anatolian region of Türkiye. These two new species are described by morphological, ecological, carpological, and phytochemical characteristics and phylogenetic analysis. *F. turcica* and *F. latialata* are morphologically distinct from *F. szowitsiana* by their habit, the stalk of the terminal umbella, and the mericarp size, as well as by the profile of their secondary metabolite markers and phylogenetic placement. The phylogenetic analyses of sequences of the internal transcribed spacer in ribosomal DNA belonging to both new taxa were conducted to reveal the evolutionary relationships of the new species. Their relationships with the other related species and proposed conservation status were reviewed. The morphological, molecular, and phytochemical evidence supported the hypothesis that *Ferula turcica* and *Ferula latialata* are two new distinct species.

**Keywords:** *Ferula*; new species; Turkey; Apiaceae; morphology; anatomy; chemotaxonomy; molecular; phylogeny

check for updates

**Citation:** Tuncay, H.O.; Akalın, E.; Doğru-Koca, A.; Eruçar, F.M.; Miski, M. Two New *Ferula* (Apiaceae) Species from Central Anatolia: *Ferula turcica* and *Ferula latialata*. *Horticulturae* **2023**, *9*, 144. https://doi.org/10.3390/horticulturae9020144

Academic Editor: Wajid Zaman

Received: 20 December 2022
Revised: 18 January 2023
Accepted: 19 January 2023
Published: 20 January 2023

**Copyright:** © 2023 by the authors. Licensee MDPI, Basel, Switzerland. This article is an open access article distributed under the terms and conditions of the Creative Commons Attribution (CC BY) license (https://creativecommons.org/licenses/by/4.0/).

## 1. Introduction

The Apiaceae family is one of the largest families among Angiosperm plants [1,2]. *Ferula* L. is the largest genus in the Apiaceae family, with approximately 213 species [3]. *Ferula* species are widespread in the temperate regions of the Euro-Asian continent, surrounded by the Canary Islands in the West, North Africa in the South, China and India in the East, and Central Europe in the North.

*Ferula* plants have been used for medicinal and culinary purposes since ancient times [4,5]. Pedanius Dioscorides described the medicinal properties of several *Ferula* resins, including asafoetida (*Ferula assa-foetida* L.), galbanum (*F. gummosa* Boiss.), sagapenum (*F. persica* Willd.), and African ammoniacum (*F. marmarica* Asch. and Taub.), in his De Materia Medica two thousand years ago [6]. Ibn-i Sina (Avicenna) described the application of the oleo-gum-resin from *Ferula foetida* in the treatment of cancerous tumors in the Canon of Medicine [7]. The resins of *Ferula* species have been used in the food and health industries as a spice, nutraceutical, and cosmeceutical in India, Iran, and Afghanistan [8,9]. Antimicrobial and anti-inflammatory activities of the essential oil and extracts of the aerial parts of *F. szowitsiana* have been reported [10,11]. The roots of *F. persica* have been used to alleviate the symptoms of diabetes in Iran and Jordan [12].

Many taxonomic studies have been conducted on the genus *Ferula*. Boissier classified the genus *Ferula* species grown in the Irano-Turanian region into three sections based on the shape of their petals and the number of their vittae: *Peucedanoides* Boiss. *Scrodosma* Bunge and *Euferula* Boiss. [13].

Korovin established the most comprehensive infrageneric classification by defining the *Ferula* genus into six subgenera and eight divisions [14]. Conversely, in the study conducted by Safina and Pimenov, *Ferula* species were examined in terms of their carpological characters, and they emphasized that the species of the *Ferula* subgenus did not form a homogeneous group [15]. A molecular study conducted by Pimenov et al. on 90 *Ferula* species yielded quite different results in comparison with Korovin's taxonomical classification [16].

According to Flora of Turkey and the East Aegean Islands, 18 *Ferula* species were listed by Peşmen in Türkiye [17]. Afterward, Sağıroğlu and Duman found four new species as a result of their revised study on *Ferula* species growing in Türkiye [18–21]. Subsequently, *F. divaricata* Pimenov and Kljuykov and *F. pisidica* Akalın and Miski were discovered in 2013 and 2020, respectively [22,23]. Followed by the addition of these novel species, the total number of *Ferula* species growing in Türkiye has reached 24.

Although based on its morphological affinity, the genus *Ferula* was accepted as a member of the tribe Peucedaneae, its classification was updated and transferred to the tribe Scandiceae based on the phylogenetic hypothesis [24]. The intrageneric phylogenetic relationships of *Ferula* are ongoing [25–28]. The sequences of the internal transcribed spacers (ITS 1, 5.8S rRNA, ITS 2) were one of the markers used in these studies. According to Panahi et al., the *Ferula* species distributed in the southwestern part of the Iran-Turanian floristic region generate a monophyletic clade and additional polytomic *F. narthex*. These lineages correspond to subgenus *Narthex* (Falc.), section *Merwia* (B. Fedtsch.) Korovin. Their results suggested that the interspecific boundaries in this section are unclear. Especially the ITS sequences of some species showed that they are identical to each other and are very close species in terms of their morphological features or were determined as synonyms. For instance, the ITS sequences of *F. gummosa* are identical to those of *F. badrakema*, *F. linczevskii*, *F. undulata*, and *F. myrioloba*.

Additionally, because of the disagreement between hypothetical phylogenetic trees based on the chloroplast DNA and ITS dataset, Panahi et al. concluded that there was reticulate evolution in section *Merwia*. They commented that this reticulate evolution resulted from hybridization and introgression, especially among Irano-Turanian species [26]. Therefore, the studies of the determination of putative new species based on the molecular data in the *Ferula* genus (especially in subgenus *Narthex*, section *Merwia*) are critical in revealing not only morphological but also phylogenetic evidence of the species and their infra-generic evolutionary relationships.

Specimens of the genus *Ferula* were collected by M. Miski and H. O. Tuncay from the shores of Tuz Lake (Konya Province) and Seyfe Lake (Kırşehir Province). Identification of the collected samples was attempted using the diagnostic key found in Flora of Turkey, as a result, the identified specimens were morphologically close to species *F. szowitsiana* DC. Thus, a detailed study was conducted, and it attempted to identify the collected species using the diagnostic keys in Flora of the U.S.S.R. and Flora Iranica [29,30]. Similarly, the collected specimens were found to be closely related to *F. szowitsiana* and *F. persica* species; however, they showed some differences from these two known species. Hence, more detailed comparative morphological, anatomical, chemical, and molecular studies were carried out to compare these potential new species with those of *F. szowitsiana* and *F. persica* species.

## 2. Results

### 2.1. *Taxonomic Treatment*

*Ferula turcica*: Akalın, Miski, and Tuncay sp. nov. (Figures 1 and 2).

**Figure 1.** (**A**,**B**) General view of *Ferula turcica*. (**C**) Sheath.

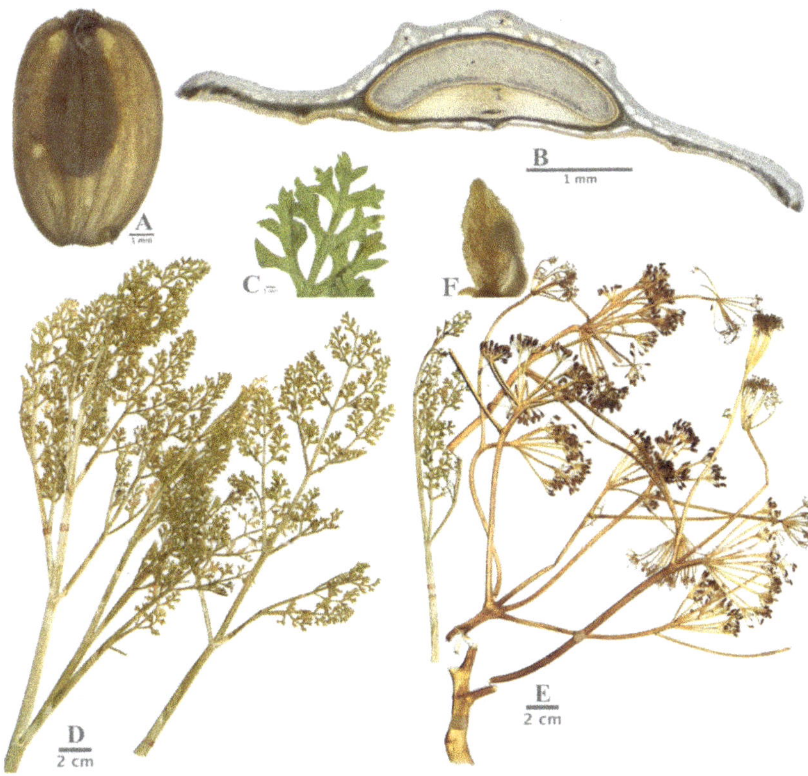


**Figure 2.** (**A**) General view of *Ferula turcica* fruit. (**B**) Cross-section of mericarp of *F. turcica*. (**C**,**D**) Basal leaf of *F. turcica*. (**E**) General view of *F. turcica* umbella. (**F**) Petals with setulose-puberulent hair.

Type: TÜRKİYE. B4 Konya: Tuz Lake, Yavşan salt pan, 910 m, 16 June 2015, M. Miski (Holotype: ISTE 116464).

Diagnosis: Caule crasso et striato, usque 100 cm; foliis basalia 5–6 pinnata, triangularia-ovata, 30–35 × 25–30 cm, dense pubescentia, segmentis ultimis pinnatisectis. Vaginae non in-flatae, 2 × 4–6 cm. Panicula laxa. Umbellis centralibus pedunculata 0.7–4 cm, umbellae lateralis raro solitaria vel 2–4 pedicellis longis, umbellis centralis (9-) 15–18 radiis 3–5 cm. Mericarpia elliptica, 5–7 mm × 9–11 mm, dorso dorsali leviter prominente et filiformi, alis lateralibus 1.2–1.9 mm latis, vittae dorsales 13–24, per valleculae 4–7 commissurales 11–14.

Description: Perennial herbs, erect, green, up to 100 cm tall, stem thick and striate. The root has a 1–3 cm width with a thick woody taproot system. A fibrous collar, which are old petioles, remains on the base of the stem. Leaves green, basal leaves 5–6 pinnate, triangular-ovate in outline, 30–35 × 25–30 cm, densely pubescent, ultimate segments pinnatisect, lobes 1–2.5 × 0.7–1 mm oblong, obtuse to acute. Sheats not inflated, 2 × 4–6 cm.

Inflorescence lax panicle, central umbels composed of fertile flowers, lateral umbels composed of sterile flowers. Central (terminal) umbella with peduncled 0.7–4 cm, lateral umbella rarely single or 2–4 on long pedicels, central umbella (9-) 15–18 rays 3–5 cm, umbellules (6-) 8–12 (−17) flowered; petals setulose-puberulent, pedicel at fruiting 0.5–1.4 mm long; sepals caducous in fruiting time.

Mericarps elliptic, 6.5 mm (5–7) width, 9.5 mm (9–11) length, dorsal ridges slightly protruding and filiform, lateral wings 1.5 mm (1.2–1.9) wide, dorsal vittae 13–24, 4–7 per vallecula, commissural 11–14 (Table 1).

**Table 1.** Comparison of the diagnostic characters of *Ferula turcica*, *F. latialata*, *F. szowitsiana*, and *F. persica*.

| Character | *F. turcica* | *F. latialata* | *F. szowitsiana* | *F. persica* |
|---|---|---|---|---|
| Stem | 70–100 cm | 80–110 cm | 50–70 cm | 70–100 cm |
| Ultimate segment of the leaf | Regular deeplobed 1–2.5 × 0.7–1 mm | Regular lobed 1–2 (−2.5) × 0.5–0.7 mm | Regular lobed 1–2 mm | Regular deeplobed |
| Hair | Densely puberulent | Puberulent | Setulose-puberulent | Pubescent |
| Ray numbers and length | (9-) 15–18 rays (3–5 cm) | 13–15 (−18) (3–5 cm) | 7–11 rays (2–5 cm) | 17–22 rays |
| Umbellules numbers | 8–12 (−17) | 8–10 (−14) | 8–12 | 15 |
| Pedicel at fruiting | 0.6–1.2 (−1.5) cm | 0.5–1 (−1.2) cm | 0.3–0.5 cm | |
| Central umbella | Peduncled 0.7–4 cm | Peduncled 0.5–1.5 (−3) cm | Shortly peduncled or sessile | Sessile |
| Shape of fruit | Elliptic | Elliptic to oblong | Elliptic to orbicular | Ovoid |
| Lateral wings | 1.2–1.9 mm | 2.5–3.9 mm | 2–4 mm | - |
| Dorsal vittae | 4–7 per vallecula | 3–5 per vallecula | (2-) 4–6 per vallecula | 5–7 per vallecula |
| Commisural vittae | 11–14 | 6–10 | 8–12 | 16–18 |
| Width of fruit average (min-max) | 6 mm (5–7 mm) | 11 mm (9–12.5 mm) | 10–13 mm | 6 mm |
| Length of fruit average (min-max) | 9.5 mm (9–11 mm) | 15.5 mm (13.5–18 mm) | 12–20 mm | 11 mm |
| Ratio of length to width of fruit | 1.6 | 1.4 | 1.4 | 1.8 |

Etymology: *F. turcica* is named after the country of Türkiye.

Phenology: Flowering time is from May to June, and fruiting is from June to July.

*Ferula latialata*: Akalın, Miski, and Tuncay sp. nov. (Figures 3 and 4).

**Figure 3.** (**A**,**C**) General view of *Ferula latialata*. (**B**) Sheath.

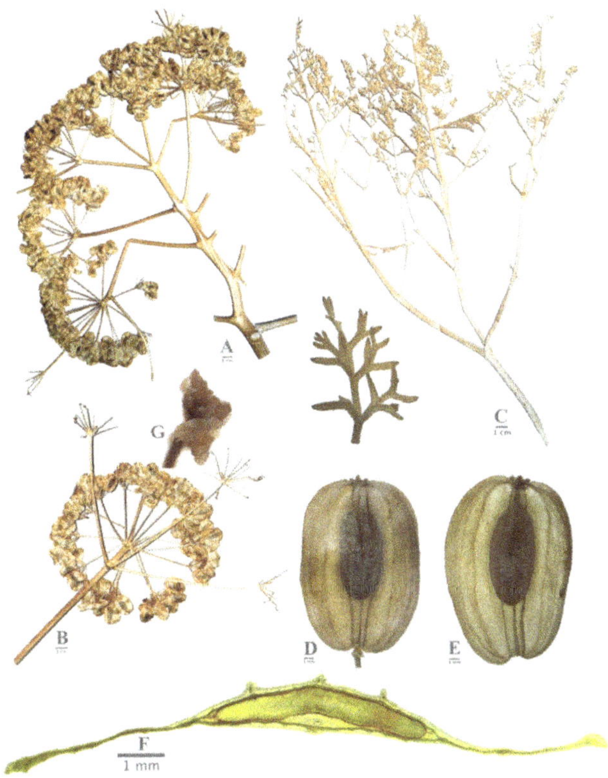


**Figure 4.** (**A**,**B**) General view of *Ferula latialata* umbella. (**C**) Basal leaf of *F. latialata*. (**D**,**E**) General view of *F. latialata* fruits. (**F**) Cross-section of mericarp of *F. latialata*. (**G**) Petals with setulose-puberulent hair.

Type: Türkiye. B5 Kırşehir: Seyfe Lake, near Yazıkınık village, 1110 m, 2 August 2021, H. O. Tuncay (Holotype: ISTE 117495).

Diagnosis: Caule crasso et striato, usque 110 cm; foliis basalia 6–7 pinnata, triangularia-ovata, 30–35 × 25–30 cm, pubescentia, segmentis ultimis pinnatisectis. Vaginae non inflatae, 2 × 4–6 cm. Panicula laxa. Umbellis centralibus pedunculata 0.5–1.5 (−2.5) cm, umbellae lateralis 1–3 (−4) pedicellis longis, umbellis centralis 13–15 (−18) radiis, 3—5 cm. Mericarpia elliptica vel oblonga, 13.5–18 mm × 9–12.5 mm, dorso dorsali leviter prominente et filiformi, alis lateralibus 2.5–3.9 mm latis, vittae dorsales per valleculae 3–5 commissurales 6–10.

Description: Perennial herbs, erect, green, up to 110 cm tall, stem thick and striate. Root 1–3 cm width with thick woody tap root system. A fibrous collar, which are old petioles, remains on the base of the stem. Leaves green, basal leaves 6–7 pinnate, triangular-ovate in outline, 30–35 × 25–30 cm, pubescent, ultimate segments pinnatisect, lobes 1–2 (−2.5) × 0.5–0.7 mm oblong, obtuse. Sheats not inflated, 2 × 4–6 cm.

Inflorescence lax panicle, central umbels composed of fertile flowers, lateral umbels composed of sterile flowers. Central (terminal) umbella with peduncled 0.5–1.5 (−2.5) cm, lateral umbella 1–3 (−4) on long pedicels, central umbella 13–15 (−18) rays 3–5 cm, umbellules 8–10 (−14) flowered; petals setulose-puberulent, pedicel at fruiting 0.5–1 mm long; sepals caducous in fruiting time.

Mericarps elliptic to oblong, 11 mm (9–12.5) width, 15.5 mm (13.5–18) length, dorsal ridges slightly protruding and filiform, lateral wings 3 mm (2.5–3.9) wide, dorsal vittae 3–5 per vallecula, commissural 6–10 (Table 1).

Etymology: The epithet name *latialata* from Latin, meaning wide, refers to the wide lateral wing in the fruit of *Ferula latialata.*

Phenology: Flowering time is from May to June, and fruiting is from June to July.

Distribution and ecology: The distribution of *Ferula turcica* (Konya) and *F. latialata* (Kırşehir) in Türkiye is shown in Figure 5. These two new species are distributed close to each other, and both are known from a single locality. Different localities have not yet been identified in field studies in similar habitats. *F. turcica* grows in halophytic soils near Tuz (salt) Lake, the second largest lake and an important source of salt in Türkiye, at an altitude of about 900 m, and its natural habitat is undisturbed. *F. turcica* grows together with *Ferula halophila* Peşmen, and they share the same habitat. The vegetation in the zone closest to the lake, which is covered with thick salt layers, consists of *Limonium lilacinum* (Boiss. and Balansa) Wagenitz, *Salicornia europaea* L., and *Halocnemum strobilaceum* Moris communities. The other new species, *F. latialata*, grows among the sunflower and wheat fields near Seyfe Lake, close to halophytic soils at an altitude of about 1100 m, and in relatively less saline soils than *F. turcica*. Species such as *Halocnemum strobilaceum* Moris, *Bassia Pilosa* (Fisch. and C.A. Mey.) Freitag and G. Kadereit, and *Camphorosma monspeliaca* L., are found in areas under the influence of the salty water of Seyfe Lake and salt marshes.

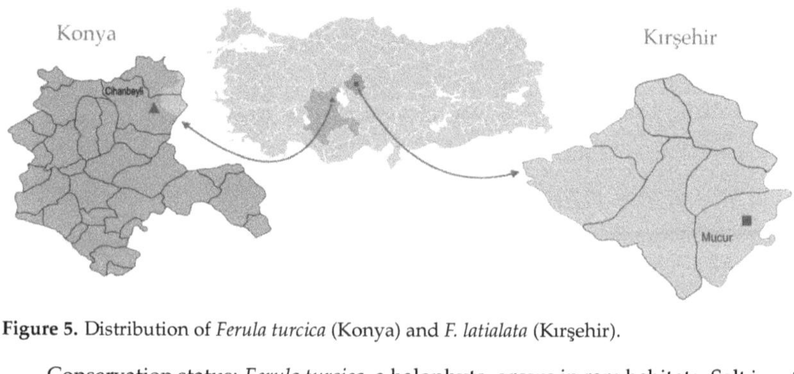


**Figure 5.** Distribution of *Ferula turcica* (Konya) and *F. latialata* (Kırşehir).

Conservation status: *Ferula turcica*, a halophyte, grows in rare habitats. Salt is extracted from Tuz Lake, where it grows, but the salt mine poses no threat to *F. turcica*. For this species, whose natural habitat is intact, its narrow distribution area and the rare soil structure in which it grows may be limiting factors. The other new species, *F. latialata*, is more threatened than *F. turcica* because it grows in agricultural areas. The number of individuals observed in the population was low. The abandonment of agricultural areas may reduce the pressure on this species and increase the number of individuals in the population. Plant breeding studies have been started for *F. latialata*, which has high environmental pressure and a very limited number of individuals. In the present records, *F. turcica* and *F. latialata* are endemic taxon to central Anatolia and are known from only one locality; therefore, they are considered as "Endangered (criterion B1)". They could also be categorized Critically Endangered (criterion B2) due to their known area of occupancy of less than 2 $km^2$ and their population size estimated to be fewer than 250 mature individuals (Criterion C). It is suggested that the species of *F. turcica* and *F. latialata* should be considered Critically Endangered (CR) according to the IUCN threat criteria [31].

### 2.2. Phenetic Analysis

Principal coordinate analysis (PCoA) based on morphological, anatomical, and chemical data was performed on *Ferula turcica*, *F. latialata*, and related species (Figure 6). The 21 morphological and chemical characters used for the analysis are given in Table S1. Coordinate two clearly separates *F. turcica*, *F. latialata*, and *F. szowitsiana* from *F. drudeana* and *F. persica*, which are located on the positive side of the axis. Moreover, coordinate one separates *F. turcica* and *F. latialata* from allied species *F. szowitsiana*, which are located on the negative side of the axis.

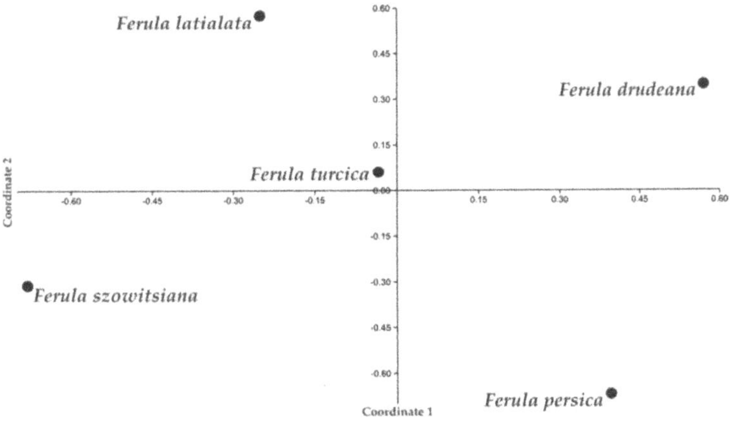


**Figure 6.** Principal coordinate analysis performed on 21 morphological, anatomical, and chemical characters of *Ferula turcica*, *F. latialata*, and related species.

### 2.3. Phylogenetic Evaluations

The phylogenetic hypothesis is presented in Figure 7. This tree is in agreement with those of Panahi et al. and Piwczyński et al. [26,27]. The genus *Ferula* is not monophyletic because of *Leutea*. The represented species of *Leutea* is nested in the *Ferula* ingroup in all the analyses, compatible with the results of Piwczyński et al. [27] and Panahi et al. [25,26]. The intra-generic classification in *Ferula* is not strongly supported in either subgenus nor are the sectional taxonomic levels by phylogenetic analysis. As discussed by Panahi, the species in nearly all the clades are located as polytomic terminals, maybe because of conspecific or nuclear-mediated introgression. Subgenus *Narthex* has numerous polytomic clades, which were determined as sections. The sampling is especially focused on subgenus *Narthex* section *Merwia* according to Panahi et al. [26]. Section *Merwia* does not generate a monophyletic clade. This section includes a clade (clade A) and two polytomic species, *F. kuhistanica* and *F. narthex*. Clade A is supported by medium posterior probability, whereas it is not supported by the bootstrap values (PP = 0.87). Clade A is divided into clades B and

C. Clade B is strongly supported by posterior probability, whereas it is not supported by the bootstrap values (PP = 0.93). This clade comprises numerous polytomic lineages. One of them, clade D, includes both of the new species, *F. turcica* and *F. latialata*. This clade is supported by strong posterior probability and moderate bootstrap values (PP = 0.99, BS-ML = 63, BS-MP = 80). *F. szowitsiana*, which was collected from the same locality (Tuzgölü, Yavşan salt pan) as *F. turcica* and is one of the accessions belonging to *F. drudeana*, are nested in clade D as polytomic lineages.

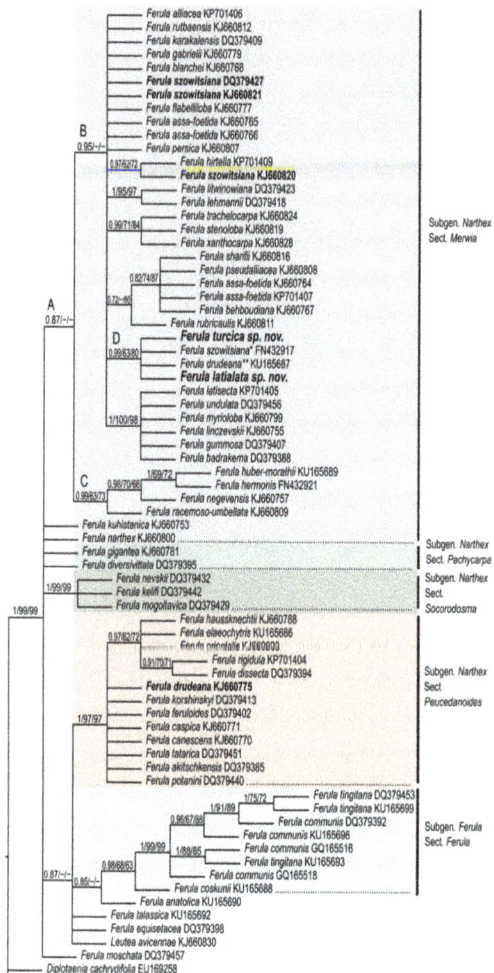


* Identified as *F. szowitsiana* in GenBank. It was collected from the same are as *F. turcica*.
** In addition to being morphologically different from *F. turcica* and *F. latialata*, it is not in the same clade as the *F. drudeana* specimen collected by Siehe.

**Figure 7.** Bayesian estimate of the phylogeny of the genus *Ferula* focused on subgenus *Narthex* based on the ITS sequence dataset. The two new species described herein, *F. turcica* and *F. latialata*, and related species, *F. szowitsiana* and *F. drudeana*, are indicated by bold letters. The supporting posterior probability and bootstrap values (the first obtained from maximum likelihood analysis; the second obtained from maximum parsimony analysis) are presented above the branches. Alphabetical designations for some clades are addressed in the main text.

## 2.4. Chemotaxonomic Characteristics

The monograph of E. Korovin [14] lists 15 *Ferula* species in section *Merwia*, and the phytochemical investigations performed on some of these species indicated that their common secondary metabolites are sesquiterpene coumarins and sulfur-containing compounds [32–37]. Therefore, all *Ferula* species belonging to section *Merwia* in Türkiye have been classified as *F. szowitsiana*. However, preliminary phytochemical studies of *F. szowitsiana* collected from three locations in central Anatolia, i.e., the Kırşehir, Konya, and Sivas provinces, indicated that their phytochemical profiles are completely different. The major sesquiterpene coumarins isolated from the roots of *F. szowitsiana* were galbanic acid (**a**), methyl galbanate (**b**), and szowitsiacoumarin B (**c**) (Figure 8), rearranged monocyclic sesquiterpene coumarin ethers and bicyclic drimane type sesquiterpene coumarin ethers [38,39]. In contrast, the major sesquiterpene coumarins of the root extract of *Ferula turcica* collected from the shores of Tuz Lake were identified as kellerin (**d**), gummosin (**e**), persicasulphide A (**f**), and persicasulphide C (**g**), bicyclic drimane type sesquiterpene coumarin ethers and sulfur-containing compounds (Figure 8). Whereas colladonin (**h**), badrakemin (**i**), badrakemin acetate (**j**), and conferol (**k**), bicyclic drimane type sesquiterpene coumarin ethers, were found as the major compounds of *Ferula latialata* (Figure 8). Umbelliprenin (**l**), the biogenetic precursor of sesquiterpene coumarins of *Ferula* species, is also present in all three *Ferula* species (Figure 8). The HPLC chromatogram profiles (Figure S1) of the dichloromethane extracts of *F. latialata* (Kırşehir), *F. turcica* (Tuz Lake), and *F. szowitsiana* (Sivas) clearly showed that *F. latialata* and *F. turcica* should be different species individually.

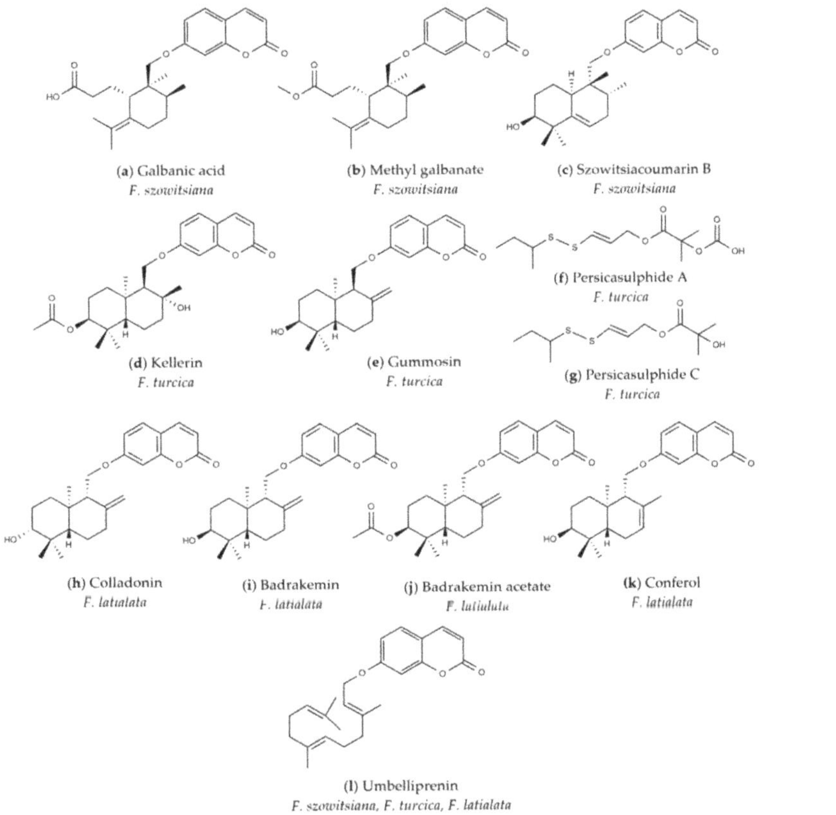


**Figure 8.** Secondary metabolites of *Ferula latialata*, *F. turcica*, and *F. szowitsiana*.

## 3. Discussion

*Ferula szowitsiana* is a species that is widely spread geographically, starting from Afghanistan, Turkmenistan, and Uzbekistan to Iran, Transcaucasus, and Türkiye. *F. persica*, which is quite close to *F. szowitsiana*, also spread in similar regions but not as widely as *F. szowitsiana*. The two new species, *F. turcica* and *F. latialata*, are morphologically close to *F. szowitsiana* and *F. persica*, which are members of subgenus *Narthex* section *Merwia* in Irano-Turanian groups. It was stated that the most prominent feature of subgenus *Narthex* is that the differentiation among the species of this subgenus is not clear [26]. When the two new species were compared to *F. szowitsiana* and *F. persica*, both new species were taller, had a greater number of rays, and had pedunculated central umbella. Compared to *F. latialata* and *F. turcica*, as the epithet name suggests, *F. latialata* has a wide lateral fruit wing and larger schizocarp fruit. In contrast, *F. turcica* has more dorsal and commissural vittae. Both new species grow in halophilous areas. It is thought that this habitat change may be one of the reasons for the differentiation from *F. szowitsiana*. The two new species, taxonomically supported by evident individual morphological, chemical, and molecular data, are considered not to be hybrids or variations of *F. szowitsiana* or *F. persica*.

The type specimen of *Ferula drudeana* was collected by Siehe (Siehe 408) [17]. Then, it was rediscovered by Sağıroğlu and Duman at the same location as the type specimen [40]. Sağıroğlu collected the specimens (Sağıroğlu 2525, GenBank barcode no: KU165687) from Kayseri. Another *F. drudeana* accession in GenBank (barcode no: KJ660775) was collected by Siehe (Siehe 163, in W). This population is nested in Section *Peucedanoides* (Figure 7, indicated by bold letters). *F. turcica* clearly differs from *F. drudeana* by puberulent petals (not glabrous), 4–7 vallecular vittae (not 2–3), and 1.2–1.9-mm wide lateral wings (not less than 0.5 mm). The other new species, *F. latialata*, has 2.5–3.9-mm wide lateral wings (not less than 0.5 mm) and setulose-puberulent petals (not glabrous). These distinct and significant morphological differences, supported by flower and fruit morphological characteristics, display that *F. drudeana* is different from both of the new species. Actually, *F. drudeana* is a remarkable species in the genus *Ferula* with its monocarpic form, thick stem, and glabrous petals. However, solving the polyphyly problem of *F. drudeana* needs more sampling and more morphological and phylogenetic studies. In clade D, the other species, *F. szowitsiana* (GenBank barcode no: FN432917), was collected from Tuzgölü, Yavşan salt pan. It is the type locality of *F. turcica*, as well. Additionally, more specimens from the Yavşan salt pan were collected by Davis (D.16670) and determined as *F. szowitsiana* [17]. On the other hand, *F. szowitsiana* was determined as a polyphyletic species (Figure 7, clade B), not only herein but also by Piwczyński et al. [27] and Panahi et al. [26]. It is clear that the Tuz gölü Yavşan populations identified as *F. szowitsiana* are genetically and morphologically different from the other collections of *F. szowitsiana* (Figure 7, clade B, Table 1). According to both our morphological and phylogenetic results, the new species could not be recognized, and therefore, these populations were misidentified as *F. szowitsiana*. As a result, the FN432917 specimen should be *F. turcica*, as well.

One of the major secondary metabolites from the root extracts of *Ferula turcica* species collected from the shores of Tuz Lake was previously isolated as the major secondary metabolite of *F. persica* [41,42]. However, the botanical features and phytochemical profile of *F. turcica* do not match those of *F. persica*. Moreover, the secondary metabolite profiles of *F. latialata* and *F. turcica* were not similar to those of *F. szowitsiana* and *F. persica*.

According to the data obtained, it is hypothesized that *Ferula persica* is at one end of the distribution of *F. szowitsiana*, and the two new species, *F. latialata* and *F. turcica*, are at the other end as a result of speciation. Aside from the observation of the morphological differences during the fieldwork, the chemical and molecular evidence supported the hypothesis that *F. turcica* and *F. latialata* are two new distinct species from *F. szowitsiana* and *F. persica*.

Identification key to *Ferula turcica*, *F. latialata*, and related species:

1. Leaf ultimate segment, linear-setaceous, dorsal vittae 2–3 per vallecula, petals glabrous ........................................................................................................................................*drudeana*

1. Leaf ultimate segment, not linear-setaceous, dorsal vittae 3–7 per vallecula, petals setulose-puberulent..........2
2. Stem 50–70 cm, rays 7–11 per umbel..........*szowitsiana*
2. Stem 70–110 cm, rays 11–22 per umbel..........3
3. Central umbella subsessile, mericarps ovoid, commissural vittae 16–18..........*persica*
3. Central umbella peduncled, mericarps elliptic or elliptic to oblong, commissural vittae 6–14..........4
4. Leaves densely puberulent, 15–18 rays, mericarps elliptic, lateral wings 1–2 mm..........*turcica*
4. Leaves puberulent, 13–15 rays, mericarps elliptic to oblong, lateral wings 2.5–4 mm..........*latialata*

## 4. Materials and Methods

The study was based on fieldwork, literature surveys, herbaria materials, and chemical and molecular studies. The new species materials were compared to the herbaria materials (Appendix A) of *Ferula* in Herbarium of Istanbul University Faculty of Pharmacy (ISTE), Royal Botanic Garden Edinburgh (E), Royal Botanic Gardens Kew (K), Moscow University Herbarium (MW), Natural History Museum Vienna (W), and Conservatoire et Jardin botaniques de la Ville de Genève (G). The voucher specimens were deposited in ISTE (*F. turcica* ISTE 116464, *F. latialata* ISTE 117495, *F. szowitsiana* ISTE116468).

Field studies were conducted by M. Miski and H.O. Tuncay in 2014, 2015, and 2021. *F. turcica* specimens were collected by M. Miski on 16 June 2015 from Konya. *F. latialata* specimens were collected from Kırşehir by M. Miski on 10 June 2014 and by H.O. Tuncay on 2 August 2021. Populations were observed during fieldwork, and a protection status was recommended according to IUCN threat criteria.

For fruit anatomy, the fruits were first submerged in a warm water–alcohol mixture (70% ethanol), and then all of the mericarps were cut by hand in the middle part with a razor. At least 40 mature fruits of *F. turcica and F. latialata* were analyzed. Samples were examined in Sartur reagent (a compound reagent of Sudan III, lactic acid, aniline, iodine, potassium iodide, water, and alcohol) [43]. Photographs were taken with an iPhone X. Measurements of the mericarps were made using ImageJ. The fruit morphology and anatomy were described using the terms of Botanical Latin [44] and Kızılarslan and Akalın [45]. The principal coordinate analysis, supported by the Gower similarity index, based on 21 morphological, anatomical, and chemical characters, was calculated and plotted with PAleontological STatistics (PAST) version 4.11 [46].

DNA extraction, amplification, sequencing, and phylogenetic analysis:

The total genomic DNAs were isolated from dried leaves of *F. turcica and F. latialata* using a GeneAll Plant SV Mini Kit (Seoul, Republic of Korea) following the protocol of the manufacturer. The complete ITS region in each genomic DNA was PCR-amplified using primers ITS4 and ITS5 [47]. Sequencing was performed by Atlas Biotechnologies (Ankara, Türkiye). The relevant marker sequences of all of the other taxa used for phylogenetic analysis were obtained from GenBank. Their NCBI barcode numbers are displayed in Figure 7. The outgroups were chosen according to Downie et al. and Panahi et al. [25,48]. ITS sequences (a total of two sequences) were obtained from the type population belonging to each of the two new species.

Representative species of the main lineages of the genus *Ferula* subgenus *Narthex* were selected according to the method of Panahi et al. and Piwczyński et al. [26,27] to determine the phylogenetic position of both new species. The aligned data matrix comprising 73 species × 618 alleles, was prepared using the MAFFT v. 7 online multi-alignment program [49] and analyzed with MrBayes v. 3.2.7a [50] with 20 million generations and a burn-in of 10% for the Bayes inference. The evolutionary substitution model was determined as TIM2+G using JModeltest 2.1.7 [51,52]. The ESS was checked using Tracer v. 1.7 [53]. The same dataset was analyzed using raxmlGUI v. 4.0b08 [54,55], with 500 runs and 1000 bootstrap replicates for the maximum likelihood approach. Additionally, the maximum parsimony tree was calculated using MEGA11 based on the same dataset with 1000 bootstrap replications [56]. The phylogenetic trees were displayed and manipulated by FigTree v. 1.4.3 and TreeGraph 2 [57]. The posterior probability and bootstrap values (>50) were added to the branches of the Bayes tree in Figure 7.

Chemical Extraction:

Plant materials were extracted by maceration and continuous Soxhlet procedures, respectively. Dichloromethane (Merck, Darmstadt, Germany) was used during the extraction, and the solvent was evaporated by a rotary evaporator with low pressure and temperature. Once the dichloromethane extract was dissolved with acetone (Merck, Darmstadt, Germany) and allowed to rest at room temperature, it was kept at 4 °C until sedimentation of the hydrocarbons occurred. The sediment of the hydrocarbons was filtered by a Nuche Erlenmayer flask under a vacuum. The extract was then dried again using a rotary evaporator (Buchi, Flawil, Switzerland) with low temperature and pressure [58,59].

Analytical High-Performance Liquid Chromatography (HPLC):

Analytical HPLC analyses were performed using a Shimadzu 10A model apparatus (Shimadzu Analytical and Measuring Instruments, Kyoto, Japan). The system comprised a pump (LC-10AD), a diode-array detector (DAD) (SPD-M10A), and an autosampler (SIL-10AD). During the process, Shimadzu LC Solutions software was used to control the system and conduct post-run analyses of the data. Millipore (Billerica, MA, USA) was used to obtain Milli-Q ultrapure water. Water (ultrapure):acetonitrile (60:40 >> 0:100) solvents were used as the mobile phases with gradient elution, and a Luna C18 (Phenomenex, CA, USA) column was used as a stationary phase. The flow rate was 0.5 mL/min, the temperature was 30 °C, and the injection volume of the samples was 10 μL. The time of the analysis was arranged for 40 min, and the mobile phase was planned as follows: 0–5 min, 40% A and 60% B; 5–30 min, 0% A, 100% B; 30–40 min, 0% A, 100% B; 40–41 min, 40% A, 60% B. A wavelength of 259 nm was used during the analyses. The method was revised according to the extract to obtain quite a separation [60].

**Supplementary Materials:** The following supporting information can be downloaded at: https://www.mdpi.com/article/10.3390/horticulturae9020144/s1, Table S1: morphological, anatomical, and chemical characters used for principal coordinate analysis; Figure S1: comparison of (**A**) *Ferula latialata*, (**B**) *F. turcica*, and (**C**) *F. szowitsiana* HPLC chromatograms.

**Author Contributions:** Conceptualization, H.O.T., E.A. and M.M.; methodology, H.O.T., A.D.-K. and F.M.E.; software, H.O.T., A.D.-K. and F.M.E.; validation, H.O.T., E.A., A.D.-K., F.M.E. and M.M.; formal analysis, H.O.T., E.A., A.D.-K., F.M.E. and M.M.; investigation, H.O.T., E.A., A.D.-K., F.M.E. and M.M.; resources, H.O.T., E.A., A.D.-K., F.M.E. and M.M.; data curation, H.O.T., E.A., A.D.-K., F.M.E. and M.M.; writing—original draft preparation, H.O.T., A.D.-K. and F.M.E.; writing—review and editing, H.O.T., E.A., A.D.-K., F.M.E. and M.M.; visualization, H.O.T., A.D.-K. and F.M.E.; supervision, E.A., M.M. and A.D.-K.; project administration, H.O.T., E.A., A.D.-K., F.M.E. and M.M.; funding acquisition, H.O.T., E.A., A.D.-K., F.M.E. and M.M. All authors have read and agreed to the published version of the manuscript.

**Funding:** This study was funded by the Scientific Research Projects Coordination Unit of Istanbul University. Project number: 30731.

**Data Availability Statement:** Not applicable.

**Conflicts of Interest:** The authors declare no conflict of interest.

## Appendix A

Additional herbarium specimens examined morphologically:

*Ferula szowitsiana*: ISTE 109444, ISTE 109420, ISTE 73511, ISTE 19932, ISTE 15377, ISTE 105388, ISTE 12865, ISTE 21102, ISTE 62970; E00428291, E00428292, E00392437, E00175642, E00175641, E00175640, E00433774, E00433776, E00467554, E00467555, E00467556, E00467557, E00467558, E00467559, E00467560, E00467560, E00467562, E00467563, E00467564, E00467565, E00467566,

E00467567, E00467568, E00467569, E00467570, E00467571, E00467572, E00467573, E00467574, E00467575, E00279094; K001097219, K001097218, K000568185; MW0744692, MW0744691.

*Ferula persica*: E00205681, E00467654, E00467655, E00467656, E00360731; K001097211; MW0744674, MW0744675, MW0744676, MW0744677, MW0744678, MW0744679, MW0700492, MW0700493, MW0754156.

*Ferula persica* var. latisecta: W1961-0001614.

## References


1. Pimenov, M.G.; Leonov, M.V.e. *The Genera of the Umbelliferae: A Nomenclator*, 2nd ed.; Royal Botanic Gardens, Kew: London, UK, 1993.
2. Hickey, M.; Clive, K. *Common Families of Flowering Plants*, 3rd ed.; Cambridge University Press: London, UK, 1997.
3. Plants of the World Online (POWO), *Ferula* L. Available online: https://powo.science.kew.org/taxon/30105171-2#publications (accessed on 30 August 2022).
4. Mohammadhosseini, M.; Venditti, A.; Sarker, S.D.; Nahar, L.; Akbarzadeh, A. The genus *Ferula*: Ethnobotany, phytochemistry and bioactivities—A review. *Ind. Crops Prod.* **2019**, *129*, 350–394. [CrossRef]
5. Iranshahy, M.; Iranshahi, M. Traditional uses, phytochemistry and pharmacology of asafoetida (*Ferula assa-foetida* oleo-gum-resin)—A review. *J. Ethnopharmacol.* **2011**, *134*, 1–10. [CrossRef] [PubMed]
6. Gunther, R.T. *The Greek Herbal of Dioscorides*, 3rd ed.; Hafner Publishing Company: London, UK, 1968.
7. Eisenman, S.W.; Zaurov, D.E.; Struwe, L. *Medicinal Plants of Central Asia: Uzbekistan and Kyrgyzstan*; Springer: New York, NY, USA, 2013.
8. Shahrajabian, M.H.; Sun, W.; Soleymani, A.; Khoshkaram, M.; Cheng, Q. Asafoetida, God's Food, a Natural Medicine. *Pharmacogn. Commn.* **2021**, *11*, 36–39. [CrossRef]
9. Kerscher, M.; Buntrock, H. Cosmetics and cosmeceuticals. In *Cosmetic Medicine and Surgery*, 1st ed.; CRC Press: Boca Raton, FL, USA, 2017; pp. 91–102.
10. Özek, G.; Özek, T.; Işcan, G.; Başer, K.H.C.; Duran, A.; Hamzaoglu, E. Composition and Antimicrobial Activity of the Oils of *Ferula szowitsiana* DC. from Turkey. *J. Essent. Oil Res.* **2008**, *20*, 186–190. [CrossRef]
11. Saghravanian, S.J.; Fereidoni, M.; Asadollahi, A. Effect of hydroalcoholic extract of *Ferula szowitsiana* DC. on paw edema in rat. *J. Kashan Univ. Med. Sci.* **2016**, *20*, 125–132.
12. Afifi, F.U.; Abu-Irmaileh, B. Herbal medicine in Jordan with special emphasis on less commonly used medicinal herbs. *J. Ethnopharmacol.* **2000**, *72*, 101–110. [CrossRef]
13. Boissier, E. *Flora Orientalis: Sive, Enumeratio Plantarum in Oriente a Graecia et Aegypto ad Indiae Fines Hucusque Observatarum*; H. Georg: Basileae, Geneva, 1872; Volume 2.
14. Korovin, E.P. *Generis Ferula (Tourn.) L. Monographia Illustrata*; Academiae Scientiarum UzRSS: Tashkent, Uzbekistan, 1947.
15. Safina, L.; Pimenov, M. Carpology of the species of type subgenus of the genus *Ferula* and some problems of their systematics. *Feddes Repert.* **1990**, *101*, 135–151. [CrossRef]
16. Pimenov, M.; Terekhin, A.; Devyatkova, G.; Baranova, Y.V. Klassifikatsiya vidov roda *Ferula* L. Umbelliferae) s pomoschchiyu yerakhicheskogo klaster-analiza. *Vopr Kibern.* **1978**, *47*, 98–113.
17. Peşmen, H. *Ferula* L. In *Flora of Turkey and the East Aegean Islands*; Davis, P.H., Ed.; Edinburgh University Press: Edinburgh, Scotland, 1972; Volume 4, pp. 440–453.
18. Duman, H.; Sağıroğlu, M. A new species of *Ferula* (Apiaceae) from south Anatolia, Turkey. *Botl. J. Linn. Soc.* **2005**, *147*, 357–361. [CrossRef]
19. Sağıroğlu, M.; Duman, H. *Ferula* parva Freyn & Bornm.(Apiaceae): A contribution to an enigmatic species from Turkey. *Turk. J. Bot.* **2006**, *30*, 399–404.
20. Sağıroğlu, M.; Duman, H. *Ferula mervynii* (Apiaceae), a distinct new species from north-east Anatolia, Turkey. *Bot. J. Linn. Soc.* **2007**, *153*, 357–362. [CrossRef]
21. Sağıroğlu, M.; Duman, H. *Ferula brevipedicellata* and *F. duranii* (Apiaceae), two new species from Anatolia, Turkey. *Ann. Bot. Fenn.* **2010**, *47*, 293–300. [CrossRef]
22. Pimenov, M.G.; Kljuykov, E.V. *Ferula divaricata* (Umbelliferae), a new species from Central Anatolia, Turkey. *Phytotaxa* **2013**, *99*, 35–39. [CrossRef]
23. Akalın, E.; Tuncay, H.O.; Olcay, B.; Miski, M. A New *Ferula* (Apiaceae) Species from Southwest Anatolia: *Ferula pisidica* Akalın & Miski. *Plants* **2020**, *9*, 740.
24. Kurzyna-Młynik, R.; Oskolski, A.A.; Downie, S.R.; Kopacz, R.; Wojewódzka, A.; Spalik, K. Phylogenetic position of the genus *Ferula* (Apiaceae) and its placement in tribe Scandiceae as inferred from nrDNA ITS sequence variation. *Plant Syst. Evol.* **2008**, *274*, 47–66. [CrossRef]
25. Panahi, M.; Banasiak, Ł.; Piwczyński, M.; Puchałka, R.; Oskolski, A.A.; Spalik, K. Phylogenetic relationships among *Dorema*, *Ferula* and *Leutea* (Apiaceae: Scandiceae: Ferulinae) inferred from nrDNA ITS and cpDNA noncoding sequences. *Taxon* **2015**, *64*, 770–783. [CrossRef]

26. Panahi, M.; Banasiak, Ł.; Piwczyński, M.; Puchałka, R.; Kanani, M.R.; Oskolski, A.A.; Modnicki, D.; Miłobędzka, A.; Spalik, K. Taxonomy of the traditional medicinal plant genus *Ferula* (Apiaceae) is confounded by incongruence between nuclear rDNA and plastid DNA. *Bot. J. Linn. Soc.* **2018**, *188*, 173–189. [CrossRef]
27. Piwczyński, M.; Wyborska, D.; Gołębiewska, J.; Puchałka, R. Phylogenetic positions of seven poorly known species of *Ferula* (Apiaceae) with remarks on the phylogenetic utility of the plastid trnH-psbA, trnS-trnG, and atpB-rbcL intergenic spacers. *System. Biodivers.* **2018**, *16*, 428–440. [CrossRef]
28. Elibol, Z.; Menemen, Y.; Sağiroğlu, M.; Duman, H. A molecular phylogenetic study on some Turkish *Ferula* L.(Apiaceae) species using nrDNA ITS sequences. *Pak. J. Bot.* **2012**, *44*, 589–594.
29. Korovin, E.P. *Ferula* L. In *Flora of the U.S.S.R.*; Shishkin, B.K., Ed.; Akademii Nauk SSSR: Moscow, Russia, 1951; Volume XVII, pp. 44–101.
30. Chamberlain, D.F.; Rechinger, K.H. *Ferula* L. In *Flora Iranica*; Rechinger, K.H., Ed.; Akademische Druck und Verlagsanstalt: Graz, Austria, 1987; Volume 162, pp. 387–426.
31. Commission, N.R.S.S.; Commission, I.S.S. *IUCN Red List Categories and Criteria*; IUCN: Grand, Switzerland, 2001.
32. Khasanov, T.K.; Saidkhodzhaev, A.I.; Nikonov, G.K. Structure and configuration of the coumarins mogoltadone and mogoltadin. *Chem. Nat. Compd.* **1974**, *10*, 20–23. [CrossRef]
33. Turabelidze, D.G.; Kemertelidze, É.P. Farnesiferol C from the roots of *Ferula szovitsiana*. *Chem. Nat. Compd.* **1976**, *12*, 589. [CrossRef]
34. Nabiev, A.A.; Malikov, V.M. Microlobin—A new coumarin from *Ferula microloba*. *Chem. Nat. Compd.* **1983**, *19*, 664–667. [CrossRef]
35. Sagitdinova, G.V.; Saidkhodzhaev, A.I.; Malikov, V.M. Structure and stereochemistry of the coumarins of *Ferula lehmannii*. *Chem. Nat. Compd.* **1983**, *19*, 672–675. [CrossRef]
36. Babekov, A.U.; Saidkhodzhaev, A.I.; Keneshov, B.M. Terpenoids of *Ferula litwinowiana*. *Chem. Nat. Compd.* **1999**, *35*, 365. [CrossRef]
37. Kasaian, J.; Asili, J.; Iranshahi, M. Sulphur-containing compounds in the essential oil of *Ferula alliacea* roots and their mass spectral fragmentation patterns. *Pharm. Biol.* **2016**, *54*, 2264–2268. [CrossRef]
38. Bagirov, V.Y.; Gasanova, R.Y.; Burma, O.I.; Ban'kovskii, A.I. Coumarins of *Ferula szovitsiana* and *F. persica*. *Chem. Nat. Compd.* **1977**, *13*, 240–241. [CrossRef]
39. Iranshahi, M.; Arfa, P.; Ramezani, M.; Jaafari, M.R.; Sadeghian, H.; Bassarello, C.; Piacente, S.; Pizza, C. Sesquiterpene coumarins from *Ferula szowitsiana* and in vitro antileishmanial activity of 7-prenyloxycoumarins against promastigotes. *Phytochemistry* **2007**, *68*, 554–561. [CrossRef]
40. Sağıroğlu, M.; Duman, H. Rediscovery of *Ferula anatolica* and *Ferula drudeana* Apiaceae from Turkey. *Biodivers Conserv.* **2011**, *4*, 191–197.
41. Iranshahi, M.; Amin, G.-R.; Amini, M.; Shafiee, A. Sulfur containing derivatives from *Ferula persica* var. *latisecta*. *Phytochemistry* **2003**, *63*, 965–966. [CrossRef]
42. Iranshahi, M.; Amin, G.; Shafiee, A. A New Coumarin from *Ferula persica*. *Pharm. Biol.* **2008**, *42*, 440–442. [CrossRef]
43. Çelebioğlu, S.; Baytop, T. Bitkisel tozların tetkiki için yeni bir reaktif. *Farmakolog* **1949**, *19*, 301.
44. Stearn, W.T. *Botanical Latin*, 4th ed.; Timber Press, Inc.: Portland, OR, USA, 2004.
45. Kızılarslan-Hançer, Ç.; Akalın, E. Apiaceae familyası meyve anatomisindeki "Vitta" terimi ve yerleşimleri. *Avrasya Terim Derg.* **2017**, *5*, 19–24.
46. Hammer, Ø.; Harper, D.A.; Ryan, P.D. PAST: Paleontological statistics software package for education and data analysis. *Palaeontol. Electron.* **2001**, *4*, 9.
47. White, T.J.; Bruns, T.; Lee, S.; Taylor, J. Amplification and Direct Sequencing of Fungal Ribosomal RNA Genes for Phylogenetics. In *PCR Protocols: A Guide to Methods and Applications*; Innis, M.A., Gelfand, D.H., Sninsky, J.J., White, T.J., Eds.; Academic Press: San Diego, CA, USA, 1990; pp. 315–322.
48. Downie, S.R.; Spalik, K.; Katz-Downie, D.S.; Reduron, J.-P. Major clades within Apiaceae subfamily Apioideae as inferred by phylogenetic analysis of nrDNA ITS sequences. *Plant Divers. Evol.* **2010**, *128*, 111. [CrossRef]
49. Katoh, K.; Rozewicki, J.; Yamada, K.D. MAFFT online service: Multiple sequence alignment, interactive sequence choice and visualization. *Brief Bioinform.* **2019**, *20*, 1160–1166. [CrossRef]
50. Ronquist, F.; Teslenko, M.; van der Mark, P.; Ayres, D.L.; Darling, A.; Höhna, S.; Larget, B.; Liu, L.; Suchard, M.A.; Huelsenbeck, J.P. MrBayes 3.2: Efficient Bayesian Phylogenetic Inference and Model Choice Across a Large Model Space. *Syst. Biol.* **2012**, *61*, 539–542. [CrossRef]
51. Guindon, S.; Gascuel, O. A simple, fast, and accurate algorithm to estimate large phylogenies by maximum likelihood. *Syst. Biol.* **2003**, *52*, 696–704. [CrossRef]
52. Darriba, D.; Taboada, G.L.; Doallo, R.; Posada, D. jModelTest 2: More models, new heuristics and parallel computing. *Nat. Methods* **2012**, *9*, 772. [CrossRef]
53. Rambaut, A.; Drummond, A.J.; Xie, D.; Baele, G.; Suchard, M.A. Posterior Summarization in Bayesian Phylogenetics Using Tracer 1.7. *Syst. Biol.* **2018**, *67*, 901–904. [CrossRef]
54. Stamatakis, A. RAxML version 8: A tool for phylogenetic analysis and post-analysis of large phylogenies. *Bioinformatics* **2014**, *30*, 1312–1313. [CrossRef]
55. Silvestro, D.; Michalak, I. raxmlGUI: A graphical front-end for RAxML. *Org. Divers. Evol.* **2012**, *12*, 335–337. [CrossRef]

56. Tamura, K.; Stecher, G.; Kumar, S. MEGA11: Molecular evolutionary genetics analysis version 11. *Mol. Bio. Evol.* **2021**, *38*, 3022–3027. [CrossRef] [PubMed]
57. Stöver, B.C.; Müller, K.F. TreeGraph 2: Combining and visualizing evidence from different phylogenetic analyses. *BMC Bioinform.* **2010**, *11*, 7. [CrossRef] [PubMed]
58. Tan, N.; Yazıcı-Tütüniş, S.; Bilgin, M.; Tan, E.; Miski, M. Antibacterial activities of pyrenylated coumarins from the roots of *Prangos hulusii*. *Molecules* **2017**, *22*, 1098. [CrossRef]
59. Tosun, F.; Beutler, J.A.; Ransom, T.T.; Miski, M. Anatolicin, a highly potent and selective cytotoxic sesquiterpene coumarin from the root extract of *Heptaptera anatolica*. *Molecules* **2019**, *24*, 1153. [CrossRef]
60. Amin, A.; Tuenter, E.; Cos, P.; Maes, L.; Exarchou, V.; Apers, S.; Pieters, L. Antiprotozoal and antiglycation activities of sesquiterpene coumarins from *Ferula narthex* exudate. *Molecules* **2016**, *21*, 1287. [CrossRef]

**Disclaimer/Publisher's Note:** The statements, opinions and data contained in all publications are solely those of the individual author(s) and contributor(s) and not of MDPI and/or the editor(s). MDPI and/or the editor(s) disclaim responsibility for any injury to people or property resulting from any ideas, methods, instructions or products referred to in the content.

*horticulturae*

MDPI

*Article*

# Seasonal Development of *Paeonia obovata* and *Paeonia oreogeton* and Their Contents of Biologically Active and Reserve Substances in the Forest-Steppe Zone of Western Siberia

**Olga V. Kalendar [1], Vera A. Kostikova [1,*], Tatiana A. Kukushkina [1], Andrey S. Erst [1], Alexander A. Kuznetsov [2], Maxim S. Kulikovskiy [3] and Olga Y. Vasilyeva [1]**

[1] Central Siberian Botanical Garden, Siberian Branch of Russian Academy of Sciences, Novosibirsk 630090, Russia
[2] Laboratory Herbarium (TK), Tomsk State University, Tomsk 634050, Russia
[3] K.A. Timiryazev Institute of Plant Physiology RAS, IPP RAS, Moscow 127276, Russia
* Correspondence: serebryakova-va@yandex.ru; Tel.: +7-(383)-339-9810

**Abstract:** *Paeonia obovata* and *Paeonia oreogeton* belong to the monotypic family Paeoniaceae. Both are popular as ornamental plants. *P. obovata* and *P. oreogeton* have been introduced into Novosibirsk Oblast (Western Siberia) from Primorye (Far East). The aim of the study was to assess their adaptability as well as the effect of seasonal developmental stages on the accumulation of secondary metabolites and reserve substances in the leaves and rhizomes under the conditions of Akademgorodok (Novosibirsk, Russia). According to long-term data (15 years), *P. obovata* and *P. oreogeton* complete the entire growth cycle here, including flowering and fruiting. Both species exhibited abundant flowering, but in the first 3 years, *P. oreogeton* did not bloom; yet under the microclimatic conditions specifically selected for this species (a more shaded area), it started to bloom and fruit yearly. A biochemical analysis (by spectrometric method) of *P. obovata* and *P. oreogeton* grown in Akademgorodok showed that the leaves accumulate higher concentrations of flavonols (*P. obovata*: 1.77%), tannins (*P. oreogeton*: 16.42%), ascorbic acid (*P. oreogeton*: 155.2 mg/100 g), and sugars (*P. obovata*: 20.85%) as compared to the roots. Peony rhizomes contain higher concentrations of protopectins (*P. oreogeton*: 13.03%), saponins (*P. obovata*: 21.06%), and starch (*P. obovata*: 30.20%) than the leaves do. These data can help to increase the levels of these natural compounds in these species. Further investigation into the dynamics of accumulation of biologically active substances in the organs of peonies will help to identify introduced plant species having high biochemical potential for the pharmaceutical industry.

**Keywords:** *Paeonia*; weather conditions; ascorbic acid; flavonol; tannin; catechin; pectin; protopectin; saponin; sugar; starch

check for updates

**Citation:** Kalendar, O.V.; Kostikova, V.A.; Kukushkina, T.A.; Erst, A.S.; Kuznetsov, A.A.; Kulikovskiy, M.S.; Vasilyeva, O.Y. Seasonal Development of *Paeonia obovata* and *Paeonia oreogeton* and Their Contents of Biologically Active and Reserve Substances in the Forest-Steppe Zone of Western Siberia. *Horticulturae* **2023**, *9*, 102. https://doi.org/10.3390/horticulturae9010102

Academic Editor: Wajid Zaman

Received: 5 December 2022
Revised: 4 January 2023
Accepted: 9 January 2023
Published: 12 January 2023

**Copyright:** © 2023 by the authors. Licensee MDPI, Basel, Switzerland. This article is an open access article distributed under the terms and conditions of the Creative Commons Attribution (CC BY) license (https://creativecommons.org/licenses/by/4.0/).

## 1. Introduction

The monotypic family Paeoniaceae F. Rudolphi includes only one genus *Paeonia* L., which is a relic of ancient mesophilic arcto-tertiary flora, which includes 33 species growing in Europe, in the Mediterranean, and in East and Southeast Asia [1,2]. Peonies are highly popular in floriculture worldwide; they are both beautifully flowering and decorative foliaceous plants owing to powerful shoots up to 1 m in height with dense foliage. Wild and cultivated peony species are also attractive as medicinal and food plants because they contain biologically active substances: e.g., carbohydrates, terpenoids, steroids, phenol-carboxylic acids, phenol glycosides, flavonoids, tannins, and vitamins [3–6]. Data on pharmacological properties of representatives of the genus *Paeonia* have revealed their antioxidant, anti-inflammatory, antitumor, antibacterial, antiviral, cardiovascular, and neuroprotective effects [7–10]. Peonies are used in conventional medicine in many countries. For example, in Chinese conventional medicine, dried root bark of *P. suffruticosa* Andrews is popular as a febricide; it is used for blood cooling and promoting blood circulation to eliminate stasis [11]. Roots of *P. lactiflora* Pall. are officially included in the Japanese Pharmacopoeia (JP17), Korean Pharmacopoeia X (KPX), European Pharmacopoeia (EP10.2), and British Pharmacopoeia (BP 2020). *P. anomala* L. is used in Russian conventional medicine [12]. Its alcohol tincture (*Tinctura paeoniae*) is used as a typical medicinal formulation and has a sedative effect; it is indicated for neurasthenia, insomnia, and dysautonomia. For preparation of a peony tincture and peony extract, pharmaceutical manufacturers use grass (shoots), rhizomes, and roots [13,14].

*Paeonia anomala* grows in steppe zones of the Caucasus and in Southern and Central Europe, China, and Siberia (Figure 1). This is a deciduous herbaceous plant with triternate leaves that are deeply incised and pinnately dissected into lanceolate and linear lobes. Flowers have different colors: purple, dark red, and reddish-pink. The ovary and fruits are covered with felted hairs, rarely almost naked. The fruits are ovoid or oval in shape. Mature fruits are straight or patulous [15,16]. Harvesting of *P. anomala* raw materials in nature causes great damage to coenopopulations because harvesters excessively dig up the largest rhizomes of mature generative plants. Many medicinal species of this genus have not been studied properly except for a few species (*P. anomala*, *P. emodi* Royle, *P. lactiflora*, *P. ostii* T. Hong & J.X. Zhang, and *P. suffruticosa*) widely used in folk medicine [4,10,17,18]. Phytochemical and pharmacological research on other *Paeonia* species may open up new medicinal resources.

**Figure 1.** Peonies: (**A**) *P. oreogeton*; (**B**) *P. obovata*; (**C**) *P. anomala*. Photo by Olga V. Kalendar.

According to the latest taxonomic studies [19,20], *P. oreogeton* S. Moore is a form of *P. obovata* Maxim.: a species native to the Far East, China, Korea, and Japan (Figure 1). This is a herbaceous plant having biternate or triternate leaves with smooth-edged wide lobes. Flowers of *P. obovata* have different colors: red, pink, purple, and pinkish-violet. *P. oreogeton* has yellow or cream-colored flowers and biternate leaves with smooth-edged wide lobes; it grows in the Far East of Russia and occurs in China, on the Korean Peninsula, and in Japan [21]. The most important advantages of these peonies are their durability in culture and high winter hardiness even in USDA (United States Department of Agriculture) zones 2 and 3. Accordingly, they are economically profitable crops that do not require constant or frequent renewal of flower beds or purchases of plants to restore plantings affected by wintering. In *P. obovata*, the roots are used most often for crude medicinal formulations in traditional Chinese medicine. It is used as a tincture for the treatment of dyspepsia and menopausal disorders. In addition, the roots of this plant are employed as a substitute for medicinal *P. lactiflora* in China for the treatment of chest pain, abdominal pain, eye redness, dysmenorrhea, amenorrhea, blood vomiting, carbuncles, and bruising and serve as antiplatelet, sedative, astringent, and antispasmodic therapeutics [13,22]. The profile of monoterpene compounds, of their glycosides (derivatives of paeoniflorin), and of tannins in the roots of *P. obovata* has been sufficiently studied. Seven proanthocyanidins have been found in the roots of this species [23,24]. Flowers contain anthocyanins: pelargonidin 3-glucoside, cyanidin 3,5- diglucoside, peonidin 3-glucoside, and peonidin 3,5-Di-*O*-$\beta$-D-glucopyranoside [25]. Seventy-seven percent of the total content of volatile compounds in *P. obovata* flowers is represented by two monoterpene alcohols. Two sesquiterpenes are also present in its flowers [26]. A relatively recent study by Bae et al. [27] showed that a methanolic extract of *P. obovata* has an antiulcerogenic activity. The other species, *P. oreogeton*, has been studied insufficiently.

In this regard, the aim of this study was to analyze seasonal development of *P. oreogeton* and *P. obovata* and their levels of biologically active and reserve substances under ex situ conditions and to determine the optimal time for harvesting of raw material in the form of leaves and rhizomes.

## 2. Materials and Methods

### 2.1. Plant Material

The study was conducted in the Central Siberian Botanical Garden, the Siberian Branch of the Russian Academy of Sciences (CSBG SB RAS, Novosibirsk, Russia); N 54.819308, E 83.102064. We transferred *P. oreogeton* and *P. obovata* (growing in the monsoon climate of the Russian Far East) to the CSBG SB RAS (Western Siberia) and analyzed *P. anomala* brought from natural Siberian habitats. *P. oreogeton* was brought from Primorye, Khasansky district (De Livron Island) N 42.694390, E 131.365044, and *P. obovata* was brought from Primorye, Khasansky district (environs of Slavyanka village); N 42.911206, E 131.338515. The plants were identified by an expert from the CSBG SB RAS (Novosibirsk, Russia). Voucher specimens were deposited in the Plant Material Storage Room in the Laboratory of Introduction of Ornamental Plants (CSBG SB RAS).

The plants were grown on experimental plots of the Laboratory of Introduction of Ornamental Plants (CSBG SB RAS); this territory has an irrigation system. Soils in these plots are gray forest soils with a bulk of density in the 0-20 cm layer: 0.8-1.18 g/cm$^3$. The content of humus in the 0-20 cm layer is 2-4%, and at a depth of 50-60 cm, no more than 0.8%. The total natural reserves of nutrients are low, and therefore organic and mineral fertilizers were applied. Winter sheltering of the plants was not carried out. Mycological analysis and identification of peony lesions caused by *Cronartium flaccidum* (Alb. et Schw.) Wint. were performed by Dr. I.G. Vorobieva. The identification of pathogens and measures to combat them are implemented by the Plant protection group in the CSBG SB RAS.

For the analysis of biologically active substances of *P. oreogeton* and *P. obovata*, the leaves and rhizomes were collected at the beginning of the growing season (May 2017) and rhizomes at its end (17 August 2017) from Collections of Living Plants Indoors and Outdoors (unique scientific unit No. 440534 of CSBG SB RAS) in a garden plot located at a forest-steppe site with gray forest soil and mean yearly temperature/precipitation of 1.8 °C/448 mm. For biochemical analysis, samples of the raw material from above-ground and underground organs were taken from mature generative plants: ontogenetic state g2. The raw material was dried and ground to obtain a representative sample for the analysis.

### 2.2. Seasonal Development Analysis

Rhythms of growth and development of peonies were studied by the phenological observation technique developed earlier [28,29]. For convenience, start dates of the main plant phenophases are presented in phenospectra. Qualitative traits of ontogenetic states were described according to a scale of ontogeny periodization [30,31]. *P. anomala* (a representative of local flora) served as a control for comparative analyses of biological characteristics.

### 2.3. Extract Preparation

Approximately 0.5 g (accurately weighed) of the raw material ground to a particle size of ~3 mm was placed into a 100 mL flask and exhaustively extracted with 70% ethanol. Extraction completeness was verified by means of a reaction with a 5% sodium hydroxide solution (until discoloration). After that, the volume of the filtered extract was measured.

Moisture was quantified via drying of a sample to constant weight at 105 °C in a thermostat. Biochemical parameters were calculated for absolutely dry weight of the raw material. All chemical analyses were performed on two biological replicates and three technical replicates [32].

### 2.4. Determination of Flavonol Contents

This procedure was performed by a spectrophotometric method based on the complexation reaction between flavonols and aluminum chloride [33]. An extract (0.1 mL) was placed into two 5 mL test tubes, 0.2 mL of a 2% ethanol solution of aluminum chloride was added into one test tube, 1–2 drops of 30% acetic acid were added into the other one, and the solution was brought to the nominal volume with 96% ethanol. The solutions were mixed, and after 40 min, optical density of the solution containing aluminum chloride was measured on an SF-56 spectrophotometer (Lomo, St. Petersburg, Russia) at a wavelength of 415 nm in a cuvette with a 1-cm light path, using a solution of acetic acid as a control. The amount of flavonols in each sample was determined by means of a calibration curve built based on rutine (Chemapol, Mumbai, MH, India).

### 2.5. Quantification of Catechins

The content of catechins was determined spectrophotometrically by the method based on the ability of catechins to produce a crimson color in a solution of vanillin in concentrated hydrochloric acid [34,35]. A 0.8 mL aliquot of an extract was placed into two test tubes. Next, 4 mL of a 1% solution of vanillin in concentrated hydrochloric acid was poured into one of them, and the volumes were adjusted to 5 mL in both tubes with concentrated hydrochloric acid. A tube without vanillin served as a control. In the presence of catechins, the sample became pink, raspberry, or orange-red. After 5 min, the intensity of colors was measured using the SF-56 spectrophotometer (Lomo, St. Petersburg, Russia) at 504 nm in a cuvette with a light path of 1 cm. The standard curve was constructed with (±)-catechin (Sigma, St. Louis, MO, USA).

### 2.6. Quantification of Tannins

This assay of tannins (hydrolyzable tannins) was performed by the method proposed by L.M. Fedoseeva [36]. An extract (10 mL) was placed into a 100 mL volumetric flask, and 10 mL of a 2% aqueous solution of ammonium molybdate was introduced. The content was brought to the nominal volume with purified water and incubated for 15 min. The intensity of the resulting color was measured using the SF-56 spectrophotometer (Lomo, St. Petersburg, Russia) at 420 nm in a cuvette having a 1-cm light path. A government standard sample of tannin (Sigma, St. Louis, MO, USA) served as a standard.

### 2.7. Quantification of Saponins

The concentration of saponins was determined by the gravimetric method. Approximately 2 g of the air-dry material was extracted with chloroform in a Soxhlet apparatus until complete discoloration to remove lipids and resins, which hamper saponin analysis. The samples were dried and extracted in a water bath at 70 °C for 30 min successively with 50%, 60%, and 96% ethanol. The combined extract was evaporated to 5 mL, and sevenfold volume of acetone was added. After 18 h, the formed precipitate was filtered off, dried at 70 °C, and weighed, and the percentage of saponins was calculated [37]. Qualitative assays (reactions) for detecting saponins were as follows: 1) foaming equal in volume and stability is achieved after shaking of the extract with an acidic or alkaline solution; 2) acetone added to the extract induces the formation of a white flocculent precipitate, which indicates the presence of triterpene saponins in samples.

### 2.8. Quantification of Ascorbic Acid

This procedure was performed by titration based on its reducing properties [32]. A raw-material sample (2–5 g), thoroughly ground up in a mortar to a homogeneous mass, was extracted with 20 mL of a 1% hydrochloric acid solution. After that, the extract was poured into a 100 mL volumetric flask and brought to the nominal volume with a 1% oxalic acid solution, which improves the stability of ascorbic acid in the extract. The extract was incubated for 5 min, then filtered and titrated until staining clear pink with the Tillmans reagent. For preparation of a 0.001N dye solution, 60 mg of 2,6-dichlorophenolindophenol was dissolved in 200 mL of warm distilled water; 4–5 drops of 0.01N sodium hydroxide were added; after 10 min of vigorous shaking, the solution was passed through a dense filter into a dry flask. A mixture of 1% hydrochloric acid and 1% oxalic acid at a ratio of 1:5 was used as a control.

### 2.9. *Quantification of Pectins and Protopectins*

Pectin substances (protopectins and pectins) were quantitated by the carbazole-free method based on specific yellowish-orange staining of uronic acids in the presence of thymol in a sulfuric acid solution [32,38]. A 0.5–1.0 g ground-up sample of the air-dry material was extracted thrice with hot 80–82% ethanol (40, 30, and 25 mL) in a boiling water bath under reflux for 20–30 min (to extract free carbohydrates, which hamper pectin analysis) and passed through a paper filter into a flask. The filtered sample was dried at 50 °C until the smell of alcohol disappeared.

#### 2.9.1. Extraction of Water-Soluble Pectin (Extract I)

Fifty milliliters of distilled water heated to 45 °C was added to the dried residue of the raw material; extraction in the water bath was performed at 45 °C for 1 h. The liquid was filtered into a 100 mL volumetric flask; after cooling, the volume was brought to the mark with water.

#### 2.9.2. Extraction of Protopectin (Extract II)

Fifty milliliters of 0.3N hydrochloric acid was poured into an extraction flask containing the residual raw material; the solution was heated for 30 min in a boiling water bath under reflux. Then, the solution was filtered into a 200 mL volumetric flask, and the extract was washed with 50 mL of hot water. The filter and the precipitate were placed into the same extraction flask filled with 50 mL of a 1% ammonium citrate solution, which was then kept in the boiling water bath for 30 min. After that, the solution was filtered into a flask containing the filtrate of the hydrochloric acid extract and washed with hot water. After cooling, the volume was brought to the mark with water.

#### 2.9.3. The Reaction with Thymol

Concentrated sulfuric acid cooled to 4 °C was added dropwise with cooling to 0.5 mL of cooled extracts I and II and thoroughly shaken; the test tubes were heated for 6 min in the boiling water bath and cooled; 0.1 mL of a 0.2% alcohol solution of thymol was added, and the solution was thoroughly mixed. After the reaction with thymol, optical density of the colored solutions was measured using the Agilent 8453 spectrophotometer at 480 nm in a cuvette with a 1-cm light path. The control was 96% ethanol. Pectins and protopectins were quantified with the help of a calibration curve plotted using galacturonic acid standards (Merck, Rahway, NJ, USA).

### 2.10. *Quantification of Carotenoids*

The concentration of carotenoids was determined in an acetone–ethanol extract [32,39]. A 0.1 g air-dried sample was ground in a mortar to a homogeneous mass during successive addition of 0.1 g of calcium carbonate (to neutralize organic acids because carotenoids are unstable in an acidic environment), 1 mL of dimethylformamide (for pigment stability), and 2 g of anhydrous sodium sulfate. Carotenoids were extracted first with acetone (20 mL: one time, and 5 mL: two times), and then the extraction was continued with 96% ethanol (5 mL: three times) to extract lycopene. Next, exhaustive extraction was continued with acetone until discoloration. The volume of the combined extract was measured. The carotenoid content was measured at 662 nm (for chlorophyll a), 644 nm (for chlorophyll b), and 440.5 nm (for carotenoids) using the SF-56 spectrophotometer (Lomo, St. Petersburg, Russia) in a cuvette with a 1-cm light path. The control was 96% ethanol. The concentration of carotenoids ($mg/dm^3$) was computed using the formula:

$$C_{car} = 4.695 \times D440.5 - 0.268 \times (5.134 \times D662 - 20.436 \times D644),$$

where D is optical density of the extract, and $C_{car}$ is the concentration of carotenoids, mg/dm$^3$.

The concentration of carotenoids (mg/100 g) was found by means of the formula:

$$X = C_{car} \times V \times V_2 \times 100/(M \times V_1 \times 1000),$$

where $C_{car}$ is the concentration of carotenoids, mg/dm$^3$; V is the volume of the initial extract, mL; $V_1$ is the volume of the initial extract used for dilution, mL; $V_2$ is the volume of the diluted extract, mL; and M is absolute dry weight of the raw material, g.

### *2.11. Quantification of Sugar and Starch*

For sugar quantification, the method proposed by A.S. Shvetsova and E.Kh. Lukyanenko was employed. It involves the reduction of potassium ferricyanide with reducing sugars to ferrocyanide in an alkaline medium. The latter causes stable blue staining with iron sulfate in the presence of gelatin [32]. An extract (0.1 mL) and an alkaline solution of potassium ferricyanide (2 mL; 1.65 g of ferricyanide and 10 g of sodium carbonate dissolved in 1 L of water) and distilled water (2 mL) were poured into 50 mL measuring tubes up to the 10 mL mark. The tubes were shaken and heated in a boiling water bath for 15 min. After that, the tubes were cooled, and 4 mL of a ferrous sulfate solution was poured into each measuring tube. To prepare the assay solution, 1 g of ferrous sulfate was dissolved in 10 mL of concentrated sulfuric acid in a 1 L flask, and water was added to bring the volume to the mark. A working solution of ferrous sulfate was prepared on the day of use via mixing of a 10% gelatin solution with the ferrous sulfate solution at a ratio of 1:20. The solution was then stirred, and the volume in the test tube was adjusted to the mark of 50 mL with distilled water. After vigorous stirring, optical density of the solution was measured using the SF-56 spectrophotometer (Lomo, St. Petersburg, Russia) at 690 nm (glucose absorption maximum) in a cuvette with a 1-cm light path. The sugar concentration was calculated by means of a calibration curve plotted for glucose (Reachem, Russia). A blank sample was used as a control.

For starch quantification, the extract was hydrolyzed with a 1% hydrochloric acid solution for 6 h [40]. After neutralization of the extract, the content of sugars was determined by the above method, and the previously determined content of sugars was subtracted. Given that 1 g of glucose corresponds to 0.89996 g of starch, the resulting difference (in %) was multiplied by 0.89996 to determine the starch concentration in an analyzed sample.

## 3. Results

### *3.1. Seasonal Development of Paeonia in the Continental Climate*

*P. oreogeton* and *P. obovata* grow naturally in Primorye with its monsoon climate. In our experiments, *P. oreogeton*, *P. obovata*, and *P. anomala*, which are representative of local flora, were grown ex situ under the conditions of the continental climate (Western Siberia). According to long-term data (15 years), *P. obovata* and *P. oreogeton* complete the entire growth cycle here, including flowering and fruiting. Both species exhibited abundant flowering, but in the first 3 years, *P. oreogeton* did not bloom; yet under the microclimatic conditions specifically selected for this species (a more shaded area), it started to bloom and fruit yearly.

Table S1 presents a comparison of the main climatic characteristics of the two climate types. Figures 2–4 illustrate phenophases of *P. oreogeton*, *P. obovata*, and *P. anomala* (control) under the conditions of the forest-steppe zone of Western Siberia during three growing seasons. During the observation period from 2007 to 2018, the 2012 growing season in Akademgorodok (Novosibirsk) was abnormally dry (Table S2), and the 2013 season was abnormally humid (Table S3).

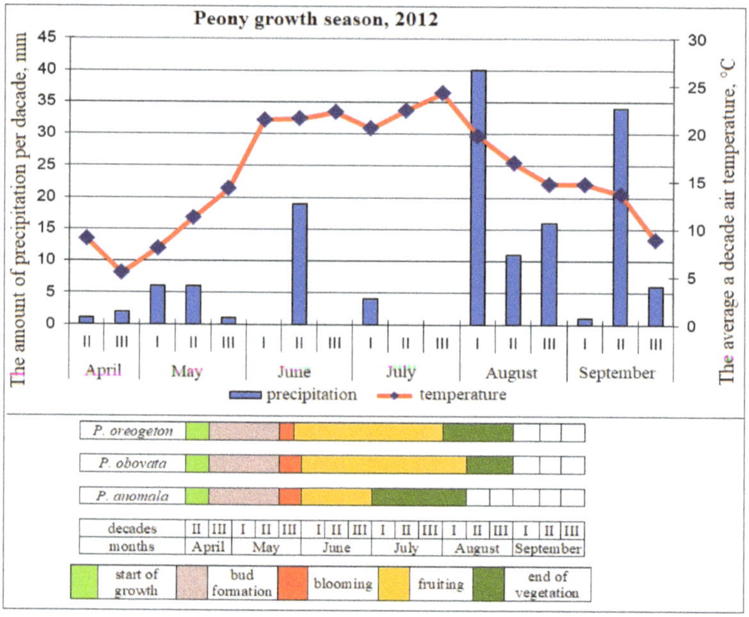


**Figure 2.** Hydrothermal conditions and peony phenophases in Western Siberia in 2012.

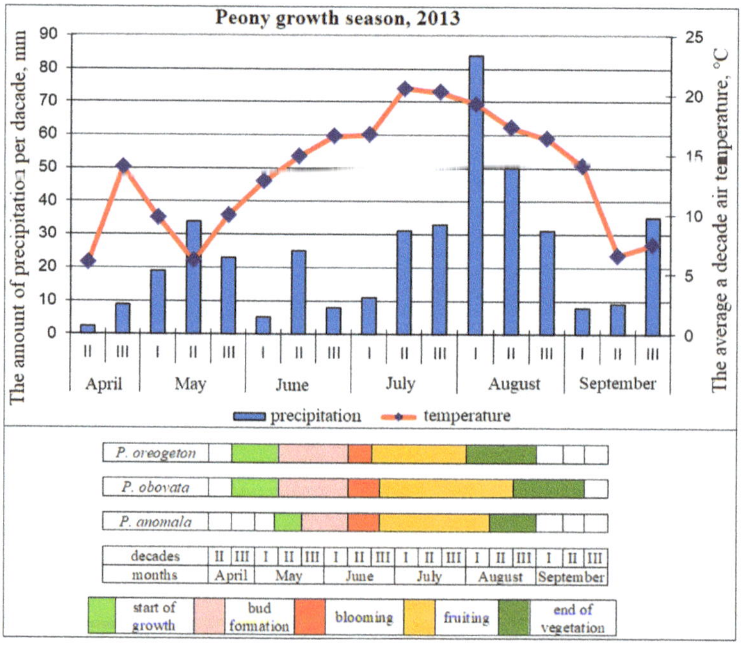


**Figure 3.** Hydrothermal conditions and peony phenophases in Western Siberia in 2013.

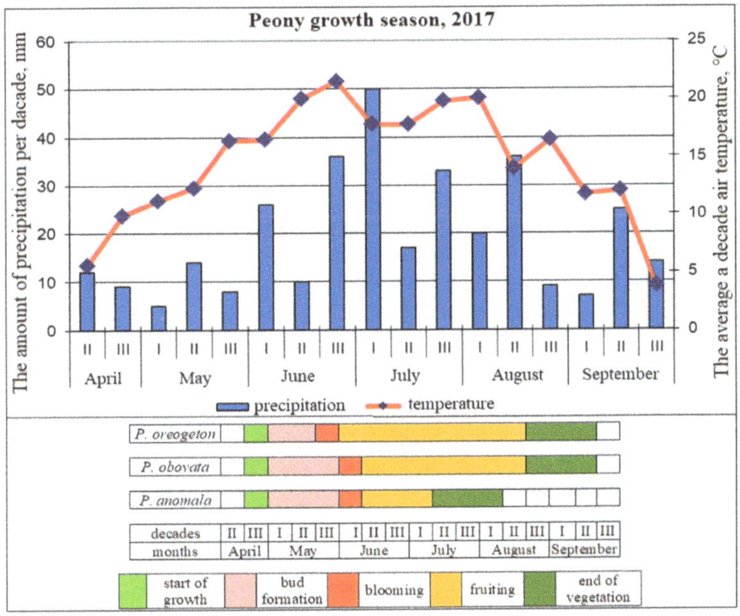


**Figure 4.** Hydrothermal conditions and peony phenophases in Western Siberia in 2017.

The examination of seasonal development and amplitude of adaptations showed that the above-ground shoots of peonies were damaged by high temperatures during moisture deficiency under abnormally dry weather conditions in 2012 (Figure 5A). Increased ambient humidity and low temperatures caused fungal diseases under abnormally humid weather conditions in 2013 (Figure 5B).

**Figure 5.** Peony shoots: (**A**) *P. anomala* shoots damaged by low humidity in Western Siberia in 2012; (**B**) *P. obovata* shoots damaged by rust in Western Siberia in 2013.

*3.2. Contents of Biologically Active and Reserve Substances in the Leaves and Roots of Paeonia*

The year 2017 was chosen as a model one to study the content of biologically active and reserve substances because this year did not show strong deviations of mean monthly hydrothermal characteristics. We measured the contents of phenolic compounds (flavonols, catechins, and tannins) and of ascorbic acid, pectin substances, saponins, and carotenoids in the leaves and rhizomes of two peony species—*P. obovata* and *P. oreogeton*—at the beginning of the growing season (Table 1). The content of flavonols that was found in the peony leaves at the beginning of the growing season was not very high. In the leaves of *P. obovata*, the content of flavonols reached 1.77%, which slightly exceeded that in the leaves of *P. oreogeton*. Flavonols were not found in rhizomes of the analyzed peonies. Carotenoids and flavonols were detectable only in peony leaves. Their level in the leaves of *P. obovata* (89.2 mg/100 g) was three-fold higher than that in *P. oreogeton* (29.1 mg/100 g). The level of tannins was sixfold higher in the leaves of *P. oreogeton* (16.42%) than in its rhizomes (2.56%). Concentrations of tannins were comparable between leaves and roots of *P. obovata*. The content of catechins was higher in the leaves of *P. oreogeton* (68.3 mg/100 g) and *P. obovata* (84.4 mg/100 g) as compared to the rhizomes. The content of ascorbic acid was fivefold higher in the peony leaves than in the peony rhizomes. The highest content of saponins was found in the leaves of *P. obovata* (21.06%), while their content in the rhizomes was lower. Levels of pectins in the peony leaves and rhizomes were not very high. Their content in *P. oreogeton* was higher in the leaves (1.89%), whereas in *P. obovata*, it was higher in the rhizomes (2.02%). The concentration of protopectins significantly exceeded that of pectins. In the underground organs of peonies, the level of protopectins was higher than that in the leaves. Protopectins mostly accumulated in the rhizomes of *P. oreogeton* (13.03%).

**Table 1.** The contents of biologically active and reserve substances in peonies at the beginning of the growing season in Western Siberia.

| Substances | *P. oreogeton* | | *P. obovata* | |
|---|---|---|---|---|
| | Rhizomes | Leaves | Rhizomes | Leaves |
| Moisture (% dry weight) | 66.86 | 80.39 | 78.17 | 74.63 |
| Flavonols (% dry weight) | no | $1.17 \pm 0.04^{b}$ | no | $1.77 \pm 0.05^{a}$ |
| Catechins (g (100 g dry weight)$^{-1}$) | $28.1 \pm 0.48^{d}$ | $68.3 \pm 1.17^{b}$ | $44.0 \pm 0.76^{c}$ | $84.4 \pm 1.45^{a}$ |
| Tannins (% dry weight) | $2.56 \pm 0.02^{d}$ | $16.42 \pm 0.14^{a}$ | $11.59 \pm 0.10^{c}$ | $12.81 \pm 0.11^{b}$ |
| Ascorbic acid (g (100 g dry weight)$^{-1}$) | $30.5 \pm 1.22^{b}$ | $155.2 \pm 6.21^{a}$ | $31.5 \pm 1.26^{b}$ | $151.8 \pm 6.07^{a}$ |
| Pectins (% dry weight) | $1.15 \pm 0.01^{b}$ | $1.89 \pm 0.06^{a}$ | $2.02 \pm 0.10^{a}$ | $0.86 \pm 0.01^{c}$ |
| Protopectins (% dry weight) | $13.03 \pm 0.26^{a}$ | $5.99 \pm 0.01^{c}$ | $7.02 \pm 0.09^{b}$ | $3.96 \pm 0.02^{d}$ |
| Saponins (% dry weight) | $13.37 \pm 0.51^{b}$ | $11.25 \pm 0.43^{c}$ | $21.06 \pm 0.80^{a}$ | $12.94 \pm 0.49^{b}$ |
| Carotenoids (g (100 g dry weight)$^{-1}$) | no | $29.1 \pm 0.29^{b}$ | no | $89.2 \pm 0.87^{a}$ |
| Sugar (% dry weight) | $5.97 \pm 0.21^{c}$ | $17.44 \pm 0.61^{b}$ | $17.22 \pm 0.60^{b}$ | $20.85 \pm 0.73^{a}$ |
| Starch (% dry weight) | $25.31 \pm 0.94^{a}$ | - | $10.51 \pm 0.39^{b}$ | - |

Note: mean values and standard deviation ($n$ = 3) are presented; no, not found; "-", not tested; means with different superscript letters (a, b, c, or d) in the same row are significantly different ($p \leq 0.05$) according to Tukey's honestly significant difference (HSD) test.

We determined contents of reserve substances (starch and sugars) in the leaves and rhizomes of Far Eastern peonies at the beginning of the growing season (Table 1). The content of sugars in *P. oreogeton* (17.44%) and *P. obovata* (20.85%) was higher in the leaves. The rhizomes of the Far Eastern peony *P. oreogeton* accumulated up to 25.31% of starch.

In addition, we studied the content of biologically active compounds and reserve substances (starch and sugars) in the rhizomes of Far Eastern peonies at the end of the growing season in Western Siberia (Figures 6 and 7). At the beginning of the growing season, protopectins accumulated mainly in the rhizomes of *P. oreogeton* (13.03%). The contents of protopectins in the rhizomes of *P. obovata* were similar when we compared the beginning (7.02%) and the end (7.62%) of the growing season (Figure 6). Contents of saponins (17.9%), catechins (44.3 mg/100 g), and sugars (7.36%) in the rhizomes of *P. oreogeton* were higher at the end of the growing season, whereas in *P. obovata* (21.06%,

44.0 mg/100 g, and 17.22%, respectively), this parameter was higher at the beginning of the growing season. At the end of the growing season, the content of saponins in *P. obovata* rhizomes dropped fivefold (to 4.49%). In *P. oreogeton* and *P. obovata*, ascorbic acid (45.6 and 50.9 mg/100 g, respectively) and starch (27.13% and 30.2%, respectively) accumulated by the end of the growing season (Figures 6 and 7). At the beginning of the growing season, the content of pectins was low and reached 1.15% in *P. oreogeton* and 2.02% in *P. obovata*; by the end of the growing season, this parameter decreased.

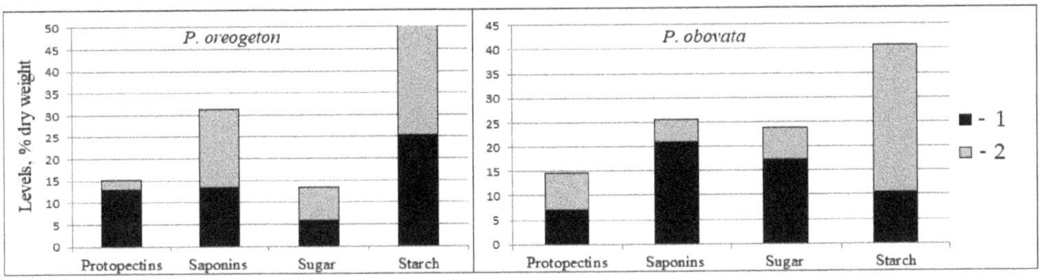


**Figure 6.** Levels of protopectins, saponins, sugars, and starch in peony rhizomes at the beginning (1) and end (2) of the growing season in Western Siberia.

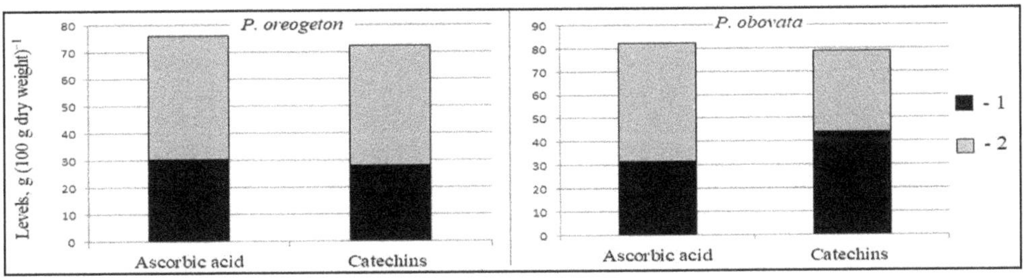


**Figure 7.** Levels of ascorbic acid and catechins in peony rhizomes at the beginning (1) and end (2) of the growing season in Western Siberia.

## 4. Discussion

The synthesis and accumulation of biologically active and reserve substances in medicinal plants are dynamic processes that change during plant ontogenesis and depend on numerous environmental factors [41–45]. Results of research by other authors [46,47] show that the concentration of biologically active substances (which in natural habitats perform an adaptive function) diminishes in some medicinal plants growing in areas strongly affected by agriculture. The accumulation of biologically active and reserve substances is influenced by hydrothermal conditions of growing seasons.

An analysis of phenophases by means of long-term phenological observation of representatives of the genus *Paeonia* grown in the forest-steppe zone of Western Siberia revealed that *P. obovata*, *P. oreogeton*, and *P. anomala* start to grow between the second decade of April and the first decade of May [48].

Under the abnormally dry weather conditions in 2012, since the first days of the second decade of April, all three species exhibited abundant growth due to early onset of the spring and a warm first decade of April: the mean 10-day temperature was 4.9 °C, and an excess of the mean monthly temperature was 4.1 °C (Table S2). Budding and flowering stages were registered at similar time points, but abundant flowering of *P. obovata* and *P. anomala* lasted longer as compared to *P. oreogeton*, which is more mesophilic (Figure 2).

It should be noted that in Primorye [49], the duration of flowering of *P. oreogeton* varies from 15 to 23 days, whereas ex situ flowering of this plant is slightly longer (by 5–6 days). At low humidity, in our study, the fruiting stage in all three species was found to be shifted by one decade, whereas in *P. anomala*, it terminated 8 days earlier. In 2012 and subsequent years, the time point of the end of the growing season for *P. anomala* was easily distinguished because in natural habitats, this plant prepares for the wintering period earlier as compared to *P. obovata* and *P. oreogeton*, whose growing seasons are longer on the Primorye territory.

A gradual temperature increase in early April in abnormally humid 2013 stimulated the mean annual growth of the Far Eastern species, but the decreased temperatures in the first and second decades of May shifted the budding and flowering stages by 10–13 days (Table S3, Figure 3). The duration of flowering lengthened by 2–6 days. Fruit ripening was longer, and fungal diseases afflicted the leaves of all species in the period from the second decade of July to the third decade of August owing to a large amount of precipitation and a cool period at the end of the summer (Figure 5b). This fact and weather conditions did not allow to harvest valuable plant raw materials.

We noticed that the growing season of 2017 (model weather conditions) is characterized by smooth temperatures and uniform precipitation levels, which were favorable for preservation of assimilating leaves on the shoots of Far Eastern peonies. Just as in coastal spring conditions, cool and humid weather induced earlier longer flowering of *P. oreogeton*. The local species *P. anomala* completed its growth at the time point optimal for preparation for wintering and differentiation of underground renewal buds. In 2017, hydrothermal conditions in the second decade of May and at the end of the second decade of August were favorable for harvesting of valuable plant materials (Figure 4).

Under drier weather conditions in 2012, the flowering stage of *P. oreogeton*, which exhibits greater mesophilicity, began early (3rd decade of May) and lasted 2–3 days less (6 days) than in abnormally humid 2013 (2nd decade of June, 8 days). Cool wet weather was similar to that during the coastal spring in 2017 (model weather conditions) and stimulated earlier (3rd decade of May) prolonged flowering (10 days) of *P. oreogeton*.

The coastal species *P. obovata* successfully adapted to the continental climate of Western Siberia, and its flowering stages turned out to be similar to those of the local species *P. anomala*. Under abnormally dry weather conditions in 2012, flowering began in the 3rd decade of May and lasted for 10 days. Abnormally humid weather conditions in 2013 caused flowering to start at the mean annual time point of flowering (2nd decade of June), which lasted for 13 days. In 2017 (model weather conditions), the flowering stage began on June 1 and lasted for 10 days. In Primorye, the period from the beginning of growth to flowering is 35–48 days for *P. obovata* [49], and during its introduction into Western Siberia (CSBG SB RAS), this period reached 28–40 days due to a more intense heat gain at the beginning of the growing season; this phenomenon is typical of the south of Siberia.

When species are grown as sources of biologically active substances, it is not recommended to harvest raw material of above-ground vegetative mass of peonies during excessively humid periods (as in 2013 for example) because leaves are affected by rust, spotting, and fungal spores, which can get into medicines and dietary supplements during industrial processing. It is better to harvest the above-ground part of these plants during the growing seasons that do not show substantial deviations in terms of hydrothermal parameters (as in 2017 for example).

The above analysis of concentrations of reserve substances and biologically active compounds indicates that there are differences in levels of biologically active and reserve substances between the two introduced species of peonies (*P. oreogeton* and *P. obovata*) grown under the conditions of the CSBG SB RAS. This can be attributed to a difference in adaptation mechanisms among the specimens collected in nature. Individual plants respond differently to the change of environmental conditions from the monsoon Far Eastern climate to the continental Siberian climate. Multifunctional adaptation of vegetative organs of peonies proceeds at morphological, anatomical, physiological, and biochemical levels; this adaptation improves active life of cultivars under specific growth conditions and facilitates selection of promising cultivars and species from various regions of Russia and abroad [4,6,50–56]. For example, the total content of flavonoids in the leaves and stems of *P. lactiflora* grown in the Botanical Garden of the Mongolian Academy of Sciences (Ulaanbaatar) decreases linearly during its growth, and the highest content is observed at the flowering stage [51]. Nonetheless, the differences in the accumulation of substances may be an invariable species trait fixed at the genetic level. To test the latter hypothesis, it is necessary to analyze accumulation of substances in plants collected under natural conditions.

Levels of some substances in the Far Eastern species *P. oreogeton* and *P. obovata* grown in Akademgorodok (Novosibirsk) are less than those in the Siberian species *P. anomala* (our data) and *P. hybrida* and the Siberian-Far Eastern species *P. lactiflora* [56]. The content of some substances in the leaves and rhizomes of the above species has been reported previously [56]. For example, the content of flavonols is higher in the leaves of *P. hybrida* (1.96%) and *P. anomala* (1.95%), the content of catechins is higher in the rhizomes of *P. lactiflora* (1287.8 mg/100 g), the content of ascorbic acid is higher in the leaves of *P. anomala* (1205.69 mg/100 g), and the content of protopectins is greater in the rhizomes of *P. anomala* (15.9%), whereas the content of reserve substances, namely, sugars, is higher in the rhizomes of *P. lactiflora* (22.06%), and the content of starch is greater in the rhizomes of *P. anomala* (31.2%). On the other hand, the concentration of tannins is higher in the leaves of *P. oreogeton* (16.42%), the concentration of saponins is greater in the rhizomes of *P. oreogeton* (21.06%), and the concentration of pectins is higher in rhizomes of *P. obovata* (2.02%). Higher contents of tannins, saponins, and pectins are most likely due to adaptation of the Far Eastern peony species to the Siberian conditions. It should be noted that the Siberian and Far Eastern species grown in Akademgorodok exhibited higher levels of some substances. Thus, species that are not used as medicinal plants in Russia can replace the official medicinal species *P. anomala*. Levels of some substances, such as catechins, sugars, tannins, saponins, and pectins, is lower in *P. anomala* than in some Siberian and Far Eastern species.

## 5. Conclusions

The study leads to the conclusion that introduced species *P. obovata* and *P. oreogeton* are promising for further research under the conditions of Akademgorodok (Novosibirsk). In this environment, *P. obovata* and *P. oreogeton* start to grow between the second decade of April and the first decade of May, go through the entire growth cycle, including flowering and fruiting, but *P. oreogeton* requires appropriate microclimatic conditions (a more shaded area). To prevent fungal spores from getting into medicines and dietary supplements during industrial processing, peony aerial shoots as a medicinal raw material should not be harvested during periods of excessive humidity.

A biochemical analysis of *P. obovata* and *P. oreogeton* grown in Akademgorodok showed that the leaves accumulate higher concentrations of flavonols (*P. obovata*: 1.77%, and *P. oreogeton*: 1.17%), tannins (*P. oreogeton*: 16.42%, *P. obovata*: 12.81%), ascorbic acid (*P. oreogeton*: 155.2 mg/100 g, *P. obovata*: 151.8 mg/100 g), and sugars (*P. oreogeton*: 17.44%, *P. obovata*: 20.85%) than the roots do. Peony rhizomes contain higher concentrations of protopectins (*P. oreogeton*: 13.03%, *P. obovata*: 7%), saponins (*P. oreogeton*: 17.90%, *P. obovata*: 21.06%), and starch (*P. oreogeton*: 27.13%, *P. obovata*: 30.20%) than the leaves do. We believe that species of the genus *Paeonia* grown in the CSBG SB RAS are a promising plant material. Further investigation into the dynamics of accumulation of biologically active substances in the organs of peonies will help to identify introduced plant species having high biochemical potential for the pharmaceutical industry.

**Supplementary Materials:** The following supporting information can be downloaded at: https://www.mdpi.com/article/10.3390/horticulturae9010102/s1, Table S1: The main climatic characteristics of the growing seasons in Novosibirsk and Vladivostok; Table S2: Main climatic characteristics of the 2012 growing season (Novosibirsk); Table S3: Main climatic characteristics of the 2013 growing season (Novosibirsk).

**Author Contributions:** Conceptualization, O.V.K., V.A.K. and O.Y.V.; methodology, O.V.K. and T.A.K.; software, A.A.K.; validation, T.A.K. and M.S.K.; formal analysis, O.V.K., O.Y.V. and V.A.K.; investigation, O.V.K., T.A.K. and O.Y.V.; resources, O.Y.V. and A.S.E.; data curation, O.V.K. and V.A.K.; writing—original draft preparation, O.V.K., O.Y.V. and V.A.K.; writing—review and editing, O.V.K., O.Y.V. and V.A.K.; visualization, A.S.E.; supervision, V.A.K.; project administration, V.A.K.; funding acquisition, O.Y.V., A.A.K. and M.S.K. All authors have read and agreed to the published version of the manuscript.

**Funding:** The study was conducted within state assignments for the CSBG SB RAS (projects No. AAAA-A21-121011290024-5 and AAAA-A21-121011290025-2), the Institute of Plant Physiology RAS (project No. 122042700045-3), and the Tomsk State University Development Program (Priority2030).

**Institutional Review Board Statement:** Not applicable.

**Informed Consent Statement:** Not applicable.

**Data Availability Statement:** Not applicable.

**Acknowledgments:** The authors thank Nikolai A. Shevchuk for comments and proofreading.

**Conflicts of Interest:** The authors declare no conflict of interest.

## References

1. Kemularia-Natadze, L.M. On the position of the family Paeoniaceae in the system of angiosperms. *Notes Syst. Geogr. Plants* **1958**, *20*, 19–28. (In Russian)
2. Hong, D.Y. Peonies of the world. In *Taxonomy and Phytogeography*; Pt. 1; Kew Publishing, and Missouri Botanical Garden Press: London, UK, 2010.
3. Yang, Y.; Li, S.S.; da Silva, J.A.T.; Yu, X.N.; Wang, L.S. Characterization of phytochemicals in the roots of wild herbaceous peonies from China and screening for medicinal resources. *Phytochemistry* **2020**, *174*, 112331. [CrossRef] [PubMed]
4. Reut, A.A.; Denisova, S.G.; Pupykina, K.A. Accumulation and distribution of biologically active substances in raw materials of some taxa of the genus *Paeonia* L. *Chem. Plant Raw Mater.* **2019**, *4*, 269–278. (In Russian) [CrossRef]
5. Wang, Y.Y.; Wang, C.Y.; Wang, S.T.; Li, Y.Q.; Mo, H.Z.; He, J.X. Physicochemical properties and antioxidant activities of tree peony (*Paeonia suffruticosa* Andr.) seed protein hydrolysates obtained with different proteases. *Food Chem.* **2021**, *345*, 128765. [CrossRef] [PubMed]
6. Tong, N.N.; Zhou, X.Y.; Peng, L.P.; Liu, Z.A.; Shu, Q.Y. A comprehensive study of three species of *Paeonia* stem and leaf phytochemicals, and their antioxidant activities. *J. Ethnopharmacol.* **2021**, *273*, 113985. [CrossRef] [PubMed]
7. Chen, Y.F.; Lee, M.M.; Fang, H.L.; Yang, J.G.; Chen, Y.C.; Tsai, H.Y. Paeoniflorin inhibits excitatory amino acid agonist and high-dose morphine-induced nociceptive behavior in mice via modulation of N-methyl-D-aspartate receptors. *BMC Compl. Altern. Med.* **2016**, *16*, 240. [CrossRef]
8. Enkhtuya, E.; Shimamura, T.; Kashiwagi, T.; Ukeda, H. Antioxidative constituents in the leaves of *Paeonia anomala* grown in Mongolia. *Food Sci. Technol. Res.* **2017**, *23*, 63–70. [CrossRef]
9. Yu, X.H.; Song, T.; Hou, X.L.; Sui, Y.; Li, Y.L.; Hu, D.; Wang, X.H.; Xiao, Z.X.; Wang, R.R.; Wang, J.; et al. Anti-depressant effect of *Paeonia lactiflora* pall extract in rats. *Trop. J. Pharmaceut. Res.* **2017**, *16*, 577–580. [CrossRef]
10. Li, P.; Shen, J.; Wang, Z.; Liu, S.; Liu, Q.; Li, Y.; He, C.; Xiao, P. Genus *Paeonia*: A comprehensive review on traditional uses, phytochemistry, pharmacological activities, clinical application, and toxicology. *J. Ethnopharmacol.* **2021**, *269*, 113708. [CrossRef]
11. *Pharmacopoeia of the People's Republic of China, Part I*; China Pharmaceutical Science and Technology Press: Beijing, China, 2020. (In Chinese)
12. *Rhizomes and Rhizomata et Radices Paeonia Anomala*; Pharmacopeial Article 42-531-98; Ministry of Health of the Russian Federation Pharmacopoeial State Committee: Moscow, Russia, 2000. (In Russian)
13. Peony Evasive Tincture 25 mL. Available online: https://www.rigla.ru/product/49409 (accessed on 22 November 2022).
14. Peony Extract, Coated Tablets. Available online: https://www.eapteka.ru/goods/id221531/ (accessed on 22 November 2022).
15. Degtjareva, G.; Efimov, S. The genetic diversity of *Paeonia anomala* (Paeoniaceae), as indicated by nuclear its and plastid ycf1 molecular markers. In Proceedings of the Northern Asia Plant Diversity: Current Trends in Research and Conservation: BIO Web of Conferences, Novosibirsk, Russia, 6–12 September 2021; Volume 38, p. 23.
16. Shulkina, T. *Ornamental Plants from Russia and Adjacent States of the Former Soviet Union: A Botanical Guide for Travelers and Gardeners*; Kew Publ.: London, UK, 2004.
17. Ahmad, M.; Malik, K.; Tariq, A.; Zhang, G.; Yaseen, G.; Rashid, N.; Sultana, S.; Zafar, M.; Ullah, K.; Khan, M.P.Z. Botany, ethnomedicines, phytochemistry and pharmacology of Himalayan paeony (*Paeonia emodi* Royle.). *J. Ethnopharmacol.* **2018**, *220*, 197–219. [CrossRef] [PubMed]
18. Zhao, M.; Wu, S.P. A review of the ethnobotany, phytochemistry and pharmacology of tree peony (Sect. Moutan). *S. Afr. J. Bot.* **2019**, *124*, 556–563. [CrossRef]

19. Hong, D.-Y.; Pan, K.-Y.; Rao, G.-Y. Cytogeography and taxonomy of the Paeonia obovata polyploid complex (Paeoniaceae). *Plant Syst. Evol.* **2001**, *227*, 123–136. [CrossRef]
20. Efimov, S.V.; Degtyareva, G.V.; Terentyeva, Y.I.; Samigullin, T.H.; Skaptsov, M.V.; Valiejo-Roman, C.M. Species complex of *Paeonia obovata* Maxim. (Paeoniaceae): Five species or just one? *Skvortsovia* **2018**, *4*, 113–114.
21. Halda, J.J.; Waddick, J.W. *The Genus Paeonia*; Timber Press: Portland, OR, USA, 2004.
22. Fang, Q.B.; Wang, D.Q.; Peng, H.S. Study on the relationship between classification, distribution and medicinal use of Sect. Moutan of the genus *Paeonia* in China. *Res. Pract. Chin. Med.* **2004**, *18*, 20–22.
23. Tanaka, T.; Kataoka, M.; Tsuboi, N.; Kouno, I. New monoterpene glycoside esters and phenolic constituents of Paeoniae Radix, and increase of water solubility of proanthocyanidins in the presence of paeoniflorin. *Chem. Pharm. Bull.* **2000**, *48*, 201–207. [CrossRef]
24. Wu, S.H.; Wu, D.G.; Chen, Y.W. Chemical constituents and bioactivities of plants from the genus *Paeonia*. *Chem. Biodivers.* **2010**, *41*, 90–104. [CrossRef]
25. Hosoki, T.; Mitsuhiro, S. Flower anthocyanins of herbaceous peony. *Shimane Daigaku Nogakubu Kenkyu Hokoku* **1991**, *25*, 11–14.
26. Zheng, W.Y.; Chen, Y.H.; Zhang, X.T.; Yu, Z.G. Analysis of Volatiles in *Paeonia obovata* Flowers by HS-SPME-GC-MS. *Chem. Nat. Comp.* **2016**, *52*, 922–923. [CrossRef]
27. Bae, J.Y.; Kim, C.Y.; Kim, H.J.; Park, J.H.; Ahn, M.J. Differences in the chemical profiles and biological activities of *Paeonia lactiflora* and *Paeonia obovata*. *J. Med. Food* **2015**, *18*, 224–232. [CrossRef]
28. Beideman, I.N. *Methods of Studying the Phenology of Plants and Plant Communities*; Nauka: Novosibirsk, Russia, 1974. (In Russian)
29. Koch, E.; Bruns, E.; Chmielewski, F.M.; Defila, C.; Lipa, W.; Menzel, A. *Guidelines for Plant Phenological Observations*; WMO/TD No. 1484; World Meteorological Organization: Geneva, Switzerland, 2009.
30. Uranov, A.A. *Ontogenesis and Age Composition of Flowering Plant Populations*; Nauka: Moscow, Russia, 1967; pp. 3–8. (In Russian)
31. Zaugolnova, L.B.; Zhukova, A.A.; Komarova, A.S.; Smirnova, O.V. *Plant Cenopopulations (Population Biology Essays)*; Nauka: Novosibirsk, Russia, 1988. (In Russian)
32. Ermakov, A.I.; Arasimovich, V.V.; Yarosh, N.P.; Peruanskiy, Y.V.; Lukovnikova, G.A.; Ikonnikova, M.I. *Methods of Biochemical Investigation of Plants*; Leningrad: Agropromizdat, Russia, 1987. (In Russian)
33. Brighente, I.M.C.; Dias, M.; Verdi, L.G.; Pizzolatti, M.G. Antioxidant activity and total phenolic content of some Brazilian species. *Pharm. Biol.* **2007**, *45*, 156–161. [CrossRef]
34. Sun, B.; Ricardo-da-Silva, J.M.; Spranger, I. Critical factors of vanillin assay for catechins and proanthocyanidins. *J. Agric. Food Chem.* **1998**, *46*, 4267–4274. [CrossRef]
35. Kukushkina, T.A.; Zykov, A.A.; Obukhova, L.A. Common cuff (*Alchemia vulgaris* L.) as a source of drugs of natural origin. In Proceedings of the Actual Problems of Creating New Drugs of Natural Origin: Materials of the VII International Congress, St. Petersburg, Russia, 3–5 July 2003; pp. 64–69. (In Russian).
36. Fedoseeva, L.M. The study of tannins of underground and aboveground vegetative organs of the *Bergenia Crassifolia* (L.) Fitsch., growing in Altai. *Chem. Plant Raw Mater.* **2005**, *2*, 45–50. (In Russian)
37. Kiseleva, A.V.; Volkhonskaya, T.A.; Kiselev, V.E. *Biologically Active Substances of Medicinal Plants of Southern Siberia*; Nauka: Novosibirsk, Russia, 1991. (In Russian)
38. Kriventsov, V.I. Carbazole-free method for the quantitative spectrophotometric determination of pectin substances. *Proc. Nikitsk. Bot. Gard.* **1989**, *109*, 128–137. (In Russian)
39. Kriventsov, V.I. *Guidelines for the Analysis of Fruits for Biochemical Composition*; GNBS: Yalta, Ukraine, 1982. (In Russian)
40. Borodova, V.Y.; Gorenkov, E.S.; Klyueva, O.A.; Malofeeva, L.N.; Megerdicheva, E.Y. *Guidelines for Chemical-Technological Variety Testing of Vegetable, Fruit and Berry Crops for the Canning Industry*; Russian Agricultural Academy: Moscow, Russia, 1993. (In Russian)
41. Zaprometov, M.N. *Fundamentals of Biochemistry of Phenolic Compounds*; Vysshaya shkola: Moscow, Russia, 1974. (In Russian)
42. Berezina, E.V.; Brilkina, A.A.; Veselov, A.P. Content of phenolic compounds, ascorbic acid, and photosynthetic pigments in *Vaccinium macrocarpon* Ait. dependent on seasonal plant development stages and age (the example of introduction in Russia). *Sci. Hortic.* **2017**, *218*, 139–146. [CrossRef]
43. Yang, L.; Wen, K.S.; Ruan, X.; Zhao, Y.X.; Wei, F.; Wang, Q. Response of plant secondary metabolites to environmental factors. *Molecules* **2018**, *23*, 762. [CrossRef]
44. Uddin, M. Environmental factors on secondary metabolism of medicinal plants. *Acta Sci. Pharm. Sci.* **2019**, *3*, 34–46. [CrossRef]
45. Asghar, M.; Younas, M.; Arshad, B.; Zaman, W.; Ayaz, A.; Rasheed, S.; Shah, A.; Ullah, F.; Saqib, S. Bioactive potential of cultivated *Mentha arvensis* L. for preservation and production of health-oriented food. *J. Anim. Plant Sci.* **2022**, *32*, 835–844.
46. Sobolevskaya, K.A. *Introduction of Plants in Siberia*; Nauka: Novosibirsk, Russia, 1991. (In Russian)
47. Canter, P.H.; Thomas, H.; Ernst, E. Bringing medicinal plants into cultivation: Opportunities and challenges for biotechnology. *TRENDS Biotechnol.* **2005**, *23*, 180–185. [CrossRef]
48. Komina, O.V. Biological Features of Some Species of the *Genus paeonia* L. When Introduced in the Forest-Steppe Zone of Western Siberia. Ph.D. Thesis, Central Siberian Botanical Garden SB RAS, Novosibirsk, Russia, 2014. (In Russian).
49. Makedonskaya, N.V. Rhythms of seasonal development and features of biology of Far Eastern pions in culture. In *Rhythms of Seasonal Development of Plants in Primorye*; DVNTS AN SSSR: Vladivostok, Russia, 1980; pp. 49–57. (In Russian)

50. Gubanenko, G.A.; Morozova, E.V.; Rubchevskaya, L.P. The influence of climatic factors on the content of flavonoids in the *Paeonia anomala* L. biomass. *Chem. Plant Raw Mater.* **2014**, *1*, 165–170. (In Russian)
51. Oyungerel, S.; Batzaya, G.; Byamba-Yondon, G.; Lyankhua, B.; Ochgerel, N.; Usukhjargal, D. Seasonal variation of some bioactive compounds and physiological characteristics in peony (*Paeonia lactiflora* Pall.). *Mong. J. Biol. Sci.* **2017**, *15*, 47–51.
52. Peng, L.P.; Cheng, F.Y.; Hu, X.G.; Mao, J.F.; Xu, X.X.; Zhong, Y.; Zhong, Y.; Li, S.-Y.; Xian, H.L. Modelling environmentally suitable areas for the potential introduction and cultivation of the emerging oil crop *Paeonia ostii* in China. *Sci. Rep.* **2019**, *9*, 3213. [CrossRef] [PubMed]
53. Calonghi, N.; Farruggia, G.; Boga, C.; Micheletti, G.; Fini, E.; Romani, L.; Telese, D.; Faraci, E.; Bergamini, C.; Cerini, S.; et al. Root Extracts of Two Cultivars of *Paeonia* Species: Lipid Composition and Biological Effects on Different Cell Lines: Preliminary Results. *Molecules* **2021**, *26*, 655. [CrossRef] [PubMed]
54. Wu, Y.; Hao, Z.; Tang, Y.; Zhao, D. Anthocyanin Accumulation and Differential Expression of the Biosynthetic Genes Result in a Discrepancy in the Red Color of Herbaceous Peony (*Paeonia lactiflora* Pall.) Flowers. *Horticulturae* **2022**, *8*, 349. [CrossRef]
55. Chernyshenko, O.V.; Rudaya, O.A.; Efimov, S.V.; Kiris, Y.N. The transpiration rate of some species' leaves of the genus *Paeonia* L., as one possible performance of their adaptation to the environment. *For. Bull.* **2017**, *21*, 78–86.
56. Kostikova, V.A.; Kalendar, O.V.; Tashev, N.A.; Erst, A.S.; Vasilyeva, O.Y. Biologically active and reserve substances of Siberian peonies. In Proceedings of the Northern Asia Plant Diversity: Current Trends in Research and Conservation: BIO Web of Conferences, Novosibirsk, Russia, 6–12 September 2021; Volume 38, p. 61.


**Disclaimer/Publisher's Note:** The statements, opinions and data contained in all publications are solely those of the individual author(s) and contributor(s) and not of MDPI and/or the editor(s). MDPI and/or the editor(s) disclaim responsibility for any injury to people or property resulting from any ideas, methods, instructions or products referred to in the content.

horticulturae

Article

# The Breeding of High-Quality Dandelions by NaCl Induced Callus Variation Combined with a Drosophila Tumor Cell Migration Test

Zhe Wu [1], Zhaojia Li [1], Wei Feng [1], Ran Meng [1], Xiuping Wang [1,*] and Chenxi Wu [2,*]

[1] Institute of Coastal Agriculture, Hebei Academy of Agriculture and Forestry Sciences, Tangshan 063299, China
[2] College of Traditional Chinese Medicine, North China University of Science and Technology, Tangshan 063210, China
* Correspondence: bhswxp@163.com (X.W.); chenxi.wu@ncst.edu.cn (C.W.)

**Abstract:** Creating high-quality and salt-tolerant plant germplasm is an effective strategy to improve the utilization of saline-alkali land. Salt-induced callus mutation was used to create dandelion germplasm and mutant dandelion calluses were obtained under NaCl concentrations of 0.8%, 1%, and 1.2% with the identification of random amplified polymorphic DNA (RAPD) markers. A new dandelion line, "Binpu 2", selected from the progenies of dandelion tissue culture plantlets that originated from callus treated under 0.8% NaCl, was evaluated in light of its morphological characteristics, bioactive components, and antitumor functions. Results showed that the plant shape of "Binpu 2"was nearly upright; its cichoric acid content was 6.7 mg/g, which was 39.6% and 36.7% higher than its mother plant and local dandelion cultivar, respectively; its leaf water extracts of 0.2 g/mL had a significant inhibitory effect on cell polarity disruption-induced cell migration without affecting drosophila normal growth, revealed by the strong inhibitory effect on tumor cell migration, the increased level of MMP1 and β-Integrin, and the reduced E-cadherin expression. Our results suggested that "Binpu 2" originated from salt-induced mutant performed better than its mother plant and processed strong antitumor function, which was suitable for amplified cultivation in saline-alkali land for food and medicinal industrial development.

**Keywords:** cell migration; dandelion breeding; functional verification; salt-induced mutant; tissue culture

check for updates

**Citation:** Wu, Z.; Li, Z.; Feng, W.; Meng, R.; Wang, X.; Wu, C. The Breeding of High-Quality Dandelions by NaCl Induced Callus Variation Combined with a Drosophila Tumor Cell Migration Test. *Horticulturae* **2022**, *8*, 1167. https://doi.org/10.3390/horticulturae8121167

Academic Editor: Wajid Zaman

Received: 2 November 2022
Accepted: 7 December 2022
Published: 8 December 2022

**Publisher's Note:** MDPI stays neutral with regard to jurisdictional claims in published maps and institutional affiliations.

**Copyright:** © 2022 by the authors. Licensee MDPI, Basel, Switzerland. This article is an open access article distributed under the terms and conditions of the Creative Commons Attribution (CC BY) license (https://creativecommons.org/licenses/by/4.0/).

## 1. Introduction

Soil salinization has become the main factor restricting agricultural development in China [1]. Cultivating salt-tolerant crops have represented the most fundamental and effective method of utilizing saline–alkali land and improving economic profit [2,3]. Studies have shown that plants growing under saline stress could accumulate more nutrients and active substances, and had better nutritional, medicinal, and other economic benefits [4].

Dandelion (*Taraxacum mongolicum* Hand.-Mazz.), a traditional herb plant used for medication, rubber, cosmetic, feed additives, as well as vegetables in recent years [2,5], could tolerate saline stress up to 1.5% of NaCl [6], which was suitable for cultivation in saline-alkali land. However, poor commercial characters of dandelion in saline-alkali land struggled to satisfy the demands of industrial production. Hence, it was urgent to develop dandelion varieties with good comprehensive performance. At present, most dandelion varieties were bred by natural selection, mainly because dandelion is a self-pollinating plant of with capitulum, leading to the hard application of common hybridization breeding [6]. Furthermore, related molecular breeding in dandelions has been hardly reported, mainly due to the complex genotype and little genome sequence information [7]. Thus, we adopted a new breeding strategy through salt-induced callus mutation to breed a high-quality dandelion variety suitable for cultivation in saline–alkali land.

Because of the medicinal use of dandelion, besides the biomass yield, a good cultivar with highly bioactive compounds and function was also vital. Thus, we evaluated antitumor function at the same time. Cancer is a worldwide fatal disease that seriously threatens human health. More than 90% of cancer patients died from tumor migration, instead of primary tumor overgrowth [8]. Dandelions have traditionally been employed in Chinese, Native American and Arabian traditional medicine for the purpose of treating diseases, including multiple types of cancers [9,10]. The anti-cancer efficacy of dandelion extracts has been previously reported, while their effects on tumor cell migration in vivo were largely unknown.

Natural malignant tumours can occur in *Drosophila melanogaster* (*D. melanogaster*). Tumours can be experimentally induced in larvae and adult flies by knocking down fly tumour suppressor genes. *D. melanogaster* has an important role in chemical genetics, helping to identify the pathways that are affected by pharmaceuticals, facilitating the design of more efficient derivatives and serving as a platform for semi-automated screens for new anticancer drugs [11,12]. Therefore, with less genome redundancy and powerful genetic tools, *D. melanogaster* was selected as the elegant system to test tumor metastasis.

The current study firstly adopted the method of salt-induced callus mutation combined with functional tests for dandelion breeding. Details of the breeding process were of great significance to enrich modern breeding practices and agriculture developments in saline-alkali land.

## 2. Materials and Methods

### 2.1. Dandelion Materials

A wild dandelion resource "Daye" (*Taraxacum mongolicum* Hand.-Mazz.) was selected as the mother plant due to high biomass yield. This resource was further bred into sister lines "Binpu 1" [13,14] and "Binpu 2" [15] by the Institute of Coastal Agriculture, Hebei Academy of Agriculture and Forestry Sciences, China. "Binpu 1" was noted for its high yield and "Binpu 2" was noted for its strong functions. Their varietal characteristics could be found from above mentioned references. This study mainly focused on the breeding process of "Binpu 2". The mother plant "Daye" was indoor cultured for 30 days at 25 °C with 16 h of illumination at 2500 lux. Then, the healthy and young leaves were taken as explants for producing callus. Related dandelion materials were shown in Figure S1.

### 2.2. Production of Dandelion Tissue Culture Plantlet

#### 2.2.1. Induction of Callus with Saline Treatment

Fresh sterilized leaves [6] were cut into 300 approximate pieces of about 1 $cm^2$ and then inoculated with solid MS medium [6] containing 0.5 mg/L of 6-BA and 0.04 mg/L of 2,4-D for 45 days in sterile tissue culture room. Once selected, the callus with the best growth, in order to obtain sufficient experimental materials of callus, was sub-cultured by solid MS medium for 5 times with 21 days for each time. Then, obtained calluses were cut into small pieces of about 0.25 $cm^2$ and treated by solid MS medium containing 85 mM, 136 mM, 170 mM, 205 mM, 256 mM, and 307 mM NaCl, respectively, for 28 days. Each NaCl treatment of solid MS medium contained 30 calluses. Subsequently, one of the best growing callus in each NaCl treatment was picked out into a NaCl-free solid MS medium individually for continuously culturing over 21 days, and then continued to be treated by NaCl treatment of solid MS media for 21 days. Above shift-NaCl process was repeated two times. Lastly, the calluses were transferred to NaCl-free solid MS medium for amplified cultivation over 2 months. Above all works were done in tissue culture room with 24 h of fluorescence illumination at 3000 lux and 25 °C.

2.2.2. Identification of Salt-Treated Callus

The calluses obtained above were identified by random amplified polymorphic DNA (RAPD) markers. Details of primers, via screening and identification techniques, can be seen from our previous research [14]. DNA extraction was processed by the DNA-secure plant genome DNA extraction kit (Tiangen Biotech Co., Ltd., Beijing, China). In total 41 random primers (Table S1) were used for RAPD identification. The polymerase chain reaction (PCR) amplification procedure involved pre-denaturation for 5 min at 94 °C, followed by 40 cycles of denaturation for 1 min at 94 °C, annealing for 1 min at 38 °C, extension for 2 min at 72 °C, and amplification for 10 min at 72 °C.

2.2.3. Regeneration of Dandelion Tissue Culture Plantlet

The calluses obtained above were placed on a solid MS medium containing 0.2 mg/L of 6-BA and 0.01 mg/L of NAA due to the occurrences of massive adventitious buds and then transferred to a 1/2 solid MS medium with free plant hormone for the continuous culturing process which was carried out 2 times with 21 days of each time. Then, calluses were transferred to a 1/2 solid MS medium containing 0.3 mg/L of NAA for culturing over 21 days. All above processes were finished in a tissue culture room with 24 h of fluorescence illumination at 3000 lux and 25 °C. The resulting tissue seedlings were acclimatized for 35 days to improve the survive rate for the field transplant. Firstly, the seedlings were gradually contacted in an external environment for 3 weeks, and the room was kept ventilated with a temperature at 23 °C, and a relative humidity of 80%. Then, these seedlings were transferred to flowerpots containing vermiculite-nutrient soil-perlite at a ratio of 1:4:3, and were continuously cultured for 2 weeks in field shelter and sprayed with a 1/4 liquid MS medium each day over two weeks. Finally, these tissue culture plantlets were transplanted to the outdoor field and their seeds were harvested separately for amplified field cultivation.

*2.3. Determination of Candidate Dandelion Lines*

The seeds from dandelion tissue culture seedlings were grown in a seedling tray in greenhouse and then transplanted to an observation nursery in saline-alkali land (Figure S1) when the seedlings grew to three leaves. The soil salt content within 20 cm depth in the observation nursery was checked by AZ 8373 TDS/salinity meter (AZ Instrument Corp., Taiwan, China) [2] and averaged value was around 0.3%. These dandelions plants were cultured for 6 weeks, and various characteristics (such as plant shape, leaf shape, and leaf color, as well as chicoric acid, caffeic acid, and the total flavonoids contents) were repeatedly investigated during the cultivation period. Seeds from these dandelion plants were harvested and propagated separately over 3 generations and finally the plants with stable characteristics were picked out as candidate dandelion lines according to the trait investigation. We finally obtained 6 such dandelion lines as marked by Y1–Y6, which meant these selected lines were originated from the calluses that under NaCl treatment of 85 mM (Y1), 136 mM (Y2), 170 mM (Y3), 205 mM (Y4), 256 mM (Y5), and 307 mM NaCl (Y6), respectively.

Of these characteristics, for determination of compounds contents, 0.5 g of dry dandelion powder was mixed with 25 mL of 50% methanol and ultrasonically extracted for 30 min at 60 °C. Then, the extract was filtered by 0.45 μm membrane for contents check. Contents of chicory acid, caffeic acid and chlorogenic acid were determined by Agilent1200 high performance liquid chromatography (Agilent Technologies Inc., Santa Clara, CA, USA), as described in [5,16]; the total flavonoids were detected by UV-2450 spectrophotometer (Shimadzu Co., Ltd., Kyoto, Japan), as described in [17].

*2.4. Anti-Tumor Cell Migration Evaluation*

2.4.1. Preparation of Dandelion Extracts

The above obtained dandelion lines, including their mother plant "Daye", which was cultivated in the field, were harvested at the same time and dried in oven at 60 °C for 2 days.

Then, 100 g of dry powder was taken to be extracted with 1000 mL of double-distilled water ($ddH_2O$) at 60 °C for 3 h. The extract that formed was vacuum filtered with a 0.45 μm filter. Then, the extract was frozen at −80 °C and freeze-dried using a lyophilizer, and reconstituted in $ddH_2O$ to obtain a final stock concentration of 0.5 g/mL. The extract was added directly to regular food from a 0.5 g/mL aqueous stock to a final concentration of 0.05, 0.1, 0.2, and 0.4 g/mL. For the control food, we used water alone.

#### 2.4.2. Fly Strains

The flies were kept on a cornmeal and agar medium at 25 °C with a 12 h light-dark cycle incubator [18]. *Drosophila* strains used $w^{1118}$ (#3605) which was obtained from the Bloomington *Drosophila* stock center. *UAS-scrib-IR* (#27424) was obtained from Vienna *Drosophila* RNAi center [18]. *ptc*-GAL4 was previously described [19]. For cell migration test, crosses were raised at 25 °C for 2 days, and then shifted to 29 °C for an additional 3 days, The third-instar larvae were dissected.

#### 2.4.3. Immunostaining

Larval discs were dissected and then fixed in 4% formaldehyde for 40 min. After several washes with 0.3% (*v*/*v*) PBST, the discs were stained with primary antibodies at 4 °C overnight and then secondary antibodies at room temperature for 2 h. The following antibodies were used: mouse anti-MMP1 (1:200, Developmental Studies Hybridoma Bank, DSHB (Iowa City, IA, USA), Cat. #3A6B4-c), mouse anti-β integrin (1:100, DSHB, Cat. #CF.6G11-c), rat anti-DE cadherin (1:100, DSHB, Cat. #DCAD2-c), goat anti-mouse cyanine-3 (1:1000, Life Technologies (Carlsbad, CA, USA), Cat. #A10521), goat anti-rat cyanine3 (1:1000, Life Technologies, Cat. #A10522). Vectashield mounting media (Vector Laboratories, Cat. #H-1500) was used for mounting.

### *2.5. Data Analysis*

The Data were presented as bar graphs or mean value ± standard divisions. The figures were created using the Origin 8.0 software (OriginLab Corporation, Northampton, MA, USA). For statistical significance, one-way ANOVA, along with Bonferroni's multiple-comparison test and Duncan's multiple range test, was applied for the drosophila test and dandelion filed evaluation. A $p$ value less than 0.05 was considered significant; ns was not significant, $p \geq 0.05$; * was significant, $p < 0.05$; ** meant $p < 0.01$; *** meant $p < 0.001$.

## 3. Results and Discussion

### *3.1. Identification of Salt-Induced Calluses*

RAPD identification showed that 3 (S358, OPA-19 and OPC-05) out of 41 random primers amplified different bands from that of the control corresponding to NaCl concentrations of 205 mM, 170 mM, and 136 mM, respectively (Figure 1), indicating that high-NaCl-concentration exceeding 136 mM could induce DNA variation, which may cause some different extend of trails variations from their mother plant, such as plant shape, leaf shape, and plant size, etc. [6]. However, dandelions in natural conditions also have chance to produce some morphological variations, and parts of reasons were ascribed to apomixis [6]. No matter what treatment method is used, for dandelion breeders, it is the premise to make trait variation between offspring and mother parent, and then continue to carry out excellent trait screening work for an ideal dandelion line. Our method of NaCl-induced calluses mutation provided the probability for subsequent dandelion breeding works.

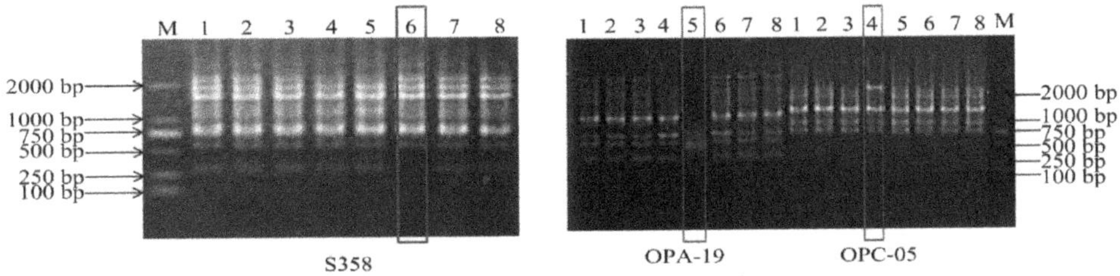


**Figure 1.** RAPD identification results. Band M was marker, number 1 to 8 were at NaCl concentrations of 0 mM, 0 mM, 85 mM, 136 mM, 170 mM, 205 mM, 256 mM and 307 mM, respectively.

### 3.2. Morphological Evaluation

As the method described for the determination of dandelion lines in the section of materials and method, we obtained 6 candidate dandelion lines and recorded 11 morphological parameters for them. Results showed that these dandelion lines showed some differences in aspects of plant shape, leaf pubescence, leaf vein color, leaf size, etc. (Table 1). Especially, we found two dandelion materials originated from that of NaCl concentrations of 136 mM (Y2) and 170 mM (Y3) were obviously advantaged than other materials in the characteristics of plant shape and whole plant size. Dandelions from Y2 appeared obviously upright growth, while Y3 showed larger size for whole plant than other materials. In addition, among the 11 morphological parameters, the purple at the base of leaf vein was stably inherited and not affected by the environment according to our long-term observations and these materials involved Y2 and Y5.

**Table 1.** Morphological evaluation of dandelion progenies.

| Index | Y1 * | Y2 | Y3 | Y4 | Y5 | Y6 | CK |
|---|---|---|---|---|---|---|---|
| Plant shape | <45° | 75–90° | 45–75° | <30° | <45° | <30° | <45° |
| Leaf pubescence | no | yes | no | no | no | only front side | no |
| Leaf margin color | green | light purple | green | green | green | green | green |
| Phyllopodium color | purple | purple | purple | purple | light purple | purple | purple |
| Leaf margin toothed | yes | light cleft | yes | yes | yes | yes | yes |
| Leaf vein | green | purple at lower | green | green | purple at lower | green | green |
| Distance of top leaf lobe to 2nd lobe | larger | smaller | middle | larger | larger | larger | larger |
| Leaf crack depth | deep | deep | deep | deep | deep | deep | deep |
| Leaf flatness | intimate smooth | intimate smooth | intimate smooth | intimate smooth | intimate smooth | intimate smooth | intimate smooth |
| Leaf length (cm) | 23.9 ± 1.4 | 22.9 ± 1.4 | 28.1 ± 1.7 | 23.3 ± 1.4 | 22.2 ± 1.3 | 23.6 ± 1.4 | 23.2 ± 1.4 |
| Leaf width (cm) | 4.7 ± 0.3 | 5.3 ± 0.3 | 6.1 ± 0.4 | 4.8 ± 0.3 | 4.5 ± 0.3 | 4.8 ± 0.3 | 4.7 ± 0.3 |

* Y1–Y6 were progenies of dandelion tissue culture plantlets originated from calluses treated at 85 mM, 136 mM, 170 mM, 205 mM, 256 mM and 307 mM NaCl, respectively; CK was female parent "Daye"; Data of leaf length or width was shown as mean ± standard deviations ($n$ = 6).

The plant shape was an important index in directly determining mechanized harvests in the future. The plant shape was divided into three criteria, namely flat growth (angle < 30°), half-upright growth (30° < angle < 75°), and upright growth (angle > 75°), based on the angle between the new leaf and ground. As plant shape of most dandelions was flat or had intimately half-upright growth, leading to that their harvest of fresh leaf was mainly relied on artificial way. Therefore, the relatively upright growth was vital trait for considering the mechanized operation in the future, which could greatly decrease the harvest cost comparing to the manual harvesting. The plant shape of dandelion lines from Y2 and Y3 were divided into upright and half-upright growth, which were named as "Binpu 2" and "Binpu 1" later. In addition, the large plant size of Y3 indicated a potentially high yield. Hence, dandelion materials from Y2 and Y3 were the focus for dandelion breeding.

### 3.3. *Yield and Bioactive Compounds Contents*

Dandelions are rich in phenolic acids and flavonoids, which have strong antioxidant and antitumor effects [20]. We further compared leaf yields and the contents of compounds from the two lines (Y2 and Y3). Overall, dandelions of Y3 showed the highest yield and Y2 had the highest compounds content (Table 2). Y2 demonstrated high contents in caffeic acid, chlorogenic acid, and cichoric acid. Of these, the cichoric acid content was 39.6%, 36.7% and 48.9% higher than its mother plant, local dandelion cultivar, and the criterion that stipulated by "the People's Republic of China Pharmacopoeia 2020", respectively. Although the total flavonoids content in Y2 were lower than in other resources, considering the relatively high fresh leaf yield, it was still considered to be a satisfied resource for high contents of bioactive compounds.

**Table 2.** Yield and bioactive compounds contents.

| Name [1] | Total Flavonoids (mg/g) | Caffeic Acid (mg/g) | Chlorogenic Acid (mg/g) | Cichoric Acid (mg/g) | Fresh Leaf Yield (t/ha) |
|---|---|---|---|---|---|
| CK | $4.51 \pm 0.25^{c}$ | $0.07 \pm 0.005^{c}$ | $0.28 \pm 0.02^{c}$ | $4.8 \pm 0.27^{b}$ | $13.2 \pm 1.14^{ab}$ |
| Y2 | $3.01 \pm 0.17^{d}$ | $0.32 \pm 0.02^{a}$ | $0.81 \pm 0.06^{a}$ | $6.7 \pm 0.38^{a}$ | $14.2 \pm 1.23^{a}$ |
| Y3 | $6.61 \pm 0.37^{b}$ | $0.06 \pm 0.004^{c}$ | $0.49 \pm 0.04^{b}$ | $3.8 \pm 0.22^{c}$ | $14.9 \pm 1.29^{a}$ |
| Anguo | $8.72 \pm 0.48^{a}$ | $0.1 \pm 0.007^{b}$ | $0.26 \pm 0.02^{c}$ | $4.9 \pm 0.28^{b}$ | $11.6 \pm 1.01^{b}$ |

[1] CK, Y2 and Y3 were referred to Table 1; 'Anguo' was the local dandelion cultivar; Fresh leaf yield has been converted by actual measured value × loss factor (0.85); Data were expressed as mean ± standard deviations ($n$ = 4), within columns, means followed by the different letters of a, b, c etc. showed significant difference according to Duncan (D) significant difference analysis ($p$ = 0.05).

The dandelion is used as traditional herbal medicine, and generally high contents of bioactive compounds are an important consideration for dandelion quality or function evaluation. Hence, based on the above results, we continued comparing the ability of anti-tumor cell migration among these dandelion lines at the same time.

### 3.4. *Tumor Cell Migration Evaluation*

#### 3.4.1. Inhibitory Dose of Dandelion Extract on Tumor Cell Migration

The mobilized UAS/GAL4 expression system was used to generate a cell migration model in *Drosophila* with third-instar larval wing epithelia. RNA interference (RNAi) mediated the knockdown of the cell polarity gene scribble (scrib) triggered by massive cell migration phenotype (Figure 2d,e) [21,22]. The green fluorescent proteins (GFPs), labeled as migrating cells, were detached from the anterior-posterior (A/P) compartment boundary and moved toward the posterior part of the wing imaginal discs (Figure 2e,e'). After quantifying this phenotype, strong increased levels in the migrating cell number, the median, or the max distance were found (Figure 2f–h). Then, we tested the inhibitory effect of different concentrations of dandelion extract (Figure 2b,c) on ptc > scrib-IR in the Drosophila cell migrating model. The results only showed the extracts at the concentration of 0.2 g/mL obviously inhibited scrib-depleted induced cell invasion without affecting the fly normal growth (Figure 2f–h).

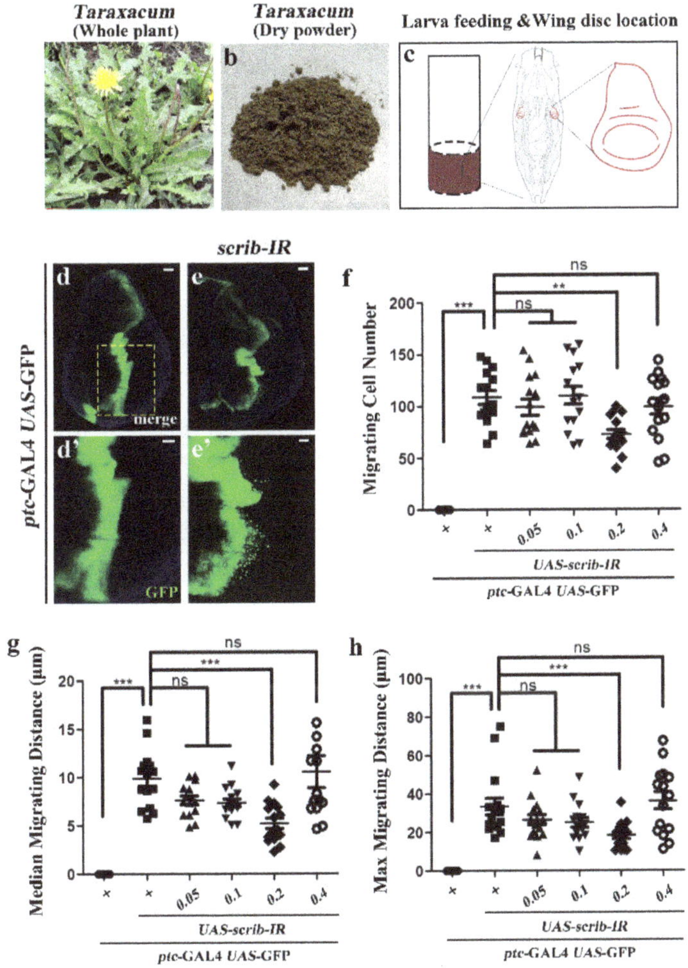


**Figure 2.** Dandelion extract inhibited cell polarity disruption-induced cell migration. Light graphs showing whole plant (**a**) or dry powder (**b**) of mother plant (CK). (**c**) Schematic view of larva feeding and the wing discs location. (**d**,**e**) Fluorescent images of 3rd instar larval wing discs. Compared with ptc-GAL4 UAS-GFP control (**d**), knockdown of scrib along the A/P boundary triggered massive cell migration in the wing pouch, which marked by green fluorescent protein (GFP) (**e**). Statistical analysis of the migrating cell number (**f**), median migrating distance (**g**) and max migrating distance (**h**) in indicated groups ($n = 20$ for each genotype). The columns from left to right were (1) ptc > GFP/+, (2) ptc > GFP/UAS scrib-IR, (3)–(6) ptc > GFP/UAS-scrib-IR larve treated with dandelion dry powder aqueous extracts at a concentration of 0.05, 0.1, 0.2 or 0.4 g/mL, respectively. Error bars indicates standard deviations. One-way ANOVA with Bonferroni multiple comparison test was used to compute $p$-values, *** $p < 0.001$, ** $p < 0.01$; ns, no significant difference. Scale bar: 40 μm (**d**,**e**), 20 μm (**d′**,**e′**).

### 3.4.2. Evaluation of Different Dandelions on Tumor Cell Migration

Based on above results, a 0.2 g/mL extract of different dandelion lines were fed to ptc > scrib-IR larvae. A loss of scrib could induce strong epithelial-mesenchymal transition (EMT) phenotypes, as revealed by the up-regulated expression of matrix metalloprotease1 (MMP1) (Figure 3a) and β-integrin (Figure S2), and a reduction in E-cadherin (a cell adhesion molecule) levels (Figure S2) [21,23,24]. By combining the statistical results of migrating cell number and migrating distances, the increased cell migration of ptc > scrib-IR was mildly suppressed by CK (mother plant) and largely inhibited by Y2 (Figure 3g–i). The up-regulated MMP1 expression was obviously suppressed by Y1, Y2, Y4, or CK, but not by Y3 (Figure 3a–f). In addition, the up-regulation of β-integrin and the reduction in E-cadherin activity were only inhibited by Y2, whereas the other tests had no effects on the β-integrin and E-cadherin alterations induced by a loss of scrib (Figure S2). Collectively, these results proved that dandelion materials from Y2 had a strong inhibitory effect on cell migration, the increased level of MMP1 and β-integrin, and the reduced expression of E-cadherin.

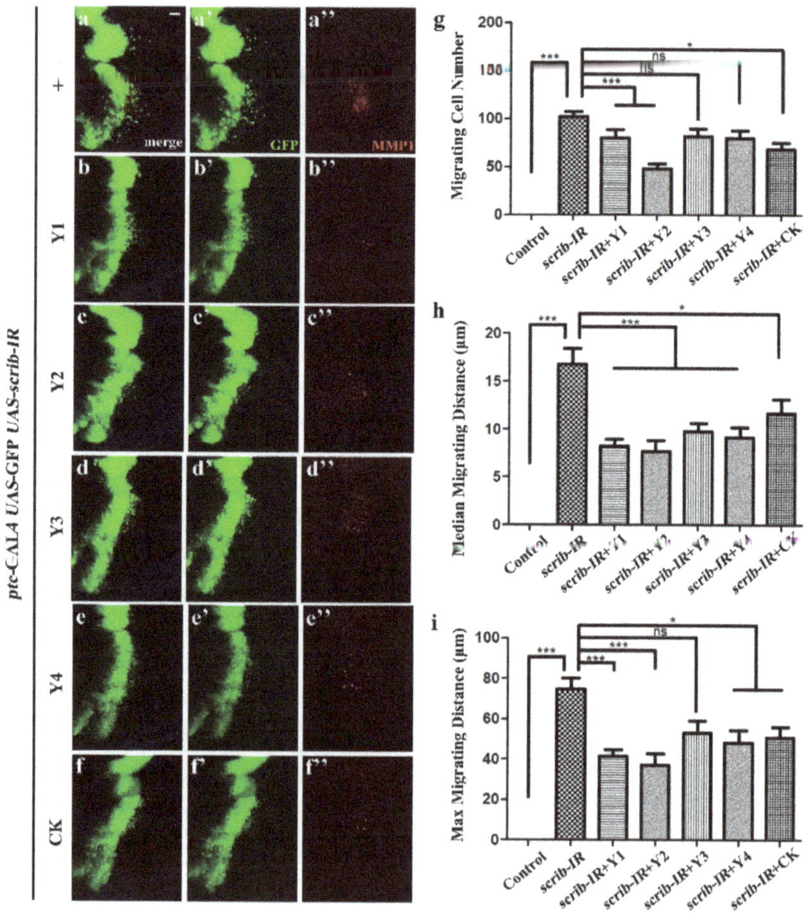


**Figure 3.** Dandelion extracts impede cell migration and MMP1 up-regulation. (**a**–**f**) Fluorescence micrographs of third instar larval wing discs were shown, anterior was to the left and dorsal up. The endogenous MMP1 expression was marked by anti-MMP1 antibody (red). Column bar graph of the migrating cell number (**g**), median migrating distance (**h**) and max migrating distance (**i**) as shown in (**a**–**f**) ($n = 20$ for each genotype), error bars indicate standard deviation. One-way ANOVA with Bonferroni multiple comparison test was used to compute $p$-values, *** $p < 0.001$, * $p < 0.05$; ns, no significant difference. Scale bar: 20 μm (**a**–**f**). CK: mother plant; Y1–Y4, represented the dandelion progenies originated from calluses that treated under NaCl concentration of 85 mM, 136 mM, 170 mM and 205 mM, respectively.

## 4. Conclusions

We created a series of dandelion progeny materials based on salt-induced callus mutation and tissue culture techniques. Combined with the results of morphological evaluation, a comparison of the main bioactive compounds, as well as an anti-tumor function evaluation of lines of Y2 and Y3, showed satisfied performance in functionality and yield. Of these, Y2 exhibited the highest phenolic acids content and strong ability of anti-tumor cell migration; thus we named Y2 as "Binpu 2". Its plant shape was upright, the fresh leaf yield was 14.2 t/ha, and the cichoric acid content was 6.7 mg/g. "Binpu 2" was suitable for mechanized harvesting and could be widely used for production in the field of food and medicine.

**Supplementary Materials:** The following supporting information can be downloaded at: https://www.mdpi.com/article/10.3390/horticulturae8121167/s1, Figure S1: Progenies of dandelion tissue culture plantlets grown on saline-alkali land; Figure S2: Dandelion extract suppresses β-integrin accumulation and reduced E-cadherin level induced by loss of scrib; Table S1: PCR primers information.

**Author Contributions:** Conceptualization and methodology, Z.W. and C.W.; software, Z.W.; validation, C.W. and R.M.; formal analysis, Z.W. and Z.L.; investigation, W.F., Z.L. and R.M.; resources, C.W. and W.F.; data curation, C.W.; writing—original draft preparation, Z.W.; writing—review and editing, Z.W. and C.W.; visualization, Z.L.; supervision, X.W.; project administration, X.W.; funding acquisition, X.W. All authors have read and agreed to the published version of the manuscript.

**Funding:** This research was funded by Hebei S & T Innovation Team of Modern Seed Industry of Traditional Chinese Medicine (21326312D-8); HAAFS Science and Technology Innovation Special Project (2022KJCXJX-BHS-4); Hebei Natural Science Foundation (H2022209027); Basic Research Funds of Hebei Academy of Agriculture and Forestry Sciences (2021010202); and Tangshan Talent Funding Project (A202203021).

**Data Availability Statement:** Not applicable.

**Acknowledgments:** We thank Lei Xue for technical support and thank the Bloomington and VDRC Fly Centers for fly stocks.

**Conflicts of Interest:** The authors declare no conflict of interest.

## References

1. Wang, X.; Xue, Z.; Lu, X.; Liu, Y.; Liu, G.; Wu, Z. Salt leaching of heavy coastal saline silty soil by controlling the soil matric potential. *Soil Water Res.* **2019**, *3*, 132–137. [CrossRef]
2. Wu, Z.; Xue, Z.; Li, H.; Zhang, X.; Wang, X.; Lu, X. Cultivation of dandelion (*Taraxacum erythropodium*) on coastal saline land based on the control of salinity and fertilizer. *Folia Hortic.* **2019**, *31*, 277–284. [CrossRef]
3. Liu, Y.; Sun, J.; Dai, X.; Zhao, Z.; Wang, X.; Zhang, G. Response of soil bacterial community structure and function under two salt-tolerant plants in a coastal saline soil area of eastern Hebei province of China. *Int. J. Phytoremediat.* **2021**, *24*, 842–854. [CrossRef] [PubMed]
4. Wu, Z.; Xue, Z.; Lu, X.; Jia, Y.; Wang, X.; Zhang, X. Salt-tolerance identification and quality evaluation of *Abelmoschus manihot* (L.) Medik. *Can. J. Plant Sci.* **2020**, *100*, 568–574. [CrossRef]
5. Xie, P.; Huang, L.; Zhang, C.; Ding, S.; Deng, Y.; Wang, X. Skin-care effects of dandelion leaf extract and stem extract: Antioxidant properties, tyrosinase inhibitory and molecular docking simulations. *Ind. Crop. Prod.* **2018**, *111*, 238–246. [CrossRef]
6. Zhang, X.; Li, Y.; Chen, H.; Shi, W. Selection and identification of salt tolerant variants of *Taraxacum officinale*. *Chin. J. Biotechnol.* **2008**, *24*, 262–271. (In Chinese)
7. Liu, Q.; Liu, Y.; Xu, Y.; Yao, L.; Liu, Z.; Cheng, H.; Ma, M.; Wu, J.; Wang, W.; Ning, W. Overexpression of and RNA interference with hydroxycinnamoyl-CoA quinate hydroxycinnamoyl transferase affect the chlorogenic acid metabolic pathway and enhance salt tolerance in *Taraxacum antungense* Kitag. *Phytochem. Lett.* **2018**, *28*, 116–123. [CrossRef]
8. Sigstedt, S.C.; Hooten, C.J.; Callewaert, M.C.; Jenkins, A.R.; Romero, A.E.; Pullin, M.J.; Kornienko, A.; Lowrey, T.K.; Slambrouck, S.V.; Steelant, W.F. Evaluation of aqueous extracts of *Taraxacum officinale* on growth and invasion of breast and prostate cancer cells. *Int. J. Oncol.* **2008**, *32*, 1085–1090. [CrossRef]
9. Nassan, M.A.; Soliman, M.M.; Ismail, S.A.; El-Shazly, S. Effect of *Taraxacum officinale* extract on PI3K/Akt pathway in DMBA-induced breast cancer in albino rats. *Biosci. Rep.* **2018**, *38*, BSR20180334. [CrossRef]
10. Menke, K.; Schwermer, M.; Felenda, J.; Beckmann, C.; Stintzing, F.; Schramm, A.; Zuzak, T. *Taraxacum officinale* extract shows antitumor effects on pediatric cancer cells and enhance mistletoe therapy. *Complement. Ther. Med.* **2018**, *40*, 158–164. [CrossRef]

11. Gonzalez, C. *Drosophila melanogaster*: A model and a tool to investigate malignancy and identify new therapeutics. *Nat. Rev. Cancer* **2013**, *13*, 172–183. [CrossRef]
12. Zhang, S.; Guo, X.; Wu, H.; Sun, Y.; Ma, X.; Li, J.; Xu, Q.; Wu, C.; Li, Q.; Jiang, C.; et al. Wingless modulates activator protein-1-mediated tumor invasion. *Oncogene* **2019**, *38*, 3871–3885. [CrossRef]
13. Wang, X.; Lu, X. A new dandelion variety "Binpu 1" with high yield, good quality and salt tolerance. *Xian Dai Nong Cun Ke Ji* **2019**, *2*, 52. (In Chinese)
14. Chen, G.; Zhang, X. Mechanism for salt tolerance of salt-tolerant mutant 'BINPU 1' of *Taraxacum mongolicum*. *Guihaia* **2021**, *41*, 1417–1424. (In Chinese)
15. Zhang, X.; Meng, R.; Feng, W.; Li, Z.; Lu, X.; Pan, X.; Wang, X. A New *Taraxacum mongolicum* Cultivar 'Binpu 2'. *Acta Hortic. Sin.* **2021**, *48*, 3025–3026. (In Chinese)
16. Wu, Z.; Li, Z.; Xue, Z.; Lu, X.; Wang, X. Optimization of extraction technology for determination of caffeic and chlorogenic acid in dandelion. *Banats J. Biotechnol.* **2020**, *11*, 26–37. [CrossRef]
17. Liu, N.; Song, M.; Wang, N.; Wang, Y.; Wang, R.; An, X.; Qi, J. The effects of solid-state fermentation on the content, composition and in vitro antioxidant activity of flavonoids from dandelion. *PLoS ONE* **2020**, *15*, e0239076. [CrossRef]
18. Ma, X.; Yang, L.; Yang, Y.; Li, M.; Li, W.; Xue, L. dUev1a modulates TNF-JNK mediated tumor progression and cell death in *Drosophila*. *Dev. Biol.* **2013**, *380*, 211–221. [CrossRef]
19. Wang, X.; Wang, Z.; Chen, Y.; Huang, X.; Hu, Y.; Zhang, R.; Ho, M.S.; Xue, L. FoxO mediates APP-induced AICD-dependent cell death. *Cell Death Dis.* **2014**, *5*, e1233. [CrossRef]
20. González-Castejón, M.; Visioli, F.; Rodriguez-Casado, A. Diverse biological activities of dandelion. *Nutr. Rev.* **2012**, *70*, 534–547. [CrossRef]
21. Ma, X.; Shao, Y.; Zheng, H.; Li, M.; Li, W.; Xue, L. Src42A modulates tumor invasion and cell death via Ben/dUev1a-mediated JNK activation in *Drosophila*. *Cell Death Dis.* **2013**, *4*, e864. [CrossRef] [PubMed]
22. Ma, X.; Li, W.; Yu, H.; Yang, Y.; Li, M.; Xue, L.; Xu, T. Bendless modulates JNK-mediated cell death and migration in *Drosophila*. *Cell Death Differ.* **2014**, *21*, 407–415. [CrossRef] [PubMed]
23. Zhao, M.; Szafranski, P.; Hall, C.A.; Goode, S. Basolateral junctions utilize warts signaling to control epithelial-mesenchymal transition and proliferation crucial for migration and invasion of *Drosophila* ovarian epithelial cells. *Genetics* **2008**, *178*, 1947–1971. [CrossRef] [PubMed]
24. Zeisberg, M.; Neilson, E.G. Biomarkers for epithelial-mesenchymal transitions. *J. Clin. Investig.* **2009**, *119*, 1429–1437. [CrossRef]

horticulturae

MDPI

Article

# Study of 15 Varieties of Herbaceous Peony Pollen Submicroscopic Morphology and Phylogenetic Relationships

Dongliang Zhang †, Anqi Xie †, Xiao Yang, Yajie Shi, Lijin Yang, Lingling Dong, Fuling Lei, Jingyue Wu * and Xia Sun *

College of Horticulture Science and Engineering, Shandong Agricultural University, Daizong Street #61, Tai'an 271018, China
* Correspondence: huixiang0813@163.com (J.W.); unsiax@sdau.edu.cn (X.S.)
† These authors contributed equally to this work.

**Abstract:** *Paeonia lactiflora* Pall. is widely used in medicine, garden applications, and as a potted ornamental. Cultivated varieties of paeonifloras suitable for cut flowers are urgently needed. In this study, the pollen morphology of *P. lactiflora* was studied and the characters of different varieties were compared, so as to provide reference for selecting suitable parents for new hybrid varieties. We examined the pollen morphology of 15 herbaceous peony varieties using scanning electron microscopy and analyzed the external pollen morphology and genetic relationship of the varieties. The pollen grains of the studied varieties were spheroidal or subspheroidal, bilaterally symmetrical monads, circular in polar view, and circular or elliptical in equatorial view. The exine of the pollen grains was observed as being relatively smooth under the light microscope, with the area around the equatorial axis having more lumina under the scanning electron microscope. The pollen grain exine sculpture was either reticular or pit type. The pollen apertures were tricolporate, arranged longitudinally, and equally spaced. The pollen grains were of two sizes: medium and small. The differences between the varieties were mainly reflected in the exine sculpture of the pollen. The closer the genetic relationship between the 15 peony varieties, the more subtle the differences in the exine sculpture. In the same cluster group, the morphological characteristics of herbaceous peony pollen were correlated with the shapes of flower and scale buds and the texture of the petals. However, the study identified no direct correlation with the cultivar type and flower color.

**Keywords:** *Paeonia lactiflora* Pall; pollen; exine sculpture; phylogenetic relationships

check for updates

**Citation:** Zhang, D.; Xie, A.; Yang, X.; Shi, Y.; Yang, L.; Dong, L.; Lei, F.; Wu, J.; Sun, X. Study of 15 Varieties of Herbaceous Peony Pollen Submicroscopic Morphology and Phylogenetic Relationships. *Horticulturae* **2022**, *8*, 1161. https://doi.org/10.3390/horticulturae8121161

Academic Editors: Wajid Zaman and Luigi De Bellis

Received: 20 October 2022
Accepted: 1 December 2022
Published: 7 December 2022

**Publisher's Note:** MDPI stays neutral with regard to jurisdictional claims in published maps and institutional affiliations.

**Copyright:** © 2022 by the authors. Licensee MDPI, Basel, Switzerland. This article is an open access article distributed under the terms and conditions of the Creative Commons Attribution (CC BY) license (https://creativecommons.org/licenses/by/4.0/).

## 1. Introduction

Pollen carries a large amount of genetic information, and the structure of pollen grains is determined by species genes. Individual pollen grains are not easily affected by the environment and have relatively stable characteristics [1–6]. Understanding the morphological characteristics of pollen is important for discussing the origin, evolution, classification, and kinship of seed plants [7,8]. Some researchers have used a scanning electron microscope (SEM) to observe the pollen morphological characteristics of ornamental plants and explain the relationship between their species or varieties [9,10]. Studies on the pollen morphology of *Paeonia lactiflora* or *Paeonia* species have also been reported [11,12].

*Paeonia lactiflora* Pall. is a species of herbaceous flowering plant in the family Paeoniaceae. It is widely used in medicine, garden applications, and as a potted ornamental. However, some characters (such as flower type, color, size, pedicel length, and upright stem) of the existing varieties are not in line with the application standards of fresh cut flowers, which restricts the production and sale of such flowers. Breeding suitable peony varieties is important for the production of fresh cut flowers. In this study, 15 varieties of peony were selected for submicroscopic morphology analysis. We used SEM to observe morphological characteristics of pollen from these 15 varieties. We then analyzed the relationship between the tested varieties based on the external morphological characteristics of the plant and exine sculpture of the pollen. The aim of the study was to provide a reference for the selection of parents of new cut flower varieties in hybrid breeding.

## 2. Materials and Methods

### 2.1. Test Materials and Sampling

The test was conducted at the Horticulture Experimental Station of Shandong Agricultural University, the Experimental Center of the College of Horticulture Science and Engineering of Shandong Agricultural University, and the College of Life Science of Shandong Agricultural University. The test materials comprised 15 varieties of herbaceous peony. The sampled peony varieties are shown in Figure 1 and detailed in Table 1. During late April to early May 2020, plant material was collected between 10:00 and 11:00 a.m. on clear days. Complete anthers with mature, but not cracked, powder were collected with tweezers and placed into a penicillin bottle filled with glutaraldehyde fixing solution. Five milliliters of the solution was then slowly drawn using a 10 mL syringe, and then pumped back in slowly. This process was repeated two or three times to ensure full contact between the solution and the anthers, and the anthers were then allowed to sink in the fixative. The bottles were then refrigerated at 0–4 °C until use.

**Figure 1.** External morphology of *Paeonia lactiflora* flowers: (**1**), Xueyuanhonghua; (**2**), Yangfeichuyu; (**3**), Fenchijinyu; (**4**), Gaoganhong; (**5**), Bingshan; (**6**), Xuefeng; (**7**), Tianshanhongxing; (**8**), Qingtianlan; (**9**), Guifeichacui; (**10**), Dafugui; (**11**), Hongfeng; (**12**), Chifen; (**13**), Hongxiuqiu; (**14**), Dongjingnvlang; (**15**), Hongfushi.

### 2.2. Electron Microscope

The pollen was dehydrated stepwise with alcohol and then evenly glued onto a circular metal platform with conductive double-sided tape using tweezers. The pollen was gold-coated in an IB-5 ion sputtering apparatus. The coated samples were scanned from different angles to observe the individual, local, and group sculpture of pollen grains and photographed using a JSM-6610LV scanning electron microscope. Twenty pollen grains from each species were selected for measurements and observations of sculpture type, polar axis length, equatorial axis length, lumina diameter, and ridge width. The values were averaged and the minimum–maximum range was recorded [13]. The submicroscopic morphology of the pollen grains was described according to Punt et al. [7].

**Table 1.** Main morphological characteristics of the varieties of *Paeonia lactiflora* in the test.

| Pollen Code | Cultivar Type | Bulbil Shape | Bud Shape | Flower Color | Petal Texture |
|---|---|---|---|---|---|
| 1 | Pavilion type | Brush-shaped | Inclined sharp-pointed peach-shaped | White | Papery |
| 2 | Crown type | Projectile-shaped | Sharp-pointed peach-shaped | White | Papery |
| 3 | Rose type | Bamboo-shaped | Inclined sharp-pointed peach-shaped | Pink | Leathery |
| 4 | Colorful-ball type | Projectile-shaped | Sharp-pointed peach-shaped | Red | Waxy |
| 5 | Crown type | Bamboo-shaped | Sharp-pointed peach-shaped | White | Papery |
| 6 | Crown type | Bamboo-shaped | Sharp-pointed peach-shaped | White | Papery |
| 7 | Rose type | Bamboo-shaped | Sharp-pointed peach-shaped | White | Waxy |
| 8 | Rose type | Brush-shaped | Sharp-pointed peach-shaped | Soft red | Papery |
| 9 | Crown type | Bamboo-shaped | Sharp-pointed peach-shaped | Pink | Papery |
| 10 | Rose type | Bamboo-shaped | Sharp-pointed peach-shaped | Red | Waxy |
| 11 | Colorful-ball type | Bamboo-shaped | Sharp-pointed peach-shaped | Red | Papery |
| 12 | Rose type | Brush-shaped | Sharp-pointed peach-shaped | Pink | Papery |
| 13 | Crown type | Bamboo-shaped | Sharp-pointed peach-shaped | Red | Waxy |
| 14 | Rose type | Bamboo-shaped | Sharp-pointed peach-shaped | Pink | Leathery |
| 15 | Colorful-ball type | Brush-shaped | Sharp-pointed peach-shaped | Burgundy | Waxy |

1, Xueyuanhonghua; 2, Yangfeichuyu; 3, Fenchijinyu; 4, Gaoganhong; 5, Bingshan; 6, Xuefeng; 7, Tianshanhongxing; 8, Qingtianlan; 9, Guifeichacui; 10, Dafugui; 11, Hongfeng; 12, Chifen; 13, Hongxiuqiu; 14, Dongjingnvlang; 15, Hongfushi.

### 2.3. Data Analysis

Based on the data of six indexes (pollen polar axis length, equatorial axis length, ratio of polar axis length to equatorial axis length, perforation diameter, number of perforations per unit area, and ridge width), systematic cluster analysis in IBM SPSS Statistics 19.0 software was used to develop a dendrogram [14].

## 3. Results and Analysis

### 3.1. External Pollen Forms

According to Erdtman's NPC classification system [8], the pollen of the tested peony varieties was classified as N3P4C5. The pollen grains were of monad type, spheroidal or subspheroidal (P/E ranging from 0.91 to 1.03), and symmetrical, except for a few pollen grains with an irregular shape (Table 2 and Figure 1). The size of normally developed grains in the same cultivar was ambiguous, but their size among different varieties was pronounced. The pollen in all tested varieties was circular in polar view and circular or elliptical in equatorial view (Table 2). The varieties with elliptical pollen in equatorial view were: Xueyuanhonghua, Yangfeichuyu, Fenchijinyu, Gaoganhong, Bingshan, Tianshanhongxing, Qingtianlan, Guifeichacui, Dafugui, Dongjingnvlang, and Hongfushi. The varieties with circular pollen in equatorial view were Xuefeng, Hongfeng, Chifen, and Hongxiuqiu.

**Table 2.** Pollen grain characteristics of *Paeonia lactiflora* varieties.

| Pollen Code | Length of Polar Axis (μm) | Length of Equatorial Axis (μm) | P/E | Perforation Number per 100 $\mu m^2$ | Lumina Diameter (μm) | Ridge Width (μm) | D/W | Shape of Two Poles | Type of Exine Sculpture |
|---|---|---|---|---|---|---|---|---|---|
| 1 | 23 (18.5–26.4) | 25.3 (21.2–28.4) | 0.91 | 140 | 0.60 | 0.41 | 1.46 | circular and elliptical | reticular |
| 2 | 24.14 (21.6–26) | 26.3 (21.6–29.4) | 0.92 | 145 | 0.60 | 0.51 | 1.18 | circular and elliptical | pit |
| 3 | 24.3 (22–27.2) | 26.4 (23.8–28.6) | 0.92 | 85 | 0.82 | 0.59 | 1.39 | circular and elliptical | reticular |
| 4 | 26.7 (22.4–29.6) | 26.9 (21.6–30.2) | 0.99 | 102 | 0.55 | 0.58 | 0.95 | circular and elliptical | pit |

**Table 2.** *Cont.*

| Pollen Code | Length of Polar Axis (μm) | Length of Equatorial Axis (μm) | P/E | Perforation Number per 100 μm$^2$ | Lumina Diameter (μm) | Ridge Width (μm) | D/W | Shape of Two Poles | Type of Exine Sculpture |
|---|---|---|---|---|---|---|---|---|---|
| 5 | 27.2 (23–31.5) | 27.8 (24.8–30.2) | 0.98 | 140 | 0.57 | 0.49 | 1.20 | circular and elliptical | pit |
| 6 | 24.3 (20–28.2) | 25.4 (23.6–26.8) | 0.96 | 170 | 0.44 | 0.51 | 0.86 | circular | pit |
| 7 | 27 (21.6–31.2) | 26.3 (21.6–29.2) | 1.03 | 70 | 0.57 | 0.69 | 0.83 | circular and elliptical | pit |
| 8 | 25 (32.1–22.2) | 25.6 (20.6–28.5) | 0.98 | 130 | 0.56 | 0.50 | 1.12 | circular and elliptical | pit |
| 9 | 25.4 (20.8–29.6) | 25.7 (19–30.8) | 0.99 | 125 | 0.48 | 0.48 | 0.86 | circular and elliptical | pit |
| 10 | 28.3 (24.4–31.6) | 27.8 (22.6–31.4) | 1.02 | 100 | 0.50 | 0.61 | 0.82 | circular and elliptical | small pit |
| 11 | 22.62 (17.8–26) | 23.9 (14.8–28.4) | 0.95 | 145 | 0.35 | 0.36 | 0.97 | circular | pit |
| 12 | 25.7 (20.8–28.87) | 26.6 (22–28.4) | 0.97 | 125 | 0.55 | 0.45 | 1.22 | circular | pit |
| 13 | 25.2 (21.8–27.8) | 25.3 (20.4–27.6) | 1.00 | 125 | 0.48 | 0.57 | 0.84 | circular | small pit |
| 14 | 22.4 (17.8–25.8) | 24.3 (17.6–29.6) | 0.92 | 160 | 0.37 | 0.42 | 0.88 | circular and elliptical | small pit |
| 15 | 24.7 (19–29.2) | 26.2 (23–29) | 0.94 | 160 | 0.49 | 0.41 | 1.2 | circular and elliptical | pit |

1, Xueyuanhonghua; 2, Yangfeichuyu; 3, Fenchijinyu; 4, Gaoganhong; 5, Bingshan; 6, Xuefeng; 7, Tianshanhongxing; 8, Qingtianlan; 9, Guifeichacui; 10, Dafugui; 11, Hongfeng; 12, Chifen; 13, Hongxiuqiu; 14, Dongjingnvlang; 15, Hongfushi.

### 3.2. Surface Ornamentation Characteristics of Pollen

The exine of the pollen grains of the 15 peony varieties was smooth at light microscope, and with a clearly defined lumina near the equatorial axis at SEM. From the equatorial axis to the poles, the diameter of the lumina gradually decreased, and the shape of the lumina openings was roughly circular or subcircular. According to the quantitative index of peony pollen morphology established by Yuan and Wang [15], the ratio of lumina diameter to ridge width was used as a reference index for the pollen exine sculpture. The exine sculpture can be divided into three types: reticular, pit, and small pit. The reticular exine type was present in Xueyuanhonghua and Fenchijinyu, the small pit type was found in Dafugui, Hongxiuqiu, and Dongjingnvlang, and the pit sculpture was present in Yangfeichuyu, Gaoganhong, Bingshan, Xuefeng, Tianshanhongxing, Qingtianlan, Guifeichacui, Hongfeng, Chifen, and Hongfushi (Table 2 and Figure 2).

### 3.3. Apertures of the Pollen Grains

The pollen apertures of the 15 herbaceous peony varieties were tricolporate, arranged longitudinally, and equally spaced. The trenches demonstrated slight and gradual narrowing toward the poles along the polar axis. The width of the pores and colpi varied slightly between the varieties. The pore membrane had protrusions in Gaoganhong, Bingshan, and Xuefeng and was without protrusions in Xueyuanhonghua, Yangfeichuyu, Fenchijinyu, Tianshanhongxing, Qingtianlan, Guifeichacui, Dafugui, Hongfeng, Chifen, Hongxiuqiu, Dongjingnvlang, and Hongfushi (Figure 2).

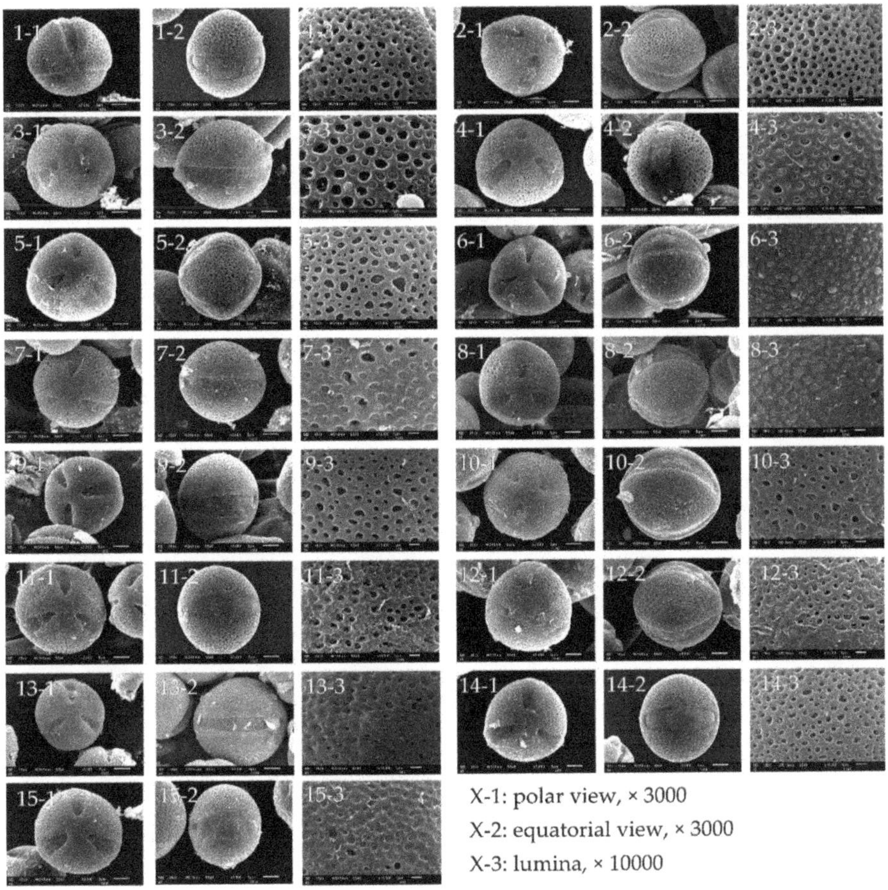


**Figure 2.** Pollen morphological characteristics of *Paeonia lactiflora* varieties: 1-1, 1-2, 1-3: Xueyuanhonghua; 2-1, 2-2, 2-3: Yangfeichuyu; 3-1, 3-2, 3-3: Fenchijinyu; 4-1, 4-2, 4-3: Gaoganhong; 5-1, 5-2, 5-3: Bingshan; 6-1, 6-2, 6-3: Xuefeng; 7-1, 7-2, 7-3: Tianshanhongxing; 8-1, 8-2, 8-3: Qingtianlan; 9-1, 9-2, 9-3: Guifeichacui; 10-1, 10-2, 10-3: Dafugui; 11-1, 11-2, 11-3: Hongfeng; 12-1, 12-2, 12-3: Chifen; 13-1, 13-2, 13-3: Hongxiuqiu; 14-1, 14-2, 14-3: Dongjingnvlang; 15-1, 15-2, 15-3: Hongfushi.

### 3.4. Pollen Grain Size

The pollen grains of the 15 peony varieties were divided into two size categories: medium (25–50 μm) and small (10–25 μm) (Table 2).

Seven varieties had a polar axis between 10 and 25 μm: Xueyuanhonghua, Yangfeichuyu, Fenchijinyu, Xuefeng, Hongfeng, Dongjingnvlang, and Hongfushi. Among them, the pollen grains of Hongfeng were the smallest (22.62 μm × 23.9 μm).

Eight varieties had a polar axis between 25 and 50 μm: Gaoganhong, Bingshan, Tianshanhongxing, Qingtianlan, Guifeichacui, Dafugui, Chifen, and Hongxiuqiu, with Dafugui having the largest pollen grains (28.3 μm × 27.8 μm).

### 3.5. Cluster Analysis

The systematic cluster analysis diagram of pollen morphology directly reflects the similarity of various units in the pollen exine sculpture (Figure 3). The closer the kinship, the earlier the aggregation. According to this, the species with close kinship can be determined, and the tested varieties can be divided into three groups.

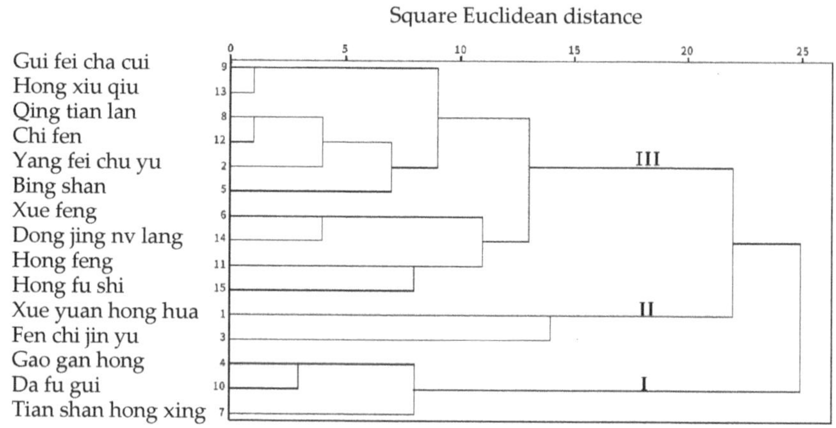


**Figure 3.** Clustering diagram of 15 varieties of *Paeonia lactiflora*.

The pollen morphology of the three varieties in Group I was nearly identical, and the length of polar axis values were mostly 26–28.5 μm. All of the pollen grain indexes of the two varieties in Group II were consistent, except for pore density, which varied slightly between the two varieties. Group III includes 10 varieties characterized by the lumina type of pollen pit, circular or subcircular pollen shape in polar view, and P/E values ranging from 0.9 to 1. The ridge width was 0.36 μm in Hongfeng and 0.45–0.55 μm in the other nine varieties.

## 4. Discussion

### *4.1. Similarities and Differences in Pollen Submicroscopic Morphological Characteristics among Herbaceous Peony Varieties*

The pollen outer wall sculpture varies with species [16,17]. The SEM revealed both similarities and differences in the pollen grains among the 15 herbaceous peony varieties that we examined. The differences were mainly reflected in the external morphology, exine sculpture, apertures, and pollen grain size. In terms of the external morphology, the pollen grains were circular in the polar view, and circular to elliptical in the equatorial view. The equatorial view can be used as a basis for analyzing the relationships among the 15 varieties.

In this study, the pollen grain exine was relatively smooth under the light microscope, and the lumina near the equatorial axis was clear under the SEM. From the equator to the poles, the diameter of the lumina gradually decreased. The lumina shape was roughly circular or elliptical. The morphological differences in the pollen grains were mainly reflected in the number of perforations per unit area, lumina diameter, ridge width, shape in polar view, and sculpture type. The number of perforations per unit area varied with cultivar. There were three basic types of sculpture: reticular, pit, and small pit. The lumina diameters and ridge widths of individual grains of the same pollen were not equal, and the size difference varied with cultivar. There were some differences between the results of this study and those of Xi [18]. The exine sculpture, shape, and size of the pollen grains among different individuals of the same species are not affected by planting location, showing strong conservation [19]. Different test materials may lead to the differences in the above conclusions.

The pollen grains in the studied varieties were tricolporate, arranged longitudinally, and equally spaced. The trenches gradually and slightly narrowed toward the poles along the polar axis. The spindle-shaped colpium was situated near the poles. The width of the colpi varied slightly among cultivar varieties. The pore membrane was either with or without protrusions. The varieties with protrusions included Gaoganhong, Bingshan, and

Xuefeng, while the other 12 varieties had no protrusions in the pore membrane. The size of the pollen grains varied by cultivar and were separated into medium (25–50 μm) and small (10–25 μm) sizes. The microstructure of the exine sculpture of the pollen grains was consistent among individuals of the same cultivar, but varied among different varieties. The results of this experiment are consistent with those reported by Jin et al. [20].

### 4.2. Kinship

Cross breeding is an important way to breed new cut flower varieties and is one of the most widely used and effective methods for breeding [21]. Therefore, understanding the morphological relationship between varieties can help to predict the values for target phenotypes when cross-breeding new varieties. According to the pollen morphological characteristics and cluster analysis diagram, the closer the genetic relationship between varieties, the more subtle the differences in pollen morphological characteristics and exine sculpture. This is consistent with the results of Hao and Ma [12]. The cluster dendrogram based on pollen morphology resolved the three varieties of Gaoganhong, Tianshanhongxing, and Dafugui in the same cluster group (Cluster I) and are therefore closely related. Their petals are waxy in texture, their scales and buds are bamboo shoot-shaped, and their flower buds are pointed peach-shaped. The varieties Xueyuanhonghua and Fenchijinyu belong to the same cluster group (Cluster II) and have the closest relationship. Their petals have a papery texture, the scales and buds are brush-shaped, and the flower buds are crooked peach-shaped. The remaining 10 varieties (Cluster III) were closely related on the basis of pollen morphology. They are characterized by pointed peach-shaped flower buds. The petals of four varieties, Xuefeng, Hongfeng, Dongjingnvlang, and Hongfushi, are leathery in texture, and their scales and buds are in the shape of bamboo shoots. The petals of the remaining six varieties are papery in texture, and the scales and buds are projectile-shaped. This indicated that the pollen morphological characteristics in Clusters I and II were consistent with the texture of the petals, the shape of the scale buds, and the shape of the flower buds. The morphological characteristics of pollen in Cluster III were consistent with the shape of the flower buds, but not with the texture of the petals or the shape of the scale buds.

## 5. Conclusions

The aim of this study was to provide a basis for the selection of parents of new cut flower varieties in hybrid breeding. The 15 tested peony varieties were clustered into three groups. The pollen morphological characteristics correlated to a certain degree with petal texture and the shape of the scales and flower buds. There was no direct correlation between the morphological characteristics of the pollen and the cultivation type or flower color. These phenotypes may be controlled by multiple genotypes or are the results of long-term natural and artificial selection, which warrants further study. In this study, we studied the pollen morphology of *P. lactiflora* and compared the characteristics of different varieties. Future research will examine the molecular biology, plant morphology, and plant physiology of the tested varieties to provide a reference for the breeding of *Paeonia* cut flowers.

**Author Contributions:** D.Z. and A.X. co-designed and undertook most of the study, analyzed the data, and wrote the paper; X.Y., Y.S., L.Y., L.D. and F.L. participated in the tests and data analysis; J.W. and X.S. co-designed the study and revised drafts of the manuscript. All authors have read and agreed to the published version of the manuscript.

**Funding:** This work was supported by the Shandong Province Improved Seed Project (No. 2021S230304-02583).

**Institutional Review Board Statement:** Not applicable.

**Informed Consent Statement:** Not applicable.

**Data Availability Statement:** The datasets used and/or analyzed during the current study are available from the corresponding author on reasonable request.

**Conflicts of Interest:** The authors declare no conflict of interest.

## References

1. Wodehouse, R.P. *Pollen Grains*; McGraw-Hill Book Co. Inc.: New York, NY, USA, 1935.
2. Agababian, V.S. Pollen morphology of the family Magnoliaceae. *Grana* **1972**, *12*, 166–176. [CrossRef]
3. Praglowski, J. *World Pollen and Spore Flora*; Almqvist & Widsell: Stockholm, Sweden, 1974; Volume 3, pp. 1–44.
4. Coimbra, S.; Costa, M. Pollen grain development is compromised in *Arabidopsis agp6 agp11* null mutants. *J. Exp. Bot.* **2009**, *60*, 3133–3142. [CrossRef] [PubMed]
5. Doyle, J.A.; Le Thomas, A. Evolution and phylogenetic significance of pollen in Annonaceae. *Bot. J. Linn. Soc.* **2012**, *169*, 190–221. [CrossRef]
6. Aguilar-Garcıá, S.A.; Figueroa-Castro, D.M.; Castañeda-Posadas, C. Pollen morphology of *Pachycereus weberi* (Cactaceae): An evaluation of variation in pollen size. *Plant Syst. Evol.* **2012**, *298*, 1845–1850. [CrossRef]
7. Punt, W.; Hoen, P.P.; Blackmore, S.; Nilsson, S.; Le Thomas, A. Glossary of pollen and spore terminology. *Rev. Palaeobot. Palynol.* **2007**, *143*, 1–81. [CrossRef]
8. Erdtman, G. *Handbook of Palynology*; Translation by Academy of Sciences in China; Science Press: Beijing, China, 1978; pp. 1–120. (In Chinese)
9. House, A.; Balkwill, K. FIB-SEM: An additional technique for investigating internal structure of pollen walls. *Microsc. Microanal.* **2013**, *19*, 1535–1541. [CrossRef] [PubMed]
10. Khan, S.U.; Zafar, M.; Ahmad, M.; Anjum, F.; Sultana, S.; Kilic, O.; Ozdemir, F.A.; Nazir, A.; Yaseen, G.; Aabidin, S.Z.U. Pollen micromorphological analysis of tribe Acacieae (Mimosaceae) with LM and SEM techniques. *Microsc. Res. Tech.* **2019**, *82*, 1610–1620. [CrossRef] [PubMed]
11. Guo, X.F.; Wang, L.Y.; Yuan, T. Studies on the pollen morphology of 4 species of wild herbaceous peony. *Sci. Silvae Sin.* **2005**, *41*, 184–188. (In Chinese)
12. Hao, L.H.; Ma, H.; da Silva, J.A.T.; Yu, X.N. Pollen morphology of herbaceous peonies (*Paeonia* L.) with different ploidy levels. *J. Am. Soc. Hortic. Sci.* **2016**, *141*, 275–284. [CrossRef]
13. Yang, S.; Zheng, Z.; Huang, K.; Li, J.; Wei, X.; Xu, Q. Southern China the main crop and vegetable and fruit modern pollen morphology and agriculture archaeology research value. *Acta Micropalaeontol. Sin.* **2012**, *29*, 80–98. (In Chinese)
14. Mert, C. Pollen morphology and anatomy of cornelian cherry (*Cornus mas* L.) varieties. *HortScience* **2009**, *44*, 519–522. [CrossRef]
15. Yuan, T.; Wang, L. Discussion on several peony pollen morphology and evolution of wild species of classification. *J. Beijing Univ.* **1999**, *21*, 17–21. (In Chinese)
16. Liu, Y.J.; Wang, J.C.; Xiong, Y.M.; Liu, R.Z.; Huang, X.F.; Yang, L.; Shen, C.G. Morphological characteristics of pollen grains from 16 germplasms of pitaya. *Fresenius Environ. Bul.* **2020**, *29*, 1522–1533.
17. Piwowarczyk, R.; Rura, K.; Krasylenko, Y.; Kasińska, J.; Pedraja, S.S. Seed micromorphology of representatives of holoparasitic Orobanchaceae genera from the Caucasus region and its taxonomic significance. *Phytotaxa* **2020**, *432*, 223–251. [CrossRef]
18. Xi, Y. The pollen morphology and exine ultrastructure of China *Paeonia*. *Acta Bot. Sin.* **1984**, *26*, 241–246. (In Chinese)
19. Humphrey, R.P. Pollen heteromorphism is pervasive in *Thalictrum* (Ranunculaceae). *Plant Syst. Evol.* **2016**, *302*, 1171–1177. [CrossRef]
20. Jin, B.; He, X.; Wu, J.; Ding, L. The relationship between morphological characteristics of pollen and cultivar evolution. *Jiangsu J. Ag. Sci.* **2005**, *21*, 63–68. (In Chinese)
21. Zhang, W.; Guan, W.L.; Li, Y.F.; Peng, L.C.; Zhang, L.; Meng, J.; Wang, J.H.; Song, J. Cytology and pollen morphology of *Bougainvillea glabra* 'Elizabeth Angus,' a cultivar with low pollen fertility. *Sci. Hortic.* **2022**, *301*, 111105. [CrossRef]

*horticulturae*

MDPI

Article

# Comparative Study on Genome Size and Phytochemical Profile of Three Potential Species of *Acacia*: Threatened and Endemic to Saudi Arabia

**Salim Khan *, Fahad Al-Qurainy, Abdulrahman Al-hashimi, Mohammad Nadeem, Mohamed Tarroum, Abdalrhaman M. Salih and Hassan O. Shaikhaldein**

Department of Botany and Microbiology, College of Science, King Saud University, P.O. Box 2455, Riyadh 11451, Saudi Arabia

* Correspondence: skhan2@ksu.edu.sa; Tel.: +966-4675865

**Abstract:** Acacias are widely distributed in tropical and subtropical regions of the world and have both economic as well as medicinal value. The estimation of genome size is very important as it changes due to the change in noncoding DNA sequence as well as genome duplication among organisms for their evolutionary aspects. Three potential species of the genus *Acacia* including *Acacia etbaica*, *Acacia johnwoodii* and *Acacia origena*, which are threatened and nearly endemic to Saudi Arabia, were collected. The present study was carried out to determine the genome size (2C DNA contents), total phenolic content (TPC), total flavonoid (TFC) and some bioactive compounds in these species for their comparison. The genome size ranged from 1.91 pg (*A. etbaica*) to 2.45 pg/2C (*A. origena*) among the *Acacia* species, which correspond to genome sizes 1843.15–2364.25 Mbp, respectively. The variation was observed in genome size within *Acacia* species as nuclei were extracted using different extraction buffers except for GB and MB01 buffers. The FTIR analysis revealed the presence of various functional groups in compounds that might be responsible for different types of phytochemicals in these *Acacia* species. Total flavonoid content (TFC) ranged from 0.647 (*A. origena*) to 1.084 mg QE/g DW (*A. etbaica*), whereas the total phenolic f content (TPC) ranged between 15.322 (*A. origena*) to 28.849 (*A. johnwoodii*) mg/g DW of GAE. HPLC analysis revealed the presence of quercetin 3-β-D-glucoside and luteolin 7-rutinoside in the leaves of all three *Acacia* species in considerable amounts, and these might have good health-promoting effects. This is our first study on genome size (2C DNA content) using flow cytometry and phytochemical profiling on these Acacias. Thus, estimated genome size and phytochemical study of these species could help to understand the biosynthesis of secondary metabolites under various genes and the evolutionary relationships among them.

**Keywords:** bioactive compounds; C value; flavonoid; FTIR; medicinal plants; phenolics; propidium iodide

check for updates

**Citation:** Khan, S.; Al-Qurainy, F.; Al-hashimi, A.; Nadeem, M.; Tarroum, M.; Salih, A.M.; Shaikhaldein, H.O. Comparative Study on Genome Size and Phytochemical Profile of Three Potential Species of *Acacia*: Threatened and Endemic to Saudi Arabia. *Horticulturae* **2022**, *8*, 994. https://doi.org/10.3390/horticulturae8110994

Academic Editor: Wajid Zaman

Received: 25 September 2022
Accepted: 18 October 2022
Published: 26 October 2022

**Publisher's Note:** MDPI stays neutral with regard to jurisdictional claims in published maps and institutional affiliations.

**Copyright:** © 2022 by the authors. Licensee MDPI, Basel, Switzerland. This article is an open access article distributed under the terms and conditions of the Creative Commons Attribution (CC BY) license (https://creativecommons.org/licenses/by/4.0/).

## 1. Introduction

*Acacia* species (approx. 1380 species) are distributed in tropical and subtropical parts of the world, including large areas of the Arabian Peninsula, deserts of Africa and the Middle East, and two-thirds are native to Australia [1–4]. *Acacia*, also known as *Acacias* (family: Fabaceae; subfamily: Mimosoideae), is considered an important shrub and tree. Different species of *Acacia* are found in Western, Northern and Eastern parts of Saudi Arabia and grow in various types of soils [5,6]. Some species of *Acacia*, viz., *Acacia johnwoodii* Boulos, *Acacia origena* Hunde and *Acacia etbaica* Schweinf, have been reported from Saudi Arabia. According to Al-Mefarrej [7], *A. origena* and *A. etbaica* are found in the Al-Baha and Aseer regions of Saudi Arabia. Acacias have various uses as these are potential sources of firewood, forage, timber, gum, fiber, tannins, folk food, and medicine; moreover, these species are used as useful for soil, environmental protection and water conservation.

Genome size is an invariable feature of an individual and is generally invariable within a species. The amount of nuclear DNA, a set of simple multiples of its basic quantity, is named 'C-values'. The amount of DNA in the unreplicated gametic nucleus of an organism is known as the 1C value (holoploid genome size) [8], which is termed genome size. Genome size (nuclear DNA content) varies approx. 2400-fold in angiosperms as a result of changes in the amount of genome duplication and noncoding DNA sequences [9]. In seed plants genome size varied from 0.13 to 254.8 pg [10,11]. Comparative studies have suggested that large genome size is maladaptive through its constraints on plant physiology in plants [12,13]. At present, duplications of whole-genome have clearly had a main effect on genome size in plants.

The use of flow cytometry (FCM) in the determination of DNA ploidy level, cell cycle analysis and assessment of nuclear DNA content have arisen as its most widespread applications [11,14]. Flow cytometry has been commonly used at the population level to identify alterations in nuclear DNA contents, i.e., cytotype and intraspecific variation, which is often one of the results of hybridization, as reported in Sorbus and Viola [15] [16,17]. Undoubtedly, this method is simple, low cost and has become a complementary means to infer output data from population genetics (i.e., allelic frequencies). There are various protocols available in the literature for the extraction of nuclei for genome size estimation as developed by researchers [18–22].

A large group of plant polyphenols known as flavonoids is found in different plant species. Flavonoids have a wide range of health-promoting properties due to their antioxidant nature. Natural flavonoids and phenolic compounds have been used to cure various human diseases [23]. These flavonoids act as shielding compounds to protect the plant against damage from high levels of solar radiation, particularly ultraviolet (UV) radiation. The flavonoid quercetin is a major constituent in plants which is found in abundance in its glycoside form [24]. Quercetin compounds have various biological and pharmacological activities, including anti-inflammatory, antiviral, antibacterial and anticarcinogenic [25–27]. Flavonoid glycosides participate in the prevention of cancer, atherosclerosis, and chronic inflammation in humans by deacceleration of oxidative degradation and treatment of amyotrophic lateral sclerosis [23,28]. The flavonoid rutin has multiple pharmacological activities and protects plants against pathogens or ultraviolet radiation and also used to inhibit the side effects of some diseases such as hypercholesteremia diabetes, and cancer treatments [29,30]. Among the various secondary metabolites, phenolics compounds have several beneficial effects to humans particularly in the cure of diseases such as neurodegenerative disorders, cancer and cardiovascular [31–34].

Based on many applications of secondary metabolites in humans, as reported in the literature, our main objective is to study the phytochemical profiling of the leaves of *Acacia* species, including *A. johnwoodii*, *A. etbaica* and *A. origena*, since a comparative phytochemical study has not been reported till date. The estimation of genome size (2C DNA content) of these *Acacia* species is important as they have a narrow geographical range, low density and small population size and grow under extreme environmental conditions. Therefore, in the present study, different extraction buffers were used for the extraction of nuclei from cotyledons of germinated seeds of Acacias for genome size estimation and a comparative study was performed among them.

## 2. Materials and Methods

### 2.1. Seeds Collection

The seeds of *Acacia* species were collected in the month of June 2021 from Abha province of Saudi Arabia. The seeds were identified by a taxonomist, and voucher specimen for *Acacia johnwoodii* (21635), *Acacia origena* (23066) and *Acacia etbaica* (23811) was deposited at the Department of Botany and Microbiology, College of Science, King Saud University, Riyadh, Saudi Arabia. After collection, the seeds were stored at room temperature, and further, they were air-dried to a minimum moisture content before storage. The collection of *Acacia* species and their information are given in (Table 1).

**Table 1.** *Acacia* species status, collection place, seed size and leaf size (plant grown in a pot in a growth chamber).

| *Acacia* Species | Accession Number | Species Status in Saudi Arabia | Collection Place | Seed Diameter (cm) (Mean ± SD) | Seed Length (cm) (Mean ± SD) | Leaf Length (cm) |
|---|---|---|---|---|---|---|
| *Acacia johnwoodii* (Boulos) Ragup., Seigler, Ebinger & Maslin | KSUSB95 | Threatened | Dhi Ain (Abha) | $0.62 \pm 0.012$ [a] | $1.1 \pm 0.100$ [a] | $4.7 \pm 0.55$ [a] |
| *Acacia etbaica* (Schweinf.) Kyal. & Boatwr | KSUSB99 | Near endemic | Wadi Khaytan (Abha) | $0.43 \pm 0.007$ [c] | $0.518 \pm 0.01$ [c] | $2.47 \pm 0.34$ [b] |
| *Acacia origena* (Hunde) Kyal. & Boatwr | KSUSB92 | Near endemic | Dhi Ain (Abha) | $0.52 \pm 0.025$ [b] | $0.630 \pm 0.01$ [b] | $1.53 \pm 0.152$ [c] |

[a,b,c] indicates significant difference according to Duncan's test ($p < 0.05$).

### 2.2. Pot Experiment

The Acacias seeds were sterilized with minor modifications, as followed by Al-Qurainy et al. [35]. The collected mature seeds were washed with tap water for 30 min, followed by doubled distilled water to remove microbes and dust. The washed seeds were treated with 50% bleach for 10 min for surface cleaning, and thereafter, they were washed with distilled water 3–4 times. The sterilized seeds were sowed in the pot containing soil consisting of peat moss and pearlite (3:1). The seedlings of all *Acacias* were raised in a growth chamber adjusted the conditions as 16-h/8-h, 25 ± 1 °C Day/night cycle. The cotyledons (20 days after sowing) were harvested for the genome size estimation. The young leaves were harvested from 3 months old plants for phytochemical profiling and bioactive compound estimation.

### 2.3. Buffer Preparation for Nuclei Extraction

Galbraith's buffer [18] (45 mM $MgCl_2$; 30 mM sodium citrate; 20 mM MOPS; 0.1% (*v*/*v*) Triton X-100; pH 7.0)

LB01 buffer [21] (15 mM Tris; 2 mM $Na_2EDTA$; 0.5 mM spermine.4HCl; 80 mM KCl; 20 mM NaCl; 15 mM β-mercaptoethanol; 0.1% (*v*/*v*) Triton X-100; pH 7.5)

MB01 [22] (20 mM MOPS; 25 mM $Na_2EDTA$; 0.7 mM spermine.4HCl; 80 mM KCl; 20 mM NaCl; 1% (*w*/*v*) PVP; 0.5% (*v*/*v*) β-mercaptoethanol; 0.2% (*v*/*v*) Triton X-100; pH 7.4

Tris-$MgCl_2$ buffer [19] (200 mM Tris; 4 mM $MgCl_2.6H_2O$; 0.5%(*v*/*v*) Triton X-100; pH 7.5)

### 2.4. Nuclei Extraction and Staining

The nuclei were extracted using the protocol as developed by authors [18–21] with slight modification in PVP (2%, *w*/*v*) and β-mercaptoethanol (0.6%, *v*/*v*) concentrations in nuclei isolation buffer. For each species, three replicates were taken for nuclei extraction as well as analysis. All necessary equipment used in the nuclei extraction was cleaned thoroughly. The whole nuclei extraction process was performed on ice. The cotyledon was used from the germinated seeds of all *Acacias* for the extraction of nuclei. The harvested cotyledons were washed with distilled water three times to remove soil, bacteria and fungi. Before extraction of the nuclei, the cotyledons were kept in doubled distilled water to get fully turgid cells for easy chopping and releasing of nuclei in the buffer. The young and fresh cotyledons of germinated seeds (25 mg for each) were chopped with a sharp razor blade into 0.4–0.5 mm size in cold nuclei isolation buffer (500 µL buffer). The released nuclei in the extraction buffer were passed through a double nylon mesh (pore size, 20 µm, purchased from Macrokun (Shijiazhuang, China). The final volume of filtered nuclei suspension was made to 750 µL. The nuclei suspension was stained with 50 µg/mL of PI dye (Propidium iodide) and incubated at 4 °C for 10 min. The same process was used for the extraction of nuclei from the standard plant (*Solanum lycopersicum*) with the same extraction buffer. Relative 2C nuclear DNA content (genome size) of each *Acacia* was estimated by comparing them with plant DNA standards (*Solanum lycopersicum*) provided by Dr. Jaroslav Dolezel.

*2.5. Flow Cytometric Analysis*

Nuclear DNA content (2C DNA content) in all *Acacia* species was calculated according to Doležel et al. [21]. A minimum of 5000 propidium iodide-stained nuclei was estimated using a flow cytometer FACS Muse cell analyzer (Sigma, St. Louis, MO, USA). A minimum flow rate (0.12 µL/s) of the capillary was set to ensure the accuracy of the results. *Solanum lycopersicum* (2C = 1.96 pg) was kindly provided by Dr. Jaroslav Dolezel, Laboratory of Molecular Cytogenetics and Cytometry, Institute of Experimental Botany, Czech Republic, for the estimation of the genome size of *Acacia* species. The histograms generated in the Muse cell analyzer were computerized, and further analysis was performed for genome size estimation. The sample 2C DNA content was calculated according to the formula:

$$2C\ DNA\ content\ (Acacias) = \frac{\text{Fluorescence mean intensity of Acacias}}{\text{Fluorescence mean intensity of standard}} \times \text{2C DNA content of standard}$$

The number of base pairs per haploid genome was calculated based on the equivalent of 1 pg DNA = 965 megabase pairs [36].

*2.6. Extraction of Phytochemicals*

Plant samples (leaves harvested from 3-month-old plants grown in pots) were washed properly and dried in the shade at room temperature (25 °C). The fine, coarse powder was made from dried leaves. 5 g of leaf powder was used for the extraction of phytochemicals in 100 mL methanol. The mixture was kept in a rotatory shaker for 12 hrs, and thereafter, it was filtered through Whatman filter paper No. 1. The filtrate was concentrated and dried under reduced pressure at 40 °C using a rotary vacuum evaporator. The concentrated samples (extract) were dissolved in methanol for analysis and kept at 4 °C until used.

2.6.1. Determination of Total Phenolic Contents (TPCs)

The total phenolic content in the methanolic leaf extract of *A. etbaica*, *A. johnwoodii* and *A. origena* was measured using Folin–Ciocalteu reagent using the method developed by Ainsworth [37]. The reaction was set up in a final volume of 2 mL (50 µL of leaf extract, 50 µL of Folin–Ciocalteu reagent and 1.9 mL of deionized water) and kept for 8 min. The above mixture was neutralized with 20% $Na_2CO_3$ solution and incubated for 30 min. The standard curve of gallic acid (1 mg/mL, dissolved in the methanol stock solution) was prepared for the estimation of TPCs in the samples. After adding all the reagents to the reaction tube and incubation, the samples were read at 765 nm wavelength as the color developed in the reaction mixture using a UV-visible spectrophotometer (SHIMADZU, UV-1800, Tokyo, Japan). The total phenolic content was estimated from the linear equation of a standard curve prepared with gallic acid ($y = 0.0017x - 0.0289$ with $R^2 = 0.9807$). The calculated TPCs were expressed as mg/g gallic acid equivalent (GAE) of dry weight sample.

2.6.2. Determination of the Total Flavonoid Content (TFCs)

Estimation of the total flavonoid content in the leaf of *A. etbaica*, *A. johnwoodii* and *A. origena* was performed using the protocol developed by Ordonez [38]. In 2 mL of Eppendorf tube, 0.2 mL of 2% $AlCl_3$ and 0.2 mL of plant extract were added and incubated at room temperature for 1 h. After incubation, deionized water (0.4 mL) was added to the above mixture. The absorbance was taken at 420 nm using a UV-visible spectrophotometer (SHIMADZU, UV-1800, Kyoto, Japan). A calibration curve was prepared using the quercetin reference standard compound. Total flavonoid content was calculated as quercetin (mg/g DW) using the following equation ($Y = 0.031x + 0.137$ with $R^2 = 0.9893$), as generated from calibration curve.

2.6.3. Estimation of Bioactive Compounds, Quercetin and Rutin in Methanolic Leaf Extract of Acacias Using HPLC

The quantification of flavonoid compounds was conducted using HPLC analysis with a UV-Vis diode array detector (DAD). The methanolic extract of the samples was analyzed using a Waters system (Agilent Technologies 1290 Infinity system, Santa Clara, CA, USA) coupled to a diode array detector with a 200–400 nm detection range and mobile phase for HPLC, as followed by Al-Qurainy et al. [35] with minor modification. The separation was performed in a ZOBRAX RX-C18 column (4.6 × 150 mm) in which the mobile phases were pumped at a flow rate of 0.800 mL/min (quercetin 3-β-D-glucoside) and 1 mL/min (luteolin 7-rutinoside) with a run time of 5 min and injection volume of 1 μL. The flavonoid standard solutions (quercetin-3-β-D-glucoside and luteolin 7-rutinoside) and samples were injected into the system using an auto-injector. Various mobile phases were checked at different flow rates with the same column temperatures in order to find a suitable separation method for the standards. Finally, the mobile phase (methanol and acetonitrile) was set up in the ratio of (65:35) for better separation of quercetin-3-β-D-glucoside, and for luteolin 7-rutinoside separation, methanol and 0.6% acetic acid ($v/v$) in HPLC grade water in the ratio of (65:35) was used. Both flavonoids (quercetin-3-β-D-glucoside and luteolin 7-rutinoside) were estimated at a wavelength (λ) = 274 nm. The standard compound, quercetin-3-β-D-glucoside and luteolin 7-rutinoside, were purchased from (Sigma–Aldrich, St. Louis, MO, USA) for estimation in samples by comparing their retention times (quercetin-3-β-D-glucoside, 1.54 min, and luteolin 7-rutinoside, 1.34 min) and absorbance spectra. Calibration curves were constructed with standard solutions (quercetin-3-β-D-glucoside) of 0.1, 0.2, 0.3 and 0.4 μg/μL and content in the samples were calculated with linear-squares $y = 1244x + 23.5$ using Microsoft Office Excel software with correlation values of ≥0.9956. Similarly, luteolin 7-rutinoside was calculated using the linear regression equations obtained from standard curve: $Y = 5260X-2$, ($R^2 = 0.9965$, Y: peak area, X: rutin content). All samples were examined in triplicate to ensure the accuracy of the results.

*2.7. Functional Groups Characterization by FTIR*

Fourier transform infrared spectrophotometer (FTIR) is the most powerful technique for the identification of the nature of chemical bonds (functional groups) present in compounds. All the Acacias extracts were analyzed in triplicates for the reproducibility of the results. The wavelength of light absorbed by the functional group present in the compound can be seen in the spectrum. The scanning was achieved at wavelength 400–4000 $cm^{-1}$, and results were shown in the % transmission analysis. The functional groups of the compounds were separated according to peak ratio (peak values in IR radiation region).

*2.8. Statistical Analysis*

The statistical analysis was carried out in SPSS software by one-way analysis of variance (ANOVA) followed by Duncan's test for the estimation of genome size and phytochemicals. The fluorescence histograms were resolved into G0/G1 (2C), S and G2/M (4C) cell cycle compartments. The mean fluorescence intensity of G0/G1 was used for the calculation of the 2C DNA content of *Acacia* species. Different letters were used on bars which represent the significant difference at $p \leq 0.05$.

## 3. Results

The morphometric traits, including seed length and diameter, were measured among three species of Acacias. Both traits of seed were found to be highest in *A.johnwoodii* than the other two species, *A. etbaica* and *A. origena* (Table 1). Seed shape was also varied among them, as given in Figure 1. The seeds were sowed in the pot in the growth chamber to get the cotyledon for genome size estimation. The morphology and size of leaves varied among the *Acacia* species (data taken in pot experiment) (Figure 1). The largest variation in leaf length was observed in *A. johnwoodii*, followed by *A. etbaica* and *A. origena*, under normal conditions. The cotyledons of germinated seeds (20 days after sowing) were used for the estimation of genome size. The excised cotyledons from germinated seeds were washed with double distilled water before chopping them into the extraction buffer. The extracted and stained nuclei (propidium iodide) were scanned with a Muse cell analyzer. The cell cycle phase ($G_0/G_1$) was found to be highest in nuclei extracted using the MBO1 buffer in all Acacias, followed by GB, Tris-$MgCl_2$ and LB01, respectively (Figures 2–4). The sharp peak of the histogram that corresponds to the G0/G1 phase (2C level) of the cell cycle was detected in all species (Figures 2–4), and the mean intensity of this phase was used for the calculation of genome size (2C DNA content). The peaks corresponding to the G2 + M (M = mitosis) phase (4C level) were also detected in all species.

**Figure 1.** *Acacia* species; morphology of seeds measured in centimeters (cm), plant grown in a pot and leaf morphology (leaf length) measured in centimeters (cm) (**a**): *A. johnwoodii*; (**b**): *A. etbaica*; (**c**): *A. origena*.

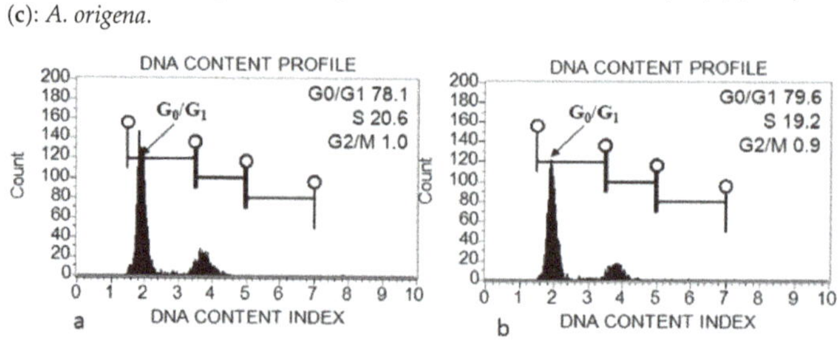


**Figure 2.** *Cont.*

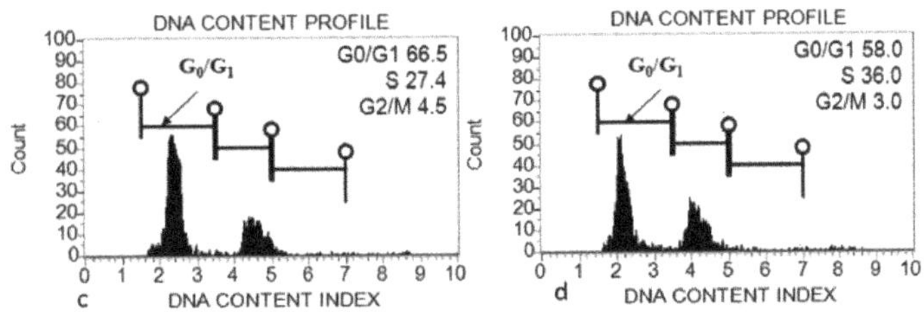


**Figure 2.** Flow cytometry (FCM) histograms of propidium iodide (PI) fluorescence intensity of nuclei prepared from cotyledon of *Acacia etbaica* using different extraction buffers. (**a**) Galbraith's buffer; (**b**) MB01 Buffer; (**c**) Tris-MgCl2 buffer; (**d**) LB01 buffer.

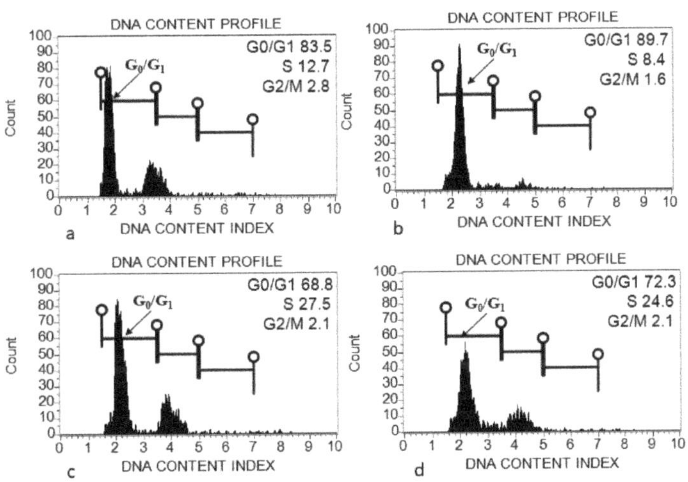


**Figure 3.** Flow cytometry (FCM) histograms of propidium iodide (PI) fluorescence intensity of nuclei prepared from cotyledon of *Acacia johnwoodii* using different extraction buffers. (**a**) Galbraith's buffer; (**b**) MB01 Buffer; (**c**) Tris-MgCl2 buffer; (**d**) LB01 buffer.

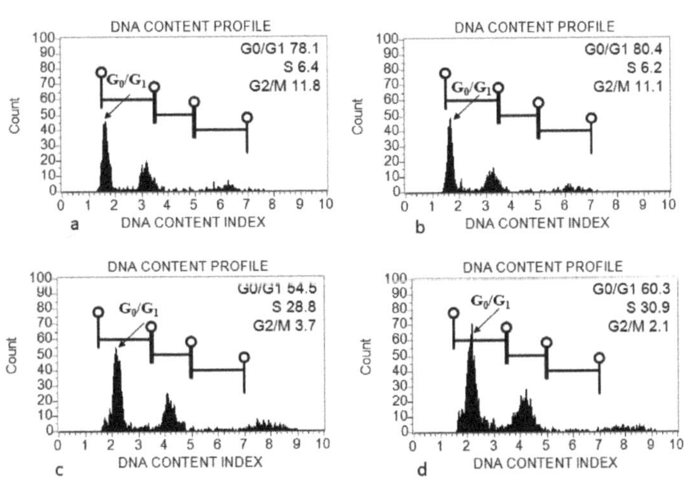


**Figure 4.** Flow cytometry (FCM) histograms of propidium iodide (PI) fluorescence intensity of nuclei prepared from cotyledon of *Acacia origena* using different extraction buffers. (**a**) Galbraith's buffer; (**b**) MB01 Buffer; (**c**) Tris-$MgCl_2$ buffer; (**d**) LB01 buffer.

The data represented in the table is the mean of three replicates ±SD. The significant differences between the *Acacia* species are shown by different letters in the table. Duncan's test at $p < 0.05$ was used for the analysis.

The genome size (2C DNA content) was compared among and within *Acacia* species in different extraction buffers. The estimated genome size using different extraction buffers ranged from 1.91 to 2.21 pg (*A. etbaica*), 2.07–2.19 pg (*A. johnwoodii*) and 2.09–2.45 pg (*A. origena*), respectively (Table 2). Among the three species of *Acacia*, the highest genome size was found to be 2.45 pg/2C in *A. origena* with LB01 buffer, whereas the lowest was 1.91 pg/2C in *A. etbaica* with GB and MB buffers, respectively. Genome size remained the same within species in MB and GB buffer in all three *Acacia* species. However, it varied among species. No significant difference was found in genome size within *A. origena*, when nuclei were extracted with buffers (GB, MB and Tris buffers). All three species showed variation in genome size with Tris-$MgCl_2$ and LB01 buffers. Significant differences were found in the genome size of *A. etbaica* and *A. johnwoodii* with Tris-$MgCl_2$ and LB01 buffers.

**Table 2.** Genome size (2C DNA content) of *Acacia* species as nuclei isolated with different nuclei extraction buffers with slight modification (0.6% β-mercaptoethanol and 2% PVP).

| Plant Species | Extraction Buffers | | | |
|---|---|---|---|---|
| | **GB Buffer (Mean ± SD)** | **MB01 Buffer (Mean ± SD)** | **Tris MgCl2 Buffer (Mean ± SD)** | **LB01 Buffer (Mean ± SD)** |
| *Acacia etbaica* (Schweinf.) Kyal. & Boatwr | $1.91 \pm 0.00^{c}$ | $1.91 \pm 0.02^{c}$ | $2.21 \pm 0.02^{a}$ | $2.11 \pm 0.02^{b}$ |
| *Acacia johnwoodii* (Boulos) Ragup., Seigler, Ebinger & Maslin | $2.19 \pm 0.04^{a}$ | $2.19 \pm 0.00^{a}$ | $2.07 \pm 0.07^{b}$ | $2.12 \pm 0.00^{ab}$ |
| *Acacia origena* (Hunde) Kyal. & Boatwr | $2.09 \pm 0.00^{a}$ | $2.09 \pm 0.01^{a}$ | $2.10 \pm 0.18^{a}$ | $2.45 \pm 0.05^{b}$ |

$^{a,b,c}$ indicates significant difference according to Duncan's test ($p < 0.05$).

### 3.1. Determination of Total Flavonoids (TFCs), Total Phenols (TPCs), and Bioactive Compounds

Total flavonoids and phenols were measured using the UV-visible spectrophotometer following the methods used by Ordonez [38]; Ainsworth [37]. The total flavonoids and phenols were calculated using the quercetin and gallic acid (GA) standard compounds curve as generated by plotting the absorbance of different concentrations of quercetin and gallic acid. The TPC content was varied in *Acacia* species, and highest was to be detected in *A. johnwoodii* (28.84 ± 1.79 mg GAE/g DW), followed by *A. etbaica* (19.49 ± 1.87 mg GAE/g DW) and *A. origena* (15.32 ± 0.94 mg GAE/g DW), respectively (Figure 5a). In contrast to TPC, the TFCs content was found to be highest in *A. etbaica* (1.084 ± 0.30 mg QE /g DW), followed by *A. johnwoodii* and *A. origena* (Figure 5b). The bioactive compounds quercetin 3-β-D-glucoside (QBDG) and luteolin 7-rutinoside were detected in HPLC in considerable amounts in all *Acacia* species (Figures 6 and 7). The quercetin 3-β-D-glucoside (QBDG), which is a flavonoid glycoside, was observed to be highest in *A. johnwoodii* (1.53 ± 0.04 mg/g DW) than the other two species of *Acacia* which had a low concentration of this compound (Figure 8a). The content of quercetin 3-β-D-glucoside content varied non-significantly between *A. etbaica* and *A. origena*. In contrast to quercetin 3-β-D-glucoside, the content of luteolin 7-rutinoside was detected in the highest concentration in *A. origena* (0.399 ± 0.01 mg/g DW), followed by *A. johnwoodii* (0.252 0.02 mg/g DW) and *A. etbaica* (0.202 ± 0.01 mg/g DW), respectively. However, a significant difference was observed in the content of luteolin 7-rutinoside among the species of *Acacia* (Figure 8b).

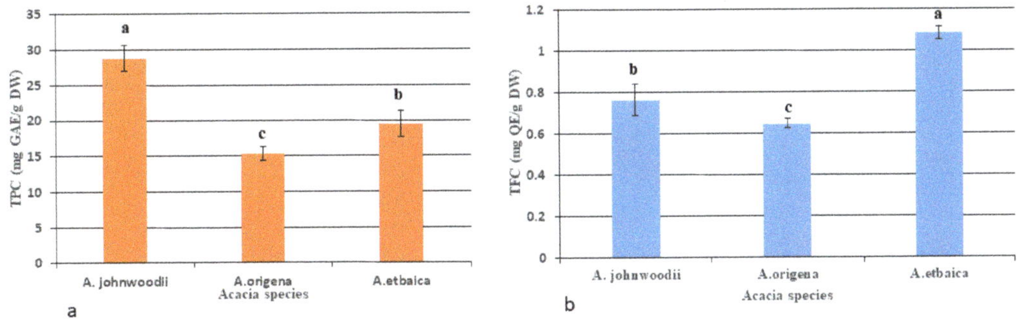


**Figure 5.** Total phenolic content (TPCs); (**a**); and flavonoid content (TFC); (**b**) in the young leaves of *Acacia* species. Values are the mean of three replicates ± S.D. Letters on bars represent the significant differences according to Duncan's test ($p < 0.05$).

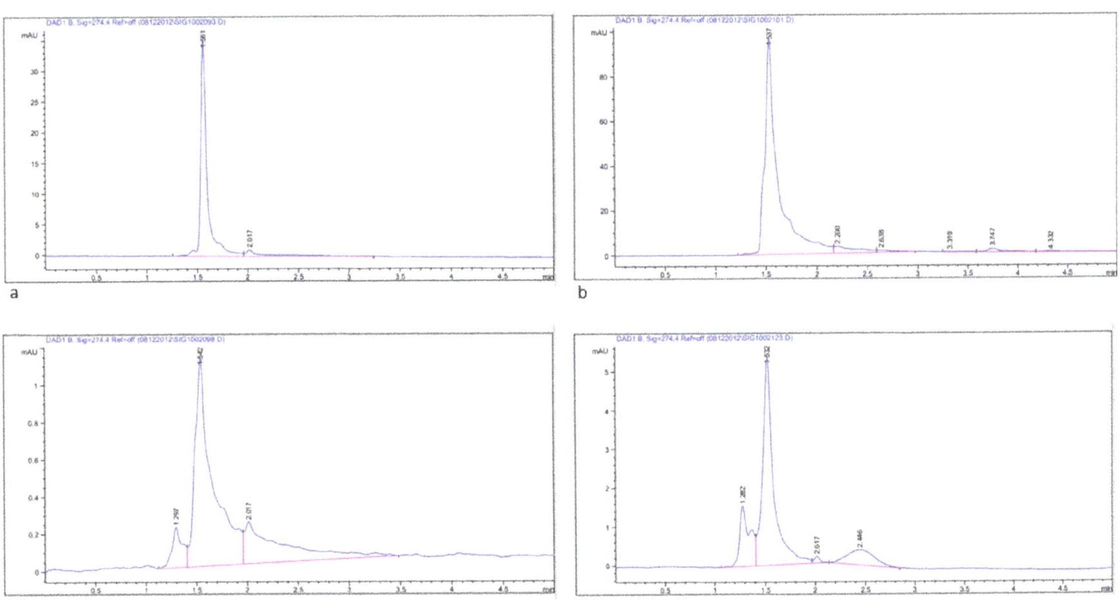

**Figure 6.** HPLC chromatogram of bioactive compound (Quercitin 3-β-D-glucoside) extracted from young leaves of *Acacia* species; (**a**) Standard compound, Quercitin 3-β-D-glucoside; (**b**) *A. johnwoodii*; (**c**) *A. origena*; (**d**) *A. etbaica*.

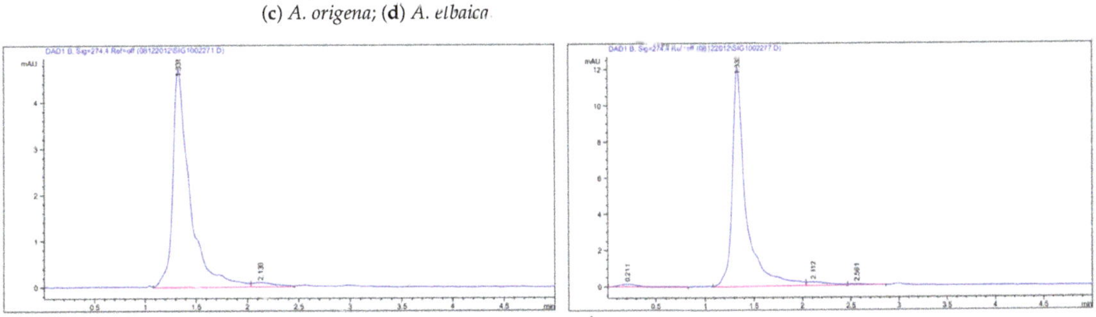

**Figure 7.** *Cont.*

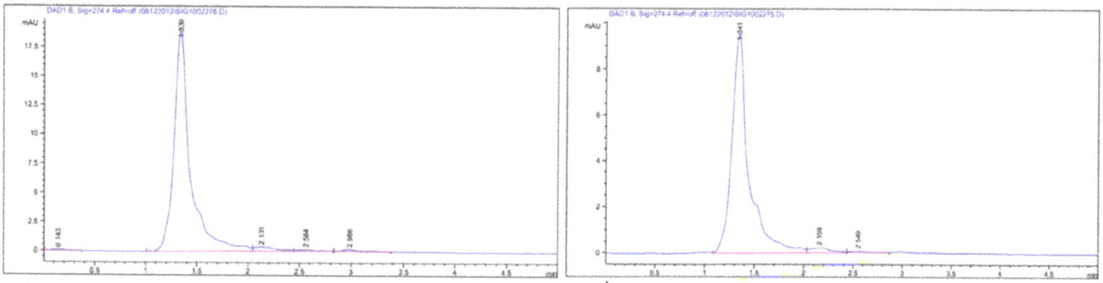


c d

**Figure 7.** HPLC chromatogram of bioactive compound (Luteolin 7-rutinoside) extracted from young leaves of *Acacia* species; (**a**) Standard compound, Luteolin 7-rutinoside; (**b**) *A. johnwoodii*; (**c**) *A. origena*; (**d**) *A. etbaica*.

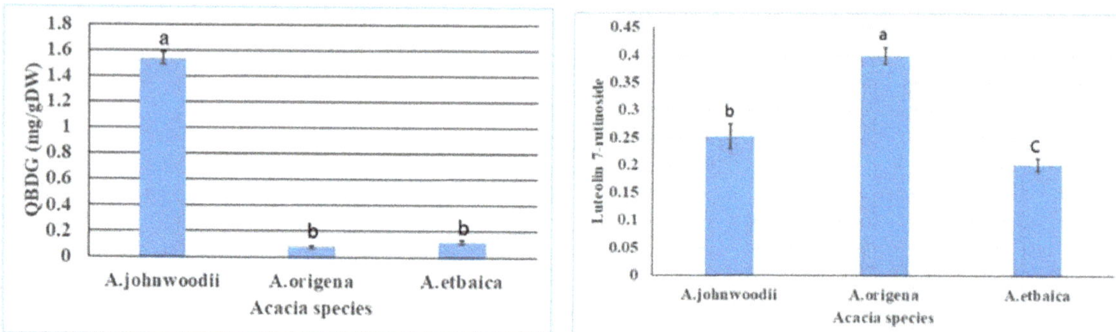


a h

**Figure 8.** Content of Quercetin 3-β-D-glucoside (QBDG) (**a**) and Luteolin 7-rutinoside (**b**) in young leaves of *Acacia* species estimated by HPLC system. Values are the mean of three replicates ± S.D. Letters on bars represent the significant differences according to Duncan's test ($p < 0.05$).

### *3.2. Fourier Transform Infrared Spectrometer (FTIR) Spectrum Analysis*

Fourier Transform Infrared Spectrometer (FTIR) spectrum analysis was used to illustrate the functional group of bio-compounds based on the peak value in the infrared region (Figure 9) and assigned their functional groups (https://www.sigmaaldrich.com/SA/en/technical-documents/technical-article/analytical-chemistry/photometry-and-reflectometry/ir-spectrum-table, accessed on 5 July 2022). The more intense band occurring at 3363.62 $cm^{-1}$, 2943.79 $cm^{-1}$, 2834.06 $cm^{-1}$, 1642.96 $cm^{-1}$, 1642.96 $cm^{-1}$, 1451.85 $cm^{-1}$, 1028.52 $cm^{-1}$ and 794.26 $cm^{-1}$ corresponding to different functional groups for related compounds in *A. origena* (Table 3).

In *A. etbaica*, the intense band was observed at 3375.47 $cm^{-1}$, 2947.10 $cm^{-1}$, 2076.89 $cm^{-1}$, 1637.10 $cm^{-1}$, 1398.89 $cm^{-1}$, 1022.61 $cm^{-1}$ and 794.46 $cm^{-1}$ corresponding to various functional groups present on different compounds (Table 4). Similarly, the peak value detected for the functional group for compounds present in *A. johnwoodii* at 3368.71 $cm^{-1}$, 2944.93 $cm^{-1}$, 2833.19 $cm^{-1}$, 1648.33 $cm^{-1}$, 1452.43 $cm^{-1}$ and 793.82 $cm^{-1}$, respectively (Table 5).

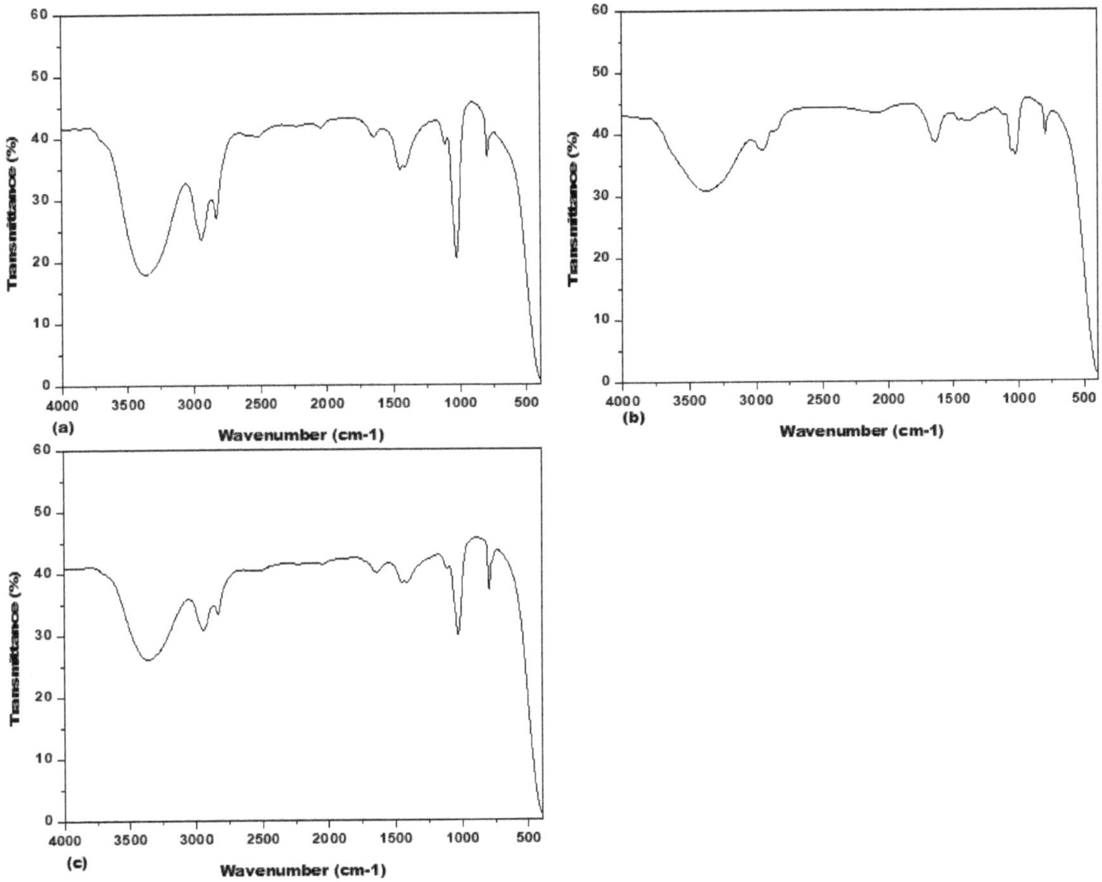


**Figure 9.** Fourier Transform Infrared Spectroscopy (FTIR) analysis for the methanolic extract of *Acacia species*: (**a**) *A. johnwoodii*; (**b**) *A. etbaica*; (**c**) *A. origena*.

**Table 3.** Fourier Transform Infrared Spectrometer (FTIR) spectrum analysis for the methanolic extract of *A. origena*.

| Frequency ($cm^{-1}$) | Intensity | Assignments | Compound Group |
|---|---|---|---|
| 794.26 | Medium | C=C bending | Alkene |
| 1028.52 | Strong | C-F stretching | Fluro compound |
| 1451.85 | Medium | C-H bending | Alkane |
| 1642.96 | medium<br>medium<br>medium | C-N stretching<br>C=C stretching<br>C–C stretching | Imine/oxime<br>Alkene<br>Conjugated alkene |
| 2834.06 | Strong, broad<br>Weak, broad | O-H stretching<br>O-H stretching | Carboxylic acid<br>Alcohol |
| 2943.79 | Strong, broad<br>Weak, broad<br>Strong, broad<br>Medium | O-H stretching<br>O-H stretching<br>N-H stretching<br>C-H stretching | Carboxylic acid<br>Alcohol<br>Amine salt<br>Alkane |
| 3363.62 | Strong, broad | O-H stretching | Alcohol |

**Table 4.** Fourier Transform Infrared Spectrometer (FTIR) spectrum analysis for the methanolic extract of *A. etbaica*.

| Frequency ($cm^{-1}$) | Intensity | Assignments | Compound Group |
|---|---|---|---|
| 794.46 | Medium | C=C bending | Alkene |
| 1022.61 | Strong | C-F stretching | Fluro compound |
| 1398.89 | Strong | S=O stretching | sulfonyl chloride |
| 1637.10 | Medium<br>Medium | C=C stretching<br>C=C stretching | cyclic alkene<br>Alkane |
| 2076.89 | Strong | N=C=S stretching | Isothiocyanate |
| 2947.10 | Weak, broad<br>Medium<br>Strong, broad | O-H stretching<br>C-H stretching<br>N-H stretching | Alcohol<br>Alkane<br>Amine salt |
| 3375.47 | Medium | N-H stretching | aliphatic primary amine |

**Table 5.** Fourier Transform Infrared Spectrometer (FTIR) spectrum analysis for the methanolic extract of *A. johnwoodii*.

| Frequency ($cm^{-1}$) | Intensity | Assignments | Compound Group |
|---|---|---|---|
| 793.82 | Medium | C=C bending | Alkene |
| 1113.21 | Strong<br>medium<br>Strong | C-F stretching<br>C-N stretching<br>C-O stretching | Fluro compound<br>amine<br>aliphatic ether |
| 1452.43 | Medium | C=H bending | Alkane |
| 1648.33 | medium<br>medium<br>medium | C-N stretching<br>C=C stretching<br>C=C stretching | Imine/oxime<br>Alkene<br>Conjugated alkene |
| 2833.19 | Strong, broad<br>Weak, broad | O-H stretching<br>O-H stretching | Carboxylic acid<br>Alcohol |
| 2944.93 | Strong, broad<br>Weak, broad<br>Strong, broad<br>Medium | O-H stretching<br>O-H stretching<br>N-H stretching<br>C-H stretching | Carboxylic acid<br>Alcohol<br>Amine salt<br>Alkane |
| 3368.71 | Strong, broad | O-H stretching | Alcohol |

## 4. Discussion

Acacias grow in dry, hot environments due to the presence of their xeromorphic structure. The high-water holding capacity of leaves and well-developed root system enable Acacias to adapt to their dry, hot environment. To analyze the plant based on such abiotic stress tolerant character, genome size and phytochemical study are necessary. Plant genome size influences stress tolerances in plants via plasticity [39]. The genome size has an impact on various parameters of the plant, such as drought tolerance, various nutritional growth, and herbivore defense traits [40,41]. Information about the genome size of any plant is very important as it has a strong correlation to many traits of the plants and their evolutionary aspects. In the present study, the seeds of all three species were collected from the same geographical region, time and same environmental conditions for genome size estimation and phytochemical study. The plants were grown in a growth chamber under optimum conditions. The morphological traits of *Acacia* species, including seed shape, size and leaf morphology, were different from each other (Figure 1). However, seed color was almost similar among the three species. The largest variation in leaf size was observed in *A. johnwoodii* than the other two *Acacia* species. Overall, the morphological traits were observed to be superior in *A. johnwoodi* than *A. origena* and *A. etbaica*. Similarly, the genome size of *A. johnwoodii* (2.19 pg/2C DNA content) was higher than *A. etbaica* (1.91 pg/2C DNA content) and *A. origena* (2.09 pg/2C DNA content) in the same extraction buffer. Variations in seed mass have been more closely associated with variations in genome size than with divergences in other ecological and morphological variables [42]. The strong relationship between DNA content, cell division rate and cell size could lead to predictable morphological differences in plants, including a negative relationship with leaf mass per unit area [43].

Identification of *Acacia* species according to phenotypic (morphological) characters is very difficult as sometimes these markers overlap with each other. Plant genomes provide knowledge on biosynthetic pathways engaged in the synthesis of phytoconstituents. Plant phenotypic expression [42,44], cell size, mitotic cell cycle and their constituent organelles [45] have been correlated to the genome size. The estimated genome size herein in *Acacia* species could be used for their identification based on observed variation among them. The genome size estimated with two buffers, namely MB01 and GB, was found to be the same within *Acacia* species and could be used for the identification of these species. In some study, genome size was stable within an individual and usually within a species [46,47]; however, there are notable exceptions where variation have been observed within species [48].

### *4.1. Genome Size (2C DNA Content) Variation within and among Acacias*

For the first time, the genome size (2C DNA content) was estimated in Acacias, including *A. johnwoodii*, *A. etbaica* and *A. origena*, which are important species found in the Kingdom of Saudi Arabia. All three species had shown genome size variation within and among when nuclei extracted with LB01 and Tris $MgCl_2$ buffer. According to extraction buffers (MB01 and GB), the genome size within *Acacia* species was found to be the same; however, it varied among the species (Table 2). The result obtained in the current study for genome size (2C DNA content) was supported by the work of other researchers who worked on various plant species of the same genus and their populations [22,49–51]. Ouarda et al. [52] found genome size variation in *Acacia tortilis* from 2.95–3.03 pg/2C in Tunisian, and those originating from Zimbabwe had consistent genome sizes (2C nuclear DNA = 1.39–1.40 pg).

Various reasons might be possible for genome size variation in studied Acacias, including biotic and abiotic factors prevailing in the natural habitat. Secondly, all three Acacias are different species that belong to the same genus, so this might be the reason for the variation in genome size. The genome size diversity in plant species is partly related to variations in chromosome number [53], and polyploidy is recognized as one of the main reasons for genome size variation [54]. Knight et al. [13] provided a literature review that covers several factors that have been linked to DNA content variation in plants, such as species diversity, latitude, altitude, precipitation temperature, generation time, seed mass, various leaf anatomical traits and growth rate. Some studies have found relationships between abiotic field conditions and genome size in plant species [45,55]. In another study, it was observed that climate seasonality and biotic interactions are potential forces shaping the evolution of genome size [56]. The ecological factors could also be important in shaping the evolutionary dynamics of genome size [57] and the origins of secondary compounds through gene duplication events [58].

Extraction buffer has an important role in genome size determination as different extraction buffers have different chemical compositions, which could help in the extraction of nuclei from different plant species in different levels of quantity and quality. The purpose of using different extraction buffers in the present study is to measure the accurate genome size of Acacias, as one nucleus extraction buffer cannot use for all plant species. In our study, genome size varied within and among *Acacia* species significantly and non-significantly in some buffers (Table 2), whereas it remained the same in some buffers. Different extraction buffers may give variation in the yield and quality of extracted nuclei, which results in variation in genome size (2C DNA content), as reported by Sadhu et al. [22].

The isolation of nuclei in pure form is very important for genome size estimation as the presence of various secondary metabolites and chlorophyll in the plant cell decreases the purity of nuclei suspension and further affects fluorochrome fluorescence. Such compounds exacerbate the problem of nuclei extraction and thus, affecting sample quality and causing problems in DNA staining [59,60]. According to Noirot et al. [59], caffeine and chlorogenic acids present in Coffea spp. hinder the PI's intercalation to the DNA. Similarly, *Lycium* species contain numerous cytosolic compounds, such as flavonoids and phenolic acids [61], which can interfere with the fluorescent staining of nuclear DNA [60,62]. Therefore, sample selection is very important for the extraction of pure nuclei. We used cotyledons of germinated seeds of Acacias for the extraction of pure nuclei, which had a low quantity of secondary metabolites and chlorophyll content as compared to the leaf sample. Extraction of nuclei from cotyledons was found to be pure as compared to leaf. Initially, we used the same buffer composition as reported in the literature for the extraction of nuclei; however, we could not get the sharp peak in FCM and hence were unable to estimate accurate genome size (data not shown). Therefore, we modified the buffer composition by adding the PVP (2% $w/v$) and 0.6% β-mercaptoethanol ($v/v$), which reduced the compounds' interaction with nuclei released from the cell and improved the purity of nuclei as indicated by sharp peaks in all Acacias (Figures 2–4). The compound β-mercaptoethanol is a reducing agent and checks the activity of phenolic compounds in the presence of another competitor, such as PVP, and breaks the hydrophobic interactions [63]. In another study, Sadhu et al. [22] used 1% of PVP in different extraction buffers, which yielded good-quality nuclei from plants of different genera of the same family from both root and shoot tissues. In earlier reports, it has been reported that PVP reduced the effect of polyphenols by changing their conformational structure, maintaining the cell compounds in a reduced state and making hydrogen bonds [63–65]. Thus, our result revealed that the two extraction buffers viz., GB and MB01, could be used for the extraction of nuclei for estimation of genome size as no variation in genome size within *Acacia* species has been observed as compared to other extraction buffers used in this study.

### 4.2. Phytochemicals Profiling in Acacias

Phytochemicals are secondary metabolites that defend the plant and are used as medicine to cure various human diseases. Phytochemical profiling was performed using the methanolic extract of young leaves of *Acacia* species for comparative study based on TFCs, TPCs and bioactive compound contents. The present study was conducted in the growth chamber for phytochemicals comparison among three *Acacia* species (*A. johnwoodii*, *A. etbaica* and *A. origena*). The leaves were harvested from 3-month-old plants for the extraction of phytochemicals in methanol. Fourier-transform infrared spectroscopy (FTIR) results showed the presence of different functional groups present on compounds in all *Acacia* species. The functional group detected on compounds were almost similar except for some unique functional groups, as aliphatic ether was detected in *A. johnwoodii* whereas isothiocyanate, cyclic alkene and sulphonyl chloride were detected in *A. etbaica*. However, functional groups, including Imine/oximes and conjugated alkenes, were not detected in *A. johnwoodii*, whereas these were detected in the other two species of *Acacia*. The obtained results show that FTIR spectroscopy is a rapid, reliable method for the comparative study of Acacias. This is one type of analytical method of fingerprinting for the identification and differentiation of plant species. Such type of fingerprinting method has been used for the identification of various plant species such as *Lycium*, some moss species, birch species and *Rhodobryum roseum* Limpr. and its adulterants [66–69].

Total flavonoids (TFCs) and phenolic contents (TPCs) were measured by colorimetric spectrophotometry, and they varied among the *Acacia* species significantly. Among the three species of Acacias, the TPCs were observed to be highest in *A. johnwoodii* (28.84 ± 1.79 mg GAE/g DW) than the other two species of *Acacia* (*A. etbaica* and *A. origena*) significantly. However, in contrast to TPCs, the TFCs were found to be highest in *A. etbaica*. This large variation in phenolic and flavonoid contents among Acacias might be responsible for different levels of medicinal properties. Many species of *Acacia* have shown medicinal properties and have been used in different countries. Several species of *Acacia* have important resources of bioactive compounds such as phenolics, flavonoids, alkaloids, saponins, terpenoids, polysaccharides and tannins [70] which are responsible for several pharmacological effects, including hypoglycemic, antibacterial, anti-inflammatory, anti-platelet aggregation, anti-cancer, anti-atherosclerotic, anti-hypertensive and analgesic properties [71]. The present result was supported by other researchers who observed variation in secondary metabolites (TFCs and TPCs) in different plant species belonging to the same genus, including *Iris* species and *Monochoria* species [72,73].

A high-performance liquid chromatographic method was used for the quantitative determination of luteolin 7-rutinoside and quercetin 3-β-D-glucoside, which are important flavonoids in plants. These bioactive compounds were extracted in 100% methanol from the young leaf of *A. johnwoodii*, *A. origena*, and *A. etbaica*, as methanol is a good solvent for the extraction of secondary metabolites. Different types of flavonoids are found in plants as quercetin-3-rutinoside, rutoside, sophorin and luteolin 7-rutinoside [74,75], and they have different pharmacological activities. The flavonoids, quercitrin and quercetin 3-β-D-glucoside (QBDG) act as a chemical chaperone for the A4V SOD1 ALS-causing mutant [76]. The concentration of both compounds, including quercetin 3-β-D-glucoside and luteolin 7-rutinoside, varied among the *Acacia* species. This variability of bioactive compound content has differentiated the *Acacia* species from each other as the highest concentration of quercetin 3-β-D-glucoside was found in *A. johnwoodii* and Luteolin 7-rutinoside in *A. origena*. The compound luteolin glycoside is found in different plant species, such as artichokes [77] and possesses a varied range of biological activities participating in the prevention and treatment of various diseases. Luteolin plays a vital role in shielding plants against ultraviolet radiation [26,78].

Based on the phytochemical study, TFCs, TPCs, luteolin 7-rutinoside and quercetin 3-β-D-glucoside, all three species of *Acacia* showed variation in a controlled environment, and these might have different pharmacological activities. The reason behind this variability of phytochemicals in these *Acacia* species might be due to the genotype variation (different species of the same genus) since no stress was imposed on the plant at any stage. However, the biosynthesis of such compounds, including phenolics, depends on the plant stage [79], growing conditions, species exclusiveness, vegetation period (altitude, climatic factors and soil properties) [80–83], and seasonal variation [84]. The variation in the content of flavonoids and phenolics has been observed in different species of Acacias and in their different parts [85,86]. Such compounds (flavonoids) could be used in other plant species effectively to analyze the evolutionary relationships of different plant families and for their authentication.

## 5. Conclusions

Estimation of genome size is very important in plant species as it makes whole genome sequencing easy, which will support carrying out a more comprehensive study of the evolution and origin. The results of the present study on genome size and phytochemical study on different species of Acacias are the first time that could be used further for molecular biology research. Our study shows that genome size (2C DNA content) varied among the species of *Acacias*. Among different nuclei extraction buffers, MB01 and GB proved to be the best buffers for extraction of nuclei for estimation of genomes size and could be used for other *Acacia* species as genome size within *Acacia* species remained the same. FTIR spectra of leaf extract revealed the presence or absence of diverse functional groups on a compound, and some are unique to *Acacia* species. The leaves of Acacias species showed the presence of quercetin 3-β-D-glucoside and luteolin 7-ritinoside, which are good antioxidant flavonoids. Further, the research could be performed in natural habitats for the screening and estimation of phytochemicals during different seasons as well as under abiotic stress conditions. More detailed studies of the phytochemical profile and extraction of individual phenolic compounds need to be carried out in the future.

**Author Contributions:** Conceptualization, S.K. and M.N.; methodology, S.K., M.N. and F.A.-Q., A.M.S.; software, M.T. and M.N. investigation, S.K.; results, S.K. and M.N.; writing—original draft preparation, S.K. and M.N.; writing—review and editing, S.K. and M.N.; supervision, F.A.-Q., A.A.-h. and H.O.S. software analysis. All authors have read and agreed to the published version of the manuscript.

**Funding:** The authors extend their appreciation to the Researchers Supporting Project No. (RSP-2021/73) at King Saud University, Riyadh, Saudi Arabia.

**Institutional Review Board Statement:** Not applicable.

**Informed Consent Statement:** Not applicable.

**Data Availability Statement:** All the data supporting the findings of this study are included in this article.

**Conflicts of Interest:** The authors declare no conflict of interest.

## References


1. Orchard, A.E.; Maslin, B.R. (1584) Proposal to conserve the name Acacia (*Leguminosae*: *Mimosoideae*) with a conserved type. *Taxon* **2003**, *52*, 362–363. [CrossRef]
2. Maslin, B.; Miller, J.; Seigler, D. Overview of the generic status of Acacia (*Leguminosae*: *Mimosoideae*). *Austral. Syst. Botany* **2003**, *16*, 1–18. [CrossRef]
3. Javed Muhammad, A.; Abdullah, M.Z.; Muhammad, N.; Ratnam, W. Detecting mislabeling and identifying unique progeny in Acacia mapping population using SNP markers. *J. For. Res.* **2017**, *28*, 1119–1127. [CrossRef]
4. Muhammad, A.J.; Ong, S.S.; Ratnam, W. Characterization of mean stem density, fibre length and lignin from two Acacia species and their hybrid. *J. For. Res.* **2018**, *29*, 549–555. [CrossRef]
5. Migahid, A.M. *Flora of Saudi Arabia*, 3rd ed.; Riyadh University Publication: Riyadh, Saudi Arabia, 1990.
6. Collenette, S. *Illustrated Guide to the Flowers of Saudi Arabia*; Scorpion: Hampshire, UK, 1985.
7. Al-Mefarrej, H. Diversity and frequency of Acacia spp. in three regions in the Kingdom of Saudi Arabia. *Afr. J. Biotechnol.* **2012**, *11*, 11420–11430. [CrossRef]
8. Greilhuber, J.; Doležel, J.; Lysák, M.A.; Bennett, M.D. The origin, evolution and proposed stabilization of the terms 'genome size'and 'C-value'to describe nuclear DNA contents. *Ann. Botany* **2005**, *95*, 255–260. [CrossRef]
9. Bennett, M.D.; Leitch, I.J. Nuclear DNA amounts in angiosperms: Targets, trends and tomorrow. *Ann. Botany* **2011**, *107*, 467–590. [CrossRef]
10. Greilhuber, J.; Borsch, T.; Müller, K.; Worberg, A.; Porembski, S.; Barthlott, W. Smallest angiosperm genomes found in Lentibulariaceae, with chromosomes of bacterial size. *Plant Biol.* **2006**, *8*, 770–777. [CrossRef]
11. Bennett, M.; Leitch, I. *Plant DNA C-Values Database*; (release 4.0, Dec. 2005); Royal Botanic Gardens: Kew, UK, 2005.
12. Vinogradov, A.E. Selfish DNA is maladaptive: Evidence from the plant Red List. *Trends Genet.* **2003**, *19*, 609–614. [CrossRef]
13. Knight, C.A.; Molinari, N.A.; Petrov, D.A. The large genome constraint hypothesis: Evolution, ecology and phenotype. *Ann. Botany* **2005**, *95*, 177–190. [CrossRef]
14. Galbraith, D.W. Cytometry and plant sciences: A personal retrospective. *Cytometry Part A J. Int. Soc. Anal. Cytol.* **2004**, *58*, 37–44. [CrossRef] [PubMed]
15. Pellicer, J.; Clermont, S.; Houston, L.; Rich, T.C.; Fay, M.F. Cytotype diversity in the Sorbus complex (*Rosaceae*) in Britain: Sorting out the puzzle. *Ann. Botany* **2012**, *110*, 1185–1193. [CrossRef] [PubMed]
16. Żabicka, J.; Migdałek, G.; Słomka, A.; Sliwinska, E.; Mackiewicz, L.; Keczyński, A.; Kuta, E. Interspecific hybridization and introgression influence biodiversity—Based on genetic diversity of Central European Viola epipsila-V. palustris complex. *Diversity* **2020**, *12*, 321. [CrossRef]
17. Levin, J.; Fay, M.F.; Pellicer, J.; Hedrén, M. Multiple independent origins of intermediate species between *Sorbus aucuparia* and *S. hybrida* (Rosaceae) in the Baltic region. *Nordic J. Botany* **2018**, *36*. [CrossRef]
18. Galbraith, D.W.; Harkins, K.R.; Maddox, J.M.; Ayres, N.M.; Sharma, D.P.; Firoozabady, E. Rapid flow cytometric analysis of the cell cycle in intact plant tissues. *Science* **1983**, *220*, 1049–1051. [CrossRef]
19. Pfosser, M.; Heberle-Bors, E.; Amon, A.; Lelley, T. Evaluation of sensitivity of flow cytometry in detecting aneuploidy in wheat using disomic and ditelosomic wheat–rye addition lines. *Cytometry J. Int. So. Anal. Cytol.* **1995**, *21*, 387–393. [CrossRef]
20. Arumuganathan, K.; Earle, E. Estimation of nuclear DNA content of plants by flow cytometry. *Plant Mol. Biol. Report.* **1991**, *9*, 229–241. [CrossRef]
21. Doležel, J.; Binarová, P.; Lucretti, S. Analysis of nuclear DNA content in plant cells by flow cytometry. *Biology* **1989**, *31*, 113–120.
22. Sadhu, A.; Bhadra, S.; Bandyopadhyay, M. Characterization of Tulbaghia violacea (Tulbaghieae, Allioideae, Amaryllidaceae) from India: A cytogenetic and molecular approach. *Nucleus* **2018**, *61*, 29–34. [CrossRef]
23. Cartea, M.E.; Francisco, M.; Soengas, P.; Velasco, P. Phenolic compounds in Brassica vegetables. *Molecules* **2010**, *16*, 251–280. [CrossRef]

24. Arts, I.C.; Sesink, A.L.; Faassen-Peters, M.; Hollman, P.C. The type of sugar moiety is a major determinant of the small intestinal uptake and subsequent biliary excretion of dietary quercetin glycosides. *Br. J. Nutr.* **2004**, *91*, 841–847. [CrossRef]
25. Di Carlo, G.; Mascolo, N.; Izzo, A.A.; Capasso, F. Flavonoids: Old and new aspects of a class of natural therapeutic drugs. *Life Sci.* **1999**, *65*, 337–353. [CrossRef]
26. Harborne, J.B.; Williams, C.A. Advances in flavonoid research since 1992. *Phytochemistry* **2000**, *55*, 481–504. [CrossRef]
27. Formica, J.; Regelson, W. Review of the biology of quercetin and related bioflavonoids. *Food Chem. Toxicol.* **1995**, *33*, 1061–1080. [CrossRef]
28. Hollman, P.C.; de Vries, J.H.; van Leeuwen, S.D.; Mengelers, M.J.; Katan, M.B. Absorption of dietary quercetin glycosides and quercetin in healthy ileostomy volunteers. *Am. J. Clin. Nutr.* **1995**, *62*, 1276–1282. [CrossRef]
29. Mehta, R.G.; Murillo, G.; Naithani, R.; Peng, X. Cancer chemoprevention by natural products: How far have we come? *Pharm. Res.* **2010**, *27*, 950–961. [CrossRef]
30. Martens, S.; Mithöfer, A. Flavones and flavone synthases. *Phytochemistry* **2005**, *66*, 2399–2407. [CrossRef]
31. Bhullar, K.S.; Rupasinghe, H. Polyphenols: Multipotent therapeutic agents in neurodegenerative diseases. *Oxidd Med. Cell. Longev.* **2013**, *2013*, 891748. [CrossRef]
32. Fernández-Arroyo, S.; Camps, J.; Menendez, J.A.; Joven, J. Managing hypertension by polyphenols. *Planta Med.* **2015**, *81*, 624–629. [CrossRef]
33. Rodrigo, R.; Libuy, M.; Feliu, F.; Hasson, D. Polyphenols in disease: From diet to supplements. *Curr. Pharm. Biotechnol.* **2014**, *15*, 304–317. [CrossRef]
34. Mocanu, M.-M.; Nagy, P.; Szöllősi, J. Chemoprevention of breast cancer by dietary polyphenols. *Molecules* **2015**, *20*, 22578–22620. [CrossRef] [PubMed]
35. Al-Qurainy, F.; Tarroum, M.; Khan, S.; Nadeem, M.; Gaafar, A.-R.Z.; Alansi, S.; Alfarraj, N.S. Genome Estimation and Phytochemical Compound Identification in the Leaves and Callus of Abrus precatorius: A Locally Endangered Plant from the Flora of Saudi Arabia. *Plants* **2022**, *11*, 567. [CrossRef]
36. Smith, J.; Heslop-Harrison, J. Nuclear DNA amounts in angiosperms. *Philos. Trans. Roy. Soc. London B* **1976**, *274*, 227–274.
37. Ainsworth, E.A.; Gillespie, K.M. Estimation of total phenolic content and other oxidation substrates in plant tissues using Folin–Ciocalteu reagent. *Nat. Protocols* **2007**, *2*, 875–877. [CrossRef] [PubMed]
38. Ordonez, A.; Gomez, J.; Vattuone, M. Antioxidant activities of *Sechium edule* (Jacq.) Swartz extracts. *Food Chem.* **2006**, *97*, 452–458. [CrossRef]
39. Meyerson, L.A.; Pyšek, P.; Lučanová, M.; Wigginton, S.; Tran, C.T.; Cronin, J.T. Plant genome size influences stress tolerance of invasive and native plants via plasticity. *Ecosphere* **2020**, *11*, e03145. [CrossRef]
40. Pyšek, P.; Skálová, H.; Čuda, J.; Guo, W.Y.; Suda, J.; Doležal, J.; Kauzál, O.; Lambertini, C.; Lučanová, M.; Mandáková, T. Small genome separates native and invasive populations in an ecologically important cosmopolitan grass. *Ecology* **2018**, *99*, 79–90. [CrossRef]
41. Meyerson, L.A.; Cronin, J.T.; Bhattarai, G.P.; Brix, H.; Lambertini, C.; Lučanová, M.; Rinehart, S.; Suda, J.; Pyšek, P. Do ploidy level and nuclear genome size and latitude of origin modify the expression of Phragmites australis traits and interactions with herbivores? *Biol. Invas.* **2016**, *18*, 2531–2549. [CrossRef]
42. Beaulieu, J.M.; Moles, A.T.; Leitch, I.J.; Bennett, M.D.; Dickie, J.B.; Knight, C.A. Correlated evolution of genome size and seed mass. *New Phytol.* **2007**, *173*, 422. [CrossRef]
43. Beaulieu, J.M.; Leitch, I.J.; Knight, C.A. Genome size evolution in relation to leaf strategy and metabolic rates revisited. *Ann. Botany* **2007**, *99*, 495–505. [CrossRef]
44. Morgan, H.D.; Westoby, M. The relationship between nuclear DNA content and leaf strategy in seed plants. *Ann. Botany* **2005**, *96*, 1321–1330. [CrossRef] [PubMed]
45. Knight, C.A.; Ackerly, D.D. Variation in nuclear DNA content across environmental gradients: A quantile regression analysis. *Ecol. Lett.* **2002**, *5*, 66–76. [CrossRef]
46. Greilhuber, J. Intraspecific variation in genome size in angiosperms: Identifying its existence. *Ann. Botany* **2005**, *95*, 91–98. [CrossRef] [PubMed]
47. Lysák, M.A.; Rostková, A.; Dixon, J.M.; Rossi, G.; Doležel, J. Limited genome size variation in Sesleria albicans. *Ann. Botany* **2000**, *86*, 399–403. [CrossRef]
48. Cullis, C.A. Mechanisms and control of rapid genomic changes in flax. *Ann. Botany* **2005**, *95*, 201–206. [CrossRef]
49. Pellicer, J.; López-Pujol, J.; Aixarch, M.; Garnatje, T.; Vallès, J.; Hidalgo, O. Detecting introgressed populations in the Iberian endemic Centaurea podospermifolia through genome size. *Plants* **2021**, *10*, 1492. [CrossRef]
50. Chen, W.; Kao, Y.; Tang, C.; Tsai, C.; Lin, T. Estimating nuclear DNA content within 50 species of the genus Phalaenopsis Blume (Orchidaceae). *Sci. Hortic.* **2013**, *161*, 70–75. [CrossRef]
51. Siljak-Yakovlev, S.; Stevanovic, V.; Tomasevic, M.; Brown, S.C.; Stevanovic, B. Genome size variation and polyploidy in the resurrection plant genus Ramonda: Cytogeography of living fossils. *Environ. Exp. Botany* **2008**, *62*, 101–112. [CrossRef]
52. El Ferchichi Ouarda, H.; Walker, D.J.; Khouja, M.L.; Correal, E. Diversity analysis of *Acacia tortilis* (Forsk.) Hayne ssp. raddiana (Savi) Brenan (Mimosaceae) using phenotypic traits, chromosome counting and DNA content approaches. *Gen. Resour. Crop Evol.* **2009**, *56*, 1001–1010. [CrossRef]

53. Bennett, M.D.; Leitch, I.J. Genome size evolution in plants. In *The Evolution of the Genome*; Elsevier: Amsterdam, The Netherlands, 2005; pp. 89–162.
54. Soltis, P.S.; Marchant, D.B.; Van de Peer, Y.; Soltis, D.E. Polyploidy and genome evolution in plants. *Curr. Opin. Gen. Dev.* **2015**, *35*, 119–125. [CrossRef]
55. Carta, A.; Peruzzi, L. Testing the large genome constraint hypothesis: Plant traits, habitat and climate seasonality in L. iliaceae. *New Phytol.* **2016**, *210*, 709–716. [CrossRef] [PubMed]
56. Cacho, N.I.; McIntyre, P.J.; Kliebenstein, D.J.; Strauss, S.Y. Genome size evolution is associated with climate seasonality and glucosinolates, but not life history, soil nutrients or range size, across a clade of mustards. *Ann. Botany* **2021**, *127*, 887–902. [CrossRef] [PubMed]
57. Guignard, M.S.; Crawley, M.J.; Kovalenko, D.; Nichols, R.A.; Trimmer, M.; Leitch, A.R.; Leitch, I.J. Interactions between plant genome size, nutrients and herbivory by rabbits, molluscs and insects on a temperate grassland. *Proc. R. Soc. B* **2019**, *286*, 20182619. [CrossRef]
58. Edger, P.P.; Heidel-Fischer, H.M.; Bekaert, M.; Rota, J.; Glöckner, G.; Platts, A.E.; Heckel, D.G.; Der, J.P.; Wafula, E.K.; Tang, M. The butterfly plant arms-race escalated by gene and genome duplications. *Proc. Natl. Acad. Sci. USA* **2015**, *112*, 8362–8366. [CrossRef]
59. Noirot, M.; Barre, P.; Duperray, C.; Louarn, J.; Hamon, S. Effects of caffeine and chlorogenic acid on propidium iodide accessibility to DNA: Consequences on genome size evaluation in coffee tree. *Ann. Botany* **2003**, *92*, 259–264. [CrossRef] [PubMed]
60. Loureiro, J.; Rodriguez, E.; DOLEŽEL, J.; Santos, C. Flow cytometric and microscopic analysis of the effect of tannic acid on plant nuclei and estimation of DNA content. *Ann. Botany* **2006**, *98*, 515–527. [CrossRef]
61. Inbaraj, B.S.; Lu, H.; Kao, T.; Chen, B. Simultaneous determination of phenolic acids and flavonoids in Lycium barbarum Linnaeus by HPLC–DAD–ESI-MS. *J. Pharm. Biomed. Anal.* **2010**, *51*, 549–556. [CrossRef]
62. Bennett, M.D.; Price, H.J.; Johnston, J.S. Anthocyanin inhibits propidium iodide DNA fluorescence in Euphorbia pulcherrima: Implications for genome size variation and flow cytometry. *Ann. Botany* **2008**, *101*, 777–790. [CrossRef]
63. Greilhuber, J.; Temsch, E.M.; Loureiro, J.C. Nuclear DNA content measurement. In *Flow Cytometry with Plant Cells: Analysis of Genes, Chromosomes and Genomes*; John Wiley & Sons: Hoboken, NJ, USA, 2007; pp. 67–101.
64. Doyle, J.J.; Doyle, J.L. A Rapid DNA Isolation Procedure for Small Quantities of Fresh Leaf Tissue. *Phytochem. Bull.* **1987**, *19*, 11–15.
65. Loureiro, J.; Rodriguez, E.; Doležel, J.; Santos, C. Two new nuclear isolation buffers for plant DNA flow cytometry: A test with 37 species. *Ann. Botany* **2007**, *100*, 875–888. [CrossRef]
66. Peng, Y.; Sun, S.; Zhao, Z.; Leung, H. A rapid method for identification of genus Lycium by FTIR spectroscopy. *Guang Pu Xue Yu Guang Pu Fen Xi Guang Pu* **2004**, *24*, 679–681. [PubMed]
67. Cao, Z.; Liu, Y.; Zhao, J. Efficient discrimination of some moss species by fourier transform infrared spectroscopy and chemometrics. *J. Spectr.* **2014**, *2014*, 191796. [CrossRef]
68. Depciuch, J.; Kasprzyk, I.; Drzymała, E.; Parlinska-Wojtan, M. Identification of birch pollen species using FTIR spectroscopy. *Aerobiologia* **2018**, *34*, 525–538. [CrossRef] [PubMed]
69. Cao, Z.; Wang, Z.; Shang, Z.; Zhao, J. Classification and identification of Rhodobryum roseum Limpr. and its adulterants based on fourier-transform infrared spectroscopy (FTIR) and chemometrics. *PLoS ONE* **2017**, *12*, e0172359. [CrossRef]
70. Kalaivani, T.; Rajasekaran, C.; Suthindhiran, K.; Mathew, L. Free radical scavenging, cytotoxic and hemolytic activities from leaves of *Acacia nilotica* (L.) Wild. ex. *Delile* subsp. indica (Benth.) Brenan. *Evid.-Based Complem. Alter. Med.* **2011**, *2011*, 274741. [CrossRef]
71. Jelassi, A.; El Ayeb-Zakhama, A.; Nejma, A.B.; Chaari, A.; Harzallah-Skhiri, F.; Jannet, H.B. Phytochemical composition and allelopathic potential of three Tunisian Acacia species. *Ind. Crops Prod.* **2016**, *83*, 339–345. [CrossRef]
72. Kaššák, P. Total flavonoids and phenolics content of the chosen genus Iris species. *Acta Univ. Agric. Silviculturae Mendelianae Brunensis* **2013**, *60*, 119–126. [CrossRef]
73. Tungmunnithum, D.; Drouet, S.; Garros, L.; Lorenzo, J.M.; Hano, C. Flavonoid Profiles and Antioxidant Potential of Monochoria angustifolia (GX Wang) Boonkerd & Tungmunnithum, a New Species from the Genus Monochoria C. Presl. *Antioxidants* **2022**, *11*, 952.
74. Kreft, S.; Knapp, M.; Kreft, I. Extraction of rutin from buckwheat (*Fagopyrum esculentum* Moench) seeds and determination by capillary electrophoresis. *J. Agri. Food Chem.* **1999**, *47*, 4649–4652. [CrossRef]
75. Shalashvili, A.; Rakviashvili, N. Rutin and luteolin 7-rutinoside from the leaves ofCitrus unshiu. *Chem. Nat. Comp.* **1984**, *20*, 621–622. [CrossRef]
76. Ip, P.; Sharda, P.R.; Cunningham, A.; Chakrabartty, S.; Pande, V.; Chakrabartty, A. Quercitrin and quercetin 3-β-d-glucoside as chemical chaperones for the A4V SOD1 ALS-causing mutant. *Protein Eng. Des. Selec.* **2017**, *30*, 431–440. [CrossRef]
77. Jiménez-Moreno, N.; Cimminelli, M.J.; Volpe, F.; Ansó, R.; Esparza, I.; Mármol, I.; Rodríguez-Yoldi, M.J.; Ancín-Azpilicueta, C. Phenolic composition of artichoke waste and its antioxidant capacity on differentiated Caco-2 cells. *Nutrients* **2019**, *11*, 1723. [CrossRef] [PubMed]
78. Antognoni, F.; Zheng, S.; Pagnucco, C.; Baraldi, R.; Poli, F.; Biondi, S. Induction of flavonoid production by UV-B radiation in Passiflora quadrangularis callus cultures. *Fitoterapia* **2007**, *78*, 345–352. [CrossRef] [PubMed]
79. Chepel, V.; Lisun, V.; Skrypnik, L. Changes in the content of some groups of phenolic compounds and biological activity of extracts of various parts of heather (*Calluna vulgaris* (L.) Hull) at different growth stages. *Plants* **2020**, *9*, 926. [CrossRef] [PubMed]

80. Rieger, G.; Muller, M.; Guttenberger, H.; Bucar, F. Influence of altitudinal variation on the content of phenolic compounds in wild populations of Calluna vulgaris, Sambucus nigra, and Vaccinium myrtillus. *J. Agric. Food Chem.* **2008**, *56*, 9080–9086. [CrossRef]
81. Yao, X.-H.; Zhang, Z.-B.; Song, P.; Hao, J.-Y.; Zhang, D.-Y.; Zhang, Y.-F. Different harvest seasons modify bioactive compounds and antioxidant activities of Pyrola incarnata. *Ind. Crops Prod.* **2016**, *94*, 405–412. [CrossRef]
82. Ribeiro, D.A.; de Macêdo, D.G.; Boligon, A.A.; Menezes, I.R.A.; de Almeida Souza, M.M.; da Costa, J.G.M. Influence of seasonality on the phenolic composition of Secondatia floribunda A. DC (Apocynaceae) during its phenological cycle. *Acta Physiol. Plant.* **2019**, *41*, 1–16. [CrossRef]
83. Yang, L.; Wen, K.-S.; Ruan, X.; Zhao, Y.-X.; Wei, F.; Wang, Q. Response of plant secondary metabolites to environmental factors. *Molecules* **2018**, *23*, 762. [CrossRef]
84. Gabr, S.; Nikles, S.; Wenzig, E.M.P.; Ardjomand-Woelkart, K.; Hathout, R.M.; El-Ahmady, S.; Motaal, A.A.; Singab, A.; Bauer, R. Characterization and optimization of phenolics extracts from Acacia species in relevance to their anti-inflammatory activity. *Biochem. Syst. Ecol.* **2018**, *78*, 21–30. [CrossRef]
85. Sulaiman, C.; Balachandran, I. Total phenolics and total flavonoids in selected Indian medicinal plants. *Indian J. Pharm. Sci.* **2012**, *74*, 258. [CrossRef]
86. Verick Purba, B.A.; Sunarti, S.; Lukmandaru, G. Phenolics content and antioxidant activity of wood extractives from three clones of Acacia hybrid (*Acacia mangium* × *Acacia auriculiformis*). *Maderas. Cien. Tecnol.* **2021**, *23*. [CrossRef]

horticulturae

Article

# Trichomes' Micromorphology and Their Evolution in Selected Species of *Causonis* (Vitaceae)

**Gaurav Parmar [1] and Wajid Zaman [2,*]**

1 National Botanical Garden, Godawari 44709, Nepal
2 Department of Life Sciences, Yeungnam University, Gyeongsan 38541, Korea
* Correspondence: wajidzaman@yu.ac.kr

**Abstract:** The Vitaceae genus *Causonis* is found in tropical to temperate climates from Asia to Australia, including the Pacific Islands. Rafinesque established the genus in 1930; however, Süssenguth classified it under *Cayratia* as a sect. *Discypharia* in 1953. The genus was resurrected in 2013 using morphological and genetic evidence. We herein provided insight into the diversity of trichomes' micromorphology of selected species of this recently reinstated genus for taxonomical implication. Simple trichomes, representing non-glandular and unbranched trichomes, are only found in *Causonis*. Trichomes vary from straight, curved, hooked, appressed, pilose, to villous in different parts like branchlets, abaxial leaf surface, and adaxial leaf surface in different species. They also vary in the same plant from the young stage to the mature stage. Most species are pubescent when young, but a few species become nearly glabrous when they are mature. Significant variations can be observed in trichomes' length between the species. Principal component analysis (PCA), based on the micromorphological traits, was carried out for the species delimitation. In Mesquite, ancestral character state reconstruction was used to examine evolutionary trends for trichomes on three different surfaces. The glabrous to sparsely pubescent state found on the branchlets and both leaf surfaces were found to be the ancestral state and, on the branchlets and both leaf surfaces, the villous hairs state was the derived state in the genus. The identification of *Causonis* species is greatly aided by trichomes morphology. Therefore, similar studies should be conducted on other Vitaceae genera to reveal the variety of trichomes found in the family.

**Keywords:** character evolution; hairs; identification; morphology; stereomicroscopy

check for updates

**Citation:** Parmar, G.; Zaman, W. Trichomes' Micromorphology and Their Evolution in Selected Species of *Causonis* (Vitaceae). *Horticulturae* **2022**, *8*, 877. https://doi.org/10.3390/horticulturae8100877

Academic Editor: Yonghua Yang

Received: 1 September 2022
Accepted: 20 September 2022
Published: 23 September 2022

**Publisher's Note:** MDPI stays neutral with regard to jurisdictional claims in published maps and institutional affiliations.

**Copyright:** © 2022 by the authors. Licensee MDPI, Basel, Switzerland. This article is an open access article distributed under the terms and conditions of the Creative Commons Attribution (CC BY) license (https://creativecommons.org/licenses/by/4.0/).

## 1. Introduction

Vitaceae is a widely known family of commercial significance owing to grapes, wines, and raisins [1]. In addition, the family also has some plants with ornamental values like *Cissus verticillata* (L.) Nicolson & C.E. Jarvis, *Parthenocissus quinquefolia* (L.) Planch., and *P. tricuspidata* (Siebold & Zuccarini) Planch. [2]. It consists of 16 genera and ca. 950 species [3]. A wide range of research is being carried out in the family from a general morphological description of new taxa (species or genus) and already described taxa or re-circumscription of taxa to different molecular level works such as phylogenetics, character evolution, and biogeography. A large number of molecular works performed on the family include phylogenetic analysis using selected chloroplast and/or nuclear markers [4–11]. With the advent of new technologies, the paradigm in the Vitaceae research has been slowly shifting from phylogenetic study to phylogenomic study to reveal long-standing phylogenetic questions like resolving the non-monophyly of taxa because of gene duplication or loss, hybridization, introgression, or incomplete lineage sorting [11–13].

*Causonis* Raf. is a recently segregated genus of Vitaceae. Wen et al. [14] resurrected the genus *Causonis* from the genus *Cayratia* Juss. (hereafter *Cayratia* s.l.) based on both molecular phylogeny and morphological evidence. Morphologically, *Causonis* do not have ventral infolds covered with a distinct membrane in seeds, while *Cayratia* s.s. have ventral infolds covered with a distinct membrane. Rafinesque created the genus in 1830 [15], although Gagnepain [16] considered them to be *Cayratia* species. Suessenguth [17] and Latiff [18] classified them as members of the *Cayratia* sect. of *Discypharia* Suess. Later, sect. *Discypharia* was treated as subg. *Discypharia* (Suess.) C.L.Li of *Cayratia* by Li [19], who was then followed by Chen, et al. [20]. Subg. *Discypharia* has a non-articulate inflorescence axis, without bracts on the inflorescence axis and ventral infolds without a distinct membrane covering, versus the articulate inflorescence axis, with bracts on the inflorescence axis and ventral infolds covered with a distinct membrane in subg. *Cayratia*.

Two species of subg. *Discypharia*, *Cayratia oligocarpa* (H. Lév. & Vaniot) Gagnep. and *C. albifolia* C.L.Li, were transferred to a new genus *Pseudocayratia* J.Wen, L.M.Lu & Z.D.Chen [21] based on the morphology in particular seed characters. Three Australian species were moved to *Causonis* by Jackes [22], namely, *C. clematidea* (F.Muell.) Jackes, *C. eurynema* (B.L.Burtt) Jackes, and *C. maritima* (Jackes) Jackes, based on endosperm shape in the seeds' cross section. The molecular phylogeny of any species requires morphological support for its distinct taxa confirmation. The identification of the genus and species of the Vitaceae has been facilitated by the use of seeds [3,10,23–25]. Gerrath et al. [26] revealed interesting findings regarding tendrils and the position of inflorescences in the Vitaceae. Trichomes, a vegetative character, are very useful and are often used in the keys for the identification of species within the genus [20,21,27]. Nonetheless, it has been poorly studied in detail in any particular genus of the family, except genus *Vitis* L. by Ma et al. [2,28].

Vitaceae is one of the families with problems in the identification of its taxa because of huge variations in its vegetative characters like leaves' architecture, tendrils' furcation, veinlets' number, veinlets' position, and sometimes trichomes [11,20,27]. These morphological characters vary from the young stage to the mature stage in the same plant. Taxa having reproductive parts as diagnostic characters are often difficult to identify in the absence of flowers or seeds. The vegetative parts like branchlets and leaves are generally easily accessible irrespective of the season or stage of growth. The identification of plants can be greatly aided by the diagnosis of plants based on these two morphological characters in the absence of floral parts or seeds, though it is not applicable every time. Trichomes can thus be incredibly helpful in identifying taxa within the Vitaceae. In addition, some *Causonis* species, such as *C. japonica* and *C. trifolia*, are highly medicinal and used to treat a variety of maladies [11]. As a result, they can be suggested for horticulture. Therefore, with an aim to unveil trichomes' diversity in different species of this horticulturally important *Causonis*, we herein (1) study different types of trichomes on stems, abaxial, and adaxial leaf surfaces for taxonomic implication in *Causonis*; (2) investigate quantitative features of trichomes on stems, abaxial, and adaxial leaf surfaces in the genus; and (3) trace evolutionary trends of trichomes on stems, abaxial, and adaxial leaf surfaces in the genus.

## 2. Materials and Methods

### 2.1. Taxon Sampling and Identification

In this study, 14 specimens representing 12 taxa of *Causonis* were studied. All of the specimens sampled were recently collected from different distribution regions of the genus, except *C. ciliifera* (Merr.) G.Parmar & L.M.Lu and *C. tenuifolia* (Wight & Arn.) G.Parmar & L.M.Lu, which were observed from the deposited specimens at the National Herbarium (PE), Institute of Botany, Chinese Academy of Sciences, Beijing. The plants were identified following Chen, et al. [20] and Jackes [27]. One additional specimen each of *C. trifolia* (L.) Mabb. & J.Wen and *C. timoriensis* var. *mekongensis* (C.Y.Wu ex W.T.Wang) G.Parmar & L.M.Lu were included in this study to show the variation in trichomes within the species. The voucher information of all studied samples is included in Table 1.

### 2.2. Microscopic Investigation

Trichomes' diversity was examined from all of the specimens at PE. All specimens were examined for variations in trichome morphology on branchlets, abaxial leaf surfaces, and adaxial leaf surfaces using a stereomicroscope with a Leica DVM6 camera. Images of trichomes on different parts were taken using a scale. The terminology used for describing trichomes follows Chen et al. [20] and Jackes [27].

**Table 1.** Voucher information of samples used in this study.

| Taxon | Voucher No. | Locality |
| --- | --- | --- |
| *Causonis ciliifera* (Merr.) G.Parmar & L.M.Lu | PE00686401 (PE) | China, Hainan, Jianfengling |
| *Causonis clematidea* (F.Muell.) Jackes | PE01966347 (PE) | Australia, Queensland, Cattle creek road |
| *Causonis corniculata* (Benth.) J.Wen & L.M.Lu | CPG09774 (PE) | China, Taiwan, Nantou |
| *Causonis daliensis* (C.L.Li) G.Parmar & L.M.Lu | VN2014116 (PE) | Vietnam, Lam Dong, Bidoup-Nui Ba |
| *Causonis fugongensis* (C.L.Li) G.Parmar & L.M.Lu | CPG33017 (PE) | China, Yunnan, Ruili |
| *Causonis japonica* (Thunb.) Raf. var. *japonica* | CPG11331 (PE) | China, Yunnan, Hekou |
| *Causonis japonica* var. *pseudotrifolia* (W.T.Wang) G.Parmar & J.Wen | CPG20403 (PE) | China, Sichuan, Guangyuan |
| *Causonis maritima* (Jackes) Jackes | AU020 (PE) | Australia, Queensland, Cairns |
| *Causonis mollis* (Wall. ex M.A.Lawson) G.Parmar & J.Wen | LA41 (PE) | Laos, Champasak, Pakse |
| *Causonis tenuifolia* (Wight & Arn.) G.Parmar & L.M.Lu. | CPG38698 (PE) | India, Kerala, Mannarkad |
| *Causonis timoriensis* var. *mekongensis* (C.Y.Wu ex W.T.Wang) G.Parmar & L.M.Lu | CPG32937 (PE) | China, Yunnan, Yingjiang |
| *Causonis timoriensis* var. *mekongensis* (C.Y.Wu ex W.T.Wang) C.L.Li | CPG18926 (PE) | China, Yunnan, Ruili |
| *Causonis trifolia* (L.) Mabb. & J.Wen | LA17 (PE) | Laos, Luang Namtha, Muang Sing |
| *Causonis trifolia* (L.) Mabb. & J.Wen | CPG23885 (PE) | Vietnam, Ninh Thuan, Ca Na |

### 2.3. Quantitative Study

Trichomes' lengths were measured for mature branchlets, abaxial leaf surfaces, and adaxial leaf surfaces, including the shortest and longest trichomes on the surface, for each species, if present. On each of the three parts of each species, six distinct trichomes were measured including the shortest and longest trichomes, and the standard deviations were calculated. Species lacking trichomes on specific parts were marked as absent. Trichomes' length of *C. trifolia* (accession number LA17) and *C. timoriensis* var. *mekongensis* (accession number CPG32937) were only recorded for the respective taxa because other species were also represented by only one accession in this study. Accessions of *C. ciliifera* and *C. tenuifolia* were studied for the types of trichomes from old collections, but they could not be studied for trichomes' length because of the limited accessibility of those samples.

### 2.4. Ancestral Character State Reconstruction

For the reconstruction of ancestral character states, trichome morphological characteristics based on the position on three different plant parts, such as trichomes on branchlets, trichomes on abaxial leaf surface, and trichomes on adaxial leaf surface, were chosen. The evolution of characters was reconstructed in Mesquite 3.61 using the chloroplast dataset. The "Trace Character History" and the Markov k-state one-parameter (Mk1) evolutionary model were employed in a maximum likelihood (ML) approach for character state reconstruction [29].

## 3. Results

### 3.1. Microscopic Investigation

The study performed on 14 different specimens of *Causonis* revealed two distinct types of trichomes: normal hairs and villous hairs. Pilose hairs, appressed hairs, hooked hairs, and curved hairs were also observed, but were very rare in occurrence (Figure 1). Specimens were found to be almost glabrous on the branchlets in *C. maritima*, *C. corniculata* (Benth.) J.Wen & L.M.Lu, *C. tenuifolia*, *C. japonica* (Thunb.) Raf. var. *japonica*, and *C. japonica* var. *pseudotrifolia* (W.T.Wang) G.Parmar & J.Wen. *Causonis timoriensis* var. *mekongensis* was found to be sparsely pubescent with simple hairs. Puberulent hairs were only found on the branchlets in *C. daliensis* (C.L.Li) G.Parmar & L.M.Lu. Villous hairs on the branchlets were observed in *C. mollis* (Wall. ex M.A.Lawson) G.Parmar & J.Wen, *C. fugongensis* (C.L.Li) G.Parmar & L.M.Lu, and *C. ciliifera*. The morphologies of the trichome on branchlets are shown in Figure 2A,B.

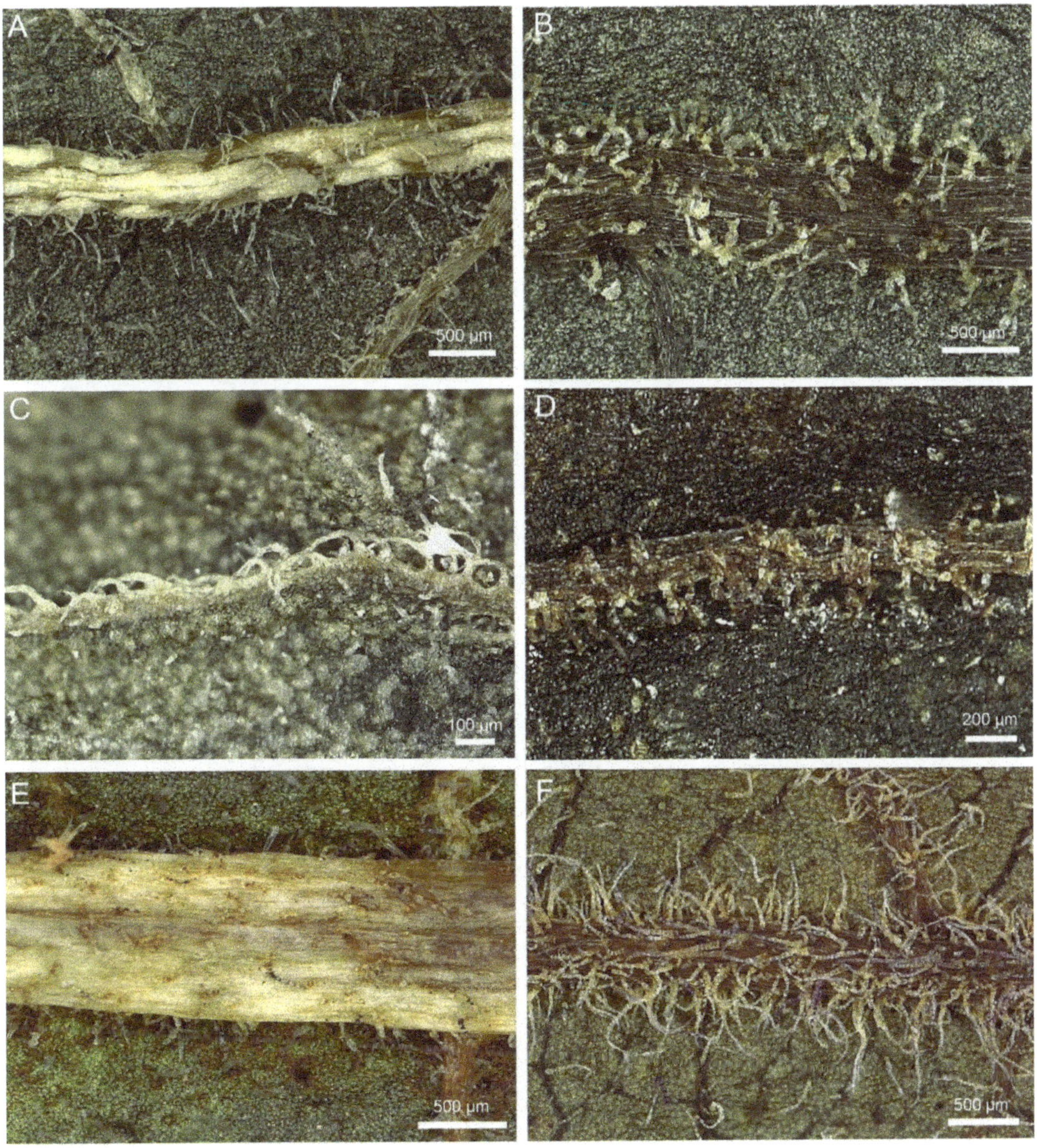


**Figure 1.** Trichomes' diversity on the midvein in leaves of *Causonis*. (**A**) Abaxial leaf surface of *C. timoriensis* var. *mekongensis* with straight hairs (CPG32937). (**B**) Abaxial leaf surface of *C. fugongensis* with curved hairs. (**C**) Adaxial leaf surface of *C. trifolia* with hooked hairs (CPG23885). (**D**) Adaxial leaf surface of *C. fugongensis* with appressed hairs. (**E**) Abaxial leaf surface of *C. timoriensis* var. *mekongensis* with pilose hairs (CPG18926). (**F**) Abaxial leaf surface of *C. mollis* with villous hairs.

(**A**)

**Figure 2.** *Cont.*

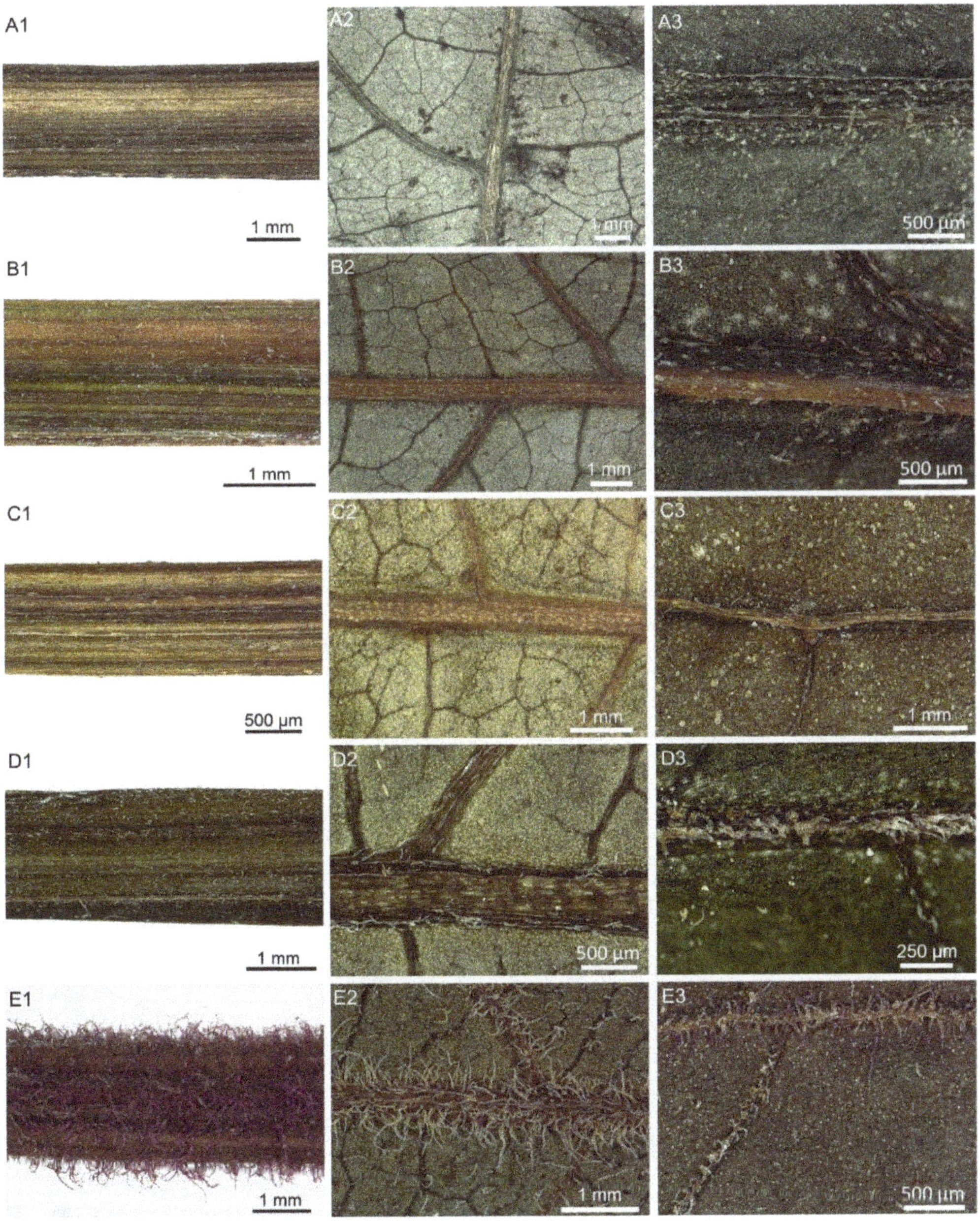


(**B**)

**Figure 2.** (**A**) Morphological variations of trichomes in *Causonis*. **A**, *C. clematidea*. **B**, *C. timoriensis* var. *mekongensis*, CPG32937. **C**, *C. trifolia*, LA17. **D**, *C. maritima*. **E**, *C. fugongensis*. Trichomes' morphologies on branchlets, abaxial, and adaxial surfaces of leaves are shown from left to right, respectively. (**B**) Morphological variations of trichomes in *Causonis*. **A**, *C. japonica* var. *japonica*. **B**, *C. japonica* var. *pseudotrifolia*. **C**, *C. corniculata*. **D**, *C. daliensis*. **E**, *C. mollis*. Trichomes' morphologies on branchlets, abaxial, and adaxial surfaces of leaves are shown from left to right, respectively.

The abaxial leaf surface had villous hairs in *C. mollis, C. fugongensis,* and *C. ciliifera,* while *C. corniculata* and *C. maritima* were glabrous on the abaxial leaf surface. The abaxial leaf surface was glabrous to sparsely pubescent in *C. japonica* var. *japonica, C. japonica* var. *pseudotrifolia,* and *C. tenuifolia,* but pubescent with straight hairs in *C. clematidea, C. daliensis,* and *C. timoriensis* var. *mekongensis.* Pilose hairs were also observed in *C. timoriensis* var. *mekongensis* and sparsely pubescent to densely pubescent with straight hairs in *C. trifolia* on the abaxial leaf surface, while *C. fugongensis* also had curved hairs on the abaxial leaf surface. The trichomes' morphologies on the abaxial leaf surfaces are shown in Figures 1 and 2A,B.

Villous hairs on the adaxial leaf surface were observed in *C. mollis* and *C. ciliifera,* sometimes confined only to the midvein. Glabrous to sparsely pubescent with simple hairs were observed in *C. clematidea, C. japonica* var. *japonica,* and *C. japonica* var. *pseudotrifolia* on the adaxial leaf surface, but pubescent with simple hairs were observed in *C. daliensis, C. trifolia,* and *C. timoriensis* var. *mekongensis.* Appressed hairs were found in *C. fugongensis* on the adaxial surface and hooked hairs on the midvein of adaxial surface in *C. maritima.* The morphological variation in trichomes used for the diagnosis of species, including trichome morphology on the adaxial leaf surface, is presented in Table 2 (Figures 1 and 2A,B).

**Table 2.** Morphology of trichomes on branchlets and leaves.

| Taxon | Trichomes on Branchlets | Trichomes on Abaxial Leaf Surface | Trichomes on Adaxial Leaf Surface |
|---|---|---|---|
| *C. ciliifera* | villous | villous | villous |
| *C. clematidea* | pubescent when young | pubescent | sparsely pubescent |
| *C. corniculata* | glabrous or sometimes sparsely pubescent | usually glabrous | usually glabrous |
| *C. daliensis* | puberulent | pubescent | pubescent |
| *C. fugongensis* | villous | villous/curved hairs | appressed hairs |
| *C. japonica* var. *japonica* | glabrous or pilose | glabrous to sparsely pubescent | glabrous to sparsely pubescent |
| *C. japonica* var. *pseudotrifolia* | almost glabrous | glabrous to sparsely pubescent | glabrous to sparsely pubescent |
| *C. maritima* | usually glabrous | usually glabrous | hooked hairs |
| *C. mollis* | villous | villous | villous |
| *C. tenuifolia* | usually glabrous | usually glabrous | usually glabrous |
| *C. timoriensis* var. *mekongensis* | sparsely pubescent or pilose | pilose or with simple hairs | pubescent |
| *C. trifolia* | sparsely pubescent | pubescent | pubescent |

### 3.2. Quantitative Study

The length of trichomes on branchlets, as well as the abaxial and adaxial leaf surfaces, wwas measured in different *Causonis* species. *Causonis corniculata* was found to have the shortest trichomes (52.716 ± 28.73 μm) on branchlets, while *C. mollis* has the longest trichomes (794.5 ± 54.32 μm), followed by *C. trifolia* (516.3 ± 466.432 μm). The observed *C. maritima* from Australia did not have any trichomes on the branchlets. On the abaxial leaf surface of *C. corniculata* and *C. maritima,* there were no trichomes. *Causonis daliensis* possessed the shortest trichomes (171.5 ± 114.414 μm) on the abaxial leaf surface. The longest trichomes on the abaxial leaf surface were observed in *C. mollis* (533 ± 318.379 μm), followed by *C. trifolia* (467.9 ± 325.638 μm). On the adaxial leaf surface, the shortest trichomes were observed in *C. daliensis* (57.316 ± 78.233 μm) and the longest in *C. trifolia* (491 ± 421.59 μm), while trichomes were absent in *C. corniculata.* The trichomes' length of different species is presented in Table 3. In addition, categorical features of trichomes on branchlets and leaves are presented in Table 4.

### 3.3. Principle Component Analysis (PCA)

Quantitative characters were employed for the PCA, and it was discovered that the first PCA variance was 89.275, while the total of the three PCA variances was 100. The eigenvalue, nevertheless, ranged from 0.3124 to 8.3719. (Table 5 and Figure 3). Based on trichomes on branchlets, abaxial leaf surfaces, and adaxial leaf surfaces, we investigated species variation. Six taxa representing five species (*C. daliensis*, *C. fugongensis*, *C. mollis*, *C. japonica* var. *japonica*, *C. japonica* var. *pseudotrifolia*, and *C. timoriensis* var. *mekongensis*) occurred in PC1, whereas PC2 was represented by four species (*C. clematidea*, *C. corniculata*, *C. maritima*, and *C. trifolia*).

**Table 3.** Quantitative features of trichomes on branchlets and leaves.

| Taxon | Trichomes' Length on Branchlets Min–Max (μm) | Trichomes' Length on Branchlets (μm) | Trichomes' Length on Abaxial Leaf Surface Min–Max (μm) | Trichomes' Length on Abaxial Leaf Surface (μm) | Trichomes Length on Adaxial Leaf Surface Min–Max (μm) | Trichomes' Length on Adaxial Leaf Surface (μm) |
|---|---|---|---|---|---|---|
| *C. clematidea* | 99.9–455.6 | 258.5 ± 164.93 | 99.9–499.9 | 268.1 ± 155.419 | 74.1–447.62 | 253.3 ± 148.996 |
| *C. corniculata* | 24.9–99.9 | 52.716 ± 28.73 | Absent | Absent | Absent | Absent |
| *C. daliensis* | 33.3–250 | 130 ± 82.235 | 53.9–307.7 | 171.5 ± 114.414 | 46.5–110.5 | 57.316 ± 78.233 |
| *C. fugongensis* | 66.7–766.7 | 360.1 ± 292.73 | 109.1–454.6 | 261.9 ± 127.304 | 87.5–212.5 | 123.1 ± 52.052 |
| *C. japonica* var. *japonica* | 66.7–177.8 | 123.3 ± 42.832 | 142.9–342.9 | 245.5 ± 97.376 | 54.6–254.6 | 141.66 ± 54.422 |
| *C. japonica* var. *pseudotrifolia* | 33.3–211.1 | 115.1 ± 67.466 | 73.7–368.4 | 212.9 ± 128.562 | 43.8–225 | 115.5 ± 72.78 |
| *C. maritima* | Absent | Absent | Absent | Absent | 83.3–266.7 | 167.23 ± 77.322 |
| *C. mollis* | 499.9–1034.5 | 794.5 ± 54.32 | 100–850 | 533 ± 318.379 | 94.1–599.9 | 216 ± 3.011 |
| *C. timoriensis* var. *mekongensis* | 49.9–233.3 | 131.9 ± 77.918 | 62.5–475 | 234.8 ± 164.124 | 80–280 | 171.1 ± 76.1 |
| *C. trifolia* | 54.6–1090.9 | 516.3 ± 466.432 | 90–870 | 467.9 ± 325.638 | 60–1020 | 491 ± 421.59 |

**Table 4.** Categorical features of trichomes on branchlets and leaves.

| Taxon | Trichomes' Length on Branchlets | Trichomes' Length on Abaxial Leaf Surface | Trichomes' Length on Adaxial Leaf Surface |
|---|---|---|---|
| *C. clematidea* | 3 | 3 | 3 |
| *C. corniculata* | 1 | 0 | 0 |
| *C. daliensis* | 2 | 3 | 1 |
| *C. fugongensis* | 2 | 3 | 2 |
| *C. japonica* var. *japonica* | 2 | 3 | 2 |
| *C. japonica* var. *pseudotrifolia* | 2 | 3 | 2 |
| *C. maritima* | 0 | 0 | 2 |
| *C. mollis* | 6 | 5 | 4 |
| *C. timoriensis* var. *mekongensis* | 2 | 3 | 2 |
| *C. trifolia* | 6 | 5 | 6 |

**Table 5.** PCA of selected species of *Causonis*.

| PC | Eigenvalue | % Variance |
|---|---|---|
| 1 | 8.37199 | 89.275 |
| 2 | 0.6934 | 7.3941 |
| 3 | 0.312386 | 3.3311 |

### *3.4. Ancestral Character State Reconstruction*

Using well-resolved chloroplast phylogeny, three morphological character states of trichomes based on their location in the *Causonis* were investigated for state evolution. According to our findings, the trichomes on branchlets in the genus were "glabrous to sparsely pubescent" in their ancestral state and "villous" in their derived state (Figure 4A). Trichomes on the abaxial leaf surface were "glabrous to sparsely pubescent" in their ancestral state, whereas "villous" was a derived state in the genus (See Figure 4B). Trichomes on the adaxial leaf surface were "glabrous to sparsely pubescent" in their ancestral state, and they were "villous" in their derived state in *Causonis* (Figure 4C).

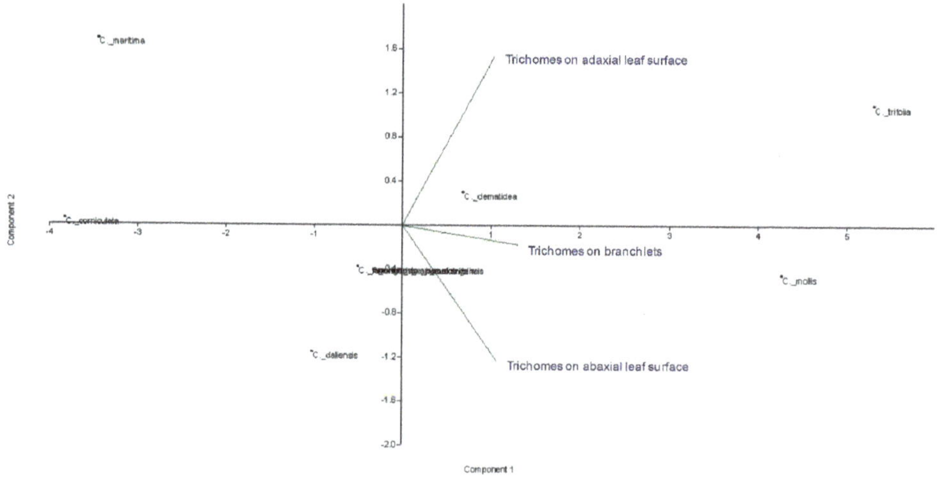


**Figure 3.** PCA of the selected taxa of *Causonis* based on trichomes on branchlets, abaxial, and adaxial leaf surfaces.

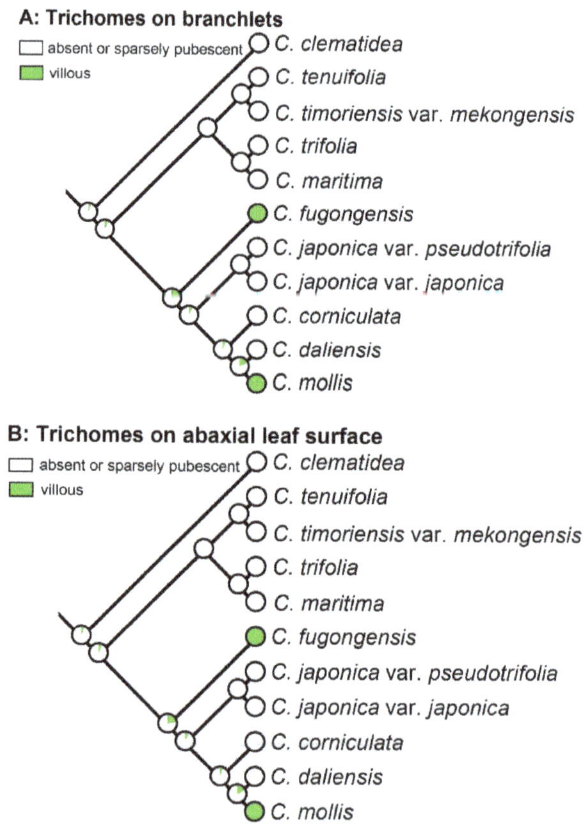


**Figure 4.** *Cont.*

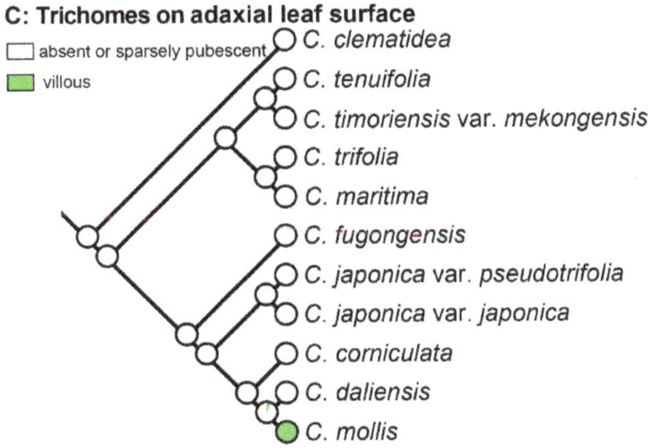


**Figure 4.** Character optimization of trichomes using the maximum likelihood approach based on the Mk1 model implemented in Mesquite. (**A**) Trichomes on branchlets. (**B**) Trichomes on abaxial leaf surface. (**C**) Trichomes on adaxial leaf surface. Pie charts show each character's ML support at its ancestral nodes.

## 4. Discussion

The morphology of vegetative parts, such as branchlets and leaves, is found to be highly helpful in identifying Vitaceae taxa [20,21,27,28]. In particular, trichomes on branchlets and leaves are highly helpful for identifying the species of *Causonis* [23,27]. Using field observations and a genus-wide screening of hundreds of herbarium specimens, branchlets were observed to be sparsely pubescent in the young/immature stage, but glabrous in the mature stage in the same plant. Sometimes, densely pubescent hairs were observed in the seedling stage or young stage in some taxa like *C. clematidea* and *C. japonica* var. *japonica*, with the morphology rarely observed in the mature stage. All of the hairs observed in the genus were simple. Glandular or ribbon-like trichomes as revealed in genus *Vitis* by Ma, et al. [28] were not found in *Causonis*. *Causonis'* non-glandular trichomes appear to be sufficient for performing its defensive and protective functions against a variety of stresses, such as serving as a mechanical barrier against low humidity, intense light and high temperatures, and insects' behaviors during oviposition and feeding [30]. Only *C. mollis*, *C. ciliifera*, and *C. fugongensis* were found to have villous hairs on their branchlets. The only species with corniculate petals are *C. ciliifera* and *C. corniculata*, but the branchlets and leaves of *C. ciliifera* are villous, while those of *C. corniculata* are nearly glabrous.

It is very challenging to identify *Causonis* species based solely on trichomes; additional characters such as petals, seeds, stipules, tendrils, or veins, besides trichomes, are needed for the identification. For example, *C. trifolia* and *C. maritima* can be easily distinguished based on trichome morphology, but an additional tendrils' morphology is needed for their confirmation. Leaves of *C. trifolia* are pubescent on both surfaces, while *C. maritima* typically have hooked hairs limited to the midvein on the adaxial surface and have a glabrous abaxial surface [27,31]. However, *C. trifolia* occasionally exhibits hooked hairs on the adaxial surface. Similarly, the abaxial leaf surface has villous hairs that limit the number of species to three: *C. ciliifera*, *C. mollis*, and *C. fugongensis*, for easy identification in the genus. *Causonis ciliifera* can be easily differentiated from the other two as having corniculate petals, considering an additional character. The appressed hairs on the adaxial leaf surface or the prominently raised veins on the abaxial leaf surface are characteristics that help identify *C. fugongensis*. On the abaxial leaf surface of *C. fugongensis*, villous hairs or curved hairs have been observed. *Causonis mollis* requires the aid of seed characters like ventral infolds shape or endosperm shape in the cross section of seed for its identification.

*Causonis corniculata* can be differentiated from *C. ciliifera*, the only another species with corniculate petals in the genus, as lacking villous hairs on branchlets and leaves, as the former generally has glabrous branchlets and leaves. *Causonis tenuifolia* and *C. corniculata* are both almost glabrous, but the former lacks corniculate petals, and the petals' morphology here is another additional character that helps in identifying both species. *Causonis timoriensis* var. *mekongensis* can be distinguished from other trifoliolate species like *C. trifolia*, *C. maritima*, and *C. japonica* var. *pseudotrifolia* by its pilose hairs on the abaxial leaf surface [20]. Without taking into account the morphological variation in trichomes, no other characters are sufficient to distinguish those species. As a result, trichomes' morphology plays a very useful role in species identification in the genus.

High variations were observed in the length of trichomes between the species. However, the length of trichomes in a species revealed a certain degree of uniformity. For example, *C. corniculata*, which lacks trichomes on the abaxial and adaxial leaf surfaces, has the shortest trichomes on branchlets, and *C. mollis* and *C. trifolia* both share the longest trichomes on all of the studied three surfaces between them. Therefore, the length of trichomes revealed a certain degree of inference in the genus.

This study showed that the glabrous to slightly pubescent trichomes are the ancestral state and the villous hairs are the derived state in the genus *Causonis*, regardless of the position of the trichomes in the genus, such as branchlets or leaves (abaxial and adaxial). It is possible that the villous hairs on branchlets and the abaxial surface of leaves have evolved at least twice, while they have evolved once on the adaxial leaf surface in the genus. This suggests that branchlets that are glabrous, sparsely pubescent, pubescent, or puberulent with normal hairs and sometimes pilose are the ancestral state of the genus. Glabrous, sparsely pubescent, or pubescent with normal hairs and sometimes pilose, hooked, curved, or appressed hairs on the leaflets' surface (abaxial and adaxial) are also the ancestral state of the genus.

## 5. Conclusions

The current study is the first comprehensive examination of the micromorphology of the trichomes in *Causonis*. Twelve taxa were studied for the morphology of trichomes on branchlets and abaxial and adaxial surfaces of leaves using a stereomicroscope. The genus was found to be represented by only simple trichomes that are non-glandular and unbranched. Different types of trichomes such as straight, curved, hooked, appressed, and villous are found in branchlets and leaves in the *Causonis*. High variations are noticed in the trichomes' length between the species. Based on the characteristics of the micromorphological trichomes, PCA is used to determine the species boundary of the taxa. Besides, we examined the character state evolution of trichomes on leaf surfaces, including the abaxial and adaxial leaf surfaces based on the chloroplast dataset using the maximum likelihood approach in Mesquite. Glabrous to sparsely pubescent hairs, irrespective of the position in plants, are the ancestral state and villous hairs are the derived state in the genus. Trichomes' morphology is found to be very useful in the identification of *Causonis* species. To uncover the whole trichome diversity within the family, a similar study should be conducted on additional genera of the Vitaceae.

**Author Contributions:** G.P. and W.Z. designed the experiments, analyzed the data, and wrote and revised the manuscript. All authors have read and agreed to the published version of the manuscript.

**Funding:** This research received no external funding.

**Institutional Review Board Statement:** Not applicable.

**Informed Consent Statement:** Not applicable.

**Acknowledgments:** Zhiduan Chen, Li-Min Lu, and Fazal Ullah provided invaluable assistance to the authors. The authors are also thankful to Viet-Cuong Dang, Jian-Fei Ye, Sun Miao, Pamela Soltis, and Douglas Soltis for their plant collections used in this study. The authors are also appreciative of the microscope facilities provided by the Institute of Botany, Chinese Academy of Sciences.

**Conflicts of Interest:** The authors declare no conflict of interest.

## References


1. Wen, J. Vitaceae. In *The Families and Genera of Vascular Plants*; Springer: Berlin, Heidelberg, 2007; Volume 9, pp. 467–479.
2. Ma, Z.-Y.; Nie, Z.-L.; Ren, C.; Liu, X.-Q.; Zimmer, E.A.; Wen, J. Phylogenomic relationships and character evolution of the grape family (Vitaceae). *Mol. Phylogenet. Evol.* **2021**, *154*, 106948. [CrossRef] [PubMed]
3. Wen, J.; Lu, L.M.; Nie, Z.L.; Liu, X.Q.; Zhang, N.; Ickert-Bond, S.; Gerrath, J.; Manchester, S.R.; Boggan, J.; Chen, Z.D. A new phylogenetic tribal classification of the grape family (Vitaceae). *J. Syst. Evol.* **2018**, *56*, 262–272. [CrossRef]
4. Ingrouille, M.J.; Chase, M.W.; Fay, M.F.; Bowman, D.; Van Der Bank, M.; Bruijn, A.D.E. Systematics of Vitaceae from the viewpoint of plastid *rbcL* DNA sequence data. *Bot. J. Linn. Soc.* **2002**, *138*, 421–432. [CrossRef]
5. Lu, L.; Wang, W.; Chen, Z.; Wen, J. Phylogeny of the non-monophyletic *Cayratia* Juss. (Vitaceae) and implications for character evolution and biogeography. *Mol. Phylogenet. Evol.* **2013**, *68*, 502–515. [CrossRef]
6. Rossetto, M.; Jackes, B.R.; Scott, K.D.; Henry, R.J. Intergeneric relationships in the Australian Vitaceae: New evidence from cpDNA analysis. *Genet. Resour. Crop Evol.* **2001**, *48*, 307–314. [CrossRef]
7. Rossetto, M.; Jackes, B.R.; Scott, K.D.; Henry, R.J. Is the genus *Cissus* (Vitaceae) monophyletic? Evidence from plastid and nuclear ribosomal DNA. *Syst. Bot.* **2002**, *27*, 522–533.
8. Rossetto, M.; Crayn, D.M.; Jackes, B.R.; Porter, C. An updated estimate of intergeneric phylogenetic relationships in the Australian Vitaceae. *Botany* **2007**, *85*, 722–730.
9. Soejima, A.; Wen, J. Phylogenetic analysis of the grape family (Vitaceae) based on three chloroplast markers. *Am. J. Bot.* **2006**, *93*, 278–287. [CrossRef]
10. Habib, S.; Dang, V.-C.; Ickert-Bond, S.M.; Zhang, J.-L.; Lu, L.-M.; Wen, J.; Chen, Z.-D. Robust phylogeny of *Tetrastigma* (Vitaceae) based on ten plastid DNA regions: Implications for infrageneric classification and seed character evolution. *Front. Plant Sci.* **2017**, *8*, 590. [CrossRef]
11. Parmar, G.; Dang, V.C.; Rabarijaona, R.N.; Chen, Z.D.; Jackes, B.R.; Barrett, R.L.; Zhang, Z.Z.; Niu, Y.T.; Trias-Blasi, A.; Wen, J. Phylogeny, character evolution and taxonomic revision of *Causonis*, a segregate genus from *Cayratia* (Vitaceae). *Taxon* **2021**, *70*, 1188–1218. [CrossRef]
12. Ma, Z.-Y.; Wen, J.; Ickert-Bond, S.M.; Nie, Z.-L.; Chen, L.-Q.; Liu, X.-Q. Phylogenomics, biogeography, and adaptive radiation of grapes. *Mol. Phylogenet. Evol.* **2018**, *129*, 258–267. [CrossRef] [PubMed]
13. Richards, Z.T.; Miller, D.J.; Wallace, C.C. Molecular phylogenetics of geographically restricted *Acropora* species: Implications for threatened species conservation. *Mol. Phylogenet. Evol.* **2013**, *69*, 837–851. [CrossRef] [PubMed]
14. Wen, J.; Lu, L.M.; Boggan, J.K. Diversity and evolution of Vitaceae in the Philippines. *Philipp. J. Sci.* **2013**, *142*, 223–244.
15. Rafinesque, C.S. *Medical Flora: Or, Manual of the Medical Botany of the United States of North America. Containing a Selection of Above 100 Figures and Descriptions of Medical Plants, with Their Names, Qualities, Properties, History, & c.: And Notes or Remarks on Nearly 500 Equivalent Substitutes*; Atkinson & Alexander: Philadelphia, PA, USA, 1830; Volume 2.
16. Gagnepain, F. Un genre meconnu: Classification des *Cissus* et *Cayratia*. *Not. Syst.* **1911**, *1*, 261–271, 306–326, 339–362.
17. Suessenguth, K. *Vitaceae*, 2nd ed.; Duncker und Humblot: Berlin, Germany, 1953; Volume 20.
18. Latiff, A. Studies in Malesian Vitaceae V. The genus *Cayratia* in the Malay peninsula. *Sains Malays* **1981**, *10*, 129–139.
19. Li, C.L. New taxa in Vitaceae from China. *Chin. J. Appl. Environ. Biol.* **1996**, *2*, 43–53.
20. Chen, Z.D.; Ren, H.; Wen, J. *Vitaceae*; Science Press and Missouri Botanical Garden Press: Beijing, China; St. Louis, MO, USA, 2007; Volume 12, pp. 390–402.
21. Wen, J.; Lu, L.M.; Hsu, T.W.; Dang, V.C.; Habib, S.; Boggan, J.K.; Okada, H.; Chen, I.J.; Chen, Z.D. *Pseudocayratia*, a new genus of Vitaceae from China and Japan with two new species and three new combinations. *J. Syst. Evol.* **2018**, *56*, 374–393. [CrossRef]
22. Jackes, B. Transfer of three species of *Cayratia* Juss., to *Causonis* Raf. (Vitaceae). *Telopea* **2020**, *23*, 69–71. [CrossRef]
23. Chen, I.; Manchester, S. Seed Morphology of Vitaceae. *Int. J. Plant Sci.* **2011**, *172*, 1–35. [CrossRef]
24. Habib, S.; Dang, V.-C.; Ickert-Bond, S.M.; Wen, J.; Chen, Z.-D.; Lu, L.-M. Evolutionary trends in *Tetrastigma* (Vitaceae): Morphological diversity and taxonomic implications. *J. Syst. Evol.* **2018**, *56*, 360–373. [CrossRef]
25. Rabarijaona, R.N.; Dang, V.C.; Parmar, G.; Liu, B.; Wen, J.; Chen, Z.D.; Lu, L.M. Phylogeny and taxonomy of *Afrocayratia*, a new genus of Vitaceae from continental Africa and Madagascar. *J. Syst. Evol.* **2020**, *58*, 1090–1107. [CrossRef]
26. Gerrath, J.M.; Posluszny, U.; Ickert Bond, S.M.; Wen, J. Inflorescence morphology and development in the basal rosid lineage Vitales. *J. Syst. Evol.* **2017**, *55*, 542–558. [CrossRef]
27. Jackes, B.R. Revision of the Australian Vitaceae, 2. *Cayratia Juss. Austrobaileya* **1987**, *2*, 365–379.
28. Ma, Z.-Y.; Wen, J.; Ickert-Bond, S.M.; Chen, L.Q.; Liu, X.Q. Morphology, structure, and ontogeny of trichomes of the grape genus (*Vitis*, Vitaceae). *Front. Plant Sci.* **2016**, *7*, 704. [CrossRef] [PubMed]
29. Maddison, W.P.; Maddison, D.R. *Mesquite: A Modular System for Evolutionary Analysis*, Version 3.51 Version 3.70. Available online: http://www.mesquiteproject.org (accessed on 1 September 2022).
30. Santos Tozin, L.R.d.; de Melo Silva, S.C.; Rodrigues, T.M. Non-glandular trichomes in Lamiaceae and Verbenaceae species: Morphological and histochemical features indicate more than physical protection. *N. Z. J. Bot.* **2016**, *54*, 446–457. [CrossRef]
31. Hsu, T.-W.; Kuoh, C.-S. *Cayratia maritima* BR Jackes (Vitaceae), a new addition to the flora of Taiwan. *Bot. Bull. Acad. Sin.* **1999**, *40*, 329–332.

MDPI
St. Alban-Anlage 66
4052 Basel
Switzerland
www.mdpi.com

*Horticulturae* Editorial Office
E-mail: horticulturae@mdpi.com
www.mdpi.com/journal/horticulturae

Disclaimer/Publisher's Note: The statements, opinions and data contained in all publications are solely those of the individual author(s) and contributor(s) and not of MDPI and/or the editor(s). MDPI and/or the editor(s) disclaim responsibility for any injury to people or property resulting from any ideas, methods, instructions or products referred to in the content.

www.ingramcontent.com/pod-product-compliance
Lightning Source LLC
Chambersburg PA
CBHW040311270926
51850CB00007BC/1045
*9783725804702*